Synthetic Polymeric Materials-Based Drug Delivery Systems for Inflammatory Diseases

This book provides a comprehensive overview of synthetic polymers and their applications in designing delivery systems for the management of inflammatory diseases. It presents introductory insights into inflammatory conditions, delves into the role of synthetic polymers, and examines diverse delivery approaches. *Synthetic Polymeric Materials-Based Drug Delivery Systems for Inflammatory Diseases* explores the potential of synthetic polymers in designing drug delivery systems for managing inflammatory diseases, including inflammatory lung diseases, inflammatory bowel diseases, and inflammatory skin diseases, as well as other conditions like cancer, neurodegenerative disorders, rheumatoid arthritis, and eye-related inflammatory conditions. It also discusses the role of synthetic polymers in modulating immune system responses in different disease conditions. Furthermore, it analyzes the 3D printing technologies employed for the preparation of drug delivery systems based on synthetic polymers. Toward the end, the book highlights the challenges and prospects of synthetic polymers in designing delivery systems for the effective management of inflammatory diseases and their clinical usage. This book is intended for researchers and professionals in the fields of pharmaceutical sciences, nanotechnology, and drug delivery systems.

Key Features

- Highlights the role of a synthetic polymer-based drug delivery system against inflammatory responses.
- Explores the cutting-edge technology of 3D printing and its application in preparing drug delivery systems based on synthetic polymers.
- Provides valuable insights into how synthetic polymers can be used to modulate immune system responses.
- Presents regulatory compliance using synthetic polymers in drug delivery systems for inflammatory diseases.
- Examines challenges associated with synthetic polymers in drug delivery systems for inflammatory diseases.

Synthetic Polymeric Materials-Based Drug Delivery Systems for Inflammatory Diseases

Edited by
Harish Dureja
Vimal Arora
Paul A. McCarron
Vandana B. Patravale
Kamal Dua

CRC Press
Taylor & Francis Group
Boca Raton London New York

CRC Press is an imprint of the
Taylor & Francis Group, an **informa** business

Designed cover image: Shutterstock

First edition published 2026
by CRC Press
2385 NW Executive Center Drive, Suite 320, Boca Raton FL 33431

and by CRC Press
4 Park Square, Milton Park, Abingdon, Oxon, OX14 4RN

CRC Press is an imprint of Taylor & Francis Group, LLC

ISBN: 978-1-032-76175-6 (hbk)
ISBN: 978-1-032-76178-7 (pbk)
ISBN: 978-1-003-47738-9 (ebk)

DOI: 10.1201/9781003477389

Typeset in Times
by SPi Technologies India Pvt Ltd (Straive)

Contents

About the Editors

Dr. Harish Dureja is presently the dean of Research and Development and the dean of the Faculty of Pharmaceutical Sciences at Maharshi Dayanand University, Rohtak; visiting professor at ARCCIM, Faculty of Health, University of Technology Sydney, Australia; and adjunct professor, School of Medical and Life Sciences, Sunway University, Kuala Lumpur, Malaysia. He has dedicated the past 25 years to teaching, research, mentoring, advising, and administration. He is actively engaged in the fields of drug delivery and regulatory science, focusing on enhancing the oral bioavailability of BCS class II drugs, developing nanoparticulate drug delivery systems, conducting in silico ADME modeling, and managing pharmaceutical regulatory affairs. Dr. Dureja holds one Indian patent and has authored over 300 research articles in various respected international and national journals. Additionally, he has authored /edited 08 books and contributed to 50 book chapters published by both national and international publishers.

Dr. Vimal Arora currently holds the position of professor and associate director at University Institute of Pharma Sciences, Chandigarh University, Punjab, India. With a career spanning 19 years, he has dedicated his expertise to the fields of pharmaceutical education and research, focusing on pharmaceutical product development and quality assurance. Dr. Arora's academic contributions include over 60 research articles, a book, 15 book chapters, 2 Indian design registrations, and 3 Indian patents (published). His work is prominently centered on the exploration of innovative techniques for designing drug delivery systems aimed at treating cancer and diabetes.

Professor Paul A. McCarron currently holds a chair in pharmaceutics and serves as the head of the School of Pharmacy and Pharmaceutical Sciences at Ulster University. As a registered pharmacist and a member of the Pharmaceutical Society of Northern Ireland, he possesses a wealth of experience in both industrial and academic pharmacy. Prof. McCarron has a specific passion for advancing drug delivery systems and excels in translating innovative delivery strategies into clinical and commercial applications. Currently, he is the company director and founder of Jenarron Therapeutics and has provided technical expertise for a range of spin-out ventures, such as Swedish Pharma. Prof. McCarron's research interests revolve around the development and characterization of nanoparticulate systems and topical formulations. Supported by substantial funding from the Research and Development Office in Northern Ireland, he successfully developed and trialed an innovative putty-type material for administering anesthesia to acute lacerations.

Dr. Vandana B. Patravale presently serves as a professor of pharmaceutics at the Institute of Chemical Technology in Mumbai, India. With over 30 years of experience in teaching and research, she has made significant contributions to her field. Dr. Patravale has published four books and 35 book chapters with international publishers. Her research primarily focuses on the development of nanocarriers, with a strong emphasis on addressing infectious diseases, cancer, and neurodegenerative disorders, many of which are linked to inflammation. Notably, she has successfully transferred more than 20 technologies to the industry.

Dr. Kamal Dua, associate professor, discipline of pharmacy, Graduate School of Health, University of Technology Sydney (UTS), has over 12 years of research experience working in the field of drug delivery targeting inflammatory diseases. Dr. Dua is also a Node Leader of Drug Delivery Research in the Centre for Inflammation at Centenary Institute/UTS, where the targets identified from the research projects are pursued to develop novel formulations as the first step toward translation into clinics. Dr. Dua's research is in two complementary areas: drug delivery and immunology, specifically addressing how these disciplines can advance one another, helping the community to live longer and healthier. This is evidenced by his extensive publication record in reputed journals. Dr. Dua's research interests focus on harnessing the pharmaceutical potential of modulating critical regulators such as interleukins and microRNAs and developing new and effective drug delivery formulations for the management of inflammation in chronic airway diseases and cancer.

1 Inflammatory Diseases

Lovedeep Singh
University Institute of Pharma Sciences, Chandigarh University, Mohali, India

Kamna, Ankita Sharma, and Manmeet Mehra
Guru Nanak Dev University, Amritsar, India

1.1 INTRODUCTION

Inflammation is a localized and transient response activated by the innate immune system to protect the organism from external damage. This response is characterized by the classic signs of rubor, tumor, calor, dolor, and laesa [1]. Inflammatory conditions are intricate disorders that occur when the body's immune system reacts to injury, infection, or other stimuli by initiating an inflammatory response [2]. The molecular cascade of inflammation is a complex phenomenon triggered by the detection of distinct molecular patterns linked to infection or tissue damage. This intricate process involves multiple pivotal regulators that orchestrate the controlled expression of proinflammatory molecules. Prolonged inflammation frequently leads to significant adverse effects on health [3]. During acute inflammatory responses, cellular and molecular mechanisms work efficiently to mitigate imminent injury or infection. However, if the acute inflammatory response is not properly regulated, it may progress to a chronic state, contributing to the onset of multiple chronic inflammatory disorders [4]. Several factors contribute significantly to the pathogenesis of inflammation, including infections and foreign irritants such as industrial chemicals [5, 6]. Research suggests that the innate immune system plays a crucial role in driving inflammation. Consequently, various immune cells, such as macrophages, mast cells, neutrophils, and lymphocytes, are significantly engaged in orchestrating inflammatory responses [1, 7].

Inflammatory stimulus triggers intracellular signaling cascades that culminate in the synthesis of various proinflammatory mediators. Essential triggers of inflammation include microbial components and cytokines such as interleukin-1β (IL-1β), interleukin-6 (IL-6) and tumor necrosis factor-α (TNF-α). These inflammatory agents induce inflammation by interaction with specific receptors [8]. The stimulation of respective receptors potentiates the essential intracellular signaling cascades, such as the mitogen-activated protein kinase (MAPK) pathway, the nuclear factor kappa-B (NF-κB) pathway, and the Janus kinase-signal transducer and activator of transcription (JAK-STAT) pathway [9, 10]. Stimulation of these cascades potentiates the activation of macrophages and adipocytes, which subsequently enhances the production of inflammatory mediators, thus aggravates inflammation [11, 12]. An exemplary instance of inflammatory disease is rheumatoid arthritis (RA), an autoimmune condition marked by persistent inflammation in the joints. RA triggers immune cell activation and the release of proinflammatory cytokines, culminating in inflammation of the synovium, destruction of cartilage, and erosion of bones [13]. Another significant inflammatory condition is inflammatory bowel disease (IBD), which includes disorders such as Crohn's disease (CD) and ulcerative colitis (UC). IBD triggers inflammation throughout the gastrointestinal tract, leading to symptoms including abdominal pain, diarrhea, and rectal bleeding. The development of IBD is attributed to a complex interplay of immune system dysregulation and environmental influences [14]. Whereas, psoriasis presents as a chronic inflammatory skin disorder characterized by red, scaly patches. It arises from immune system dysregulation, prompting excessive proliferation of keratinocytes and inflammation within the skin [15]. This chapter explores the various types of inflammatory diseases, detailing the complex interactions of inflammatory mediators within these conditions. It examines how different mediators contribute to disease progression and highlights their roles in the pathophysiology of inflammatory responses.

1.2 ACUTE VS. CHRONIC INFLAMMATORY DISEASES

Acute inflammation is a short-term protective response to harmful stimuli, such as infections or injuries, aimed at eliminating the cause and restoring tissue homeostasis. It involves the release of inflammatory cytokines and reactive oxygen species (ROS), recruiting immune cells to the site of damage [16]. Acute inflammation initiates swiftly, often within minutes to hours, and generally subsides within a few days. It is marked by distinctive signs and symptoms and is predominantly characterized by a cellular infiltrate primarily consisting of neutrophils. The redness associated with acute inflammation is attributed to augmented blood flow to the affected region, which is a consequence of vasodilation [17]. Acute inflammation serves as a protective

DOI: 10.1201/9781003477389-1

mechanism that functions to remove the underlying cause of cellular injury, eliminate necrotic cells and tissues, and initiate the process of tissue repair [17]. An example of an acute inflammatory disease is acute appendicitis, where inflammation of the appendix leads to severe abdominal pain, requiring prompt medical intervention [18]. Acute inflammation arises from various triggers such as infections, irritants, burns, traumatic brain injuries, and allergens. Acute inflammation typically occurs as a short-term response to injury or infection, characterized by redness and swelling [19]. However, chronic inflammation persists over time, often due to ongoing exposure to irritants or unresolved infections. This prolonged state can lead to significant health problems, including cardiovascular diseases, diabetes, and cancer. As inflammation continues, it damages tissues and organs, contributing to the progression of these chronic diseases and impairing the body's ability to function properly [12] (Figure 1.1). In chronic conditions, excessive ROS production and cytokine release can sustain the inflammatory response, disrupting normal tissue function and driving disease progression [16]. Chronic inflammation develops slowly over days and can persist for years. The condition exhibits less pronounced traditional clinical manifestations and is characterized by a cellular infiltrate predominantly consisting of monocytes/macrophages and lymphocytes [20]. Chronic inflammation can be induced by prolonged exposure to toxic chemicals and environmental agents, such as cigarette smoke [20, 21].

Systemic inflammation differs significantly between acute and chronic forms. Acute systemic inflammation is marked by a rapid, critical impact on microcirculation, manifesting in severe, often life-threatening conditions such as sepsis and acute trauma [21]. Chronic systemic inflammation develops over a prolonged period, with less immediate but persistent effects. Acute systemic inflammation exhibits a phased pattern with varying degrees of severity, necessitating timely intervention to prevent the most critical phases, such as secondary phlogogenic stroke and the depressive phase. Chronic systemic inflammation, observed in severe chronic diseases, often mirrors a prolonged pre-SI state seen in intensive care. However, in conditions like systemic lupus erythematosus (SLE), chronic systemic inflammation can reach hyperergic levels similar to acute systemic inflammation [22]. Furthermore, Abudukelimu et al. utilized the predictive dynamic model to simulate the extracellular network involved in inflammation to study the transition of acute inflammation to a chronic state based on antigen levels. The computational analysis revealed that when acute inflammation is sustained by a continuous influx of cytokines or other inflammatory agents above a specific threshold, it can escalate into a more severe inflammatory state characterized by significantly elevated TNF-α levels. Once this intense inflammation reaches a steady state, even if the influx of inflammatory agents decreases to very low levels, TNF-α levels remain high. This finding suggests that substantial exposure to inflammatory stimuli can transition the inflammatory response from acute to chronic. Thus, delineating that the inflammatory state of a network can vary between high and low levels based on its prior exposure highlights a mechanistic basis for the transition from acute to chronic inflammation [23]. The comparison of acute and chronic inflammation is listed in Table 1.1.

1.3 TYPES OF INFLAMMATORY DISEASES

1.3.1 Autoimmune Inflammatory Diseases

Autoimmune diseases are chronic conditions caused by a breakdown in immune tolerance to self-antigens, leading to damage that can be either systemic or organ-specific. Autoimmune inflammatory diseases are disorders in which the immune system mistakenly attacks the body's own tissues, causing chronic inflammation and damage. These

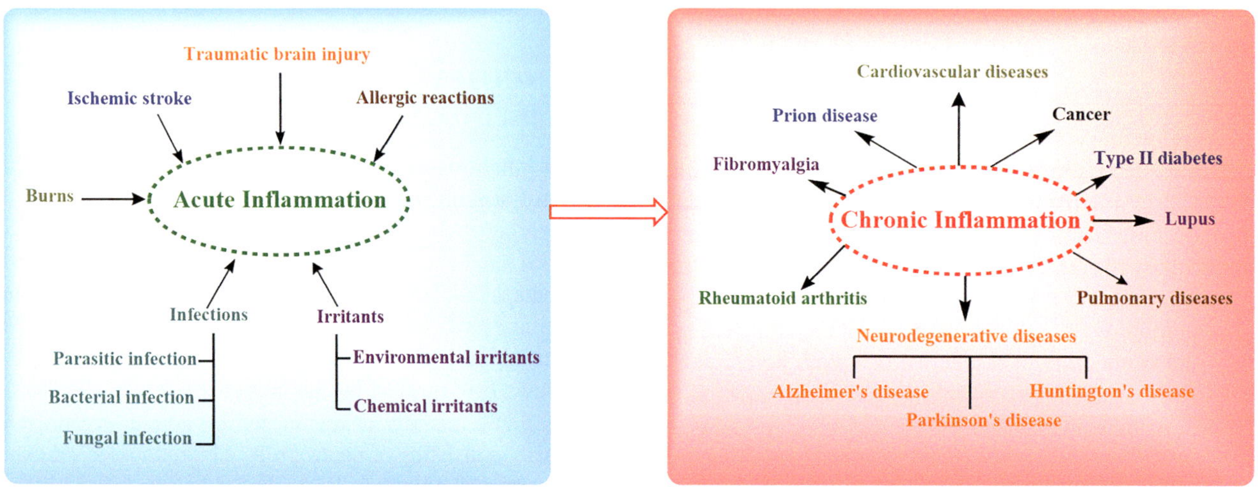

FIGURE 1.1 Overview of the triggers and outcomes of acute and chronic inflammation. Acute inflammation arises as a defense mechanism in response to external stimuli. However, when the inflammatory response is prolonged, it transitions into chronic inflammation, which is closely linked to the development of various pathological conditions.

TABLE 1.1
Comparison of Acute and Chronic Inflammation [17, 36, 37]

Feature	Acute Inflammation	Chronic Inflammation
Duration	Short-term (minutes to days)	Long-term (weeks, months, or years)
Onset	Rapid (immediate to hours)	Slow, progressive
Main Cells Involved	Neutrophils	Macrophages, lymphocytes, and plasma cells
Cause	Usually due to infections, trauma, or tissue damage	Persistent infections, prolonged exposure to irritants, or autoimmune responses
Signs & Symptoms	Redness, swelling, heat, pain, and loss of function	Often subtle or asymptomatic, with ongoing tissue damage and repair
Tissue Damage	Minimal, typically resolves with healing	Significant, with the potential for fibrosis or tissue destruction
Outcome	Resolution with healing or progression to chronicity	Ongoing inflammation, fibrosis, tissue remodeling, or organ dysfunction
Vascular Changes	Increased permeability, vasodilation	Minimal; new blood vessel formation (angiogenesis)
Inflammatory Mediators	Histamine, prostaglandins, leukotrienes, and cytokines	Cytokines, growth factors and ROS
Role	Protective response to eliminate pathogens and damaged cells	A sustained response, can lead to tissue dysfunction

diseases can affect various organs and systems, leading to diverse clinical manifestations. Key examples include RA, SLE, and multiple sclerosis (MS) [24, 25].

1.3.1.1 Rheumatoid Arthritis

RA is a chronic autoimmune disease marked by symmetrical joint inflammation, initially affecting smaller joints and progressively involving larger ones. As the disease advances, it can extend to other organs, including the skin, eyes, heart, kidneys, and lungs [26, 27]. This condition often leads to joint damage, including the destruction of bone and cartilage, along with the weakening of tendons and ligaments [27]. RA is commonly characterized by morning joint stiffness persisting for more than 30 minutes, accompanied by symptoms such as fatigue, fever, weight loss, and swollen, tender joints. Rheumatoid nodules may form under the skin. Onset generally occurs between ages 35 and 60, but juvenile RA can affect children under 16, lacking rheumatoid factor [28, 29]. In RA, the immune system erroneously targets the synovial membrane, which lines the joints, leading to inflammation and subsequent thickening of the synovium. The inflammatory cascade triggers the release of multiple proinflammatory mediators, which subsequently amplify the inflammatory response and exacerbate tissue damage [30]. RA occurs more commonly in females than in males, with women being two to three times more likely to develop the condition [26]. RA progresses through several stages. In the early stage, patients may experience joint stiffness, swelling, and mild inflammation without significant damage [31]. As the disease advances to the moderate stage, inflammation increases, leading to cartilage damage and noticeable pain, swelling, and reduced mobility. In the severe stage, joint destruction becomes more evident, causing erosion of bones, greater pain, deformities, and loss of function. By the end-stage, joints may become completely fused or nonfunctional, accompanied by chronic pain, disability, and a significant impact on daily activities [32]. The pathophysiology of RA encompasses a multifaceted interaction of genetic predispositions, environmental influences, and immunological mechanisms.

1.3.1.1.1 Genetic Factors

Genome-wide analyses of single nucleotide polymorphisms in RA patients have pinpointed the human leukocyte antigen-DR beta 1 (HLA-DRB1) gene as a significant susceptibility gene [33]. Other identified genes linked to RA susceptibility include PTPN22, CTLA4, STAT4, TNFAIP3, CCL21, and PADI4 [34]. In Japanese individuals, two haplotypes of PADI4 have been found, one of which is disease-susceptible and produces stable mRNA [35]. Moreover, anti-cyclic citrullinated peptide (anti-CCP) antibodies exhibit high specificity for RA and are linked to an elevated risk of bone and cartilage degradation. Environmental factors such as smoking, gingivitis, and gut microbiota can influence the epigenome, leading to the demethylation of histones and deoxyribonucleic acid (DNA) and promoting the transcription of proinflammatory cytokines [13, 31].

1.3.1.1.2 Environmental Factors

Environmental factors like smoking, periodontal disease, and gut microbiota dysbiosis are known to impact the development of RA [38]. These factors can alter the epigenome, resulting in changes in gene expression that enhance inflammation. For instance, smoking is linked to the citrullination of proteins, a post-translational modification that generates new antigens targeted by the immune system in RA [39].

1.3.1.1.3 Proinflammatory Cytokines

The inflammatory milieu in RA is maintained by various proinflammatory mediators such as IL-1, IL-6, IL-15, IL-21 and TNF-α. These mediators trigger signaling cascades such as NF-κB and JAK/STAT, which intensify the inflammatory response and contribute to joint destruction [40].

1.3.1.1.4 TNF-α and IL-1

It is well established that TNF-α and IL-1 are crucial cytokines in chronic joint inflammation and associated cartilage and bone erosion. Early studies involving the injection of recombinant cytokines into rabbit and rodent knee joints identified TNF-α as an inflammatory mediator and IL-1 as a key player in arthritis and cartilage destruction. TNF-α alone showed limited destructiveness but enhanced the destructive effects of IL-1 synergistically [41]. Further research using TNF-α transgenic models demonstrated that overexpression of TNFα, even without functional T and B cells, induced arthritis, indicating TNFα's arthritogenic potential [42]. TNF-α serves as a pivotal cytokine in RA, instigating inflammation through the stimulation of diverse signaling cascades, notably the NF-κB pathway. This activation prompts the transcription of genes associated with inflammation, thereby stimulating the synthesis of additional proinflammatory cytokines and adhesion molecules that sustain the inflammatory process [13]. Importantly, arthritis could be triggered by membrane-bound TNF-α (mTNFα), suggesting that effective therapies should target cell surface-bound TNF-α using antibodies or soluble receptors [43]. Blocking IL-1 receptors also effectively prevented arthritis, underscoring IL-1 as the secondary mediator responsible for joint damage [44]. IL-1 triggers the production of matrix metalloproteinases (MMPs), enzymes responsible for cartilage degradation, and promotes the differentiation and activation of osteoclasts, cells involved in bone resorption. Additionally, IL-1 heightens the inflammatory process by stimulating the proinflammatory mediators expression [45].

1.3.1.1.5 IL-6

IL-6 is synthesized by multiple cell types, including T and B lymphocytes, macrophages, and fibroblast-like synoviocytes (FLS), within the synovial tissue of patients with RA [46, 47]. IL-6 amplifies synovial inflammation, promotes cartilage degradation, and accelerates bone erosion in RA by uplifting the level of other proinflammatory mediators and facilitating osteoclast differentiation. Therapeutic interventions targeting IL-6 signaling, such as IL-6 receptor antagonists like tocilizumab, have demonstrated efficacy in managing RA, leading to clinical amelioration and impeding the progression of joint damage [31].

1.3.1.1.6 IL-15

IL-15 plays a crucial role in the stimulation and proliferation of T lymphocytes and natural killer (NK) cells. In RA, there is an observed increase in IL-15 concentrations within the synovial fluid and affected joint tissues. This heightened IL-15 presence facilitates the persistence and activation of T cells and NK cells in the synovial environment, thereby exacerbating the inflammatory milieu characteristic of RA [48, 49]. IL-15 stimulates macrophages to produce TNF-α and IL-1β, which play a key role in joint destruction in RA. IL-15 within the synovial tissue exacerbates the inflammatory process by perpetuating macrophage activation and fostering a persistent inflammatory environment [50]. IL-15 also affects B-cell function by promoting their survival and differentiation [51].

1.3.1.1.7 IL-21

IL-21 stimulates B cells and NK cell activation boosts antibody production, and its effects are mediated through a receptor predominantly expressed in CD4+ T cells, macrophages, and dendritic cells in individuals with RA [52]. Evidence suggests that IL-21 enhances the migration and invasion of rheumatoid arthritis fibroblast-like synoviocytes (RA-FLS). Research utilizing murine models has further highlighted the critical role of IL-21 in RA. In these experimental models, the application of the IL-21 receptor Fc fusion protein (IL-21RFc), which acts as an antagonist of IL-21, led to a marked decrease in the concentrations of IL-6 and IL-17 [53]. Additionally, Xing et al. reported that IL-21 substantially enhances the migration and invasion of RA-FLS. This enhancement is potentially mediated through the upregulation of MMPs including MMP-2, MMP-3, MMP-9, and MMP-13, as well as intercellular adhesion molecule-1 (ICAM-1) and cadherin-11 [54].

1.3.1.2 Systemic Lupus Erythematosus

SLE is a chronic autoimmune condition characterized by persistent inflammation and immune-related damage to various organ systems, including the skin, mucous membranes, joints, muscles, blood cells, and kidneys [55]. According to the 2019 report from the European Alliance of Associations for Rheumatology and the American College of Rheumatology, approximately 90% of individuals with SLE are female. About 40% of SLE patients develop lupus nephritis and roughly 10% of those with lupus nephritis progress to end-stage kidney disease within 10 years [55]. In SLE, the immune system becomes excessively active, producing autoantibodies that attack nuclear antigens. This leads to the production of immune complexes that accumulate within tissues, triggering complement activation and subsequent inflammation [56]. A characteristic feature of SLE is the heightened activity of B cells alongside a breakdown in B-cell tolerance. Immune complexes with nucleic acid autoantigens can activate endosomal toll-like receptors (TLRs), exacerbating SLE inflammation [57]. Additionally, this condition is marked by increased plasma cells and the synthesis of autoantibodies, which are typically non-pathogenic unless generated within an inflammatory environment. In such an environment, these autoantibodies can switch to pathogenic isotypes [58]. Proinflammatory cytokines play a pivotal role in the stimulation of T cells and the maturation of dendritic cells. Additionally, they facilitate extramedullary hematopoiesis, leading to an increase in innate immune cells, and drive the synthesis of acute-phase proteins. Autoantibodies may aggregate in tissues like the kidney glomeruli, activating myeloid effector cells through Fcγ and complement receptors, contributing to tissue damage. The advancement and severity of the disease are affected by the interaction of genetic predispositions, environmental conditions, dietary factors, and stress [58]. SLE, significant autoantibodies encompass anti-nuclear antibodies, anti-double-stranded DNA antibodies and anti-Smith antibodies [59]. Type I interferons (IFNs), especially IFN-α, play a pivotal role in this process. These cytokines are overproduced in SLE patients and enhance the presentation of autoantigens and the production of autoantibodies, thereby sustaining the inflammatory cycle [60].

1.3.1.3 Multiple Sclerosis

MS is a long-term inflammatory disease affecting the central nervous system (CNS), distinguished by the gradual degradation of myelin, gliosis, neuro-axonal damage, and persistent inflammatory processes [61]. This condition is characterized by a complex interplay of inflammatory and neurodegenerative mechanisms, leading to various clinical impairments such as motor, sensory, and cognitive dysfunctions [62, 63]. This condition involves an excess buildup of macrophages/microglia and lymphocytes within the CNS (both white and gray matter), resulting in the demyelination and damage of axons [64]. Although the precise origin of MS remains uncertain, it is commonly understood that inflammation within the CNS is considered the principal source of damage in MS [65]. Studies have demonstrated that proinflammatory mediators, including different interleukins, play a substantial role in exacerbating inflammation and causing tissue damage in MS [66]. The literature indicates that TNF levels are elevated in the cerebrospinal fluid of patients with MS [67, 68]. Similar observations have been reported in animal models of MS. Additionally, elevated levels of TNF have been identified in both central and peripheral demyelinating lesions, as well as in the serum and cerebral plaques of individuals with MS [69]. Furthermore, a study conducted on testing the potential of Atrosab, a novel premier TNFR1 blocker, in animal models of MS, specifically experimental autoimmune encephalomyelitis (EAE), revealed that Atrosab reduced symptom severity of MS [70]. Moreover, IL-17 can also promote inflammation and tissue damage within the CNS by activating other immune cells and inducing the production of proinflammatory molecules. This inflammatory cascade contributes to the destruction of myelin and neuronal tissue, exacerbating MS symptoms and disease progression [71]. Cell line research revealed that human astrocytes when exposed to IL-17, generate IL-6, a cytokine known to sustain the transformation of CD4 naïve cells into T-helper-17 (Th17) cells [72]. Furthermore, the study confirmed the importance of IL-17-triggered astrocyte activation in the development of EAE, as evidenced by the reduced severity of EAE in animals with inhibited IL-17 signaling in astrocytes [72].

1.3.2 Infectious Inflammatory Diseases

Infectious inflammatory diseases occur when infections trigger the immune system, resulting in inflammation. These diseases are triggered by a variety of agents, such as bacteria, viruses, and parasites [12]. The inflammatory response elicited by these infections may lead to tissue damage, thereby exacerbating disease symptoms and advancing its progression. Examples of infectious inflammatory diseases include asthma, tuberculosis (TB), UC, and CD. The immune response to the infection, while aimed at controlling and eliminating the pathogen, can also result in significant inflammation and damage to the affected tissues [73].

1.3.2.1 Tuberculosis

Inflammatory mediators are key contributors to the development of TB. Following infection with *Mycobacterium tuberculosis* (Mtb), the host immune system triggers the release of numerous cytokines and inflammatory molecules. Notably, TNF-α and interleukins like IL-1β are critical for the formation of granulomas, organized cellular structures that help contain and limit the spread of the infection [74]. Granulomas are central to TB pathology, serving both to contain the bacteria and as sites of chronic inflammation. Key mediators like eicosanoids can drive granuloma dissociation and exacerbate tissue damage. Eicosanoids, such as prostaglandin E2 (PGE2), are involved in balancing the immune response by counteracting type I interferons, which can suppress protective cytokines like IL-1α and IL-1β. Modulating these mediators, such as by employing nonsteroidal anti-inflammatory drugs (NSAIDs) or intervening in lipid metabolism pathways, has demonstrated promise in mitigating tuberculosis-related pathology and enhancing therapeutic efficacy [75]. Additionally, the immune response in TB involves a delicate equilibrium between proinflammatory and anti-inflammatory signals. Proinflammatory cytokines such as IL-1β and IL-6, are crucial for managing Mtb infection; however, their unregulated activity can lead to tissue damage [76, 77]. Conversely, anti-inflammatory cytokines such as IL-10 and transforming growth factor-beta (TGF-β) play a role in mitigating tissue damage. However, their activity may also suppress vigorous immune responses against bacterial pathogens, potentially facilitating the persistence and progression of the disease [78].

1.3.2.2 Inflammatory Bowel Diseases

IBD, including both CD and UC, is a significant contributor to chronic gastrointestinal disorders worldwide, characterized by varied clinical manifestations and a complex underlying etiology [79]. UC is a primary form of IBD, a persistent inflammatory condition affecting the colonic mucosa, marked by recurrent periods of flare-ups and remission [80]. UC is defined by ongoing, superficial inflammation localized to the mucosal and submucosal layers of the colon, while CD is characterized by discontinuous, transmural inflammation that can involve any segment of the gastrointestinal tract [81]. Inflammatory cells, which secrete various cytokines, are crucial to the pathogenesis of UC. M cells in the intestinal epithelium detect foreign antigens and stimulate immune cells within Peyer's patches. These patches are primarily composed of B cells, T cells (including follicular helper T cells), and dendritic cells [82]. When M cells capture bacterial antigens from the intestinal lumen, a range of immune cells produces immunoglobulin

A (IgA). This IgA binds to the antigens on pathogenic bacteria, thereby aiding in infection prevention. Dendritic cells present these bacterial antigens to naïve T cells in the mesenteric lymph nodes. Upon antigen recognition, these naïve T cells differentiate into various helper T cell subsets. These subsets release both anti-inflammatory cytokines, which enhance barrier function and facilitate inflammation resolution, and proinflammatory cytokines, which are thought to play a role in chronic inflammation and tissue damage [81, 83]. Patients with UC exhibit abnormal neutrophil activation, as documented in studies [84]. Neutrophils play a pivotal role in tissue damage by secreting proteolytic enzymes, including MMPs and neutrophil elastases, as well as ROS [85]. The accumulation of activated neutrophils leads to modifications in cryptic structures and the formation of crypt abscesses, driven by an unbalanced enzymatic response, and increased production of TNF-α and IL-1β, along with the release of calprotectin [86]. In UC, macrophages expressing elevated levels of TLR-2 along with the CX3C motif chemokine receptor tend to accumulate at sites of inflammation [87]. These macrophages demonstrate an increased ability to secrete proinflammatory cytokines. The excessive activation of macrophages contributes to the worsening of fibrosis driven by myofibroblasts through the production of factors such as TGF-β1, connective tissue growth factor, and fibroblast activating protein [88].

1.3.3 Allergic Inflammatory Diseases

Allergic inflammatory disorders are a group of conditions marked by an inappropriate immune reaction to generally benign environmental substances, termed allergens. The development of these disorders is influenced by a multifaceted interaction involving genetic susceptibility, environmental exposures, and immune system dysfunction. Common manifestations of such conditions include asthma [89].

1.3.3.1 Asthma

Asthma is a long-term inflammatory condition that can cause gradual and possibly irreversible decreases in lung function for some individuals. The inflammation in asthma occurs through a series of interactions between inflammatory cells and resident cells, leading to a chain of events that perpetuate chronic inflammation and the symptoms of the disease. These symptoms include increased inflammation, spasms of the airway smooth muscles (bronchospasm), secretion of mucus in the airways, swelling and narrowing of the airways, and damage to the bronchial epithelium [90]. Environmental triggers such as allergens and pollutants cause the airway to narrow, which leads to reversible, episodic bronchoconstriction along with hyperresponsiveness and obstruction of the airway, leading to symptoms such as chest tightness, shortness of breath, recurrent wheezing episodes and coughing [91, 92]. Other factors that might play a pathological role in asthma development are microbiome, chemical, exposure, stress, and dietary change [92]. Conventionally, asthma is classified as extrinsic and intrinsic asthma. Extrinsic asthma is caused by allergens, also called allergic asthma whereas intrinsic asthma is exacerbated by factors such as obesity, aspirin treatment, exercise, cold, stress, or pulmonary infection. Extrinsic asthma results in T-helper type 2 (Th-2) cell inflammation which is further subdivided into Th2 high and Th-2 low inflammation. In Th2 high asthma there is increased blood eosinophil counts, whereas Th2 low asthma involves neutrophils [93].

Asthma is a persistent inflammatory condition affecting the airways, distinguished by the presence of type 2 cytokines, including IL-4, IL-5, and IL-13. These cytokines contribute to airway eosinophilia, increased mucus production, bronchial hyperresponsiveness, and elevated immunoglobulin E levels. Nevertheless, approximately 50% of asthma patients do not exhibit an exaggerated type 2 immune response. "Type 2-low" asthma is distinguished by different immune features, including airway neutrophilia, systemic inflammation often linked with obesity, or, in some cases, minimal signs of immune activation [94]. The transcription factor GATA-binding protein 3 (GATA3) plays a critical role in guiding the differentiation of naïve T cells into Th2 cells and is crucial for the regulation of Th2 cytokine production [95]. Evidence suggests a rise in the population of GATA3+ T cells within the airways of stable asthmatic individuals [96]. Furthermore, IL-4 is crucial in guiding the differentiation of Th2 cells from naïve Th-0 cells and may be instrumental in the initial allergen sensitization [97]. In concert with IL-13, IL-4 triggers the production of IgE by B cells [98]. When allergens bind to IgE molecules on mast cells and basophils, which are attached to high-affinity receptors, these cells release significant amounts of IL-4 and IL-13 [99, 100], thus intensifying type-2 airway inflammation.

1.4 INFLAMMATION AND NEURODEGENERATIVE DISEASES

Substantial evidence underscores the critical role of neuroinflammation in both the initiation and perpetuation of progressive neurodegenerative processes [101]. Microglia and astrocytes are essential for modulating inflammatory processes within the CNS [102]. Proinflammatory cytokines, as well as remnants from pathogens or damaged cells, initiate the activation of quiescent microglia. This activation leads to the production of proinflammatory mediators such as interleukins, TNF-α, COX-2, nitric oxide (NO), and various proteases, which contribute to detrimental effects observed in neurodegenerative diseases [102, 103]. Cyclooxygenase (COX) expression in the brain correlates with inflammatory processes that play a crucial role in neurodegenerative conditions [104]. COX-2 is accountable for generating ROS. These species, in turn, stimulate NF-κB activation, thereby enhancing the synthesis of proinflammatory mediators such as IL-6, TNF-a, NO and PGE2 [105].

1.4.1 Alzheimer's Disease

Alzheimer's disease (AD) is a progressive neurodegenerative disorder primarily characterized by cognitive impairment and memory loss [106]. In 2022, a report by the World Health Organization (WHO) documented that the global prevalence of dementia affects over 55 million individuals. This figure is projected to rise to 139 million by 2050 [107]. According to the Alzheimer's Association, as of the latest report, approximately 6.7 million individuals aged 65 and older in the United States are affected by AD. In 2019, AD ranked as the 6th leading cause of death, and it was the seventh leading cause in 2020 and 2021. The number of family caregivers has declined, yet their caregiving hours have increased, valued at $339.5 billion in 2022 [108]. According to projections, the worldwide prevalence of dementia is expected to increase from 57.4 million individuals in 2019 to 152.8 million by 2050. Despite this increase, age-standardized prevalence rates remain stable, with a 0.1% global change [109].

The defining pathological characteristics of AD encompass the deposition of amyloid-beta (Aβ) plaques and tau protein tangles within the brain, accompanied by neuroinflammation and the loss of neurons [106]. Over the past decade, research has highlighted the role of neuroinflammation as a significant factor in AD. Chronic neuroinflammation, characterized by the sustained activation of glial cells and the excessive release of deleterious molecules, is regarded as a fundamental aspect of AD. Initially, microglial activation helps clear Aβ and tau but can later exacerbate disease progression due to prolonged inflammation. Reactive astrocytes also contribute to the inflammatory process [110]. In AD, the deposition of Aβ peptide kick-starts a range of cerebral neuroinflammatory responses through the activation of microglia. Activated microglia have the potential to adversely affect surrounding brain tissue by stimulating the release of proinflammatory mediators such as IL-1β, IL-6, and TNF-α [111] (Figure 1.2). Proinflammatory cytokines contribute to the upregulation of amyloid precursor protein (APP) production and its subsequent proteolytic cleavage, which results in increased generation of Aβ42 peptide [112]. Additionally, in AD, reactive astrocytes surround amyloid plaques and exhibit elevated levels of intermediate filaments, notably glial fibrillary acidic protein (GFAP) [113, 114]. Moreover, evidence indicates that Aβ enhances the expression of inflammatory enzymes, including COX-2 and inducible nitric oxide synthase (iNOS) (Figure 1.2). Furthermore, current research suggests that neuroinflammation may contribute to the hyperphosphorylation of tau proteins, which subsequently leads to the formation of neurofibrillary tangles [115]. Conversely, the presence of these tangles can intensify neuroinflammation by interacting with microglia, thereby promoting the release of neuroinflammatory mediators [116]. Like Aβ, tau oligomers have been shown to initiate an inflammatory response [117]. Research indicates that these oligomers stimulate the release of high mobility group box 1 (HMGB1), a protein commonly associated with AD. This protein can activate receptors, including receptors for advanced glycation end products and TLR-4, which subsequently promote the level of proinflammatory mediators [117, 118].

1.4.2 Parkinson's Disease

Parkinson's disease (PD) is a progressive neurodegenerative disorder characterized by the gradual deterioration of dopaminergic neurons in the substantia nigra (SN) region of the brain. This neuronal loss leads to a decrease in dopamine levels, a neurotransmitter crucial for motor regulation, which manifests as the hallmark motor symptoms associated with the disease [119]. As per the report of WHO, the prevalence of PD has doubled globally over the past 25 years, with over 8.5 million people affected by 2019. The rates of disability and mortality associated with PD are increasing more rapidly than any other neurological condition. In 2019, PD accounted for 5.8 million disability-adjusted life years (DALYs), representing an 81% increase since 2000. Additionally, PD was responsible for 329,000 deaths in 2019, more than doubling the mortality rate observed in 2000. This sharp escalation underscores the increasing burden of PD on global health [120]. The role of inflammation in the onset of Parkinsonian symptoms has been a topic of considerable suspicion and investigation for many years [121]. In 1988, McGeer and colleagues first established a link between inflammation and PD through a postmortem study, which revealed activated microglia in the SN of PD patients [122]. Dopaminergic neurons are vulnerable to stressors, and inflammation is acknowledged as a risk factor for neuronal dysfunction and cell death in prevalent neurodegenerative disorders [123]. Evidence delineates that microglial cell activation induces the activation of the TLR2-MyD88 pathway, which increases inflammation and proinflammatory cytokines. These cytokines further directly act on dopaminergic neurons and alpha-synuclein via the NF-κB pathway [124]. Inflammation also increases oxidative stress, leading to elevated ROS levels, which decrease the susceptibility of dopaminergic neurons. TLR-4 receptor may regulate inflammation [125]. Activation of TLR-4 initiates the activation of NF-κB through the MyD88-dependent signaling cascade, leading to the upregulation of the NOD-like receptor family, pyrin domain containing 3 (NLRP3), along with the release of multiple proinflammatory cytokines [126] (Figure 1.2).

Numerous studies have demonstrated that elevated concentrations of proinflammatory cytokines, including TNF-α, IL-1β, and IL-6, are associated with the advancement of neurodegenerative processes. Reduced levels of these cytokines may be initiated by microglial cell activation, which promotes apoptotic cell death and phagocytosis of DA neurons [127]. The degeneration of dopaminergic neurons is associated with the CX3CL1-CX3CR1 signaling pathway. This axis plays a protective role against dopaminergic neuron loss when microglial activation is absent. However, in

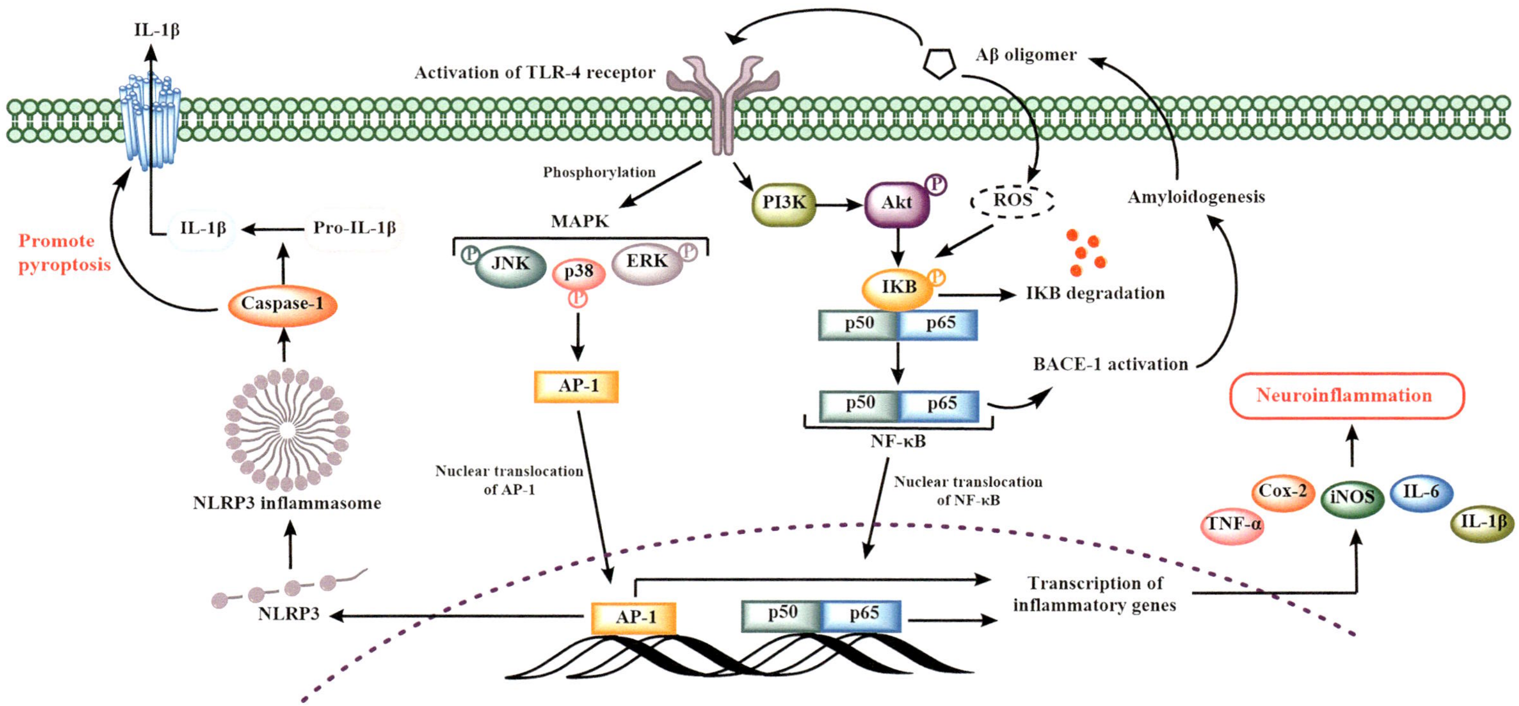

FIGURE 1.2 Schematic depiction of TLR-4-mediated neuroinflammation. TLR-4 activation recruits adaptor proteins, initiating downstream signaling. The TLR-4/MAPK pathway phosphorylates MAP kinases (ERK, JNK, p38), activating AP-1 to induce proinflammatory cytokines. Concurrently, the TLR-4/NF-κB pathway leads to IκB degradation and NF-κB nuclear translocation, promoting inflammatory gene transcription. Together, these pathways drive neuroinflammation and neuronal damage, common in neurodegenerative diseases.

the presence of microglial activation, there is a reduction in CX3CL1-CX3CR1 expression, which leads to the death of dopaminergic neurons in SN [128]. Approximately 17 monogenic genes have been implicated in the pathogenesis of PD. Of these, eight are specifically expressed in astrocytes. These include PARK7, SNCA, PLA2G6, ATP13A2, LRRK2, GBA, PINK1, and PARK2 [129]. Several of these genes are implicated in inflammatory processes. DJ-1, associated with Parkinsonism, is predominantly expressed in astrocytes compared to neurons. Genetic mutations affecting these genes disrupt TLR-3/4-mediated endocytosis, modify the production of proinflammatory cytokines, and elevate the generation of ROS [130]. These alterations result in the addition of extracellular alpha-synuclein, leading to the generation of proinflammatory mediators and an increase in the expression of ICAM. This overexpression of alpha-synuclein results in the loss of dopaminergic neurons. In the PINK1/PARKIN gene, mutation leads to mitophagy, and the accumulation of damaged mitochondria leads to neurodegeneration [131]. Another pathway, aside from TNF-α, is IFN-γ, also called type II interferon. The level of IFN-γ increases significantly with alpha-synuclein, implicating a direct link with progressive dopaminergic neuronal death [132]. The binding of IFN-γ to its receptor IFNGR leads to the activation of leucine-rich repeat kinase 2 (LRRK2), which plays roles in microglia and dopaminergic neurons. Activated LRRK2 suppresses c-Rel expression, promotes TAU oligomer formation leading to cell death, and increases inflammatory cytokine membrane receptors [133, 134].

1.4.3 Huntington's Disease

Huntington's disease (HD) is a neurodegenerative condition marked by a gradual decline in cognitive abilities, motor skills, and psychiatric health [135]. This condition arises from a genetic alteration in the huntingtin gene (HTT), which results in the synthesis of a mutant variant of the huntingtin protein (mHTT) [136]. Besides the pathological clumping of mHTT, other mechanisms, such as inflammation, appear to contribute to the development and advancement of HD. Changes in inflammation may be detectable even before the emergence of the typical symptoms associated with HD. Evidence delineates that in pathological circumstances like HD, microglia get activated before astrocytes. In neurodegenerative lesions, the activated microglia release neurotrophic or neurotoxic mediators. A clinical and positron emission tomography investigation has linked microglial activity to the severity of HD. Thus, controlling microglial activation might be a tempting treatment approach for several neurological conditions, including HD [137]. Microglial cell activation induced the activation of TLR-4, which increased the proinflammatory cytokines. Inflammation may be regulated by the TLR-4 receptor. Activation of TLR-4 initiates the NF-kB signaling cascade through the MyD88-dependent mechanism, leading to the upregulation of NLRP3 and the production of multiple proinflammatory cytokines [126]. TLR-4 functions analogously to HTT mRNA and is prominently expressed in microglia. Elevated levels of mHTT RNA are associated with upregulation of proinflammatory cytokines including TNF-α, IL-1β, and IL-6 (refer to Figure 1.2). Various genes upregulated in mHTT microglia, such as Tlr2, Cd14, Fcgrl, Clec4d, Adora3, Tlr9, and Tnfrsf1b, contribute to the increased expression of IL-6 and TNF-α [138]. HTT protein does not initiate autophagy and mitophagy, but mutant HTT binds abnormally with fusion protein dynamin-related protein 1 (Drpl) with mitochondrial fragmentation [131]. Mutant HTT acts as an inflammatory stimulus for cells. HTT also increased the MCP-1 and murine chemokine. Activation of IkappaB (IkB) kinase leads to the production of NF-β and its translocation to the nucleus and promotes the release of proinflammatory cytokines [139].

1.5 FUTURE PROSPECTS

Looking ahead, further research is needed to unravel the precise mechanisms underlying the transition from acute to chronic inflammation and its impact on various disease states. Investigating the molecular pathways involved in cytokine signaling and immune cell activation could provide insights into potential therapeutic targets. For autoimmune diseases, understanding the interplay between genetic predispositions and environmental triggers will be crucial for developing targeted interventions. In the realm of infectious diseases, a deeper exploration into the modulation of inflammatory mediators could lead to novel strategies for reducing tissue damage and improving treatment outcomes. Similarly, in allergic inflammatory diseases, elucidating the mechanisms behind different asthma phenotypes and their response to various triggers will be essential for personalized treatment approaches. Neurodegenerative diseases present an urgent need for innovative therapeutic strategies aimed at mitigating neuroinflammation and protecting neuronal integrity. Exploring the role of microglial activation and identifying pathways that regulate neuroinflammatory responses may offer new avenues for treatment. Overall, advancing our understanding of inflammation across these diverse contexts will be instrumental in developing effective therapeutic strategies and improving patient outcomes.

1.6 CONCLUSION

In summary, inflammation is a multifaceted physiological process with significant implications across various disease contexts. Acute inflammation serves as an essential protective response aimed at addressing immediate threats and initiating repair, as exemplified by conditions such as acute appendicitis. However, when inflammation becomes chronic, it transitions into a more complex pathological state characterized by sustained immune activation and tissue damage. Autoimmune inflammatory diseases like RA, SLS, and MS illustrate how the immune system's dysregulation can lead to persistent inflammation and damage to self-tissues. These diseases underscore the role of genetic, environmental, and cytokine-mediated factors in perpetuating chronic inflammation. Infectious inflammatory diseases, such as tuberculosis and IBDs, further highlight the dual nature of inflammation as both a defensive mechanism and a source of tissue damage. The balance between proinflammatory and anti-inflammatory signals is critical for controlling infections and preventing chronic inflammation. Allergic inflammatory diseases, such as asthma, add another layer of complexity, demonstrating how environmental allergens and immune dysregulation contribute to chronic airway inflammation and dysfunction. Neurodegenerative diseases, including AD, PD, and HD, exemplify how neuroinflammation plays a pivotal role in disease progression. Chronic neuroinflammation contributes to neuronal damage and disease symptoms, revealing the intricate relationship between immune responses and neurodegeneration.

FUNDING

The authors affirm that no financial support, grants, or other forms of assistance were provided in the development of this chapter.

CONFLICT OF INTEREST

All authors affirm the absence of any conflicts of interest related to this study.

ETHICAL APPROVAL

This chapter does not include any research involving human participants or animal subjects conducted by the authors.

REFERENCES

1. Bansal P., Verma D., Singh Tamber S. et al. (2023) An update on pathogenesis of inflammatory disorders with its management. *RPS Pharmacy and Pharmacology Reports* 2(2):11.
2. Barnes P. J. (2009) The cytokine network in chronic obstructive pulmonary disease. *American Journal of Respiratory Cell and Molecular Biology* 41(6):631–638.
3. Ahmed A. U. (2011) An overview of inflammation: mechanism and consequences. *Frontiers in Biology* 6(4):274–281.
4. Zhou Y., Hong Y., Huang H. (2016) Triptolide attenuates inflammatory response in membranous glomerulo-nephritis rat via downregulation of NF-κB signaling pathway. *Kidney and Blood Pressure Research* 41(6):901–910.
5. Bidian C., Filip G. A., David L. et al. (2022) Viburnum opulus fruit extract-capped gold nanoparticles attenuated oxidative stress and acute inflammation in carrageenan-induced paw edema model. *Green Chemistry Letters and Reviews* 15(2):320–336.
6. Adegbaju O. D., Otunola G. A., Afolayan A. J. (2020) Anti-inflammatory and cytotoxic evaluation of extracts from the flowering stage of Celosia argentea. *BMC Complementary Medicine and Therapies* 20:1–7.
7. Akira S., Uematsu S., Takeuchi O. (2006) Pathogen recognition and innate immunity. *Cell* 124(4):783–801.
8. Kaminska, B. (2005) MAPK signalling pathways as molecular targets for anti-inflammatory therapy—from molecular mechanisms to therapeutic benefits. *Biochimica et Biophysica Acta (BBA)-Proteins and Proteomics* 1754(1–2):253–262.
9. Hendrayani, Siti-Fauziah (2016) The inflammatory/cancer-related IL-6/STAT3/NF-κB positive feedback loop includes AUF1 and maintains the active state of breast myofibroblasts. *Oncotarget* 7(27): 41974.
10. Lai J. L., Liu Y. H., Liu, C. et al. (2017) Indirubin inhibits LPS-induced inflammation via TLR4 abrogation mediated by the NF-kB and MAPK signaling pathways. *Inflammation* 40:1–12.
11. Henríquez-Olguín C., Altamirano F., Valladares D. et al. (2015) Altered ROS production, NF-κB activation and interleukin-6 gene expression induced by electrical stimulation in dystrophic mdx skeletal muscle cells. *Biochimica et Biophysica Acta (BBA)-Molecular Basis of Disease* 1852(7):1410–1419.
12. Chen L., Deng H., Cui H. et al. (2018) Inflammatory responses and inflammation-associated diseases in organs. *Oncotarget* 9(6):7204.
13. Mcinnes I.B., Schett G. (2017) Pathogenetic insights from the treatment of rheumatoid arthritis. *The Lancet* 389(10086):2328–2337.
14. Neurath M.F. (2017) Current and emerging therapeutic targets for IBD. *Nature Reviews Gastroenterology & Hepatology* 14(5):269–278.
15. Nestle F.O., Kaplan D.H., Barker J. (2009) Mechanisms of disease: psoriasis. *New England Journal of Medicine* 361(5):496–509.
16. Kiss A.L. (2022) Inflammation in focus: the beginning and the end. *Pathology and Oncology Research* 27:1610136.
17. Medzhitov R. (2008) Origin and physiological roles of inflammation. *Nature* 454(7203):428–435.
18. Bleakley C., McDonough S., MacAuley D. (2004) The use of ice in the treatment of acute soft-tissue injury: a systematic review of randomized controlled trials. *The American Journal of Sports Medicine* 32(1):251–261.
19. Lin Y.J., Anzaghe M., Schülke S. (2020) Update on the pathomechanism, diagnosis, and treatment options for rheumatoid arthritis. *Cells* 9(4):880.
20. Furman D., Campisi J., Verdin E. et al. (2019) Chronic inflammation in the etiology of disease across the life span. *Nature Medicine* 25(12):1822–1832.
21. Arroyo V., Angeli P., Moreau R. (2021) The systemic inflammation hypothesis: towards a new paradigm of acute decompensation and multiorgan failure in cirrhosis. *Journal of Hepatology* 74(3):670–685.
22. Zotova N., Zhuravleva Y., Chereshnev V. et al. (2023) Acute and chronic systemic inflammation: features and differences in the pathogenesis, and integral criteria for verification and differentiation. *International Journal of Molecular Sciences* 24(2):1144.
23. Abudukelimu A., Barberis M., Redegeld F.A. et al. (2018) Predictable irreversible switching between acute and chronic inflammation. *Frontiers in Immunology* 9:1596.
24. Kivity S., Agmon-Levin N., Blank M., et al. (2009) Infections and autoimmunity–friends or foes?. *Trends in Immunology* 30(8):409–414.
25. Pollard K.M., Hultman P., Kono D.H. (2010) Toxicology of autoimmune diseases. *Chemical Research in Toxicology* 23(3):455–466.

26. Guo Q, Wang Y, Xu D et al. (2018) Rheumatoid arthritis: pathological mechanisms and modern pharmacologic therapies. *Bone Research* 6(1):15.
27. Lee JE, Kim IJ, Cho MS et al. (2017) A case of rheumatoid vasculitis involving hepatic artery in early rheumatoid arthritis. *Journal of Korean Medical Science* 32(7):1207–1210.
28. Picerno V, Ferro F, Adinolfi A et al. (2015) One year in review: the pathogenesis of rheumatoid arthritis. *Clin Exp Rheumatol* 33(4):551–558.
29. Bullock J, Rizvi SA, Saleh AM et al. (2019) Rheumatoid arthritis: a brief overview of the treatment. *Medical Principles and Practice*. 27(6):501–507.
30. Yap H.Y., Tee S.Z.Y., Wong M.M.T. et al. (2018) Pathogenic role of immune cells in rheumatoid arthritis: implications in clinical treatment and biomarker development. *Cells* 7(10):161.
31. Smolen JS, Aletaha D, Barton A et al. (2018) *Rheumatoid Arthritis*. Nature reviews. Dis Primers 418001.
32. O'Neil LJ, Alpízar-Rodríguez D, Deane KD (2024) Rheumatoid arthritis: The continuum of disease and strategies for prediction, early intervention, and prevention. *The Journal of Rheumatology* 51(4):337–349.
33. Wysocki T, Olesińska M, Paradowska-Gorycka A (2020) Current understanding of an emerging role of HLA-DRB1 gene in rheumatoid arthritis–from research to clinical practice. *Cells* 9(5):1127.
34. Kurko J, Besenyei T, Laki J et al. (2013) Genetics of rheumatoid arthritis—a comprehensive review. *Clinical Reviews in Allergy and Immunology* 45:170–179.
35. Klareskog L, Padyukov L, Lorentzen J et al. (2006) Mechanisms of disease: genetic susceptibility and environmental triggers in the development of rheumatoid arthritis. *Nature Clinical Practice Rheumatology* 2(8):425–433.
36. Ferrero-Miliani L, Nielsen, O.H., Andersen P.S. et al. (2007) Chronic inflammation: Importance of NOD2 and NALP3 in interleukin-1β generation. *Clinical and Experimental Immunology* 147(2), 227–235.
37. Serhan C.N., Ward P.A., Gilroy, D.W. (2005) *Fundamentals of Inflammation*. Cambridge University Press.
38. Voigt L.F., Koepsell T.D., Nelson J.L. et al. (1994) Smoking, obesity, alcohol consumption, and the risk of rheumatoid arthritis. *Epidemiology* 5(5):525–532.
39. Catrina A I, Ytterberg A J, Reynisdottir G et al. (2014) Lungs, joints and immunity against citrullinated proteins in rheumatoid arthritis. *Nature Reviews Rheumatology* 10(11):645–653.
40. Mcinnes I.B., Schett G (2011) The pathogenesis of rheumatoid arthritis. *N Engl J M* 365(23):2205–2219.
41. Van de Loo A.A., Van den Berg W.B. (1990) Effects of murine recombinant interleukin 1 on synovial joints in mice: measurement of patellar cartilage metabolism and joint inflammation. *Annals of the Rheumatic Diseases* 49(4):238–245.
42. Keffer J, Probert L, Cazlaris H, Georgopoulos S et al. (1991). Transgenic mice expressing human tumour necrosis factor: a predictive genetic model of arthritis. *The EMBO Journal* 10(13):4025–4031.
43. Waetzig G.H., Rosenstiel P, Arlt A, et al. (2005) Soluble tumor necrosis factor (TNF) receptor-1 induces apoptosis via reverse TNF signaling and autocrine transforming growth factor-β1. *The FASEB Journal* 19(1):91–93.
44. Dinarello C A, Simon A, Van Der Meer J W (2012) Treating inflammation by blocking interleukin-1 in a broad spectrum of diseases. *Nature Reviews Drug Discovery* 11(8):633–652.
45. Dinarello C A (2011) Interleukin-1 in the pathogenesis and treatment of inflammatory diseases. *Blood, The Journal of the American Society of Hematology* 117(14):3720–3732.
46. Kondo N, Kuroda T, Kobayashi D (2021) Cytokine networks in the pathogenesis of rheumatoid arthritis. *International Journal of Molecular Sciences* 22(20):10922.
47. Srirangan S, Choy E.H. (2010) The role of interleukin 6 in the pathophysiology of rheumatoid arthritis. *Therapeutic Advances in Musculoskeletal Disease* 2(5):247–256.
48. Mueller A.L., Payandeh Z, Mohammadkhani N et al. (2021) Recent advances in understanding the pathogenesis of rheumatoid arthritis: new treatment strategies. *Cells* 10(11):3017.
49. Horwitz D.A., Gray J.D., Ohtsuka K (1999) Role of NK cells and TGF-β in the regulation of T-cell-dependent antibody production in health and autoimmune disease. *Microbes and Infection* 1(15):1305–1311.
50. Ruckert R, Brandt K, Ernst M et al. (2009) Interleukin-15 stimulates macrophages to activate CD4+ T cells: a role in the pathogenesis of rheumatoid arthritis?. *Immunology* 126(1):63–73.
51. Benito-Miguel M, García-Carmona Y, Balsa A et al. (2012) IL-15 expression on RA synovial fibroblasts promotes B cell survival. *PLoS One* 7(7):40620.
52. Habib T, Nelson A, Kaushansky K et al. (2003) IL-21: a novel IL-2–family lymphokine that modulates B, T, and natural killer cell responses. *Journal of Allergy and Clinical Immunology* 112(6):1033–1045.
53. Young D.A., Hegen M, Ma H.L.M. et al. (2007) Blockade of the interleukin-21/interleukin-21 receptor pathway ameliorates disease in animal models of rheumatoid arthritis. *Arthritis & Rheumatism* 56(4):1152–1163.
54. Xing R, Jin Y, Sun L et al. (2016) Interleukin-21 induces migration and invasion of fibroblast-like synoviocytes from patients with rheumatoid arthritis. *Clinical & Experimental Immunology* 184(2):147–158.
55. Siegel C.H., Sammaritano L.R. (2024) Systemic lupus erythematosus: A review. *JAMA* 331(17):1480–1491.
56. Herrada A.A., Escobedo N, Iruretagoyena M et al. (2019) Innate immune cells' contribution to systemic lupus erythematosus. *Frontiers in Immunology* 10: 449737.
57. Fillatreau S, Manfroi B, Dörner, T (2021) Toll-like receptor signalling in B cells during systemic lupus erythematosus. *Nature Reviews Rheumatology* 17(2):98–108.
58. Gottschalk T A, Tsantikos E, Hibbs M L (2015) Pathogenic inflammation and its therapeutic targeting in systemic lupus erythematosus. *Frontiers in Immunology* 6:166622.
59. Crispín J C, Liossis S N C, Kis-Toth K et al. (2010) Pathogenesis of human systemic lupus erythematosus: recent advances. *Trends in Molecular Medicine* 16(2):47–57.
60. Banchereau J & Pascual V (2006) Type I interferon in systemic lupus erythematosus and other autoimmune diseases. *Immunity* 25(3):383–392.
61. Haase S & Linker R A (2021) Inflammation in multiple sclerosis. *Therapeutic Advances in Neurological Disorders* 14:17562864211007687.
62. Reich D.S., Lucchinetti C.F., Calabresi P.A. (2018) Multiple sclerosis. *The New England Journal of Medicine* 378(2):169–180.
63. Thompson A.J., Baranzini S.E., Geurts J et al. (2018) Ciccarelli. O multiple sclerosis *Lancet* 391:1622–1636.
64. Mao P, Reddy PH (2010) Is multiple sclerosis a mitochondrial disease?. *Biochimica et Biophysica Acta (BBA)-Molecular Basis of Disease* 1802(1):66–79.

65. Trapp B.D., Nave K.A. (2008) Multiple sclerosis: an immune or neurodegenerative disorder. *Annu. Rev. Neurosci.* 31:247–269.
66. Palle P, Monaghan K.L., Milne S.M. et al. (2017) Cytokine signaling in multiple sclerosis and its therapeutic applications. *Medical Sciences* 5(4):.23.
67. ShariefM.K.,HentgeR(1991)Associationbetweentumornecrosis factor-α and disease progression in patients with multiple sclerosis. *New England Journal of Medicine* 325(7):467–472.
68. Rossi S, Motta C, Studer V et al. (2014) Tumor necrosis factor is elevated in progressive multiple sclerosis and causes excitotoxic neurodegeneration. *Multiple Sclerosis Journal* 20(3):304–312.
69. Kemanetzoglou E, Andreadou E (2017) CNS demyelination with TNF-α blockers. *Current Neurology and Neuroscience Reports* 17:1–15.
70. Williams S.K., Fairless R, Maier O et al. (2018) Anti-TNFR1 targeting in humanized mice ameliorates disease in a model of multiple sclerosis. *Scientific Reports* 8(1):13628.
71. Milovanovic J, Arsenijevic A, Stojanovic B et al. (2020) Interleukin-17 in chronic inflammatory neurological diseases. *Frontiers in Immunology* 11:947.
72. Elain G, Jeanneau K, Rutkowska A et al. (2014) The selective anti-IL17A monoclonal antibody secukinumab (AIN457) attenuates IL17A-induced levels of IL6 in human astrocytes. *Glia* 62(5):725–735.
73. Cassell G H (1998) Infectious causes of chronic inflammatory diseases and cancer. *Emerging Infectious Diseases* 4(3):475.
74. Sasindran S.J., Torrelles J.B. (2011) Mycobacterium tuberculosis infection and inflammation: what is beneficial for the host and for the bacterium? *Frontiers in Microbiology* 2:9225.
75. Ndlovu H, Marakalala M.J. (2016) Granulomas and inflammation: host-directed therapies for tuberculosis. *Frontiers in Immunology* 7:28728.
76. Baguma R, Mbandi S K, Rodo M J et al. (2021) Inflammatory determinants of differential tuberculosis risk in pre-adolescent children and young adults. *Frontiers in Immunology* 12:639965.
77. Muefong C N, Sutherland J S (2020) Neutrophils in tuberculosis-associated inflammation and lung pathology. *Frontiers in Immunology* 27;11:962.
78. Ashenafi S, Loreti M G, Bekele A et al. (2023) Inflammatory immune profiles associated with disease severity in pulmonary tuberculosis patients with moderate to severe clinical TB or anemia. *Frontiers in Immunology* 14:1296501.
79. Eltantawy N, El-Zayyadi I A E H, Elberry A A et al. (2023) A review article of inflammatory bowel disease treatment and pharmacogenomics. *Beni-Suef University Journal of Basic and Applied Sciences* 12(1):35.
80. Ungaro R, Mehandru S, Allen P.B. et al. (2017) JF Colombel Ulcerative colitis. *The Lancet* 389:32126.
81. Neurath M.F. (2019) Targeting immune cell circuits and trafficking in inflammatory bowel disease. *Nature Immunology* 20(8):970–979.
82. Kobayashi N, Takahashi D, Takano S et al. (2019) The roles of Peyer's patches and microfold cells in the gut immune system: relevance to autoimmune diseases. *Frontiers in Immunology* 10:479654.
83. Nakase H (2020) Treatment of inflammatory bowel disease from the immunological perspective. *Immunological Medicine* 43(2):79–86.
84. Hansberry D R, Shah K, Agarwal P et al. (2017) Fecal myeloperoxidase as a biomarker for inflammatory bowel disease. *Cureus* 9(1):E1004.
85. Drury B, Hardisty G, Gray R D et al. (2021) Neutrophil extracellular traps in inflammatory bowel disease: pathogenic mechanisms and clinical translation. *Cellular and Molecular Gastroenterology and Hepatology* 12(1):321–333.
86. Yui S, Nakatani Y, Mikami M (2003) Calprotectin (S100A8/S100A9), an inflammatory protein complex from neutrophils with a broad apoptosis-inducing activity. *Biological and Pharmaceutical Bulletin* 26(6):753–760.
87. Bain C C, Scott C L, Uronen-Hansson H et al. (2013) Resident and pro-inflammatory macrophages in the colon represent alternative context-dependent fates of the same Ly6Chi monocyte precursors. *Mucosal Immunology* 6(3):498–510.
88. Yao H, Tang, G (2022) Macrophages in intestinal fibrosis and regression. *Cellular Immunology* 381:104614.
89. Galli S J, Tsai M, Piliponsky A M (2008) The development of allergic inflammation. *Nature* 454(7203):445–454.
90. Lemanske Jr, R.F. (2000) Inflammatory events in asthma: an expanding equation. *Journal of Allergy and Clinical Immunology* 105(6):S633–S636.
91. Bush A (2019) Pathophysiological mechanisms of asthma. *Frontiers in Pediatrics* 7:68.
92. Mims J.W. (2015) Asthma: definitions and pathophysiology. *In International Forum of Allergy and Rhinology* 5:S2–S6.
93. Habib N, Pasha M A, Tang D D (2022) Current understanding of asthma pathogenesis and biomarkers. *Cells* 11(17):2764.
94. Lambrecht B.N., Hammad H, Fahy J.V. (2019) The cytokines of asthma. *Immunity* 50(4):975–991.
95. Ho I.C., Pai S.Y. (2007) GATA-3-not just for Th2 cells anymore. *Cell Mol Immunol* 4(1):15–29.
96. Caramori G, Lim S, Ito K et al. (2001) Expression of GATA family of transcription factors in T-cells, monocytes and bronchial biopsies. *European Respiratory Journal* 18(3):466–473.
97. Wills-Karp M (2004) Interleukin-13 in asthma pathogenesis. *Immunological Reviews* 202(1):175–190.
98. Froidure A, Mouthuy J, Durham S.R. et al. (2016) Asthma phenotypes and IgE responses. *European Respiratory Journal* 47(1):304–319.
99. Stone K.D., Prussin C, Metcalfe D.D. (2010) IgE, mast cells, basophils, and eosinophils. *Journal of Allergy and Clinical Immunology* 125(2):S73–S80.
100. Komlósi Z.I., van de Veen W, Kovács N et al. (2022) Cellular and molecular mechanisms of allergic asthma. *Molecular Aspects of Medicine* 85:100995.
101. Solleiro-Villavicencio H, Rivas-Arancibia S (2018) Effect of chronic oxidative stress on neuroinflammatory response mediated by CD4+ T cells in neurodegenerative diseases. *Frontiers in Cellular Neuroscience* 12:114.
102. Kwon H.S., Koh S.H. (2020) Neuroinflammation in neurodegenerative disorders: the roles of microglia and astrocytes. *Translational Neurodegeneration* 9(1): 42.
103. Cai Y, Liu J, Wang B et al. (2022) Microglia in the neuroinflammatory pathogenesis of Alzheimer's disease and related therapeutic targets. *Frontiers in Immunology* 13:856376.
104. Moussa N, Dayoub N (2023) Exploring the role of COX-2 in Alzheimer's disease: Potential therapeutic implications of COX-2 inhibitors. *Saudi Pharmaceutical Journal* 31(9): 101729.
105. Ramalho T.C., Rocha M.V., da Cunha E.F. et al. (2009) The search for new COX-2 inhibitors: a review of 2002–2008 patents. *Expert Opinion on Therapeutic Patents* 19(9):1193–1228.

106. De Ture M.A., Dickson D.W. (2019) The neuropathological diagnosis of Alzheimer's disease. *Molecular Neurodegeneration* 14(1):32.
107. WHO (2022a) Dementia. https://www.who.int/news-room/fact-sheets/detail/dementia. Accessed 10 Apr 2023
108. Alzheimer's disease facts and figures. *Alzheimers Dement*. 2023 Apr;19(4):1598–1695. doi: 10.1002/alz.13016. Epub 2023 Mar 14. PMID: 36918389.
109. Nichols E, Steinmetz JD, Vollset SE et al. (2022) Estimation of the global prevalence of dementia in 2019 and forecasted prevalence in 2050: an analysis for the global burden of disease study 2019. *The Lancet Public Health* 7(2):e105–25.
110. Chen Y, Yu Y (2023) Tau and neuroinflammation in Alzheimer's disease: interplay mechanisms and clinical translation. *Journal of Neuroinflammation*. 20(1):165.
111. Wang W.Y., Tan M.S., Yu J.T. (2015a) Role of pro-inflammatory cytokines released from microglia in Alzheimer's disease. *Annals of Translational Medicine* 3(10): 136.
112. Alasmari F, Alshammari M A, Alasmari A F (2018) Neuroinflammatory cytokines induce amyloid beta neurotoxicity through modulating amyloid precursor protein levels/metabolism. *BioMed Research International* 2018(1):3087475
113. Kamphuis W, Middeldorp J, Kooijman L et al. (2014) Glial fibrillary acidic protein isoform expression in plaque related astrogliosis in Alzheimer's disease. *Neurobiology of Aging* 35(3):492–510.
114. Franciosi S, Choi H B, Kim S U et al. (2005) IL-8 enhancement of amyloid-beta (Aβ1-42)-induced expression and production of pro-inflammatory cytokines and COX-2 in cultured human microglia. *Journal of Neuroimmunology* 159(1–2):66–74.
115. Metcalfe M.J., Figueiredo-Pereira M.E. (2010) Relationship between tau pathology and neuroinflammation in Alzheimer's disease. *Mount Sinai Journal of Medicine: A Journal of Translational and Personalized Medicine: A Journal of Translational and Personalized Medicine* 77(1):50–58.
116. Nilson A.N., English K.C., Gerson J.E. et al. (2017) Tau oligomers associate with inflammation in the brain and retina of tauopathy mice and in neurodegenerative diseases. *Journal of Alzheimer's Disease* 55(3):1083–1099.
117. Laurent C, Buée L, Blum D (2018) Tau and neuroinflammation: what impact for Alzheimer's disease and tauopathies. *Biomedical Journal* 41(1):21–33.
118. Gaikwad S, Puangmalai N, Bittar A et al. (2021) Tau oligomer induced HMGB1 release contributes to cellular senescence and neuropathology linked to Alzheimer's disease and frontotemporal dementia. *Cell Reports* 36(3): 109419.
119. Ramesh, S, Arachchige A.S.P.M. (2023) Depletion of dopamine in Parkinson's disease and relevant therapeutic options: A review of the literature. *AIMS Neuroscience* 10(3):200.
120. WHO (2022b), Parkinson's disease. https://www.who.int/news/item/14-06-2022-launch-of-who-s-parkinson-disease-technical-brief
121. Lutters B, Foley P, Koehler P.J. (2018) The centennial lesson of encephalitis lethargica. *Neurology* 90(12):563–567.
122. McGeer PL, Itagaki S, Boyes BE et al. (1988) Reactive microglia are positive for HLA-DR in the substantia nigra of Parkinson's and Alzheimer's disease brains. *Neurology*. 38(8):1285
123. Jayanti S, Dalla Verde C, Tiribelli C et al. (2023) Inflammation, dopaminergic brain and bilirubin. *International Journal of Molecular Sciences* 24(14):11478.
124. Dutta D, Jana M, Majumder M et al. (2021) Selective targeting of the TLR2/MyD88/NF-κB pathway reduces α-synuclein spreading in vitro and in vivo. *Nature Communications* 12(1):5382.
125. Zhao Z, Ning J, Bao X.Q. et al. (2021) Fecal microbiota transplantation protects rotenone-induced Parkinson's disease mice via suppressing inflammation mediated by the lipopolysaccharide-TLR4 signaling pathway through the microbiota-gut-brain axis. *Microbiome* 9:1–27.
126. Yang J, Wise L, Fukuchi K.I. (2020) TLR4 cross-talk with NLRP3 inflammasome and complement signaling pathways in Alzheimer's disease. Frontiers in Immunology 11:526840.
127. Nagatsu T, Sawada, M (2005) Inflammatory process in Parkinson's disease: role for cytokines. *Current Pharmaceutical Design* 11(8):999–1016.
128. Wang Q, Liu Y, Zhou J (2015b) Neuroinflammation in Parkinson's disease and its potential as therapeutic target. *Translational Neurodegeneration* 4:1–9.
129. Booth H D, Hirst W D, Wade-Martins R (2017) The role of astrocyte dysfunction in Parkinson's disease pathogenesis. *Trends in Neurosciences* 40(6):358–370.
130. Waak J, Weber S.S., Waldenmaier A et al. (2009) Regulation of astrocyte inflammatory responses by the Parkinson's disease-associated gene DJ–1. *The FASEB Journal* 23(8):2478–2489.
131. Quinn P.M., Moreira P.I., Ambrósio A.F. et al. (2020) PINK1/PARKIN signalling in neurodegeneration and neuroinflammation. *Acta Neuropathologica Communications* 8(1):189.
132. Pajares M, I. Rojo A, Manda G et al. (2020) Inflammation in Parkinson's disease: mechanisms and therapeutic implications. *Cells* 9(7):1687.
133. Gardet A, Benita Y, Li C et al. (2010) LRRK2 is involved in the IFN-γ response and host response to pathogens. *The Journal of Immunology* 185(9):5577–5585.
134. Russo I, Bubacco L, Greggio E (2014) LRRK2 and neuroinflammation: partners in crime in Parkinson's disease. *Journal of Neuroinflammation* 11:1–9.
135. Bano D, Zanetti F, Mende Y et al. (2011) Neurodegenerative processes in Huntington's disease. *Cell Death and Disease* 2(11):228–228.
136. Tabrizi S.J., Ghosh R, Leavitt B.R. (2019). Huntingtin lowering strategies for disease modification in Huntington's disease. *Neuron* 101(5):801–819.
137. Jang M, Cho I.H. (2016). Sulforaphane ameliorates 3-nitropropionic acid-induced striatal toxicity by activating the Keap1-Nrf2-ARE pathway and inhibiting the MAPKs and NF-κB pathways. *Molecular Neurobiology* 53:2619–2635.
138. Hickman S, Izzy S, Sen P et al. (2018) Microglia in neurodegeneration. *Nature Neuroscience* 21(10):1359–1369.
139. Fatoba O, Ohtake Y, Itokazu T et al. (2020) Immunotherapies in Huntington's disease and α-Synucleinopathies. *Frontiers in Immunology* 11:517415.

2 Synthetic Polymers for Drug Delivery Systems in Inflammatory Diseases

S. Lakshmana Prabu, M. Vijayalakshmi, S. Surendhar, S. Meganathan, A. Umamaheswari, and A. Puratchikody**
University College of Engineering - BIT Campus, Anna University, Tiruchirappalli, India

2.1 INTRODUCTION

The word "inflammation" originates from the Latin word – *inflammare*, which means burn. Any damage in the body can cause a number of chemical changes in the affected area [1]. Inflammation is a complex biological response instigated by cellular and tissue complications ascending from an imbalance between regulatory signals and inflammatory process. It contributes to the body's immune defense against external chemicals, tissue injury and a variety of pathogenic pathogens [2].

During inflammation, both physical and chemical stimuli are generated as a result of tissue necrosis [3]. Macrophages play a key proinflammatory role by producing various mediators as part of the initial immune defense. These include reactive nitrogen and oxygen species, prostaglandins, tumor necrosis factor-α (TNF-α), and cytokines such as interleukin-1β (IL-1β) and interleukin-6 (IL-6) [4].

Excessive reactive oxygen species (ROS) generation and inflammation can harm cells, which can produce number of diseases, DNA damage, mutations, and other issues. Increasing evidence has exemplified the outcome of dysregulation of inflammatory pathways contributing to many manifestations of chronic illnesses. Numerous illnesses, including rheumatism, encephalitis, pneumonia, esophagitis, cancer, heart issues, and fibrosis, are linked to inflammation and discomfort [5].

2.2 POLYMERS

Polymers are the materials that are synthesized or extracted from various natural sources. With a range of delivery methods, biopolymers have gained popularity as a biomaterial that interacts with the biological system to release therapeutic chemicals [6]. Polymers can react to environmental elements such as pH, electric field, magnetic field, ionic strength, temperature, chemical, and biological stimuli in order to deliver the targeted therapeutic components. These biopolymers are extensively used in several applications due to their biocompatibility and biodegradability [7].

Over the years, researchers have explored various therapeutic approaches to combat inflammation. Among the various therapeutic approaches, synthetic polymers have been considered an emerging, promising avenue. Synthetic polymers are macromolecules constructed from repeating units of smaller monomers. It is possible to construct these polymers to have certain characteristic properties like controlled release, biocompatibility, and biodegradability. Furthermore, because of their adaptable physiochemical characteristics, drugs can be delivered specifically to inflammatory areas, increasing their effectiveness and reducing systemic toxicity.

2.3 SYNTHETIC POLYMERS

Synthetic polymers used for the drug delivery of inflammatory diseases can be broadly classified into two categories. They are

1. hydrophilic polymers and
2. hydrophobic polymers.

Synthetic polymer distribution of anti-inflammatory drugs has drawn a lot of interest as a potentially effective strategy in the field of therapeutic drug delivery. Inflammation is decreased by hydrophilic polymers with antioxidant qualities, such as polyvinyl alcohol (PVA), polyethylene glycol (PEG), and PEI. These polymers also modify immunological responses. Hydrophobic polymers, such as polylactic acid (PLA) and poly-lactic-co-glycolic acid (PLGA), provide controlled release and targeted delivery of anti-inflammatory drugs. The versatility of synthetic polymers allows for precise control over drug release kinetics, site-specific delivery, and protection of drugs from degradation. These polymers offer potential solutions to the challenges such as

* **Dr. A. Puratchikody**, contributed equally to this work, all correspondence and requests should be directed to **Dr. A. Umamaheswari**.

DOI: 10.1201/9781003477389-2

poor bioavailability, short half-life, and systemic toxicity by traditional anti-inflammatory agents.

Further research and development are required to optimize the use of synthetic polymers in anti-inflammatory applications. However, the progress made thus far suggests that synthetic polymers hold tremendous promise in revolutionizing in the field of anti-inflammatory therapeutics.

2.3.1 Hydrophilic Polymers

Hydrophilic polymers such as polyvinyl alcohol (PVA), polyethylene glycol (PEG), and polyethyleneimine (PEI) have demonstrated promising potential in modulating inflammatory responses. These polymers serve various functions, including reducing inflammation by scavenging reactive oxygen species (ROS) and altering immune cell activity.

2.3.1.1 Polyethylene Glycol

The potential of PEG to scavenge ROS and shield cells from oxidative damage has led to many research efforts. Its PEG's antioxidant characteristic that aids in its anti-inflammatory impact by decreasing the proinflammatory cytokine production and inhibiting leukocyte activation [8]. Hydrogels based on PEG have been used for encapsulating anti-inflammatory drugs, which provides sustained release of the drugs and improved therapeutic activity [9].

2.3.1.2 Polyvinyl Alcohol

PVA has demonstrated anti-inflammatory properties by inhibiting the TNF-α and IL-1β as inflammatory mediators [10]. PVA-based nanoparticles loaded with anti-inflammatory drugs have shown enhanced drug stability and targeted delivery drugs to the inflammatory sites [11].

2.3.1.3 Polyethyleneimine

PEI is a hydrophilic polymer that has shown anti-inflammatory activity. It inhibits the expression of NF-κB, a transcription factor responsible for the synthesis of proinflammatory cytokines [12]. PEI-based nanoparticles have been utilized for the targeted delivery of anti-inflammatory drugs, providing improved efficacy **polymer, polypropylene glycol (PPG), reducing** systemic toxicity [13].

2.3.2 Hydrophobic Polymers

Hydrophobic polymers include PLGA and PLA. These two polymers have anti-inflammatory properties and are widely used in drug delivery systems. In addition, they provide protection of drugs from their degradation, act as carriers to delivery anti-inflammatory drugs, and provide controlled drug release. PLA and PLGA, in particular, have been extensively studied for their biocompatibility and biodegradability. These polymers can be used to generate nanoparticles, microparticles, and scaffolds that enable targeted delivery and prolonged release of anti-inflammatory medicines [14]. PLA-based nanoparticles loaded with anti-inflammatory drugs have shown sustained release profiles and improved therapeutic efficacy in various inflammatory conditions, such as rheumatoid arthritis (RA) [15], as shown in Figure 2.1. PLGA-based microparticles have been exploited for the delivery of anti-inflammatory drugs to treat disorders like inflammatory bowel disease and asthma, which showed promising results [16].

2.3.2.1 Chitosan

Chitosan is a copolymer of β-(1-4)-linked d-glucosamines and N-acetyl-d-glucosamines produced from chitin. Chitin is found in the exoskeleton of crustaceans, such as lobsters, crabs and shrimp. Partial deacetylation of chitin produces chitosan. Based on the degree of the deacetylation process (40%–98%), different molecular weights (50–200 kDa) and viscosities are produced. Due to its remarkable properties, such as biocompatibility, biodegradability, and nontoxicity, binding ability with mammalian cells and microbial cells shows antimicrobial activity. It is widely used in medicine, the food industry, agriculture, and cosmetics preparations. In recent years, the anti-inflammatory property of chitosan has gained significant attention in both the research and medical fields [17] the structure of chitosan is shown in Figure 2.2.

Chitosan, a natural biopolymer derived from chitin, possesses significant anti-inflammatory properties. It has the ability to control the immune response, prevent the release of inflammatory mediators, suppress inflammatory cell activity, and promote tissue repair. Several research studies have provided evidence supporting the anti-inflammatory effects of chitosan in different inflammatory conditions. Continued study in this area could result in the advancement of chitosan-based therapeutics for the inflammatory illnesses treatment.

2.3.2.1.1 Anti-Inflammatory Activity

2.3.2.1.1.1 Modulation of Immune Response Chitosan has been shown to modify the immune response, leading to subsequent reduction in inflammation. Hu et al. developed chitosan nanoparticles and reported that they could upregulate the production of anti-inflammatory cytokines like interleukin-10 (IL-10) and downregulate the pro-inflammatory cytokines like tumor necrosis factor-alpha (TNF-α) and interleukin 1-beta (IL-1β) production. These findings demonstrated that chitosan has anti-inflammatory activity by regulation of immune response [18].

2.3.2.1.1.2 Inhibition of Inflammatory Mediators Numerous inflammatory mediators, such as prostaglandins, leukotrienes, and nitric oxide (NO), are inhibited by chitosan. Patrulea and team developed chitosan nanoparticles that could reduce the production of leukotriene B4 (LTB4) and prostaglandin E2 (PGE2) in lipopolysaccharide (LPS)-induced inflammation. Also it inhibited the activity of inducible nitric oxide synthase (iNOS), an enzyme important for the NO production. This NO plays a significant role in the inflammatory response [19].

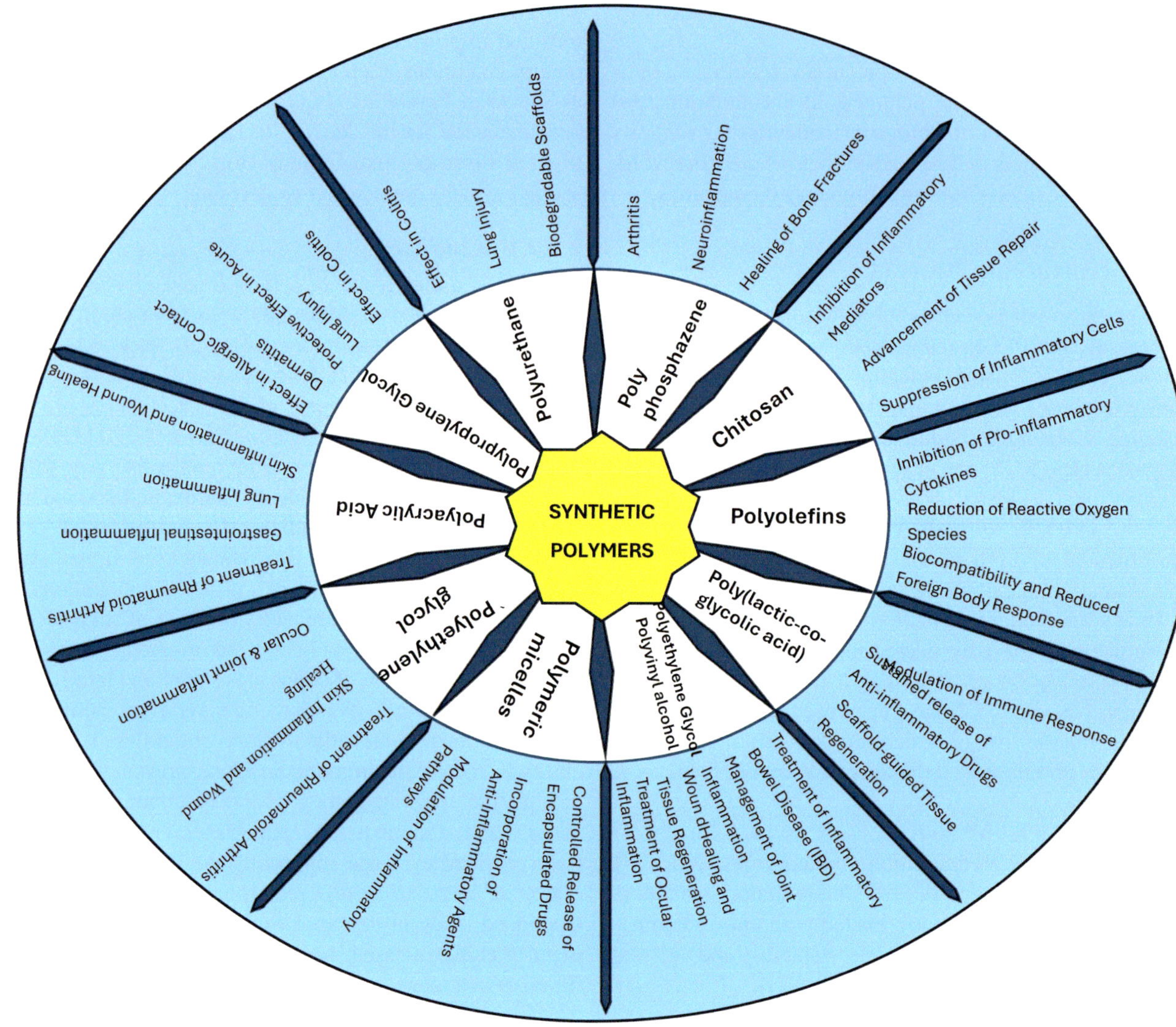

FIGURE 2.1 Synthetic polymers for inflammation.

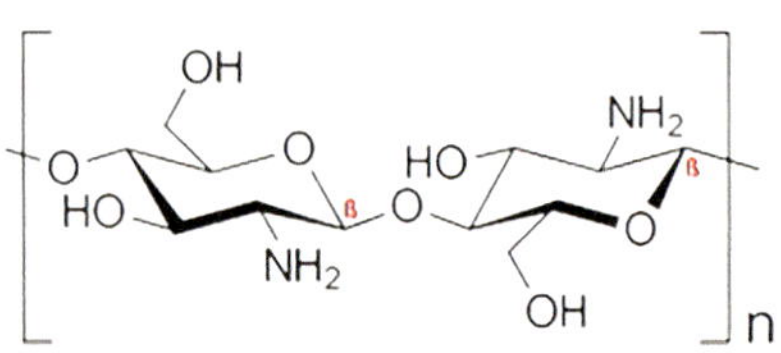

FIGURE 2.2 Structure of chitosan.

2.3.2.1.1.3 Suppression of Inflammatory Cells Chitosan has been reported to suppress the stimulation and migration of inflammatory cells, such as macrophages and neutrophils. Neutrophils are known to release various inflammatory mediators and enzymes that contribute to tissue damage during inflammation. Chitosan nanoparticles have been found to inhibit neutrophil activation and reduce the release of ROS and myeloperoxidase (MPO) [20]. It also inhibits macrophage migration, which is crucial for the recruitment of inflammatory cells to the site of inflammation [21].

2.3.2.1.1.4 Advancement of Tissue Repair Further, chitosan has been reported to support wound healing and tissue repair. Research indicates that it can augment the synthesis of growth factors, including transforming growth factor-beta (TGF-β), which are essential for tissue regeneration [22]. Furthermore, chitosan has been found to stimulate the multiplication and migration of fibroblasts, the key cells involved in the synthesis of extracellular matrix components required for tissue repair [23].

2.3.2.1.1.5 Research Evidence Numerous research studies have provided evidence supporting the anti-inflammatory property of chitosan. Zhang et al. used the mouse model of acute lung produced by lipopolysaccharide (LPS) and examined the anti-inflammatory properties of chitosan. The administration of chitosan was found to considerably reduce lung inflammation by lowering proinflammatory cytokine levels and preventing inflammatory cells from penetrating the lung tissue [22]. Another study examined the effect of chitosan on anti-inflammatory effects in a rat model of colitis. The study findings indicated that chitosan significantly reduced colonic inflammation by reducing the inflammatory cytokines level and improved histopathological changes in the colon [24]. Alarcón-Alarcón C and coworkers determined that chitosan gel reduced gingival inflammation and improved periodontal health in patients with periodontitis [25].

2.3.2.2 Polyolefins

Polyolefins are a versatile group of polymers made from basic olefins, such as ethylene and propylene. They are widely used due to their excellent physical and chemical properties, including high tensile strength, chemical resistance, thermal stability, and low cost. In recent years, the anti-inflammatory properties of polyolefins have garnered significant attention in various biomedical applications, the structure of polyolefins in Figure 2.3.

Polyolefins are widely used synthetic polymers with anti-inflammatory properties, making them promising candidates for use in tissue engineering, drug delivery, and implantable devices. They inhibit proinflammatory cytokines, scavenge ROS, and modulate immune responses. Numerous investigational studies have supported the anti-inflammatory potential of polyolefins.

2.3.2.2.1 Anti-inflammatory Activity

Based on the type and content of the polyolefin, there are a number of potential inflammatory response mechanisms that could account for the anti-inflammatory activity of polyolefins.

2.3.2.2.2 Inhibition of Proinflammatory Cytokines

Polyolefins modulate the production and release of IL-6, TNF-α and IL-1β as proinflammatory cytokines. Proinflammatory cytokines, such as IL-6, IL-1β, and TNF-α in macrophages, were markedly lowered by the nanocomposite containing polyolefins [26].

2.3.2.2.3 Reduction of Reactive Oxygen Species

Polyolefins have antioxidant properties that enable them to scavenge and reduce the ROS levels produced during inflammatory processes. This ROS is essential for initiating and propagating the inflammatory reactions. By lowering ROS levels in macrophages, a study on polyolefin nanofibers showed their anti-inflammatory effect [26].

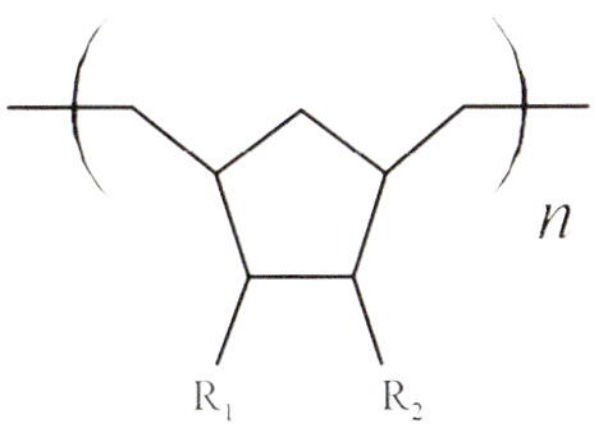

FIGURE 2.3 Structure of polyolefins.

2.3.2.2.4 Modulation of Immune Response

Polyolefins have been found to modulate the immune response, leading to an anti-inflammatory effect. Polyethylene and polypropylene as polyolefins noticed to reduce the immune cells like lymphocytes and monocytes production, thereby suppressing the immune response and subsequent inflammation [27].

2.3.2.2.5 Biocompatibility and Reduced Foreign Body Response

Polyolefins are well-known for their excellent biocompatibility, which reduces the foreign body response and inflammation at the interface between biomaterial and tissue. The ability of polyolefins to integrate into biological tissues without eliciting a response is crucial for their use in implantable devices and tissue engineering scaffolds. In vivo techniques are used to investigate the biocompatibility of polyethylene-based implants, and minimal inflammation at the implant sites indicates the anti-inflammatory nature of polyolefins [28].

2.3.2.2.6 Research Evidence

Various research studies support the anti-inflammatory properties of polyolefins. Studies have shown that polyethylene fibers can promote wound healing and reduce inflammation by decreasing inflammatory cell infiltration and proinflammatory cytokine levels in a rat model [29].

Tang F. et al. examined the potential of polypropylene nonwoven fabrics as anti-inflammatory materials in a rat model by subcutaneous inflammation. The outcomes showed that the infiltration of inflammatory cells and the expression of proinflammatory cytokines in the inflammatory tissues were considerably decreased by using polypropylene nonwoven fabrics [30]. Furthermore, Biggs et al. explored the anti-inflammatory properties of polyolefin-based meshes used for hernia repair. The findings demonstrated its activity by decreasing the inflammatory response in the presence of polyolefin-based meshes [31].

2.3.2.3 Poly-Lactic-Co-Glycolic Acid

PLGA is a biodegradable copolymer made up of glycolic acid and lactic acid, the structure of PLGA is given in Figure 2.4. Due to its biocompatibility, adjustable degradation rate, and controlled release properties, it is widely used in biomedical applications like tissue engineering, delivery of drug and regenerative medicine. Recently, the anti-inflammatory property of PLGA has gained significant attention.

A variety of characteristics of PLGA, a biodegradable copolymer, support its anti-inflammatory activity. The

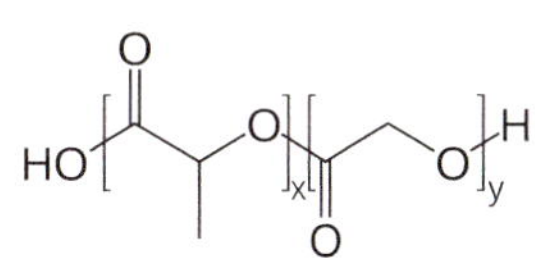

FIGURE 2.4 Structure of PLGA.

sustained release of anti-inflammatory drugs, modulation of immune response and scaffold-guided tissue regeneration make PLGA a versatile material in the field of regenerative medicine and delivery of drugs. Outcome of the study indicates the ability of PLGA-based systems to reduce inflammation and promote tissue healing in various biomedical applications. Further research and development in this area hold great promise for the utilization of PLGA as an effective anti-inflammatory material.

2.3.2.3.1 PLGA and Inflammation

PLGA possesses several inherent characteristics that contribute to its anti-inflammatory effects. These mechanisms are crucial for modulating the inflammatory response and promoting tissue regeneration.

2.3.2.3.2 Sustained Release of Anti-Inflammatory Drugs

PLGA is well-known for its ability to encapsulate and sustain the release of drugs, including anti-inflammatory agents. A sustained and localized release of anti-inflammatory medications can be accomplished by integrating nonsteroidal anti-inflammatory medicines (NSAIDs) or corticosteroids into PLGA-based systems. This targeted drug delivery allows more effective suppression of inflammation at the site of action. Arcos and team developed PLGA microspheres loaded with the NSAID ibuprofen and examined for tissue healing in animal model and demonstrated that the microspheres significantly reduced the inflammation in the tissues [32].

2.3.2.3.3 Modulation of Immune Response

Reducing the expression of proinflammatory cytokines and increasing the expression of anti-inflammatory cytokines in immune cells, including macrophages and dendritic cells, can modify the immune response. Vasita et al., developed PLGA nanoparticle and demonstrated that developed nanoparticles could differentiate the macrophages in anti-inflammatory phenotype, reducing the secretion of proinflammatory cytokines [33].

2.3.2.3.4 Scaffold-guided Tissue Regeneration

PLGA-based scaffolds can facilitate tissue regeneration by providing a suitable environment for cell infiltration, proliferation, and differentiation. This characteristic, which helps to reduce inflammation and encourage tissue repair, is essential during the inflammatory stage of wound healing. PLGA-based nanocomposite scaffold was developed that demonstrated anti-inflammatory properties by downregulating proinflammatory gene expression in human mesenchymal stem cells [34].

2.3.2.3.5 Research Evidence

An experiment was conducted by Park et al. on the treatment of inflammatory bowel disease by adding dexamethasone to a PLGA-based system. As a result, the findings showed that in an animal model of colitis, a PLGA-based system reduced proinflammatory cytokines production [35].

Numerous research studies have highlighted the anti-inflammatory potential of PLGA-based systems. Al Samri et al. made an investigation of PLGA-based hydrogel loaded with diclofenac sodium in a rat model, which resulted in hydrogel showing significant reduction in inflammatory markers and improved healing of wounds [36]. PLGA scaffold loaded with the anti-inflammatory peptide Ac2-26 was developed by Gagliardi et al. and examined in a mouse model. The PLGA scaffold increased the release of Ac2-26 and significantly reduced the inflammation and inhibited the angiogenesis [37].

2.3.2.4 Polyethylene Glycol

PEG is a multipurpose polymer that finds extensive application in a range of industries, such as food, cosmetics, and pharmaceuticals. Its anti-inflammatory qualities have drawn attention in the field of inflammation research. The structure of PEG is given in Figure 2.5.

PEG exhibits significant anti-inflammatory properties through its modulation of immune cell function, inflammatory mediators, inhibition of leukocyte activation and migration, stabilization of cell membranes, and antioxidant activity. These mechanisms enable PEG to alleviate inflammation and considered it as a potential candidate for various applications as a therapeutic agent. Its use in treating inflammatory bowel disease, managing joint inflammation, wound healing, and ocular inflammation showed promising results. However, further research and clinical studies are needed to validate the efficacy and safety. Inflammation is a complex biological response to injury or infection. It is characterized by fever, discomfort, edema, and redness. Acute inflammation serves as a defense mechanism, but persistent inflammation can cause tissue damage and the emergence of a number of illnesses, including inflammatory bowel disease, cardiovascular disease, and RA. Therefore, finding effective anti-inflammatory agents is of great importance in managing these conditions.

2.3.2.4.1 Modulation of Immune Cells

Many immune cells, such as lymphocytes, macrophages, and neutrophils have been demonstrated to have their functions modulated by PEG. When it comes to the beginning and development of inflammation, these immune cells are crucial.

Zhang et al. demonstrated that PEG can attenuate neutrophil activation and migration. Neutrophils are the first responders to inflammation and release proinflammatory mediators. PEG treatment led to a reduction in neutrophil infiltration and subsequently decreased tissue damage in a

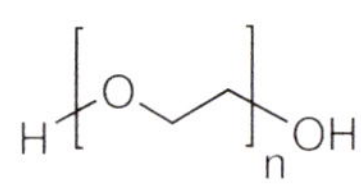

FIGURE 2.5 Structure of PEG.

mouse model of inflammation. Additionally, PEG has been found to alter macrophage phenotype. Macrophages have different phenotypes, with M1 macrophages promoting inflammation and M2 macrophages promoting tissue repair and resolution of inflammation. PEG treatment has been reported to shift the macrophage phenotype toward M2 anti-inflammatory state, thereby reducing inflammation and tissue damage [38].

2.3.2.4.2 *Inhibition of Inflammatory Mediators*

PEG has been shown to prevent the production of various inflammatory mediators, including prostaglandins, cytokines, and chemokines. These mediators play a crucial role in the instigation and maintenance of inflammation. The Wang team demonstrated that PEG can inhibit the production of prostaglandin E2 (PGE2), a potent proinflammatory mediator. Enzyme cyclooxygenase-2 (COX-2) produces the PGE2 and plays a vital role in inflammation and pain. PEG reduced COX-2 expression and subsequently decreased PGE2 production, leading to reduced inflammation and pain in a mouse model. PEG was evaluated for its anti-inflammatory activity in a mouse model of colitis, and the results revealed that PEG suppressed the proinflammatory cytokines production. These cytokines are crucial in regulating the immune response and are elevated in various inflammatory diseases [39, 40].

2.3.2.4.3 *Inhibition of Leukocyte Activation and Migration*

PEG can interfere with adhesion molecules involved in leukocyte-endothelial cell interactions and ultimately inhibit leukocyte activation and migration. This phenomenon could potentially reduce the inflammatory cells at the inflammation site during the inflammatory response [41].

2.3.2.4.4 *Stabilization of Cell Membranes*

Another mechanism through which PEG exhibits anti-inflammatory effects is its ability to stabilize cell membranes. PEG acts as a barrier on the cell surface, preventing the efflux of intracellular contents by stabilizing cell membranes that may contribute to inflammatory reactions. Also, PEG can limit the release of proinflammatory molecules from damaged cells [42].

2.3.2.4.5 *Reduction of Oxidative Stress*

An imbalance between the generation of ROS and antioxidant defenses is known as oxidative stress. It can cause tissue damage and has a major impact on inflammation. PEG's antioxidant potential have been shown to decrease oxidative stress. The general anti-inflammatory properties of PEG are partly attributed to this antioxidant characteristic.

Juarez-Moreno K and co-workers demonstrated that PEG treatment could scavenge ROS and inhibit lipid peroxidation. Lipid peroxidation is a chain reaction that occurs when ROS attack polyunsaturated fatty acids, leading to the generation of toxic by-products. PEG treatment reduced lipid peroxidation, thereby protecting cells from oxidative damage in a cell culture model [43]. Liu and team established that PEG treatment can increase the expression of antioxidant enzymes, such as catalase (CAT) and superoxide dismutase (SOD). These enzymes play a crucial role in neutralizing ROS and preventing oxidative damage. PEG treatment upregulated the expression of SOD and CAT, leading to oxidative stress reduction and inflammation in a mouse model [44].

2.3.2.4.6 *Therapeutic Applications of PEG as an Anti-Inflammatory Agent*

Given its diverse mechanisms of action, PEG holds considerable promise as a therapeutic agent for various inflammatory conditions. Here, we discuss some potential therapeutic applications of PEG as an anti-inflammatory agent:

2.3.2.4.7 *Treatment of Inflammatory Bowel Disease (IBD)*

Crohn's disease and ulcerative colitis are examples of IBDs that are characterized by persistent inflammation in the gastrointestinal system. Further PEG was investigated for treating this IBD and the study results outlined that PEG showed significant reduction of IBD by modulating the inflammatory mediators and reduce leukocyte migration [45].

Liu and team performed a comparative study between IBD drug and PEG-based nanoparticles for its IBD effect and the study results highlighted those nanoparticles showed better efficacy than the pure drug [44].

2.3.2.4.8 *Management of Joint Inflammation*

Inflammatory joint diseases like RA and osteoarthritis are associated with significant pain and disability. PEG-based hydrogels and nanoparticles have shown significant anti-inflammatory effect in the affected joints by providing localized relief [46]. Further, PEG can be used as a lubricating agent to reduce friction within the joint space, thereby alleviating pain and inflammation. Studies have demonstrated the potential benefits of PEG-based interventions in managing joint inflammation [47].

2.3.2.4.9 *Wound Healing and Tissue Regeneration*

Inflammation is an integral part of the wound healing process, but excessive or prolonged inflammation can impede healing. Due to its anti-inflammatory properties, PEG plays a pivotal part in regeneration of tissue and wound healing processes. PEG-based dressings have been developed to provide a moist wound environment. These dressings have been shown to accelerate wound healing and enhance tissue regeneration [48]. Additionally, PEG hydrogels can serve as scaffolds for tissue engineering applications, facilitating the repair and regeneration of damaged tissues [49].

2.3.2.4.10 *Treatment of Ocular Inflammation*

Ocular inflammation, such as uveitis and dry eye disease, can lead to vision impairment and discomfort. PEG-formulated eye drops have demonstrated anti-inflammatory effects by reducing cytokine levels and promoting tear film stability [50]. The ocular surface lubrication provided by

PEG can alleviate symptoms associated with dry eye disease and reduce inflammation in the eye. PEG-based interventions hold promise in ocular inflammation treatment and related conditions.

2.3.2.5 Polymeric Micelles

In an aqueous solution, amphiphilic block copolymers spontaneously self-assemble to form nanoscale structures known as polymeric micelles, as shown in Figure 2.6. Because these micelles can efficiently solubilize and transport hydrophobic medicines, they have drawn a lot of attention in the field of drug delivery systems. The anti-inflammatory properties of polymeric micelles have also been investigated recently.

Polymeric micelles have arisen as promising candidates in the treatment of inflammation due to their unique properties and ability to encapsulate or incorporate anti-inflammatory agents. By modulating the release of drugs or incorporating specific anti-inflammatory agents, micelles can effectively reduce inflammation and provide sustained therapeutic effects. Additionally, the targeted drug delivery capabilities of micelles offer the potential for localized treatment and reduced systemic side effects.

2.3.2.5.1 Polymeric Micelles in Drug Delivery

Polymeric micelles are specifically used as drug carriers. The core of these micelles can encapsulate hydrophobic drugs, protecting them from degradation and enhancing their solubility. Additionally, the outer shell consists of hydrophilic polymer blocks, allowing for stealth-like behavior, prolonged circulation time, and reduced clearance by the mononuclear phagocyte system (MPS). [51]

Polymeric micelles can be engineered to selectively accumulate at the site of inflammation, such as inflamed tissues or tumor tissues with enhanced vascular penetrability. The improved penetrability and retention (EPR) effect allows micelles to passively accumulate in the inflammation areas due to leaky vasculature and dysfunctional lymphatics. As a result, these polymeric micelles have the potential to deliver the drug in the targeted site and reduced systemic toxicity [52].

2.3.2.5.2 Anti-Inflammatory Properties of Polymeric Micelles

Polymeric micelles have shown promise as anti-inflammatory agents in various preclinical models. Studies have demonstrated that by modulating the release of encapsulated drugs or by incorporating specific anti-inflammatory agents, polymeric micelles can effectively reduce inflammation and provide therapeutic benefits.

2.3.2.5.3 Controlled Release of Encapsulated Drugs

One approach to harness the anti-inflammatory properties of polymeric micelles is by incorporating anti-inflammatory drugs or agents into the core of the micelles. Incorporating the drugs into the micelles can reduce the side effects and provides sustained therapeutic action. Danhier et al. performed a comparative study between pure drug (dexamethasone) and drug in polymeric micelles and evaluated for its anti-inflammatory potential. The study results demonstrated that polymeric micelles significantly reduced inflammation compared to free dexamethasone because of its prolonged and controlled release of the drug substance from the micelles [53].

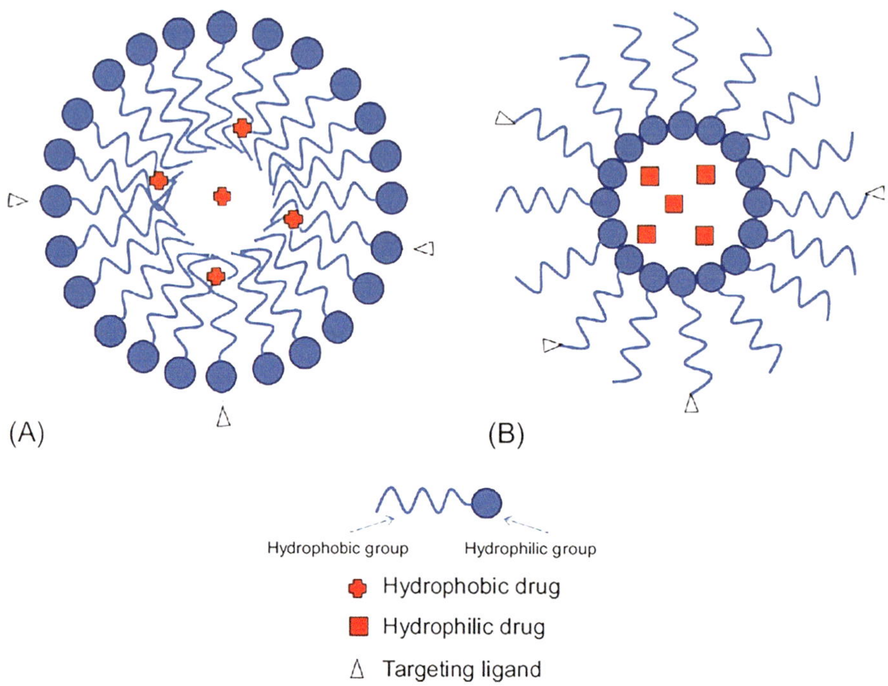

FIGURE 2.6 Structure of polymeric micelles.

2.3.2.5.4 *Incorporation of Anti-Inflammatory Agents*

In addition to encapsulating anti-inflammatory drugs, polymeric micelles can also be designed to incorporate anti-inflammatory agents directly into the polymer structure. This approach allows for sustained release of the drugs and provides a continuous anti-inflammatory effect.

Polymeric micelles loaded with curcumin, a natural anti-inflammatory compound derived from turmeric was developed and examined for its anti-inflammatory effect. This micelle efficiently encapsulated the curcumin and exhibited sustained release over time. The curcumin-loaded micelles demonstrated potent anti-inflammatory effects, with a substantial reduction in inflammation compared to free curcumin. The researchers suggested that the sustained release of curcumin from the micelles contributed to the enhanced therapeutic efficacy [54].

2.3.2.5.5 *Modulation of Inflammatory Pathways*

Polymeric micelles exert their anti-inflammatory effect at the cellular level by modulating the inflammatory pathways. Incorporating specific molecules or peptides that target inflammatory receptors or signaling pathways. Polymeric micelles containing a peptide inhibitor were developed that specifically target interleukin-1 receptor-associated kinase 4 (IRAK-4) and evaluated in an animal model. IRAK-4 plays a crucial role in the activation of factor-kappa B (NF-κB) pathway is one of the inflammatory signaling pathways. These developed micelles released the IRAK-4 in a sustained manner and reduced the inflammation by blocking the inflammatory signaling pathway [55].

2.3.2.6 Polyvinyl Alcohol

PVA is a versatile synthetic polymer (structure given in Figure 2.7) that has gained significant attention in various fields, including pharmaceuticals and biomedical applications. It is known for its excellent film-forming properties, biocompatibility and biodegradability. In recent years, PVA has also shown promising anti-inflammatory properties, making it a candidate for inflammatory treatment.

PVA is a versatile synthetic polymer that has shown promising anti-inflammatory properties due to its unique physicochemical properties and biological characteristics. The anti-inflammatory activity of PVA can be attributed to various mechanisms, including its role in modulating key signaling pathways, and it's scavenged free radical and immune response.

2.3.2.6.1 *PVA as an Anti-Inflammatory Agent*

2.3.2.6.1.1 *Modulation of Signaling Pathways*

Inflammation is a complex biological process regulated by various signaling pathways. Among these mitogen-activated protein kinase (MAPK) and nuclear factor-kappa B (NF-κB) signaling pathways are two pathways involved in the process of inflammation. PVA has been shown to modulate these pathways and inhibit the proinflammatory cytokines production [56]. Lee and team investigated the anti-inflammatory activity of PVA in LPS inspired macrophages. The study results significantly suppressed the MAPK and NF-κB signaling pathways activation and inhibited the pro-inflammatory cytokines production [57].

2.3.2.6.1.2 Reduction of Oxidative Stress Increased oxidative stress, which produces free radicals and ROS, is frequently linked to inflammation. PVA possesses antioxidant properties, enabling it to scavenge ROS and protect cells from oxidative damage. Anti-inflammatory and anti-oxidant effect of PVA was examined in the colitis model. PVA administration significantly reduced colonic injury and inflammation by decreasing markers of lipid peroxidation and oxidative stress in the colonic tissue. The study results highlighted its therapeutic potential in IBDs [58].

2.3.2.6.1.3 Modulation of Immune Response PVA has also shown immunomodulatory effects, influencing the immune response during inflammation. It regulates the expression and release of immune mediators, which play a crucial part in the progression of inflammatory diseases.

Kim et al. investigated the immunomodulatory effects of PVA in a RA model. PVA treatment significantly reduced joint inflammation, cartilage destruction, and bone erosion. It also decreased the proinflammatory cytokines production and increased the secretion of anti-inflammatory cytokines. The researchers suggested that PVA exerts its immunomodulatory effects by balancing the inflammatory and anti-inflammatory cytokine profiles, thereby attenuating the inflammatory response in RA [59].

2.3.2.7 Polyethylene Oxide (PEO)

Polyethylene oxide (PEO) is a water-soluble linear polymer known for its diverse applications in various fields, including medicine, the chemical structure of PEO is given in Figure 2.8. In recent years, PEO has garnered attention for its potential as an anti inflammatory agent.

PEO exhibits significant anti-inflammatory properties through its modulation of inflammatory mediators, inhibition of leukocyte activation and migration, reduction of oxidative stress and promotion of tissue repair. These mechanisms enable PEO as a potential candidate for various therapeutic applications. It is used in treating IBD, skin inflammation and wound healing, ocular inflammation, and joint inflammation. However, further research and clinical examinations are required to validate the efficiency and

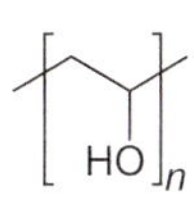

FIGURE 2.7 Structure of PVA.

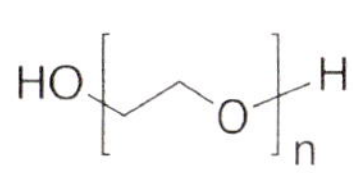

FIGURE 2.8 Structure of PEO.

safety of PEO-based anti-inflammatory interventions. The diverse nature of PEO's anti-inflammatory properties positions it as a promising avenue for future therapeutic developments.

2.3.2.7.1 *Modulation of Inflammatory Mediators*

Numerous inflammatory mediators like adhesion molecules, chemokines and cytokines have been demonstrated to be modulated by PEO. Also, it can prevent the expression of proinflammatory cytokines like tumor necrosis factor-alpha (TNF-α), interleukin-6 (IL-6), and interleukin-1 beta (IL-1β) [60].

2.3.2.7.2 *Inhibition of Leukocyte Activation and Migration*

PEO can interfere with adhesion molecules involved in leukocyte-endothelial cell interactions, ultimately inhibiting the activation and migration of leukocytes. This process reduces the inflammatory response by reducing the involvement of inflammatory cells to the site of inflammation [61].

2.3.2.7.3 *Reduction of Oxidative Stress*

During inflammation oxidative stress has a critical role and PEO possesses antioxidant properties that can counteract oxidative damage. PEO acts as a scavenger of ROS and inhibits lipid peroxidation. These antioxidant activities contribute to its overall anti-inflammatory effects and reduce oxidative stress-induced inflammation.

2.3.2.7.4 *Advancement of Tissue Repair*

PEO aids in repair of tissue and regeneration by angiogenesis promotion and re-epithelialization. It enhances the endothelial cell relocation and proliferation and keratinocytes, promoting the restoration of damaged tissues [62].

2.3.2.7.5 *Therapeutic Applications of PEO as an Anti-Inflammatory Agent*

PEO's multifaceted anti-inflammatory mechanisms make it a promising candidate for various therapeutic applications.

2.3.2.7.6 *Treatment of Inflammatory Bowel Disease*

IBD, including conditions such as Crohn's disease and ulcerative colitis, is characterized by chronic inflammation in the gastrointestinal tract. PEO can be used as a potential treatment option for IBD due to its inflammatory mediators modulating ability and reduce leukocyte migration [63].

2.3.2.7.7 *Skin Inflammation and Wound Healing*

Inflammatory skin conditions like dermatitis and psoriasis are associated with significant discomfort and impaired barrier function. PEO-based formulations have been investigated for their anti-inflammatory effects in skin inflammation. Studies have shown that PEO can reduce proinflammatory cytokine levels and alleviate symptoms of inflammatory skin diseases [64]. PEO-based dressings and hydrogels have exhibited potential in wound healing by enhancing tissue regeneration and accelerating wound closure [65].

2.3.2.7.8 *Ocular Inflammation*

Ocular inflammation, such as uveitis and dry eye disease, can lead to vision impairment and discomfort. PEO-based eye drops have demonstrated anti-inflammatory effects by reducing cytokine levels and promoting tear film stability. Also, PEO can act as a lubricating agent, reducing friction and inflammation on the ocular surface [66].

2.3.2.7.9 *Joint Inflammation*

Both RA and osteoarthritis are associated with joint inflammation and pain. PEO-based interventions have shown their potential in the management of joint inflammation. PEO hydrogels can provide sustained release of anti-inflammatory drugs to the affected joints, reducing inflammation and pain [67]. PEO-based lubricants can improve joint mobility and alleviate symptoms associated with joint inflammation [68].

2.3.2.8 Polyacrylic Acid

Polyacrylic acid (PAA) is a synthetic polymer (structure given in Figure 2.9) with various applications in industries ranging from textiles to pharmaceuticals. In recent years, PAA has expanded attention for its potential as an anti-inflammatory agent. PAA possesses significant anti-inflammatory properties through its modulation of inflammatory mediators, inhibition of leukocyte activation and migration, reduction of oxidative stress, and promotion of tissue repair.

These mechanisms enable PAA to attenuate inflammation and considered as a probable candidate for various therapeutic applications. It is used in treating RA, gastrointestinal inflammation, lung inflammation, and various skin infections. However, further research and clinical studies are required to validate the efficiency and safety of PAA-based anti-inflammatory interventions.

PAA may be a good fit for a number of therapeutic uses because of these mechanisms that allow it to reduce inflammation. RA, lung, gastrointestinal, and various skin infections are among the conditions it is used to treat. However, further research and clinical studies are needed to validate the efficacy and safety of PAA-based anti-inflammatory interventions.

2.3.2.8.1 *Modulation of Inflammatory Mediators*

PAA has been shown to modulate several inflammatory mediators like chemokines, cytokines and adhesion molecules. It can inhibit the proinflammatory cytokines production [69].

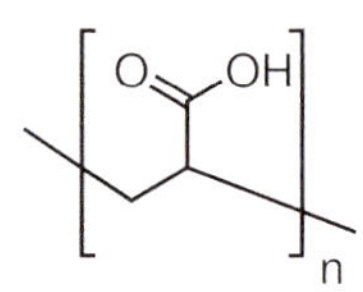

FIGURE 2.9 Structure of polyacrylic acid.

2.3.2.8.2 *Inhibition of Leukocyte Activation and Migration*

PAA can interfere with adhesion molecules involved in leukocyte-endothelial cell interactions, ultimately inhibiting the activation and migration of leukocytes. This phenomenon reduces the inflammatory cells recruitment at the inflammation site, thereby mitigating the inflammatory response [70].

2.3.2.8.3 *Reduction of Oxidative Stress*

Inflammation is largely caused by oxidative stress, and PAA has antioxidant qualities that can prevent oxidative damage. Inhibiting lipid peroxidation and acting as a scavenger of ROS, PAA lowers the oxidative stress brought on by inflammation [71].

2.3.2.8.4 *Promotion of Tissue Repair*

PAA stimulates angiogenesis and re-epithelialization, which helps in tissue regeneration and repair. It promotes the healing of injured tissues by increasing the migration and proliferation of keratinocytes and endothelial cells [72].

2.3.2.8.5 *Therapeutic Applications of PAA as an Anti-Inflammatory Agent*

PAA's diverse anti-inflammatory mechanisms make it a promising candidate for various therapeutic applications.

2.3.2.8.6 *Treatment of Rheumatoid Arthritis*

RA is linked to inflammation and damage to the joints. Interventions based on PAA have been demonstrated to encouraging the outcomes in RA treatment. Studies have demonstrated that intra-articular injection of PAA hydrogels can reduce joint inflammation and ameliorate the symptoms of RA. The ability of PAA to modulate inflammatory mediators and inhibit leukocyte migration makes it a potential therapeutic option for management of RA [73].

2.3.2.8.7 *Gastrointestinal Inflammation*

Gastrointestinal inflammation, such as gastritis and colitis, can cause significant discomfort and impairment of normal digestive function. PAA-based formulations have demonstrated anti-inflammatory effects in gastrointestinal inflammation. PAA coated mesalamine nanoparticles have shown enhanced anti-inflammatory efficacy and improved drug delivery to the inflamed gastrointestinal tract [74].

2.3.2.8.8 *Lung Inflammation*

Lung inflammation, such as chronic obstructive pulmonary disease (COPD) and asthma, can lead to respiratory symptoms and impaired lung function. PAA-based interventions have shown potential in the management of lung inflammation. Lentka and team investigated the effect of PAA nanoparticles in an animal model with a respiratory inflammatory condition, and the study results demonstrated that the nanoparticle reduced inflammation and improved the lung function [75].

2.3.2.8.9 *Skin Inflammation and Wound Healing*

Inflammatory skin conditions, such as dermatitis and psoriasis, can cause significant discomfort and impaired skin barrier function. PAA-based formulations have demonstrated anti-inflammatory effects in skin inflammation. PAA hydrogels and creams have been shown to reduce inflammatory cytokine levels and alleviate symptoms of inflammatory skin diseases. Furthermore, PAA can promote wound healing by enhancing tissue regeneration and accelerating wound closure [76].

2.3.2.9 Polypropylene Glycol

A synthetic PPG belongs to the family of polyols. It is made up of repeating units of propylene oxide, as shown in Figure 2.10 and is used in many different industries, such as personal care, cosmetics and pharmaceuticals. PPG has garnered interest lately due to its possible anti-inflammatory qualities, which makes it a viable option for the creation of innovative anti-inflammatory treatments. PPG, a versatile synthetic polymer, exhibits promising anti-inflammatory properties.

PPG's antioxidant activity, which aids in scavenging ROS and lowering oxidative stress, may be responsible for its anti-inflammatory properties. In addition, PPG's anti-irritant properties contribute to its ability to mitigate inflammation by acting as a protective barrier on the skin or mucosal surfaces by enhancing the skin's hydration and moisture retention. They also inhibit the proinflammatory mediator's production, downregulate NF-κB signaling pathway activation, and promote the production of anti-inflammatory cytokines.

Studies investigating PPG's anti-inflammatory properties have demonstrated its effectiveness in mitigating inflammation in various models, including allergic contact dermatitis, acute lung injury, and colitis. Considering the PPG's anti-inflammatory property, it is an innovative candidate for anti-inflammatory treatments. To fully understand the precise processes underlying PPG's anti-inflammatory activity and its potential uses in a range of inflammatory disorders, more research is necessary.

2.3.2.9.1 *Research Studies*

2.3.2.9.1.1 Effect in Allergic Contact Dermatitis Choi et al. examined the anti-inflammatory efficacy of PPG in a mouse model having allergic contact dermatitis (ACD). The study results outlined that topical application of PPG inhibited the development of ACD and reduced ear thickness, edema formation, and inflammatory cell infiltration. They emphasized that the PPG's anti-inflammatory effect might be due to down regulation of proinflammatory mediators [76].

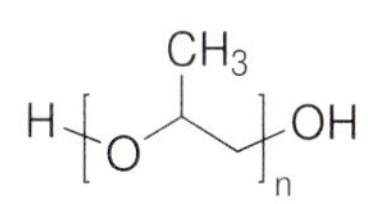

FIGURE 2.10 Structure of PPG.

2.3.2.9.1.2 Protective Effect in Acute Lung Injury Using lipopolysaccharide (LPS) in an induced acute lung injury (ALI) mouse model, the protective effect of PPG against ALI was examined. The findings showed that the injection of PPG considerably reduced lung tissue damage and inhibited the release of proinflammatory cytokines and monocyte chemoattractant protein-1 (MCP-1). The study's findings outlined that PPG exhibited an anti-inflammatory effect by preventing NF-κB activation, which involved in controlling the inflammatory responses, to provide its anti-inflammatory action [77].

2.3.2.9.1.3 Effect in Colitis A study conducted by Malik et al. investigated the potential of PPG in attenuating colitis in an animal model. Through a reduction in colonic tissue damage, a decrease in inflammatory cell infiltration, and a suppression of proinflammatory cytokine production, the observed data demonstrated that PPG therapy considerably reduced the severity of colonic inflammation. They said that the NF-κB signaling pathway may be inhibited, ROS may be activated and lowered, and anti-inflammatory cytokine production, such as interleukin-10 (IL-10), may be upregulated to produce its anti-inflammatory effects [77].

2.3.2.10 Polyurethane

Polyurethane (PU) is a polymer (structure given in Figure 2.11) that finds applications in different fields like medicine, electronics, automotive, and textiles. PU-based materials are known for their durability, flexibility, and biocompatibility, making them suitable for use in medical devices and implants. The potential anti-inflammatory qualities of PU have been brought to light by recent research, which makes it a compelling option for the creation of anti-inflammatory treatments.

PU is a polymer with several applications that demonstrate anti-inflammatory properties by regulating the activity of immune cells (neutrophils and macrophages), suppressing proinflammatory cytokines, and interfering with inflammatory signaling pathways (NF-κB and MAPK pathways). It has been demonstrated that PU induces the synthesis of anti-inflammatory cytokines while inhibiting the production of proinflammatory cytokines, such as tumor necrosis factor-alpha (TNF-α), interleukin-6 (IL-6), and interleukin-1 beta (IL-1β). Research studies investigating PU's anti-inflammatory properties have demonstrated its effectiveness in wound healing, lung injury, and bone regeneration models.

PU is a viable option for the development of anti-inflammatory therapies because of its anti-inflammatory capabilities, especially in the field of regenerative medicine. To fully understand the intricate mechanisms underlying PU's anti-inflammatory activity and investigate its potential uses in a range of inflammatory disorders, more research is necessary.

2.3.2.10.1 Research Studies on PU

2.3.2.10.1.1 Effect in Wound Healing Anti-inflammatory property in wound healing was investigated in diabetic wounds in rats by dressing the wounds with PU. The results disclosed that the PU dressing significantly reduced the inflammation in the wounds. They proposed possible mechanisms by which the anti-inflammatory impact could be attributed to an increase in the production of anti-inflammatory cytokines and decrease in proinflammatory cytokines production [78].

2.3.2.10.1.2 Lung Injury Anti-inflammatory effect of PU nanoparticles was investigated in mouse model having ALI, whereas the lung injury was induced by LPS. The results proved that PU nanoparticles significantly inhibited the activation of NF-κB pathway and reduced the production of proinflammatory cytokines, such as IL-6 and TNF-α in lung tissues. Also, it attenuated the damage of lung tissue and decreased the infiltration of inflammatory cells. The study suggested that PU nanoparticles can be considered as a therapeutic compound for lung inflammatory diseases [79].

2.3.2.10.1.3 Biodegradable Scaffolds Mattu et al., stated a hypothesis that PU-based biodegradable scaffold is having anti-inflammatory property in a rat model of a calvarial bone defect. The team consolidated that the PU-based scaffold have potential application in bone tissue engineering by decreasing the appearance of proinflammatory cytokines at the defect site. Also decreased the inflammatory cells infiltration and promoted the polarization of macrophages [80].

2.3.2.11 Polyphosphazenes

Polyphosphazenes are a unique type of polymers composed of alternating nitrogen and phosphorus as a backbone, with different side groups linked to the phosphorus atoms, as shown in Figure 2.12.

These polymers have a wide range of uses, such as biomedical coatings, tissue engineering, and drug delivery. In recent times, polyphosphazenes have garnered noteworthy interest owing to their exceptional biocompatibility and adjustable characteristics. Furthermore, recent studies indicate that polyphosphazenes may have anti-inflammatory qualities, which makes them a viable option for a number of medicinal uses.

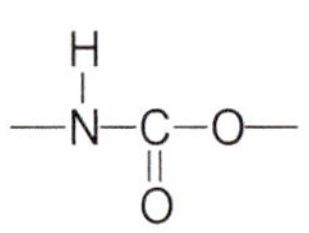

FIGURE 2.11 Structure of PU.

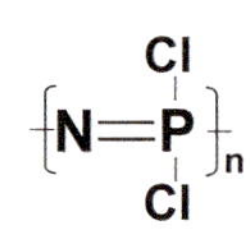

FIGURE 2.12 Structure of polyphosphazenes.

Polyphosphazenes, a unique type class of polymers having different variety of applications, possess anti-inflammatory properties by suppressing proinflammatory cytokines, modulation of immune cell responses (macrophages and neutrophils), and regulation of inflammatory signaling pathways. These mechanisms contribute to the overall attenuation of the inflammatory response.

Research studies investigating the anti-inflammatory properties of polyphosphazenes have highlighted their effectiveness in various inflammatory conditions, including arthritis, neuroinflammation, and bone fracture healing. Polyphosphazenes are a prospective option for the development of therapeutic strategies that target inflammatory disorders because of their anti-inflammatory activity. To understand the intricate mechanisms underlying polyphosphazenes' anti-inflammatory effects and maximize their potential for use in certain inflammatory illnesses, more research studies are necessary.

2.3.2.11.1 Research Studies on Polyphosphazenes' Anti-Inflammatory Properties

2.3.2.11.1.1 Arthritis Anti-inflammatory properties of polyphosphazene nanoparticles' was studied in a RA induced rat model. The results showed that the particles dramatically decreased the proinflammatory cytokines production in the inflamed joints. Also, the histological analysis revealed the synovial inflammation reduction and joint destruction. The result study suggested that polyphosphazenes hold promise as a therapeutic strategy for RA due to their anti-inflammatory properties [81].

2.3.2.11.1.2 Neuroinflammation Anti-inflammatory effect of polyphosphazene nanoparticles was assessed by microglial activation in a mouse model having neuroinflammation, where the neuroinflammation was induced by LPS. Attenuation of microglial cells is a key mediator of neuroinflammation. The results exhibited that nanoparticles significantly reduced the microglial cells activation and reduced the proinflammatory cytokines release in the brain tissue. Furthermore, treatment with polyphosphazene nanoparticles improved behavioral deficits associated with neuroinflammation. The study indicated that polyphosphazenes have potential as therapeutic agents for neuroinflammatory disorders [82].

2.3.2.11.1.3 Healing of Bone Fractures Anti-inflammatory property of polyphosphazene-based scaffolds in a rat model of bone fracture by assessing regeneration of bone was investigated. The results demonstrated that fabricated polyphosphazene scaffolds reduced the production of proinflammatory cytokine at the fracture site. Also, it stimulated the anti-inflammatory cytokines secretion including IL-10. Histological analysis outlined the enhanced bone healing by decreasing the inflammation in the fracture site. The study suggested that polyphosphazene-based scaffolds have potential applications in promoting bone regeneration by modulating the inflammatory response [83].

2.4 FUTURE PERSPECTIVES AND CONCLUSION

Inflammation is a crucial component of the body's immune response, aimed at protecting tissues from injury or infection. With advancements in material science, the use of synthetic polymers in anti-inflammation therapy has gained significant attention. In this chapter, the future perspective of studying synthetic polymers in anti-inflammation therapy is explored.

The use of synthetic polymers as carriers for anti-inflammatory drugs offers several advantages. It is possible to create polymers to have particular physical characteristics, like site-specific targeting and regulated release kinetics. PLGA was investigated for drug delivery applications. Shen et al. explained that the PLGA-based nanoparticle systems is an effective carrier for as anti-inflammatory medication. These nanoparticles exhibited excellent drug encapsulation efficiency and sustained release, leading to improved therapeutic outcomes compared to conventional drug delivery systems [84]. Synthetic polymers also hold the potential to modulate the immune response, allowing for targeted anti-inflammatory therapy. For instance, conjugation of anti-inflammatory drugs to polymers can enhance their stability and prolong their circulation time in the body. Furthermore, specific modifications can be made to the polymer structure to trigger the release of anti-inflammatory agents in response to inflammation markers. This approach allows for precise control over the site and duration of drug release, minimizing adverse effects. A study by Zhang et al. explored polymer-based systems for the modulation of immune responses in inflammatory diseases, highlighting the potential of these approaches for future therapeutics [85]. Synthetic polymers can also be designed to specifically target inflammatory mediators, providing a novel strategy for anti-inflammatory therapy. For example, molecularly imprinted polymers (MIPs) have gained consideration due to their capability to selectively recognize and bind target molecules. Through the integration of specific molecular recognition sites within the polymer matrix, MIPs can sequester inflammatory mediators, preventing their interaction with cells and subsequent inflammatory responses. A study by Gao et al. explained the successful creation of MIP-based systems for the selective removal of inflammatory cytokines, highlighting their potential for combating inflammation-associated diseases [86].

Inflammation is a complex process involving multiple pathways and signaling molecules. Synthetic polymers offer the opportunity to develop combination therapies, incorporating multiple anti-inflammatory drugs or agents to target different aspects of the inflammatory response simultaneously. This approach can enhance the efficacy of therapy and decrease the drug resistance risk. A recent review article by Hu et al. discussed the potential of combination therapies using synthetic polymers for anti-inflammatory treatment, providing insights into future directions for research and development in this area [87].

The future of studying synthetic polymers in anti-inflammation therapy holds immense potential. With the ability to act as drug delivery systems, modulate immune responses, regulate inflammatory mediators, and enable combination therapies, synthetic polymers offer innovative approaches for combating chronic inflammation. However, additional research studies are needed to optimize the polymer properties, stability, and biocompatibility to ensure their efficacy and safety in clinical applications. The integration of interdisciplinary fields, including material science, pharmacology, and immunology, will propel the development of novel synthetic polymer-based anti-inflammatory therapeutics, leading to improved patient outcomes.

REFERENCES

1. Ward MA, Georgiou TK. Thermoresponsive polymers for biomedical applications. *Polymers*. 2011 Aug 3;3(3):1215–42.
2. Twaites BR, De Las Heras Alarcón C, Lavigne M, Saulnier A, Pennadam SS, Cunliffe D, Górecki DC, Alexander C. Thermoresponsive polymers as gene delivery vectors: Cell viability, DNA transport and transfection studies. *Journal of Controlled Release*. 2005 Nov 28;108(2–3):472–83.
3. Mahapatro A, Singh DK. Biodegradable nanoparticles are excellent vehicle for site directed in-vivo delivery of drugs and vaccines. *Journal of Nanobiotechnology*. 2011 Dec;9:1–1.
4. Mishra B, Vuppu S, Rath K. The role of microbial pullulan, a biopolymer in pharmaceutical approaches: A review *Journal of Applied Pharmaceutical Science* 2011 Aug 30(Issue):45–50.
5. Eliyahu H, Barenholz Y, Domb AJ. Polymers for DNA delivery. *Molecules*. 2005 Jan 31;10(1):34–64.
6. Pack DW, Hoffman AS, Pun S, Stayton PS. Design and development of polymers for gene delivery. *Nature Reviews Drug Discovery*. 2005 Jul 1;4(7):581–93.
7. Dang JM, Leong KW. Natural polymers for gene delivery and tissue engineering. *Advanced Drug Delivery Reviews*. 2006 Jul 7;58(4):487–99.
8. Jang YH, Blanco M, Creek J, Tang Y, Goddard WA. Wax inhibition by comb-like polymers: Support of the incorporation– perturbation mechanism from molecular dynamics simulations. *The Journal of Physical Chemistry B*. 2007 Nov 22;111(46):13173–9.
9. Roodbandi AS, Naderi H, Hashenmi-Nejad N, Choobineh A, Baneshi MR, Feyzi V. Technical report on the modification of 3-dimensional non-contact human body laser scanner for the measurement of anthropometric dimensions: Verification of its accuracy and precision. *Journal of Lasers in Medical Sciences*. 2017;8(1):22.
10. Nagaraj K, Kristiansen LV, Skrzynski A, Castiella C, Garcia-Alonso L, Hortsch M. Pathogenic human L1-CAM mutations reduce the adhesion-dependent activation of EGFR. *Human Molecular Genetics*. 2009 Oct 15;18(20):3822–31.
11. DeRouen TA, Wiesenbach C. The summer institute in clinical dental research methods: Still going and growing after twenty years. *Journal of Dental Education*. 2012 Nov;76(11):1408–15.
12. Giannocco G, Oliveira KC, Crajoinas RO, Venturini G, Salles TA, Fonseca-Alaniz MH, Maciel RM, Girardi AC. Dipeptidyl peptidase IV inhibition upregulates GLUT4 translocation and expression in heart and skeletal muscle of spontaneously hypertensive rats. *European Journal of Pharmacology*. 2013 Jan 5;698(1–3):74–86.
13. Gindes L, Archiron R. Tetralogy of Fallot: Evaluation by 4D spatiotemporal image correlation. *Ultrasound in Obstetrics and Gynecology*. 2008 Sep 1;32(4): 598–9.
14. Wu Y, Liu DH, Zhang XX. Thymidine kinase 2 gene compound heterozygous mutation leads to mitochondrial DNA depletion syndrome-2: A case report. Zhonghua er ke za zhi= Chinese *Journal of Pediatrics*. 2018 May 1;56(5):381–2.
15. Mankaï A, Sakly W, Thabet Y, Achour A, Manoubi W, Ghedira I. Anti-Saccharomyces cerevisiae antibodies in patients with systemic lupus erythematosus. *Rheumatology International*. 2013 Mar;33:665–9.
16. Koskinen CA, Lindström UÅ. An envisioning about the caring in listening. *Scandinavian Journal of Caring Sciences*. 2015 Sep;29(3):548–54.
17. Elieh-Ali-Komi D, Hamblin MR. Chitin and chitosan: Production and application of versatile biomedical nanomaterials. *International Journal of Advanced Research*. 2016 Mar;4(3):411.
18. Hu Q, Sun W, Wang C, Gu Z, et al. Targeted uptake and stimulated release of magnetic nanoparticles in tumor extracellular pH microenvironment. *Journal of the American Chemical Society*, 2015, 137(36), 11461–4.
19. Gindes L, Archiron R. Tetralogy of Fallot: Evaluation by 4D spatiotemporal image correlation. *Ultrasound in Obstetrics and Gynecology*. 2008 Sep 1;32(4):598–9.
20. Kang ML, Cho CS, Yoo HS. Application of chitosan microspheres for nasal delivery of vaccines. *Biotechnology Advances*. 2009 Nov 1;27(6):857–65.
21. Tran TV, Phuong TH, Tran NQ, Nguyen CK, Nguyen DH. Polymeric chitosan based nanogels as a potential platform for dual targeted drug delivery in cancer therapy. *International Journal of Nanotechnology*. 2018;15(1–3):188–98.
22. Zhang Y, Cheng X, Wang J, Wang Y, Shi B, Huang C, Yang X, Liu T. Novel chitosan/collagen scaffold containing transforming growth factor-β1 DNA for periodontal tissue engineering. *Biochemical and Biophysical Research Communications*. 2006 May 26;344(1):362–9.
23. Kosik-Kozioł A, Costantini M, Mróz A, Idaszek J, Heljak M, Jaroszewicz J, Kijeńska E, Szöke K, Frerker N, Barbetta A, Brinchmann JE. 3D bioprinted hydrogel model incorporating β-tricalcium phosphate for calcified cartilage tissue engineering. *Biofabrication*. 2019 May 3;11(3):035016.
24. Zheng J, Yuan X, Cheng G, Jiao S, Feng C, Zhao X, Yin H, Du Y, Liu H. Chitosan oligosaccharides improve the disturbance in glucose metabolism and reverse the dysbiosis of gut microbiota in diabetic mice. *Carbohydrate Polymers*. 2018 Jun 15;190:77–86.
25. Alarcón-Alarcón C, Inostroza-Riquelme M, Torres-Gallegos C, Araya C, Miranda M, Sánchez-Caamaño JC, Moreno-Villoslada I, Oyarzun-Ampuero FA. Protection of astaxanthin from photodegradation by its inclusion in hierarchically assembled nano and microstructures with potential as food. *Food Hydrocolloids*. 2018 Oct 1;83:36–44.
26. Li J, Zhang X, Lv F, Liu L, Liu J, et al. Biocompatibility and anti-inflammatory properties of size-controlled polyethylene nanocomposites for potential bone tissue engineering

applications. *ACS Applied Materials & Interfaces*, 2018, 10(35), 29436–47.

27. Bures R, Skopalik J, Tomankova KB, Homola T, et al. Polyethylene and polypropylene microparticles embedded in human immune system cell lines reduce immune response. *Journal of Biomaterials Applications*, 2015, 31(5), 610–23.
28. Kweh SWK, Cheong XL, Kwong CH, Abdul Rahman N, et al. Comparison of in vivo biocompatibility of polyethylene-based implants through urine analysis. *Journal of Applied Polymer Science*, 2020, 137(18), 48654.
29. Yun YR, Kim IG, Koh YH, Kim HW, et al. Enhanced wound healing effect of polyethyleneimine-grafted polyethylene fibers by simultaneous delivery of basic fibroblast growth factor and small hairpin RNA targeting IL-6 in a rat pressure ulcer model. *International Journal of Nanomedicine*, 2019, 14, 3113–27.
30. Tang F, Fan K, Wang K, Bian C. Atractylodin attenuates lipopolysaccharide-induced acute lung injury by inhibiting NLRP3 inflammasome and TLR4 pathways. *Journal of Pharmacological Sciences*. 2018 Apr 1;136(4):203–11.
31. Biggs MJ, Sutherland AD, Murariu M, et al. A preliminary investigation into the etiology and prevention of bacterial attachment to hernia repair meshes. *Journal of the Mechanical Behavior of Biomedical Materials*, 2016, 61, 124–35.
32. Arcos N, Olmo R, Tapia C, López-Noriega A, et al. An Ibuprofen loaded Mg-HPMO hybrid material with anti-inflammatory properties. *European Journal of Pharmaceutical Sciences*, 2017, 102, 73–80.
33. Vasita R, Shanmugam K, & Katt DS. Improved immune response from PLGA-based nanocomposite scaffolds for tissue engineering applications. *International Journal of Nanomedicine*, 2017, 12, 2955–72.
34. Gaharwar AK, Peppas NA, & Khademhosseini A. Nanocomposite hydrogels for biomedical applications. *Biotechnology and Bioengineering*, 2014, 111(3), 441–53.
35. Park JY, Kwon OH, Kim TH, Baek KJ, et al. Immobilization of dexamethasone in PLGA microspheres for the treatment of inflammatory bowel disease. *Materials Science and Engineering: C*, 2020, 106, 110172.
36. Al Samri K, Albrecht W, Vermonden T, Colella F, et al. Mucoadhesive thermosensitive hydrogels: Glycol chitosan-graft-poly(N-isopropylacrylamide) and PLGA-based systems for sustained delivery of diclofenac in dysphagic patients. *Carbohydrate Polymers*, 2019, 204, 39–52.
37. Gagliardi M, Liu X, Veronese C, Fong CY, et al. Dexamethasone loaded PLGA micro and nanoparticles for the prevention of inflammatory corneal angiogenesis. *Investigative Ophthalmology & Visual Science*, 2018, 59(9), 3764–75.
38. Zhang Y, Zhang L, Chen Y, Zhang Y, et al. Anti-inflammatory effects of polypropylene nonwoven fabrics through the regulation of macrophage polarization in rat model of subcutaneous inflammation. *Journal of Biomedical Materials Research Part A*, 2018, 106(2), 502–11. https://doi.org/10.1002/jbm.a.36347
39. Nicholson TA, Sagmeister M, Wijesinghe SN, Farah H, Hardy RS, Jones SW. Oligonucleotide therapeutics for age-related musculoskeletal disorders: Successes and challenges. *Pharmaceutics*. 2023 Jan 10;15(1):237.
40. Yamamoto C, Aoyagi K, Hayashi Y, Morita I, Sakisaka S. The effect of polyethylene glycol-electrolyte on dextran sulfate sodium-induced colitis in rats. *Gastrointestinal Endoscopy*. 2005 Apr 1;61(5): AB268.
41. Cappelli E, Taylor R, Cevasco M, Abbondandolo A, et al. Involvement of XRCC1 and DNA ligase III gene products in DNA base excision repair. *The Journal of Biological Chemistry*, 1997, 272(38), 23970–5. https://doi.org/10.1074/jbc.272.38.23970
42. Zhou HH, Wang JL, Zhao X. Effect of denitrifying biofilm development on the surface configuration and chemical composition of PBS polymer. Huan Jing ke Xue= *Huanjing Kexue*. 2006 Dec 1;27(12):2507–10.
43. Juarez-Moreno K, Ayala M, Vazquez-Duhalt R. Antioxidant capacity of poly (ethylene glycol) (PEG) as protection mechanism against hydrogen peroxide inactivation of peroxidases. *Applied Biochemistry and Biotechnology*. 2015 Nov;177:1364–73.
44. Liu TH, Beckman JS, Freeman BA, Hogan EL, Hsu CY. Polyethylene glycol- conjugated superoxide dismutase and catalase reduce ischemic brain injury. Am J Physiol. 1989 Feb;256(2 Pt 2):H589-93.
45. Thomas TL, Bazemore SA, Kornbrust DJ. Single-dose hemodynamic toxicity and pharmacokinetics of a partial phosphorothioate anti-HIV oligonucleotide (AR177) after intravenous infusion to cynomolgus monkeys. *The Journal of Pharmacology and Experimental Therapeutics*, 1996, 278(3), 1306–12.
46. Li X, Luo R, Chan CP, Wu X, Zhang S, Hua W. The clinical correlates and prognostic impact of QRS prolongation in patients with dilated cardiomyopathy: A single-center cohort study. *International Journal of Cardiology* 2014 Apr 1;172(3):e475–7.
47. Aaron SE, Hunnicutt JL, Embry AE, Bowden MG, Gregory CM. POWER training in chronic stroke individuals: Differences between responders and nonresponders. *Topics in Stroke Rehabilitation* 2017 Oct;24(7):496–502
48. Peter KM. Veterinarians as leaders in animal welfare. *Journal of the American Veterinary Medical Association*. 2010 Mar 1;236(5):509.
49. Fagan MJ, Connelly CD, Williams BS, Fisher ES. Integrating team training in the pediatric life support program: An effective and efficient approach?. JONA: *The Journal of Nursing Administration*. 2018 May 1;48(5):279–84.. https://doi.org/10.1097/NNA.0000000000000613
50. Cox PM, Williamson MS, Nijsse FJ, Huntingford C. Cox et al. Reply. *Nature*. 2018 Nov 1;563(7729): E10–5.
51. Xu W, Ling P, Zhang T. Polymeric micelles, a promising drug delivery system to enhance bioavailability of poorly water-soluble drugs. *Journal of Drug Delivery*. 2013;2013(1):340315.
52. Ghezzi M, Pescina S, Padula C, Santi P, Del Favero E, Cantù L, Nicoli S. Polymeric micelles in drug delivery: An insight of the techniques for their characterization and assessment in biorelevant conditions. *Journal of Controlled Release*. 2021 Apr 10;332:312–36.
53. Danhier F, Ansorena E, Silva JM, Coco R, Le Breton A, Préat V. PLGA-based nanoparticles: An overview of biomedical applications. *Journal of Controlled Release*. 2012 Jul 20;161(2):505–22.
54. Yang X, Li Z, Wang N, Li L, Song L, He T, Sun L, Wang Z, Wu Q, Luo N, Yi C. Curcumin-encapsulated polymeric micelles suppress the development of colon cancer in vitro and in vivo. *Scientific reports*. 2015 May 18;5(1):10322.
55. Hu Q, Sun W, Wang C, Gu Z, et al. Targeted uptake and stimulated release of magnetic nanoparticles in tumor extracellular pH microenvironment. *Journal of the American Chemical Society*, 2015, 137(36), 11461–4.

56. Lawrence T. The nuclear factor NF-kappaB pathway in inflammation. *Cold Spring Harb Perspect Biol*. 2009 Dec;1(6):a001651.
57. Lee H, Kim CS, Ko JK. et al. Polyvinyl alcohol attenuates Toll-like receptor 4 signaling and NF-kappa B-mediated gene expression in macrophages. *Experimental and Molecular Medicine*, 2006, 38(3), 276–84.
58. Youssef SS, Borai IH, El-Sawy MM, Abuelwafa, AA. Up-regulation of antioxidant polyvinyl alcohol by treatment with vitamin C and zinc sulfate in a rat model of acetic acid-induced colitis. *Experimental and Therapeutic Medicine*, 2020, 20(1), 345–54.
59. Kim YJ, Kim HJ, Kim JI, et al. Polyvinyl alcohol attenuates rheumatoid arthritis by modulating immune dysregulation in mice. *Journal of Microbiology and Biotechnology*, 2019, 29(7), 1147–54.
60. Patil PS, Saklani A, Mehta S. Need for standardized endoscopic tattooing agents. *Indian Journal of Gastroenterology*. 2014 Jul;33:392.
61. Partinen M, Gislason T. Basic Nordic Sleep Questionnaire (BNSQ): A quantitated measure of subjective sleep complaints. *Journal of Sleep Research*. 1995 Jun;4:150–5.
62. Shahid SK. Novel anti-Helicobacter pylori therapies. *Pharmaceutical patent analyst*, 2014, 3(4), 411–27.
63. Long S, Xu Y, Wang C, Li C, Liu D, Piao X. Effects of dietary supplementation with a combination of plant oils on performance, meat quality and fatty acid deposition of broilers. *Asian-Australasian journal of animal sciences*. 2018 Nov;31(11):1773.
64. Lin B, and Looi A. Orbital lymphoma. *Ophthalmology*, 2007, 114(7), 1423.
65. González-Burguera I, Ricobaraza A, Aretxabala X, Barrondo S, del Caño GG, de Jesús ML, Sallés J. Highly efficient generation of glutamatergic/cholinergic NT2-derived postmitotic human neurons by short-term treatment with the nucleoside analogue cytosine β-d-arabinofuranoside. *Stem Cell Research*. 2016 Mar 1;16(2):541–51.
66. Xiao P, Fink M, Boccara AC. Full-field spatially incoherent illumination interferometry: A spatial resolution almost insensitive to aberrations. *Optics Letters*. 2016 Sep 1;41(17):3920–3.
67. Caudron F, Barral Y. A super-assembly of Whi3 encodes memory of deceptive encounters by single cells during yeast courtship. *Cell*. 2013 Dec 5;155(6):1244–57.
68. Jiao X, Yu W, Qian J, Chen Y, Wei P, Fang W, Yu G. ADAM-17 is a poor prognostic indicator for patients with hilar cholangiocarcinoma and is regulated by FoxM1. *BMC Cancer*. 2018 Dec;18:1–1.
69. Adlaf EM, Hamilton HA, Wu F, Noh S. Adolescent stigma towards drug addiction: Effects of age and drug use behaviour. *Addictive Behaviors*. 2009 Apr 1;34(4):360–4.
70. Myers B, Dittmeyer K, Greenwood-Van Meerveld B. Involvement of amygdaloid corticosterone in altered visceral and somatic sensation. *Behavioural Brain Research*. 2007 Jul 19;181(1):163–7.
71. Díez CM, Gaut BS. The jury may be out, but it is important that it deliberates: A response to Besnard and Rubio de Casas about olive domestication. *New Phytologist*. 2016 Jan 1;209(2):471–3.
72. Mologni L. Current and future treatment of anaplastic lymphoma kinase-rearranged cancer. *World Journal of Clinical Oncology*. 2015 Oct 10;6(5):104.
73. Bernstein V, Kolodney E. Postcollision multifragmentation in fullerene-surface impact: Microscopic insights via molecular dynamics simulations. *The Journal of Chemical Physics*. 2016 Jul 28;145(4): 044303.
74. Lentka Ł, Smulko J, Kotarski M, Granqvist CG, Ionescu R. Non-gaussian resistance fluctuations in gold-nanoparticle-based gas sensors: An appraisal of different evaluation techniques. *Sensors*. 2017 Apr 3;17(4):757.
75. Yamashita Y, Ohashi J, Hirai Y, Choi YL, Kaneda R, Fujiwara SI, Arai Y, Akutsu M, Tsutsumi C, Miyazaki Y, Usuki K. Gene expression profiles of CD133-positive fractions predict the survival of individuals with acute Myeloid Leukemia. *Cancer Genomics & Proteomics*. 2006 May 1;3(3–4):169–81.
76. Choi KH, Kim J, Kwon OS, Kim MJ, Ryu YH, Park JE. Is heart rate variability (HRV) an adequate tool for evaluating human emotions? –A focus on the use of the International Affective Picture System (IAPS). *Psychiatry Research*. 2017 May 1;251:192–6.
77. Wang WF, Yang CY, Wu YF. SVM-based classification method to identify alcohol consumption using ECG and PPG monitoring. *Personal and Ubiquitous Computing*. 2018 Apr;22:275–87.
78. Morales-González M, Díaz LE, Dominguez-Paz C, Valero MF. Insights into the design of polyurethane dressings suitable for the stages of skin wound-healing: A systematic review. *Polymers (Basel)*. 2022 Jul 24;14(15):2990.
79. Liu W, Li S, Wang B, Peng P, Gao C. Physiologically responsive polyurethanes for tissue repair and regeneration. *Advanced NanoBiomed Research*. 2022 Oct;2(10):2200061.
80. Sartori S, Chiono V, Tonda-Turo C, Mattu C, Gianluca C. Biomimetic polyurethanes in nano and regenerative medicine. *Journal of Materials Chemistry B*. 2014;2(32):5128–44.
81. Su X, Jing X, Jiang W, Li M, Liu K, Teng M, Ma Y, Wang D, Meng L, Zhang Y, Ji W. Polyphosphazene nanodrugs for targeting delivery and inflammation responsive release of curcumin to treat acute lung injury by effectively inhibiting cytokine storms. *Colloids and Surfaces B: Biointerfaces*. 2023 Sep 1;229:113446.
82. Huang Y, Du Z, Wei P, Chen F, Guan B, Zhao Z, Zhang X, Cai Q, Mao J, Leng H, Yang X. Biodegradable microspheres made of conductive polyorganophosphazene showing antioxidant capacity for improved bone regeneration. *Chemical Engineering Journal*. 2020 Oct 1;397:125352.
83. Jin GW, Rejinold NS, Choy JH. Polyphosphazene-based biomaterials for biomedical applications. *International Journal of Molecular Sciences* 2022 Dec 15;23(24):15993.
84. Shariati A, Chegini Z, Ghaznavi-Rad E, Zare EN, Hosseini SM. PLGA-based nanoplatforms in drug delivery for inhibition and destruction of microbial biofilm. *Frontiers in Cellular and Infection Microbiology* 2022 Jun 21;12:926363.
85. Zhang J, Zhang M, Tang K, Verpoort F, Sun T. Polymer-based stimuli-responsive recyclable catalytic systems for organic synthesis. *Small*. 2014 Jan 15;10(1):32–46.
86. Gao M, Gao Y, Chen G, Huang X, Xu X, Lv J, Wang J, Xu D, Liu G. Recent advances and future trends in the detection of contaminants by molecularly imprinted polymers in food samples. *Frontiers in Chemistry*. 2020 Dec 1;8:616326.
87. Hu CM, Zhang L. Nanoparticle-based combination therapy toward overcoming drug resistance in cancer. *Biochemical Pharmacology*. 2012 Apr 15;83(8):1104–11.

3 Drug Delivery Approaches in the Management of Inflammatory Diseases

Gopal Bajad, Shristi Arya*, Soham Loharkar, Saptarshee Bhattacharjee, Dipak Panchal, and Soyal S. Sayyed*
National Institute of Pharmaceutical Education and Research, Hyderabad, India

Shivani Sharma
Hindu College Pharmacy, Sonipat, India

Bhavana Bojja, Darshana Tijare, Jitendra Kumar, Chandanapalli Dinesh Kumar, Jaya Narhire, and Pawan Devangan
National Institute of Pharmaceutical Education and Research, Hyderabad, India

Ashish Baldi
Maharaja Ranjit Singh Punjab Technical University, Bathinda, India

Jitender Madan
National Institute of Pharmaceutical Education and Research, Hyderabad, India

3.1 INTRODUCTION

Inflammation has traditionally been acknowledged as a defensive strategy of all organisms in response to any injury or infection. However, it is progressively becoming indisputable that chronic inflammation is unanimously linked to diseases associated with lifespan and affluence, such as obesity, neurodegenerative, cardiovascular diseases, and cancer (1,2,3,4). The expanding prevalence of these conditions in recent decades brings up important queries like what aspects of human physiology make us prone to these diseases and why chronic inflammation connects diseases from different origins. Although the reasons for the close affiliation of inflammation and disease in extreme cases are well comprehended, it is evident in the case of unregulated inflammatory responses leading to conditions like tissue damage and sepsis (5, 6). The primary attributes of inflammation and inflammatory processes contribute to disease susceptibility but are still elusive and also the features of the target tissues and/or organs that make them susceptible to inflammatory pathologies. An inflammatory response can be initiated by multiple harmful stimuli, such as injury and infection, which lead to intricate processes involving different cell types and molecular mediators. This response can be dissembled into four common components: inflammatory mediators, inducers and target tissues. Together, these components work to enhance the elimination of harmful stimuli, adaptation to a noxious state, and restoration of tissue homeostasis (7). However, modification of target tissue functional states underlies the pathological potential of the inflammatory response. This alteration/modification can lead to various adaptive benefits but also comes with baggage, as it impairs the normal function of the affected tissues and organs. Modification or alteration of target tissue can be achieved via proper utilization of various drug delivery approaches employing suitable delivery routes.

* Authors contributed equally.

DOI: 10.1201/9781003477389-3

A summary of various routes of administration for targeting various types of inflammatory disorders are depicted in Figure 3.1.

Oral administration is the most common method for treating localized gastrointestinal inflammatory conditions due to its noninvasive nature, ease of implication, and ability to target local inflammation sites (8). There are multiple delivery systems for the treatment of inflammatory diseases, which include stimuli-sensitive and carrier-based delivery systems. Stimuli-sensitive systems encompass time-based, pH-based, enzyme-based, and pressure-based mechanisms (9). Polymeric nanoparticles (NPs), lipid-based nanovesicles, and hybrid systems are all examples of nanocarrier-based delivery systems. In designing oral drug delivery systems, one must consider factors such as intestinal transit time, gastrointestinal pH, disease site, gut microbiota, enzymes, and drug solubility in the colon, as these influence drug release and therapeutic efficacy (10). However, since the majority of pain and inflammatory issues involve the skin's surface, topically administering nonsteroidal anti-inflammatory drugs (NSAIDs) has several benefits. Studies have indicated that administering NSAIDs directly to the site of inflammation can alleviate discomfort and swelling (11). Several advantages come with topical delivery of these medications over oral administration, including self-administration, low systemic side effects, noninvasiveness, no hepatic metabolism and no risk of gastrointestinal irritation (12). On the other hand, parenteral drug administration integrates a range of delivery systems, such as NPs, liposomes, microemulsions, and suspensions. Reduced first-pass metabolism and improved efficacy are two major benefits of injectable approach (13). Parenteral routes of administration are used to treat autoimmune encephalitis, neuromyelitis optica, multiple sclerosis (MS), chronic inflammatory demyelinating polyneuropathy and other neurological disorders. Inflammatory diseases encompass asthma, chronic obstructive pulmonary disease (COPD), pulmonary fibrosis, and rheumatoid arthritis-induced interstitial lung disease (RA-ILD), are often caused by the natural immune response. Lung epithelial cells produce inflammatory mediators, leading to mucous secretion and inflammation (14). Conventional drug delivery for inflammatory lung diseases through oral and parenteral routes has drawbacks, such as inadequate concentration of a drug at the targeted site and systemic distribution leading to adverse effects. Furthermore, chronic inflammatory disease requires long-term therapy, which involves a risk of resistance and poor therapeutic efficacy. Thus, inhalation delivery can improve drug concentration at the site of action and reduce systemic exposure, potentially improving therapeutic outcomes. If administered intravenously, a wide spectrum of nanomedicines that are not polyethylene glycol (PEG) PEGylated are likely to

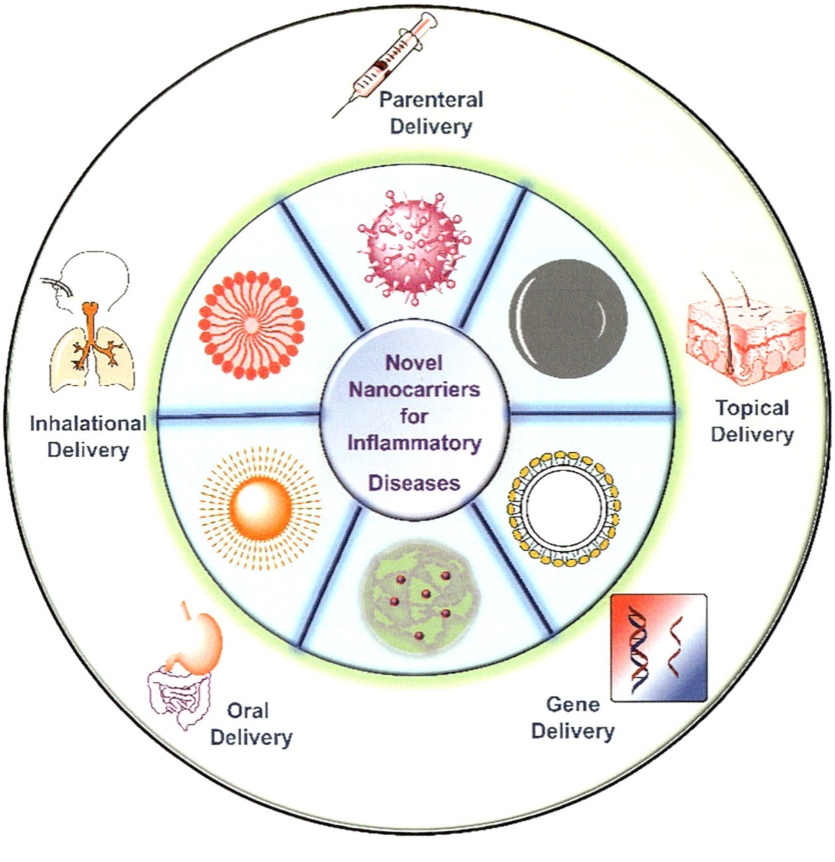

FIGURE 3.1 Various delivery approaches for the management of inflammatory disorders.

get absorbed and destroyed by the immune system. Biomimetic NPs, a groundbreaking delivery strategy, is a hybrid nanostructure made of immune cells, erythrocytes, and platelet cells that have an outer layer that resembles a cell membrane (15, 16). Liposomes made with cell membrane proteins and NPs coated with cell membranes fall under such a category. This method prolongs the NPs' circulation time and improves their targeting capability by imitating the biological properties of the source cells. Moreover, the intrinsic attributes of the membrane covering make biomimetic NPs less likely to be identified by the immune system. In virtue of their versatile NP cores and distinct biodistribution properties generated from various cell types, biomimetic nanomedicines present a promising new strategy. However, there are still significant issues that need to be amended, like possible immunological reactions, complement activation upon delivery, and difficulties in scaling up and sterilizing the biomimetic nanodevices (17).

3.2 ORAL DELIVERY APPROACHES FOR THE MANAGEMENT OF INFLAMMATORY DISEASES

Inflammatory diseases encompass a wide range of conditions, primarily affecting the stomach and intestinal regions, where oral delivery approaches have proven effective in managing such types of disorders. Notable examples of such diseases include inflammatory bowel diseases (IBD), celiac diseases, peptic ulcers, etc. Within the category of IBD, two prominent conditions are Crohn's disease (CD) and ulcerative colitis (UC) (18). Despite both falling under the same umbrella of IBD, CD and UC have distinct characteristics. From mouth to anus, CD can affect any part of the gastrointestinal (GI) tract, often leading to a wide range of symptoms and complications. In contrast, UC is confined to the rectal and colonic regions, causing continuous inflammation and ulceration of the colon lining (19). Furthermore, the chronic nature of these inflammatory diseases necessitates ongoing research into more efficient and patient-friendly treatment modalities, with oral drug delivery systems offering promising advancement in this field (20).

Oral administration is a frequently utilized method for treating localized gastrointestinal conditions (21). Noninvasive nature, ease of administration, and ability to target specific sites of inflammation are inherent benefits of oral drug delivery (8). Achieving a localized action of the drug is necessary to treat IBDs despite systemic administration, and the oral route facilitates this site-specific action (22). Various delivery approaches are available for treating inflammatory diseases, including stimuli-sensitive delivery and carrier-based delivery systems. The stimuli-sensitive delivery systems encompass time-based, pH-based, enzyme-based, and pressure-based systems (9). Meanwhile, nanocarrier-based delivery systems include polymeric NPs, lipid-based vesicular systems, and hybrid drug delivery systems. Several factors must be considered when designing an oral drug delivery system, as they influence the drug release and its therapeutic efficacy. These factors include the disease site, pH of the gastrointestinal lumen, intestinal transit time, gut microflora and their enzymes, and drug dissolution in the colon (23).

3.2.1 Site-Specific Drug Delivery Systems

The primary clinical symptoms of IBD are restricted to the intestine; hence, oral administration is the recommended mode of treatment. Furthermore, because the colon is involved in IBD therapy, oral administration methods that target the colon are preferred. Oral medication delivery systems that are colon targeted usually capitalize on the physiological characteristics of the colon, including its pH, transit duration, microflora, osmotic potential, and pressure. Microflora-activated systems, pH-dependent systems, pressure-based systems, time-dependent systems, prodrugs, and osmotic-triggered systems are majorly employed for site-specific delivery (Table 3.1) (24). Drug release from pH-dependent release devices is enabled by the use of pH-sensitive polymers that dissolve upon the changes in the pH of the GI tract. Prodrugs are chemical substances that lack pharmacological activity; instead, they are converted into active agents and carrier molecules by bacterial enzymes in the colon through linker cleavage. Drugs are transported to a specific area of the GI tract at predetermined time points using time-dependent delivery devices that provide the drug release after a lag period. The greater intestinal pressure brought on by peristaltic activity causes pressure-controlled delivery mechanisms to collapse. The release of active substances is regulated by a semipermeable membrane in osmotic-controlled delivery systems (25) (Figure 3.2).

The delivery mechanisms mentioned above are probably impacted by mucosal inflammation since they can concurrently alter GI motility, induce diarrhea, and alter intestinal volume and pH. The area between the colon and the rectum typically has a pH of 7.0 ± 0.7. However, in patients with UC, this pH ranges from 2.7 to 5.5; in individuals with CD, it averages 5.3 (26). Additionally, patients with IBD have altered normal intestinal microbiota, which may have an impact on targeting systems mediated by microflora-activated systems (27). Compared to healthy controls, patients with IBD have altered GI tract transit times (28). Furthermore, alterations in gastrointestinal motility reduce the efficacy of osmotic and pressure-regulated drug delivery systems in terms (29).

3.2.2 Carrier-Based Oral Delivery System in the Management of Inflammatory Diseases

3.2.2.1 Nanoparticle-Based Delivery System

Nanocarrier-based delivery systems are considered the most effective delivery systems as they enhance the

TABLE 3.1
Different Site-specific Delivery Approaches for Oral Administration in the Management of Inflammatory Diseases

Delivery Strategy	Drug Release Mechanism	Comments	Case Study	Reference
Prodrugs	The bond between the drug and carrier is cleaved by the gut microflora enzymes, gut enzymes.	Site-specific delivery is achieved by Prodrugs. The drug molecules are protected from degradation in unspecific regions where drug release is undesired.	The prodrug of budesonide (budesonide-beta-D-glucuronide), which is hydrolyzed by colon-specific beta-glucuronidase, is designed for treating ulcerative pancolitis.	(30)
pH-dependent systems	The polymers that show solubility at a specific pH of the gastro intestinal tract (GIT) are employed.	The broad range of pH differences in GIT provides the advantage of pH-sensitive release systems. The pH is liable to changes caused by feeding, fasting, and disease conditions.	pH-sensitive edgragit-S_{100} coated mesalazine and curcumin-loaded liposomes for treating UC.	(31)
Time-dependent systems	The drug is released after a predetermined lag time. The lag period depends on the GIT transit times.	The transit time may be inconsistent, different from person to person, and affected by different disease conditions.	Ketoprofen time-based release tablets for colon-specific delivery were prepared. The ketoprofen core was primarily coated with a kollidon-90, an intermediate coating of hydroxy propyl methyl cellulose (HPMC) that swells and gives a lag phase release; the outer enteric coating, which dissolves at pH > 5, was made with Eudragit.	(32)
Microflora-activated systems	The drug is released by the fermentation of non-starch polysaccharide coating/matrix by anaerobic bacteria.	This approach is best suited for colon targeting as non-starch polysaccharides are degraded by colon microflora. The release of the drug is very slow, as polysaccharide degradation takes time.	Poly[bi(o-carboxyphenyl) adipate-polyethylene glycol] anhydrides—P(BOCA-PEG)—the polymeric conjugate of 5-aminosalicylic acid (5-ASA) was synthesized. The azo linkage was degraded by the colon bacteria, which provided a site-specific release.	(33)
Pressure-based systems	The increase in pressure due to peristalsis movements triggers the release of drugs from systems that are liable to pressure changes.	The intraluminal pressure in the small intestine is buffered by the presence of a large amount of fluids. This approach is mostly used for colon targeting as reabsorption of water occurs and peristalsis movements can increase pressure.	The pressure-controlled colon delivery capsule (PCDC) of 5-aminosalicylic acid (5-ASA) was prepared by coating the Ethyle cellulose into the gelatin capsule with 2.0 mm diameter pores for treating UC.	(34)
osmotic-controlled delivery systems	When the osmogene comes in contact with the gut fluid, there develops an osmotic pressure difference, and the drug releases from a semipermeable membrane.	There are chances of dose dumping and hypersensitive reactions due to the implantation of the device.	Microbial triggered colon-targeted osmotic pump (MTCT-OP) of budesonide for treating IBD was developed. Chitosan was used as a pore-forming agent, with cellulose acetate as the semipermeable membrane and PEG 400 as a plasticizer. Whole system was enterically coated with Eudragit EL-100.	(35)

therapeutic efficacy of drugs, provide more localized action, and prevent unnecessary drug exposure (36). NPs are artificially made particles with sizes in the nanoscale range, generally between 1 to 100 nm (37). Various polymers are used in the preparation of NPs, including poly(lactic-co-glycolic-acid) (PLGA), polyethylene glycol (PEG) and Eudragit, etc. (38). PLGA and PEG aid in site-specific targeting and protect the drug from degradation in the acidic conditions of the stomach. Coating NPs with Eudragit makes them pH-sensitive and stable, allowing for drug release in an optimal environment at the appropriate pH.

NPs-based drug delivery involves several approaches, including ligand-mediated targeting, size-mediated targeting, redox-mediated targeting and surface-charge targeting (10). Inflammatory diseases exhibit an enhanced permeability and retention effect (EPR)– i.e., enhanced permeability and retention effect similar to that observed in tumor

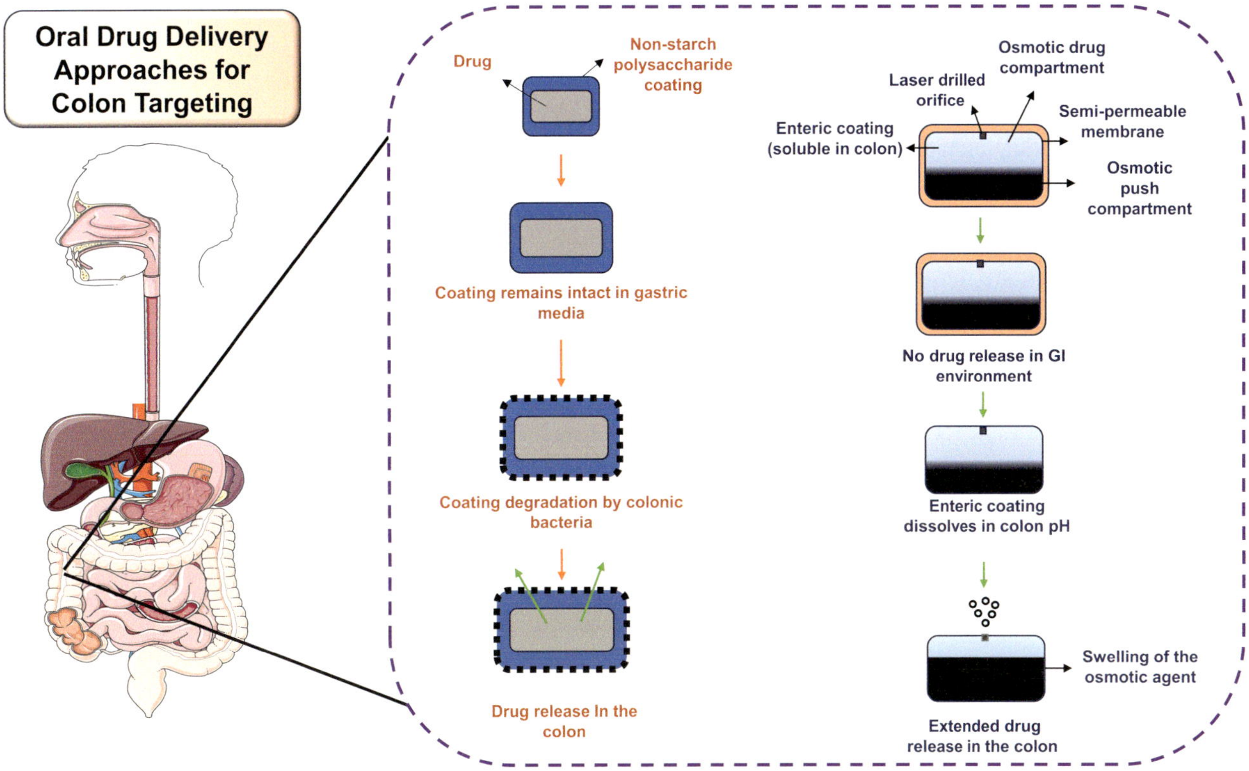

FIGURE 3.2 Schematic illustration of colon-targeted delivery systems utilizing osmotic-controlled systems.

vasculature (39). Due to the EPR effect, NPs demonstrate increased permeability, making their size a critical factor in size-mediated targeting (40). In IBD, mucosal composition changes, allowing electrostatic interactions to target inflamed regions. Thus, surface-charged NPs can be used for targeting. Both positively and negatively charged NPs are employed in surface-charge-mediated targeting. In IBD, colonic mucins with negatively charged carbohydrate moieties are present, which can be targeted with cationic NPs due to their interaction with oppositely charged molecules (41). Despite enhanced intestinal motility in IBD, cationic NPs can still be retained in the gastrointestinal tract for extended periods (42, 43). Additionally, inflamed regions exhibit increased levels of positively charged proteins, such as eosinophil cationic proteins and transferrin, enhancing the affinity of anionic NPs for inflamed tissue (44). Furthermore, IBD-inflamed tissues show a ten-to-hundred-fold increase in reactive oxygen species (ROS) concentration (45, 46). Therefore, radical-containing NPs can be used to deliver drugs to inflamed sites. For instance, Vong et al. developed nitroxide radical-containing NPs that help suppress inflammation by scavenging ROS. This system is advantageous, as it acts only on superficial lesions, preventing deeper penetration into the mucosa and thereby reducing systemic adverse effects (47). Ligand-mediated NPs target inflammatory markers such as CD98 Fab, Ly6c, and CD44, which are highly expressed in inflammatory conditions. Naserifar et al. prepared PLGA NPs loaded with resveratrol (PLGA-RSV) with a folate ligand surface coating (FA-PLGA-RSV) for enhanced targeting and improved permeation. PLGA-RSV and FA-PLGA-RSV showed a 61% and 99% increase in the permeability rates through the Caco-2 monolayer, respectively. Both PLGA-RSV and FA-PLGA-RSV significantly inhibited inflammation and reduced neutrophil and lymphocyte accumulation (48).

3.2.2.2 Lipid-Based Vesicular System

Liposomes are small vesicles with a spherical form that primarily comprise various phospholipids and cholesterol (49). Liposomes work effectively with both lipophilic and hydrophilic drugs (50). Liposome's surface modification boosts their site-specific action, which is crucial in treating IBD. Zhao et al. synthesized lactoferrin-modified liposomes to deliver patchouli alcohol a sesquiterpene alcohol to macrophages in the colon for anti-inflammatory benefits. The findings showed reduced disease activity index and decreased body weight (51). Wang et al. designed curcumin-loaded liposomes; based on *in vivo* research, it was observed that formulation lowered UC symptoms such as diarrhea, bloody stools, and weight loss. Moreover, studies also demonstrated that it could prevent colon tissue damage and shortening caused by dextran sulfate sodium salt. The treated mice depicted a reduction in the concentration of various inflammatory mediators, such as interleukin-6 (IL-6), tumor necrosis factor-α (TNF-α), malondialdehyde (MDA), and colonic myeloperoxidase (MPO) (52).

3.2.2.3 Hybrid Drug Delivery Systems

Hybrid drug delivery systems (HDD) employs the excellent properties of two sizes – nano- and micro-size delivery systems – at the same time, and they are challenging to prepare, but together, they offer the benefits of both systems (53). HDD systems leverage the unique physicochemical properties of different delivery methods, providing a wide range of desirable characteristics within a single system. However, HDD systems have some limitations, including low productivity, high energy consumption, expensive setup, and lengthy process duration (54). Zang et al. developed a nano in micro-encapsulated hybrid delivery system where PLGA encapsulates berberine and is further entrapped in an Eudragit FS30D matrix. The dual polymeric coating offers the benefits of a dual-release profile, providing both sustained and immediate release. The dissolution of the Eudragit FS30D matrix results in immediate action, while the PLGA NPs ensure sustained release. The micro- and nano- encapsulated hybrid delivery system (MNEHDS) mediated delivery demonstrated an alleviating effect on pain by downregulating mitogen-activated protein kinase (MAPK) and NF-κB. This system reduced dosing frequency while enhancing therapeutic efficacy (55).

3.3 PARENTERAL DELIVERY APPROACHES FOR THE MANAGEMENT OF INFLAMMATORY DISEASES

The parenteral drug delivery system (PDDS) is extensively used in inflammatory diseases that are administrated via various injection routes, such as intramuscular (IM), subcutaneous (SC), intravenous (IV), intradermal, and intraarticular (56). PDDS consists of various dosage forms, including liposomes, microemulsion, suspension, and NPs. This system offers several advantages over the conventional delivery system, including reduced first-pass metabolism and increased drug efficacy (57). Parenteral routes are utilized to treat inflammatory diseases such as brain disorders, osteoarthritis (intraarticular), other neurological disorders like chronic inflammatory demyelinating polyneuropathy, MS, neuromyelitis optica, autoimmune encephalitis, Guillain-Barré (GBS) syndrome.

Arthritis, characterized by inflammation and swelling of the joints, is present in two forms: rheumatoid arthritis (RA) and osteoarthritis (OA) (58). Arthritis affects most people worldwide, though the exact causes remain unclear. Various parenteral novel carriers were developed, such as long-acting parenteral system (liposomes and lipid emulsion), very long act (polymeric micelles, carbon nanotubes, and NPs of iron oxide), and ultra-long act (nanosphere, microsphere, and polymer-drug conjugates) to treat arthritis (59). Synthetic polymers such as polyvinyl alcohol (PVA), PEG, and PLGA employed in the above systems show biocompatibility and biodegradability (60).

Liposomes are bilayer vesicular systems; cationic liposomes are important for anti-inflammatory properties as they inhibit the formation of TNF-α and Nitric oxide as pro-inflammatory agents (61). Cationic lipids such as sphingosine have the property of reducing protein kinase C (PKC) activity because of their structural similarity to PKC and sphingosine (62). Researchers developed dexamethasone (DXM)-loaded liposomes that have been used to treat RA through the parenteral route (63). Liposomes were shown to have a small size of 58 nm; it was followed by transport by the site via passive diffusion and produced an effect on the inflamed joint. Polymeric liposomes encapsulated with dexamethasone were studied against inflammatory disease. Polymeric liposomes were prepared using PEG as a polymer, which overcomes several disadvantages, including drug leakage, rapid uptake of RES, and problems related to blood circulation. These liposomal formulations improved the therapeutic potency by inhibiting inflammatory activators. A pharmacokinetic study of the PEGylated liposomes experimented on rats (Sprague-Dawley) exhibited a two-fold improvement in the area under curve (10,226 μg/L h) as compared to nonpolymeric liposomes (64).

Lipid emulsion is another vital drug delivery system for inflammatory hydrophobic drugs given through the parenteral route. The emulsion is a heterogeneous system in which the oil phase (lipid) is dispersed in an aqueous phase by using an emulsifier to stabilize the system, and the size of the parenteral lipid emulsion is less than 500 nm (65, 66). Lipid emulsions were formulated for the corticosteroid drugs for intravenous administration, for example, the incorporation of triamcinolone acetonide palmitate (TAP) into oil in water emulsion. The emulsion improved the pharmacodynamic and pharmacokinetic properties of the drug in addition to improved solubility (67).

Polymeric micelles consist of a hydrophilic shell and hydrophobic core and a size range of 10 to 100 nm (68) for arthritic disease. PEGylated micelles of size less than 50 nm modified the pharmacokinetics profile, and 94.2% drug loading was reported. This formulation (PEG-PCL-DXM) inhibited mRNA-containing cytokines such as interleukin-1β (IL-1β) and TNF-α (69). Triamcinolone (TA) incorporated into the hyaluronic acid (HA) micelles was studied to treat arthritics via injection. HA is used to target CD44 receptors that are abundant in synovial fluid. HA-based micellar formulation of TA reduced the systemic toxicity, and increased therapeutic effect. The formulation also depicted a sustained release for 76 h (70).

Microemulsion is used in parenteral delivery over another delivery system due to its improved solubility and bioavailability, reduced side effects, and increased residence time in blood (71, 72). Nonionic surfactants are mostly used to prepare the microemulsion as they demonstrate sustained release, biocompatibility, and increased solubilization. Surfactants must be sterilizable, non-pyrogenic non-irritating and non-haemolytic. The FDA approved several surfactants for parenteral administration (73, 74). Recognized surfactants for parenteral use are synthetic surfactants, including Solutol® HS 15, poloxamer 188, Span 60, Tween 80, and natural surfactants like lecithin (75, 76).

Parenteral microemulsions exhibit several advantages, such as increased cellular uptake and RES clearance (77). Some research groups described Kolliphor HS® 15 as a safe, synthetic surfactant used parenterally in the range of 50% to 65% (w/w) (78, 79). According to Pineros et al., a lysine clonixinate-loaded microemulsion formulation was prepared for parenteral administration (intramuscular). The microemulsion comprised Labrafil® as the oil phase, Tween 80 and Lauroglycol® as surfactants, and water as aqueous phase. Formulations controlled drug release due to the presence of an oil phase that increases the viscosity and increment in droplet size, which slows the *in vitro* release (80). Sulistio et al. prepared hydrogel of diclofenac drug-conjugated synthetic polyurethane polymer for parenteral administration. This system increased the drug efficacy and reduced the GIT and coronary side effects. An *in vivo* study observed the zero order in drug release. The hydrogel depicted a sustained release of diclofenac for two weeks, significantly higher than the half-life of diclofenac (5h). The fluorescence image resulted in significant retention of polymer in the joints. This drug delivery system was used in the long-term treatment of osteoarthritis (81).

MS is an autoimmune inflammatory disease primarily affecting the central nervous system, such as the brain and spinal cord. MS is a common neurodegenerative disease for young adults and it is characterized by relapse damage, inflammation, gliosis, and the volume of brain loss (82, 83). MS is found to be more commonly seen in Europe and Australia as compared to other parts of the globe; it is also reported in Western countries and North America (84). Parenterally approved drugs for MS include Copaxone, natalizumab, alemtuzumab, daclizumab, and mitoxantrone (85). Copaxone (glatiramer acetate) is an injectable drug for MS, and it is a mixture of salt amino acids such as L-glutamic acid, L-alanine, L-tyrosine, and L-lysine. Copaxone exerts a mechanism through cytokine changes and proinflammatory balance (86).

Alemtuzumab is a monoclonal antibody with a ligand for the CD52 receptor and CD52 consisting of monocytes and granulocytes (87). Phase III clinical trials of alemtuzumab studied the parenteral administration of the drugs three times a week (88, 89). Another monoclonal antibody, daclizumab, utilized for MS disease via parenteral administration, targets the CD25 (IL-2) receptor with expressed T cells (90). Autoimmune encephalitis (AE) is an immune-responsive disorder showing changes in behavior, seizures, and impairment in memory (91, 92). First-line treatments for AE are intravenous steroids, plasma exchange, and intravenous immunoglobulins.

Gout is an inflammatory disease related to monosodium urate present in joints and triggers the release of inflammatory cytokines (93). Inflammatory drugs administrated parenterally have significant benefits, including a decreased concentration of drugs, lower drug interaction, reduced systemic side effects, and direct targeting of gout joints. Intraarticularly injected drugs for gout disease enhance the therapeutic activity and reduce the cost of the drug (94). Mohamed et al. developed a colchicine-based nanoemulsion for intraarticular delivery. Nanoemulsion consists of isopropyl myristate (IPM), water, ethanol, Span 20, and Tween 80 as oil phase, aqueous phase, cosurfactant, and surfactant, respectively. Emulsions were prepared by using the water titration method. The diameter of the emulsions ranged from 103.26 to 222 nm with a polydispersity index (PDI) of 0.21 to 0.60, a zeta potential of 34 mV, and an entrapment efficiency of 75.65%. Nanoemulsion was intraarticularly injected into albino mice, and biodistribution was studied. It was observed that colchicine-loaded nanoemulsion reduced inflammation of knees as compared to colchicine solution. The formulation showed prolonged retention time (95). Some drugs used for gout treatment are administered parenterally, which include rasburicase, pegloticase, canakinumab, anakinra, rilonacept, and genakumab (96). GBS is a frequently occurring inflammatory disease caused by flaccid muscle weakness, areflexia, and polyneuropathy (97). The annual GBS case statistics are reported to be around 0.5 to 2 patients per 100,000 people which men were more affected than females (98). Patients with GBS need to prevent viral infection, physical treatment, and autonomic dysfunction (99). The first-choice drug treatment for GBS is intravenous immunoglobulin. Some researchers have studied the effects of the parenteral combination of methylprednisolone with immunoglobulin for GBS. Chronic inflammatory demyelinating polyneuropathy (CIDP) is a neuropathy inflammatory disease associated with sensory loss, pain, and limb weakness. Treatment for CIDP is steroids and immunoglobulin. Drugs were used for CIDP, including prednisone, azathioprine, methotrexate, cyclosporine A, azathioprine, rituximab, tacrolimus, interferon-α, interferon-β, natalizumab, etc. (100, 101). Researchers studied CIPD treated with monthly methylprednisolone (iv) for five years. It observed improvement in patients in the first, fourth, and fifth years, so treatment is more effective for a long time (102). Börü ÜT et al. comparative study of immunoglobulin (iv) and methylprednisolone administration via parenterally for CIPD disease of six months. Immunoglobulin shows adverse effects and inefficacy compared to methylprednisolone (103).

3.4 TOPICAL APPROACHES FOR THE MANAGEMENT OF INFLAMMATORY DISEASES

Topical drug delivery is a safer route for treating localized inflammatory skin conditions. In order to enhance therapeutic results, promote treatment adherence, and minimize adverse effects, an optimal topical administration formulation is therefore required for proper selection and development (104–106). The major objectives of treatment for inflammatory skin diseases are quality of life enhancement and symptom management. Corticosteroids, vitamin D3 analogues, methotrexate, cyclosporine, and other disease-modifying anti-rheumatic medications, as well as recently developed biologically targeted medications that target the IL-23/Th-17 axis, such as brodalumab andrisankizumab, are among the therapy choices for inflammatory skin disorders. The topical administration of those molecules can overcome many severe side effects resulting from systemic therapy. For instance, Methotrexate is known to cause bone marrow suppression, cirrhosis, hepatic fibrosis, and gastrointestinal and mouth ulcers. When treating psoriasis, topical use of methotrexate may help prevent the systemic side effects of oral methotrexate. The immune system's response to harmful stimuli, like infections, damaged cells, toxic chemicals, or radiation, is known as inflammation (107). It takes action by eliminating harmful stimuli and starting the healing process (108). During an acute inflammatory response, cells and molecules work together to prevent harm or infection. Uncontrolled acute inflammation can cause chronic inflammatory disorders (109). Inflammation causes redness, swelling, heat, discomfort, and loss of function in tissues due to vascular, immunological and inflammatory cell responses to infection or damage. During inflammation, the microcirculatory system undergoes alterations such as vascular permeability, leukocyte recruitment, and inflammatory mediator release (110, 111).

3.4.1 Skin and the Barriers in the Skin Epidermis

The body's biggest organ is the skin. The skin is made up of layers of epidermis, dermis, and hypodermis and covers the entire outer surface of body. Epidermis is the outermost layer of the skin. The dermis, which is in the middle, has sweat glands, different types of connective fibers, and sensory receptors. The hypodermis, on the other hand, is the subcutaneous layer of adipose tissue and serves as an anchor for the other two skin layers. Treatments utilizing basic topical formulations, including lotions and creams, are less effective because of the skin's capacity to function as a barrier (112). The stratum corneum, the outermost layer of the epidermis, acts as a barrier between the internal cells and the external environment, shielding them from physical, chemical, and mechanical harm as well as dehydration and infection. The skin also acts as a defense barrier against external threats such as bacteria, poisons, and UV radiation; thus, skin barriers are essential components of the body's defense system (113, 114). Some conditions like cuts, burns, and eczema may all disrupt the skin barrier's protective function, making it more susceptible to infections and other skin issues. Maintaining skin barrier integrity is critical for overall skin health, which may be accomplished through optimal hydration, skincare, and protection from external stresses (115). Various skincare products, such as moisturizers and barrier repair creams, are designed to protect and reinforce the skin barrier, keeping it healthy and robust (113). Topical drug delivery is the application of medicine directly to the skin or mucous membranes for localized therapy. It entails administering pharmaceuticals such as creams, ointments, gels, lotions, patches, and sprays to the afflicted part of the body, where the drug is absorbed via the skin or mucous membranes to reach the intended site of action (116).

3.4.2 Commonly Used Therapeutics for Topical Inflammation

Various therapeutics including NSAIDs, topical immunomodulators, antibiotic, or antifungal, moisturizer for soothing the inflamed skin, corticosteroids, antiseptics, antipruritic, etc., are used in different inflammatory skin disorders. These therapeutics are delivered in various forms such as ointments, gels, creams and lotions. Use of these medications depends on condition being treated, severity of condition or inflammation. Some of them are discussed in the following sections.

3.4.2.1 Topical Nonsteroidal Anti-Inflammatory Drugs

NSAIDs including diclofenac, ibuprofen, and naproxen can be formulated as topical creams or gels. They work by inhibiting the production of prostaglandins, chemicals in the body that induce inflammation. These topical NSAIDs can help to relieve localized pain and inflammation in conditions including arthritis and tendonitis (117, 118).

3.4.2.2 Topical Corticosteroids

Corticosteroid creams and ointments are commonly used to treat inflammation and irritation caused by conditions including eczema, psoriasis, and allergies. They act by suppressing the immune system and reducing inflammation. To avoid side effects such as skin thinning, topical corticosteroids should be used cautiously and under medical supervision (119).

3.4.2.3 Topical Calcineurin Inhibitors

Tacrolimus and pimecrolimus are two topically applied calcineurin inhibitors. They are suggested for the treatment of atopic dermatitis and eczema, among other inflammatory skin disorders. These drugs reduce inflammation and the response of the immune system (120, 121).

3.4.2.4 Topical Immunomodulators

These medications reduce inflammation by changing the immune system. They are used to address conditions such as atopic dermatitis. Two such are tacrolimus and crisaborole (122).

3.4.2.5 Topical Antibiotics

Topical antibiotics may be recommended to treat bacterial infections that produce inflammation in order to both control the infection and lessen inflammation (123).

3.4.2.6 Topical Antipruritic

Topical antipruritic drugs result in the decrease of the itching or irritation brought on by inflammatory skin disorders. Ingredients like menthol, pramoxine, lidocaine, etc., could be present (124, 125).

3.4.3 Delivery Approaches for the Management of Topical Anti-Inflammatory Diseases

For topical anti-inflammatory disorders therapy, several delivery techniques are being investigated with the major goal of therapeutic effectiveness, safety, and patient compliance. Anti-inflammatory medication is still often used topically in the form of creams, ointments, gels, and lotions. These formulations are appropriate for controlling localized inflammation since they are simple to apply and distribute directly to the afflicted region (Figure 3.3) (126–129).

3.4.3.1 Nanoparticle-Based Delivery Systems

The formulation and development of nanoparticulate delivery systems, for the encapsulation and administration of anti-inflammatory drugs is made possible by nanotechnology (130). These NPs can improve treatment results by preventing drug degradation, increasing drug penetration into the skin, and delivering sustained release (131). Alsaidan et al. developed and optimized hydrocortisone loaded-lipid-polymer hybrid NPs for topical targeting the inflammatory disease and *in vitro* and *in vivo* assessment efficacy of the formulation. Polymer-lipid hybrid NPs were prepared by phospholipid and polycaprolactone (PCL) following the nanoprecipitation method. Histological analysis following topical administration of the improved formulation, demonstrated increase in epidermal thickness. In an ear rosacea model produced by croton oil, the improved formulation demonstrated encouraging anti-inflammatory effects *in vivo*. These results suggest that using lipomers, an outstanding anti-inflammatory drug, may be a viable way to increase hydrocortisone acetate's topical efficacy against inflammatory illnesses. All of these findings confirm that hydrocortisone may be effectively delivered to the skin via lipid polymer hybrid NPs to treat inflammatory skin diseases (132).

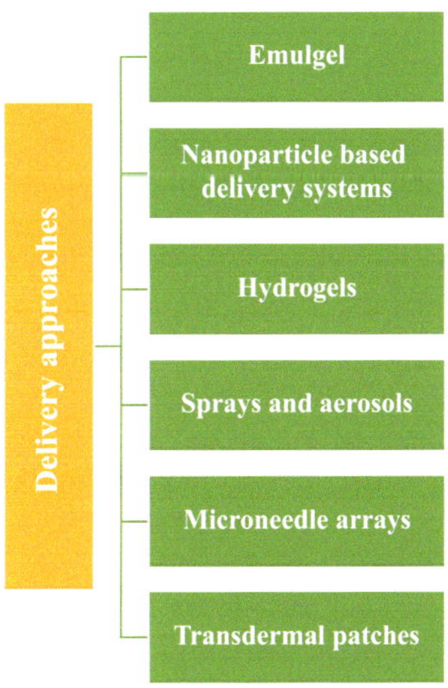

FIGURE 3.3 Advanced delivery strategies for topical or transdermal applications.

3.4.3.2 Transdermal Patches

Adhesive patches called transdermal patches are used to apply medications topically so that they may be absorbed systemically. Transdermal patches can be made to administer anti-inflammatory medications locally to the site of inflammation, offering prolonged release and reducing systemic adverse effects, even though they are often employed for systemic administration (133). Trivedi et al. formulated and assessed dexketoprofen trometamol containing transdermal patches. The patches were prepared by the solvent casting method using varying concentrations of Eudragit RS 100, HPMC, and ethyl cellulose polymers. The synthesized compositions' physical properties were all the same. The polymer combination (HPMC: EC in a 4:1 ratio) in formulation F6 demonstrated a maximum release of 85.77% in 24 hours. The resulting data was fitted to the Higuchi and Peppas model, which was the zero-order model. According to the findings, a transdermal patch containing dexketoprofen and trometamol can be developed to provide more therapeutic effects (134).

3.4.3.3 Microneedle Arrays

Tiny needles used in microneedle arrays are minimally invasive devices that pierce the epidermis to create micropores for the administration of drugs. Anti-inflammatory medication-loaded microneedles can be used topically to target certain inflammatory areas and improve drug penetration into the skin. Jungyoon et al. developed triamcinolone acetonide (TA)-loaded biocompatible and biodegradable candle flame shape microneedles (candlelit-DMN) for treating inflammatory disease. When Candlelit-DMN was inserted into human skin and co-administration of controlled release of TA was executed in the target site of the skin layer, the ex vivo results demonstrated the uniform distribution of TA from the TA-candlelit-DMN across the skin. The researchers inserted Candlelit-DMN into the *in vivo* skin inflammation mouse model, efficiently reducing inflammation by decreasing cytokine gene expression and inflammatory cell infiltration (135).

3.4.3.4 Sprays and Aerosols

Anti-inflammatory medication may be applied to the skin conveniently and effectively with the help of sprays and aerosols. These formulations may be sprayed as small mists or droplets, which enables quick and even medication absorption into the skin while covering the afflicted region uniformly. Narumon Changsan et al. formulated and assessed polymeric film forming spray containing *Centella asiatica* extract, which is used to heal acute wounds. The spray product included the following contents: 0.20±0.02 mg/ml of asiatic acid, 0.16±0.01 mg/ml of madecassic acid, 0.32±0.03 mg/ml of madecassoside, and 0.10±0.00 mg/ml of madecassic acid. The viscosity was 33 ± 4cP, and the pH was 5.5±0.01. When the product was preserved at 30±2°C and 40±2°C with 75±5% relative humidity, it remained stable for six months. According to the findings, a polymeric film spray containing *Centella asiatica* was successfully developed to improve wound healing (136).

3.4.3.5 Hydrogels

Hydrogels are water-based semisolid formulations that can be loaded with anti-inflammatory medicines for topical administration (137). These formulations hydrate the skin, improve medication penetration, and stick effectively to the skin surface, resulting in extended contact with the afflicted region and sustained drug release (138). Kolipaka et al. developed and optimized oxicam-loaded hydrogel to treat RA. They were prepared hydrogel using carbopol, poloxamer, water, and alcohol as the solvents. The formulation showed sustained release of the drug and good physical stability. The hydrogel formulations are designed and optimized by design of experiment (DoE) and show potential as a successful transdermal drug delivery formulation (139).

3.4.3.6 Emulgel

Emulgel is a novel drug delivery system that intends to improve the therapeutic efficacy of lipophilic drugs. Lipophilic formulations have several disadvantages, such as limited oral bioavailability, variable absorption, and low solubility. The goal of emulgel, a combined preparation of many systems, is to address these challenges (140). The emulgel prepared by adding nanoemulsion to the gel enhances the drug's stability, permeability, and regulated release. The advantages of emulgel in accomplishing targeted distribution, a safe profile, easy administration, and not undergoing first-pass metabolism or gastrointestinal degradation have all contributed to its growing attention. These properties make emulgel a more promising carrier for the topical drug delivery system (141). Rao et al. prepared glucosamine sulfate potassium chloride (GSPC) containing emulgel for treating osteoarthritis. The optimum microemulsion contains 60 mg of GSPC, 15% oil (oleic acid), 20% surfactant (Tween 20), and 35% cosurfactant (propylene glycol). The formulation of glucosamine sulfate potassium chloride-containing emulgel was achieved by adding the optimal microemulsion formulation in a 1:1 (w/w) ratio to gel made by carbopol 934. Consequently, it can be said that GSPC emulgels can be used topically in OA as a controlled release of the drug and effectively across the deeper layer of skin (142).

3.5 INHALATION DELIVERY APPROACHES FOR THE MANAGEMENT OF INFLAMMATORY LUNG DISEASES

Inflammatory lung diseases like asthma, COPD, pulmonary fibrosis, RA-ILD, allergic rhinitis, etc., are often the results of natural immune response. The epithelial cells of the lungs produce inflammatory mediators, including reactive oxygen radicals, cytokines such as TNF-α, IL-1ß, granulocyte/macrophage colony-stimulating factor, and platelet-activating factor, which stimulates the production of inflammatory cells and eicosanoids that further stimulates the goblet cells for mucous secretion and inflammation (143).

Asthma is the most common chronic lung disease characterized by airways inflammation making it difficult to breathe. Symptoms include shortness of breath, coughing, wheezing, and chest tightness. According to the World Health Organization's (WHO) 2019 report, approximately 262 million people were affected by asthma, causing 45500 deaths worldwide. Asthma can be successfully managed by inhaled corticosteroids such as Beclomethasone dipropionate, fluticasone propionate, budesonide, mometasone furoate, and $ß_2$ adrenergic receptor agonists such as formoterol, salmeterol, salbutamol, etc. Another common lung inflammatory disease is COPD, causing 3.23 million deaths in 2019, as per a WHO report. COPD is generally characterized by chronic peripheral progressive narrowing of airways, which leads to shortness of breath and chronic cough. Inhalation delivery of short as well as long-acting bronchodilators alone or in combination with corticosteroids reduces inflammation and relaxes the airways, improving the quality of life. Conventional drug delivery for inflammatory lung diseases includes primarily oral and parenteral delivery, which comprises some drawbacks, mostly leading to inadequate concentrations of drugs at the target site. Systemic distribution of drugs administered by these routes has a low therapeutic index and often leads to extrapulmonary adverse events. Moreover, chronic inflammatory disease requires long-term therapy, so there is always the risk of drug resistance and poor therapeutic efficacy. Inhalation delivery systems for the management of pulmonary inflammatory diseases can increase drug therapeutic concentration directly at the site of action, reducing off-target inadequacies.

Inhalation delivery is preferred for local and systemic inflammatory diseases for its numerous advantages over traditional delivery systems (Figure 3.4). It allows easy absorption of drugs due to the large surface area of thin alveolar epithelial cell membranes with the presence of a high number of capillaries of high permeability and high blood flow in the lungs (144). The absorption from the lungs ensures high bioavailability with a rapid onset of action as first-pass metabolism is avoided. Inflammatory diseases localized within the pulmonary system can be passively targeted by inhalation delivery as high drug concentrations can be achieved at the site of action. Lower drug doses are required to provide sufficient local and systemic drug concentrations, which will reduce the side effects of potent drugs (145). Transport and deposition of inhaled particles in the respiratory tract depend on a number of factors such as particle attributes: size, shape, density, charge, inertia, hygroscopicity; patient factors: gender, age, disease state, breathing pattern, tidal volume, mucociliary clearance (146). The deposition of inhaled particles in the lungs occurs by three mechanisms, viz. impaction, sedimentation, and Brownian diffusion. Large particles have a mass median aerodynamic diameter (MMAD) >5μ deposit in the upper respiratory tract by impaction mechanism; the sedimentation mechanism applies to particles having MMAD 1-6μ, which deposit

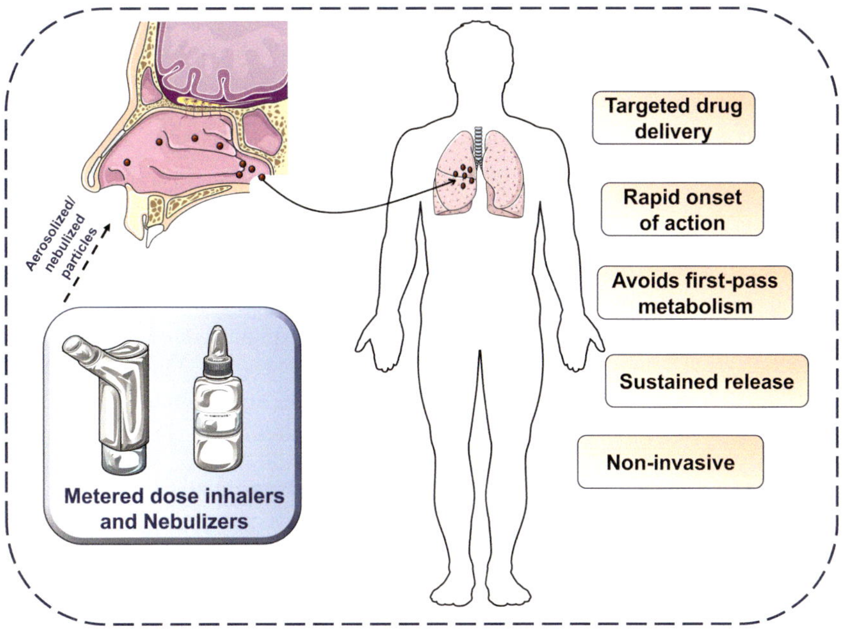

FIGURE 3.4 Inhalational delivery systems and their associated advantages.

particles in small airways and alveolar cavities by gravitational settling and fine particles having MMAD <1μ follows Brownian diffusion (147, 148).

3.5.1 Inhalation Devices

Administration of pulmonary formulation requires inhalation devices, including dry powder inhalers (DPI), metered dose inhalers (MDI), nebulizers, and soft mist inhalers (SMI).

3.5.1.1 Dry Powder Inhalers

DPIs are devices that deliver dry powder formulation of an active drug substance into deeper parts of the lungs with the help of the patient's respiration. Dry powder formulations are loose agglomerates of micronized drug substances of MMAD<5μ or micronized drug substances adhered on lactose carriers (149). The general design of DPIs consists of a mouthpiece, capsule or dose chamber, grid, needles attached with pressure buttons, and air inlet (150). DPIs are of three types, viz. single and multiple-use DPIs, and powder-assisted DPIs. Single-use DPIs devices are activated by breath in which capsule of powder formulation is placed in between the needles fixed with pressure buttons – e.g., Rotahaler® (GSK), Breezhaler® (Novartis). Multidose DPIs are either a self-measured dose reservoir system, e.g., turbuhaler® (AstraZeneca), or the dispense of individual premeasured doses into blisters, strips, dimples, and tubes. Powder-assisted DPIs incorporated with battery-driven impellers assist in inhalation, making them suitable for patients with impaired lung function. e.g., MicroDose® (MicroDoseTherapeutx) (151).

3.5.1.2 Metered Dose Inhalers

MDIs, also known as pressurized MDIs (pMDI), are devices that use pressurized propellants to deliver a specific amount of dose in aerosolized form. The drug is either dissolved or dispersed in the propellants. Previously, chlorofluorocarbons (CFC) were used as propellants, but due to their ozone-depleting effect now, they are replaced with novel propellants such as hydrofluoroalkanes. The design of MDIs consists of a formulation vial, actuator, metering valve, dose counter, and mouthpiece. Key parameters impacting the performance of MDIs include aerodynamic particle size distribution (APSD) (<5μ), dose uniformity, chemical and physical stability, and extent of leaching. MDIs don't need the help of the patient's respiration to deliver the dose; hence, they are very useful for patients with impaired lung function – for example, Beclovent® and Advair® (152).

3.5.1.3 Nebulizers

Nebulizers convert liquid formulation into aerosolized mist using mechanical or thermal energy to deposit the dose into the lungs. These devices are commonly used by patients who have difficulty using MDIs and DPIs or young children aged 4–5 years. The basic advantage of using a nebulizer is simple aqueous solutions and process-sensitive proteins, peptides, and biological medications can be used (153, 154). There are three types of nebulizers, i.e., Jet nebulizer (e.g., LC Sprint®, PARI GmbH, Germany), Ultrasonic nebulizer (e.g., PolyGreen KN-9210®, PolyGreen, Germany) and vibrating mesh nebulizers (e.g., MicroAir®, Omron Healthcare, Netherland) depending on the energy used for aerosolization.

3.5.2 Formulation Approaches for Inhalational Delivery

3.5.2.1 Direct Inhalation of Dry Powder

Formulation in the form of dry powder is meant to be administered via DPIs. Direct inhalation of dry powder is dependent on its particle size distribution. For effective aerosolization as well as deposition, particle size is required to be in the range of 1–5 μ. To assess the quality of effective dry powder, some parameters are kept in check, such as MMAD, emitted dose (ED), fine particle fraction (FPF), geometric standard deviation (GSD) and median diameter (Dv50) (155, 156). There are various techniques for the production of dry powder for inhalation, such as micronization by supercritical fluid drying, jet milling, spray-freeze drying and spray-drying (157). Lactose is the most commonly used carrier in the dry powder formulations, which acts as a diluent as well as a flow aid – e.g., Lactohale100, Lactohale200, Lactohale300.

3.5.2.2 Carrier-Based Approaches

Direct inhalation of dry powder may cause localized side effects – e.g., direct inhalation of corticosteroids may cause candidiasis and dyspnea, as well as systemic side effects (158). So, the specificity of inhaled particles needs to be enhanced. Also, there has been a need to increase residence time to decrease the rate of administration in order to improve patient compliance. So, formulating inhalation products in the form of polymeric and lipid-based carrier systems will be a possible method to achieve these targets.

3.5.2.3 Polymeric Microparticles (MPs)

Inhalable microparticles offer several potential features such as drug targeting, controlled release of drug, protection of active ingredients against degradation, and reduction of side effects. The potential toxicity of polymers and maintenance of requisite particle size distribution limits the pulmonary use of microparticles for inhalation. For the deepest deposition in the lungs, particle size should be between 1 and 5 μm (159). In order to enhance pulmonary residence time and respirable fraction, Naikwade et al. prepared budesonide sustained release microparticles using chitosan for pulmonary administration for the treatment of COPD and asthma. Budesonide is a potent corticosteroid that inhibits inflammatory symptoms. Chitosan is a polysaccharide with mucoadhesive properties and is used extensively as a sustained-release formulation for pulmonary delivery. Physicochemical characterization of microparticles revealed the MMAD and GSD for conventional, porous particle formulations, and microspheres were 2.75 μm, 4.30 μm, 4.60 μm and 2.56, 2.54, 1.75, respectively. A pharmacokinetic study in rats exhibited that developed formulations have one to fourfold enhanced bioavailability and enhanced lung deposition of budesonide with extended $t_{1/2}$ from 9.5 to 14h when compared to conventional formulations (160). Moreover, Sakagami et al. formulated mucoadhesive microspheres of Beclomethasone (BMS) for powder inhalation in order to prolong drug action and reduce drug dosage. Aqueous and ethanolic solutions of hydroxypropyl cellulose (HPC) containing BMS (drug: polymer = 1:4) were spray-dried, producing amorphous and crystalline BMS-incorporated HPC microspheres. The guinea pigs were employed to administer the as powder aerosols, followed by a comparison of BMS retention in the lungs and its eosinophil infiltration inhibitory duration. As a result, crystalline BMS microspheres showed an eosinophil infiltration inhibitory effect for 24 hours as compared to that of the amorphous counterpart (6 hours), even at a very low dose of 0.25 mg/kg. Therefore, HPC microspheres of crystalline BMS are attributed to prolonged lung retention even at much lower doses (161).

3.5.2.4 Nanoparticles

The dry powder NPs show significant potential to bypass natural defenses such as mucociliary clearance, enzymatic degradation, and alveolar macrophages. These NPs effectively target the organ/tissue by active or passive mechanisms, resulting in higher drug accumulation at the diseased site, reducing frequent dosing, and improving patient compliance. Buhecha et al. engineered budesonide and theophylline co-encapsulated poly (lactic acid) (PLA) NPs for pulmonary drug administration via the inhalation route. Budesonide is a widely used corticosteroid by inhalation route for the management of asthma and COPD, while theophylline serves as a potent bronchodilator. Research findings have indicated that the formulation is optimal for inhalation, with a particle size distribution ranging from 190 to 400 nm and −10 to −16 mV zeta potential. FPF was found to be 48% and 75% for budesonide and theophylline, respectively. Cytotoxicity assays demonstrated that the inhaled NPs are biocompatible with airway epithelial cells, and permeability studies revealed the sustained release profile for the 24 hours of the encapsulated therapeutics. From this observation, it was concluded that PLA NPs are appropriate medication delivery vehicles for combination therapy (162).

3.5.2.5 Nanocomposite Microparticles

The major challenge with inhalational NPs is their small size (<0.5μm), which can result in them being easily exhaled before reaching their target. While the use of microparticles in 1–5 μm range achieves deep lung deposition and susceptibility to opsonization via alveolar route. Researchers are exploring new approaches that combine the advantage of both nano and micro sizes to enhance pulmonary drug delivery, such as nanocomposite, where NPs are held together by the carriers. Nanocomposites consist of drug-loaded NPs with various carriers (Sugars), which turn into microparticles in the 1–5 μm range suitable for lung targeting. After reaching the alveolar surface, sugar dissolves, and nanocomposite disintegrates into NPs and shows their therapeutic efficacy (163). Mohamed et al. formulated miRNA (miR-146a) encapsulated nanocomposite microparticles (NCMPs) for the treatment of COPD. Researchers formulated miRNA-loaded NPs using a single emulsion method (o/w). Further, nanocomposite microparticles are prepared by spray-drying technique using various ratios of mannitol and l-leucine as excipients. The ratio of l-leucine to mannitol (25:75w/w) shows a high percentage yield and low moisture content. The aerosolization performance showed ED of 81.81 ± 3.0%, high FPF of 51.33 ± 2.9%, and MMAD of ≤5 μm, suggesting deposition in the respirable region of the lungs. *In vitro* study confirmed that after spray-drying, miR-146a encapsulated NCMPs retained their biological functionality, effectively reducing IRAK1 and TRAF6 gene expression in A549 cells. This indicates the potential of miR-146a encapsulated NCMPs for pulmonary drug delivery to manage COPD (164).

3.5.2.6 Inhalational Liposomes

Lipid nanocarriers for pulmonary administration have been studied extensively because of use of natural phospholipids for their preparation. There are potential benefits in using liposomes for aerosol delivery, such as serving as a matrix for the solubilization of poor solubility drugs, a reservoir for sustained pulmonary release to sustain therapeutic drug levels or a means of facilitating intracellular delivery of therapeutic agents. Moreover, liposome-based pulmonary medication delivery has the potential to lessen systemic harmful drug effects and avoid local discomfort (165).

3.6 GENE THERAPY APPROACHES FOR THE MANAGEMENT OF INFLAMMATORY DISEASES

In past years, significant progress has been achieved in the realm of designing gene delivery technology. Gene delivery offers enhanced therapeutic efficacy while circumventing off-target adverse effects. Small interfering RNAs (siRNAs) are a promising approach for silencing specific genes associated with inflammation, apart from the prime obstacle of delivering it into the cytosol of the immune cell. Multiple mediators were considered for developing gene nano-therapies. For instance, siRNA targeting TNF-α was encapsulated in mannose-modified trimethyl chitosan-cysteine NPs. These NPs were internalized by rat peritoneal exudate cells *in vitro*, and orally administered TNF-α siRNA-loaded NPs achieved effective TNF-α knockdown at even low doses (166). In a similar way, cationic lipid nanoemulsion were anticipated for targeting the brain with anti-TNF-α siRNA. The nanoemulsion was intranasally administered in an LPS-induced neuroinflammation rat model, resulted in greater accumulation in the brain and subsequent downregulation of TNF-α (167).

Among different inflammatory diseases, RA is the most prevalent type of chronic arthritis, marked by inflammation of the synovial tissue, and leads to damage to the affected area. The presence of inflammation and hypertrophy of the soft tissues neighboring the synovial joints are usual manifestations of RA. Autoantibodies like rheumatoid factors (RFs) and antibodies targeting collagen type II are common. However, anti-citrullinated protein antibodies (ACPAs) are now recognized as more reliable indicators for diagnosing RA than RF tests (168). The pathophysiology of RA is shaped by fibroblast-like synoviocytes (FLS). Targeting FLS for treatment can restore synovial balance and prevent bone and cartilage degeneration. To overcome the resistance of RA FLS to apoptosis, using viral vectors to deliver the proapoptotic gene PUMA has shown promise, but has limitations. As per studies conducted, the ROS-sensitive polymers show potential for nonviral gene therapy to regulate RA FLS (169).

MS is another chronic autoimmune inflammatory disease that primarily affects young adults by targeting the central nervous system. Despite extensive research, the exact pathology of MS remains elusive. As the disease progresses, compartmentalized inflammation in the central nervous system (CNS) leads to ongoing inflammatory and degenerative changes, ultimately driving the progression of the disease. However, studies strongly suggest the role of genetic and environmental factors. Obesity, smoking, and Epstein-Barr virus infections are the prominent risk factors for developing MS. The immunomodulatory capability of recent MS therapies backs the idea of MS being an autoimmune disease (170). In animal models of neurodegenerative diseases, brain-derived neurotrophic factors (BDFN) show a neuroprotective effect. However, when administered, its short half-life and hindrance by the blood brain barrier (BBB) restrict its effectiveness in the CNS. In a study, genetically engineered bone marrow stem cells (BMSCs) were developed to deliver BDNF into the CNS of an animal model with autoimmune encephalomyelitis, a condition that mimics MS, to analyze the impact of BDNF. The mice receiving BDNF-engineered BMSCs exhibited a significant delay in the onset of autoimmune encephalomyelitis and a fall in overall clinical severity compared to the mice that received BMSCs transfected with an empty vector lacking the BDNF gene. This method has the potential to be a novel therapeutic strategy for delivering the BDNF genes or different therapeutic proteins in the CNS (171).

Psoriasis also fits within the spectrum of inflammatory skin conditions identified by scaly red patches. When observed under the microscope, psoriatic skin shows abnormal keratinization and influx of different inflammatory cells, indicating an inflammatory response. Recent research has shown that the inflammation in psoriatic skin can lead to systemic inflammatory disorders, increasing the risk of cardiovascular disease and bone fractures in psoriasis patients (172). Patients suffering from psoriasis exhibit increased expression of TNF-α. The RNA interference (RNAi), mediated by siRNA, enables the elimination of some genes. However, the implication necessitates the use of carriers that can provide protection to the siRNA and deliver it to the cellular target. Accordingly, a delivery system for siRNA was developed using hybrid polymer-lipid nanoparticles (PLNs), which were then combined with photochemical internalization. This process involves photoactivation of the photosensitizer TPPS2a to enhance the endosomal escape of TNF-α siRNA into the cytoplasm. The goal was to utilize this system as a topical formulation for treating psoriasis. The PLNs were made up of 2% Compritol, 1.5% Poloxamer 188, and 0.1% poly(allylamine hydrochloride). They had an average size of 142 nm, a zeta potential of +25 mV, and could efficiently encapsulate TPPS2a and siRNA. Furthermore, there was no cellular toxicity evident in these compounds, and they exhibited high cellular internalization. The PLNs carried siRNA and TPPS2a into the skin, as determined by *in vitro* distribution tests. The efficacy of the treatment to silence TNFα in psoriatic lesions was confirmed in a hairless mouse model. The results showed a reduction in redness and scaling of the skin, indicating the potential of the developed treatment for psoriasis (173).

Inflammatory myopathies, or myositis, are a diverse group of rare diseases that can affect multiple organs and systems, such as the muscles, skin, lungs, and joints, falling in the category of inflammatory diseases. The randomized controlled trials offer evidence for treatment in certain patient subsets, the rarity and diversity of these diseases make it challenging to enroll enough patients with similar symptoms and to design appropriate studies (174). Other acknowledged autoimmune conditions like Graves' disease (GD) are characterized by a variety of clinical presentations. One of the main characteristics of GD is the presence of thyrotropin receptor autoantibodies (TRAb), that activate the TSH receptor and cause an overabundance of thyroid hormone to be produced (175). A multitude of endogenous, genetic, environmental, and epigenetic variables are involved in the intricate pathophysiology of GD (176). There is significant evidence from epidemiological, familial, and twin research that GD is largely inheritable and has a strong genetic component. Based on data from twin studies, it appears that genetic variables are important, explaining around 70% of the chance of getting GD (177). In a comparable vein, lymphocyte infiltration of parenchyma, increased thyroid volume, and the development of antibodies specific to thyroid antigens are thought to be the hallmarks of Hashimoto thyroiditis (HT), an autoimmune thyroid disease. HT is regarded as an autoimmune thyroid illness (AITD), along with GD. HT is currently the most common cause of hypothyroidism; also, people with HT have an increased risk of developing malignant neoplasms and cardiovascular disorders. The condition known as "struma lymphomatosa" is caused by lymphocytes invading thyroid tissue and increasing the gland's bulk. The exact etiology of HT is not clear; however, genetic, environmental factors and epigenetic changes are thought to have a crucial role (178).

Inflammatory conditions like RA and MS are distinguished by the presence of inflammation that causes substantial tissue damage. Conventional treatments, such as nonsteroidal anti-inflammatory drugs (NSAIDs) and biologics, in most cases, do not offer sustained relief, and that may lead to serious adverse effects. Gene therapy in the recent past emerged as a prosperous alternative, which utilizes the administration of therapeutic genes to alter the immune response and lower the inflammation on the molecular scale. Gene therapy involves the introduction of specific genes into the cell of the patient to generate therapeutic proteins or emend the genetic abnormalities. In the realm of inflammatory diseases, the objective is to deliver genes that can carry anti-inflammatory cytokines, immunomodulatory proteins, or other therapeutic agents directly to the target tissue. This pin-point delivery method has the potential to provide long-lasting therapeutic benefits while reducing systemic side effects. The fascinating use of gene therapy in the treatment of inflammatory diseases involves the utilization of vectors, specifically viral vectors, to administer anti-inflammatory genes. The adeno-associated viruses (AAVs) are highly preferred for their low immunogenicity and capability to sustain long-term gene expression. For instance, in the treatment of RA, AAVs have been used to transfer genes that encode anti-inflammatory IL-10 directly to the affected joints. Preclinical research has shown that this method can effectively mitigate inflammation and lower joint damage in animal models (179). Another innovative approach involves the use of gene editing tools like CRISPR/Cas9 to modify or alter genes associated with inflammation. CRISPR/Cas9 has the ability to precisely alter the genomic sequence in order to downregulate the proinflammatory genes or bolster the anti-inflammatory pathways. In the context of IBD, CRISPR/Cas9 has been employed to deactivate genes that activate the inflammatory response in gut epithelial cells, resulting in decreased inflammation and enhanced gut function in experimental models (180). Gene therapy holds promise in the treatment of inflammatory diseases; however, it faces significant challenges. The key concerns boil down to efficient, targeted delivery of the therapeutic genes to specific tissues and ensuring the long-term safety of gene editing tools to avoid off-target effects. Moreover, regulatory and ethical considerations of gene therapy need to be addressed carefully in order to achieve clinical translation. Continued progress in designing vectors, gene editing tools, and delivery approaches is foreseen to fix the present obstacles, clearing the way for the practical implication of gene therapy in the treatment of chronic inflammation.

3.7 CONCLUSION AND FUTURE PERSPECTIVES

Continuous advancement in biomedical sciences helps understand and elucidate the underlying pathophysiological mechanisms, diagnosis, and treatment of many inflammatory and infectious diseases. Irrespective of the rapid technological development, treating chronic, acute, and autoimmune inflammatory diseases causes a significant reduction in patient's quality of life. The poor prognosis of autoimmune inflammatory diseases remains a major endangering factor for human health. Due to indeterminacy in pathogenesis, no specific drug is used as a therapeutic agent. An in-depth molecular understanding of pathogenesis led to the development of new targeted therapies in recent years, which have attracted attention in autoimmune inflammation. Polymeric implants and injectables like particulates or hydrogels are excellent platforms for significant local and long-term drug delivery in treating acute and chronic inflammation. Polymeric implants may not be compatible biologically or mechanically with tissues or organs. Therefore, injectable formulations are preferred for local delivery to joints, sinuses, or the colon. Despite these various applications, the empty polymeric carriers are responsible for causing new inflammation and need to be removed physically or degraded synchronously with drug release at the site of action. Advancements in different drug delivery approaches have recently been perceived using stimuli-responsive polymers, nanocarriers, and biological vectors as drug delivery agents for treating inflammatory conditions by delivering drugs to their respective sites with the appropriate amount of drug. Utilization of these approaches can circumvent the drug dose-related side effects and toxicity. Molecular or microenvironmental changes in physiology like low pH, increased ROS, and over-expressed enzymes are observed in several inflammatory conditions like RA and inflammatory vascular diseases. Regardless of these extensive applications in inflammatory conditions, only a few bio-responsive drug delivery approaches are intensively studied. Selecting an optimal biochemical signal according to the inflammation condition is crucial in developing a rational stimuli-responsive drug delivery. Considering pH-sensitive drug delivery systems, some tissues like the stomach and sub-cellular organelles like endosomes might be recognized as off-targets for drug release rather than the inflammatory site, resulting in developing side effects and might also impair *in vivo* efficacy.

Conventional therapeutic interventions result in low therapeutic efficiency, non-specific targeting, unforeseen side effects, and drug resistance, that are resolved by aiding nanotechnology. Nanotechnology provides versatile methods for diagnosis and management of infectious and inflammatory diseases. Surface modification and adjustment of the dose of the nanocarriers overcome some drawbacks that lead to site-specific targeting, reducing toxicity and increasing stability and biocompatibility. Recent trends include utilizing artificial intelligence (AI) to assess and predict the toxicity of nanomaterials. AI provides critical tools for interactions understanding between nanomaterials and various biological systems, resulting in cost-effective nano-delivery systems.

ACKNOWLEDGMENT

Authors are highly thankful to the Department of Pharmaceuticals, Ministry of Chemicals and Fertilizers, Government of India, for providing financial assistance.

REFERENCES

1. Wyss-Coray T, Mucke L. (2002) Inflammation in neurodegenerative disease—a double-edged sword. *Neuron* 35:419–32.
2. Libby P. (2006) Inflammation and cardiovascular disease mechanisms. *Am J Clin Nutr*. 83:456S–460S.
3. Hotamisligil GS, Erbay E. (2008) Nutrient sensing and inflammation in metabolic diseases. *Nat Rev Immunol*. 8:923–34.
4. Trinchieri G. (2012) Cancer and inflammation: an old intuition with rapidly evolving new concepts. *Annu Rev Immunol*. 30:677–706.
5. Nathan C, Ding A. (2010) Nonresolving inflammation. *Cell*. 140:871–82.
6. Gubala V, Harris LF, Ricco AJ, Tan MX, Williams DE. (2012) Point of care diagnostics: status and future. *Anal Chem*. 84:487–515.
7. Medzhitov R. (2010) Inflammation 2010: New Adventures of an Old Flame. *Cell*. 140:771–6.
8. Hua S. (2020) Advances in Oral Drug Delivery for Regional Targeting in the Gastrointestinal Tract - Influence of Physiological, *Pathophysiol Pharmaceut Factors. Front Pharmacol*. 28:11.
9. Moroz E, Matoori S, Leroux JC. (2016) Oral delivery of macromolecular drugs: Where we are after almost 100 years of attempts. *Adv Drug Deliv Rev*. 101:108–21.
10. Hua S, Marks E, Schneider JJ, Keely S. (2015) Advances in oral nano-delivery systems for colon targeted drug delivery in inflammatory bowel disease: Selective targeting to diseased versus healthy tissue. *Nanomedicine*. 11:1117–32.
11. Devangan P, Bajad G, Loharkar S, Wadate N, Gollapalli S, Dutta T, et al. (2024) Quality-By-Design Strategy for Developing Novel Herbal Products. In: Jain NK, Bajwa N, editors. *Introduction to Quality by Design (QbD)*: From Theory to Practice Springer Nature Singapore, pp. 263–95.
12. Jyothi VGSS, Pawar J, Fernandes V, Kumar R, Singh C, Singh SB, et al. (2022) Film forming topical dermal spray of meloxicam attenuated pain and inflammation in carrageenan-induced paw oedema in Sprague Dawley rats. *J Drug Deliv Sci Technol*. 70:103195.
13. Lautenschläger C, Schmidt C, Fischer D, Stallmach A. (2014) Drug delivery strategies in the therapy of inflammatory bowel disease. *Adv Drug Deliv Rev*. 71:58–76.
14. Moldoveanu B, Otmishi P, Jani P, Walker J, Sarmiento X, Guardiola J, et al. (2009) Inflammatory mechanisms in the lung. *J Inflamm Res*. 2.1–11.
15. Carmona-Ribeiro AM. (2012) Preparation and characterization of biomimetic nanoparticles for drug delivery. *Methods Mol Biol*.; 906: 283–94.
16. Hu CMJ, Fang RH, Wang KC, Luk BT, Thamphiwatana S, Dehaini D, et al. (2015) Nanoparticle biointerfacing by platelet membrane cloaking. *Nature*. 526:118–21.
17. Brusini R, Varna M, Couvreur P. (2020) Advanced nanomedicines for the treatment of inflammatory diseases. *Adv Drug Deliv Rev*. 157:161–78.
18. Baumgart DC, Sandborn WJ. (2007) Inflammatory bowel disease: clinical aspects and established and evolving therapies. *The Lancet*. 369:1641–57.
19. Vuyyuru SK, Kedia S, Sahu P, Ahuja V. (2022) Immune-mediated inflammatory diseases of the gastrointestinal tract: Beyond Crohn's disease and ulcerative colitis. *JGH Open*. 6:100–11.

20. Lou J, Duan H, Qin Q, Teng Z, Gan F, Zhou X, et al. (2023) Advances in oral drug delivery systems: Challenges and Opportunities. *Pharmaceutics*. 15:484.
21. Hua S. (2020) Advances in oral drug delivery for regional targeting in the gastrointestinal tract - influence of physiological, pathophysiological and pharmaceutical factors. *Front Pharmacol* 11: 11.
22. Li X, Lu C, Yang Y, Yu C, Rao Y. (2020) Site-specific targeted drug delivery systems for the treatment of inflammatory bowel disease. *Biomedicine & Pharmacotherapy*. 129:110486.
23. Hua S, Marks E, Schneider JJ, Keely S. (2015) Advances in oral nano-delivery systems for colon targeted drug delivery in inflammatory bowel disease: Selective targeting to diseased versus healthy tissue. *Nanomedicine*. 11:1117–32.
24. Yang L, Chu JS, Fix JA. (2002) Colon-specific drug delivery: new approaches and in vitro/in vivo evaluation. *Int J Pharm*. 235:1–15.
25. Arya S, Bahuguna D, Bajad G, Loharkar S, Devangan P, Khatri DK, et al. (2023) Colloidal therapeutics in the management of traumatic brain injury: Portray of biomarkers and drug-targets, preclinical and clinical pieces of evidence and future prospects. 230
26. Philip A, Philip B. (2010) Colon targeted drug delivery systems: A review on primary and novel approaches. *Oman Med J*. 25:70–8.
27. Swidsinski A, Loening-Baucke V, Theissig F, Engelhardt H, Bengmark S, Koch S, et al. (2007) Comparative study of the intestinal mucus barrier in normal and inflamed colon. *Gut*. 56:343–50.
28. Nguyen TLA, Vieira-Silva S, Liston A, Raes J. (2015) How informative is the mouse for human gut microbiota research? *Dis Model Mech*. 8:1–16.
29. Hebden, Blackshaw, Perkins, Wilson, Spiller. (2000) Limited exposure of the healthy distal colon to orally-dosed formulation is further exaggerated in active left-sided ulcerative colitis. *Aliment Pharmacol Ther*. 14:155–61.
30. Cui N, Friend DR, Fedorak RN. (1994) A budesonide prodrug accelerates treatment of colitis in rats. *Gut*. 35:1439–46.
31. Aib S, Iqbal K, Khan N, Khalid S, Adnan M, Umair SM, et al. (2022) pH-sensitive liposomes for colonic co-delivery of mesalazine and curcumin for the treatment of ulcerative colitis. *J Drug Deliv Sci Technol*. 72:103335.
32. Gazzaniga A, Iamartino P, Maffione G, Sangalli ME. (1994) Oral delayed-release system for colonic specific delivery. *Int J Pharm*. 108:77–83.
33. Cai Q, Zhu KJ, Zhang J. (2005) Salicylic acid and PEG-contained polyanhydrides: Synthesis, characterization, and in vitro salicylic acid release. *Drug Deliv*. 12:97–102.
34. Hua Z, Kimura G, Ito Y, Mawatari S, Shimokawa T, Yoshikawa H, et al.(1999) Technology to obtain sustained release characteristics of drugs after delivered to the colon. *J Drug Target*. 6:439–48.
35. Liu H, Yang XG, Nie SF, Wei LL, Zhou LL, Liu H, et al. (2007) Chitosan-based controlled porosity osmotic pump for colon-specific delivery system: Screening of formulation variables and in vitro investigation. *Int J Pharm*. 332:115–24.
36. Alshawwa SZ, Kassem AA, Farid RM, Mostafa SK, Labib GS.(2022) Nanocarrier drug delivery systems: Characterization, limitations, future perspectives and implementation of artificial intelligence. *Pharmaceutics*. 14:883.
37. Murthy SK. (2007) Nanoparticles in modern medicine: state of the art and future challenges. *Int J Nanomedicine*. 2:129–41.
38. Badwaik HR, Kumari L, Nakhate K, Verma VS, Sakure K. (2019) Phytoconstituent plumbagin: Chemical, biotechnological and pharmaceutical aspects. *Studies in Natural Products Chemistry* Volume 63. 415–60.
39. Tolstanova G, Deng X, French SW, Lungo W, Paunovic B, Khomenko T, et al. (2012) Early endothelial damage and increased colonic vascular permeability in the development of experimental ulcerative colitis in rats and mice. *Laboratory Investigation*. 92:9–21.
40. Youshia J, Lamprecht A. (2016) Size-dependent nanoparticulate drug delivery in inflammatory bowel diseases. *Expert Opin Drug Deliv*. 13:281–94.
41. Schmidt C, Lautenschlaeger C, Collnot EM, Schumann M, Bojarski C, Schulzke JD, et al. (2013) Nano- and microscaled particles for drug targeting to inflamed intestinal mucosa—A first in vivo study in human patients. *Journal of Controlled Release*. 165:139–45.
42. Antoni L. (2014) Intestinal barrier in inflammatory bowel disease. *World J Gastroenterol*. 20:1165.
43. Han HK, Shin HJ, Ha DH. (2012) Improved oral bioavailability of alendronate via the mucoadhesive liposomal delivery system. *European Journal of Pharmaceutical Sciences*. 46:500–7.
44. Yasmin F, Najeeb H, Shaikh S, Hasanain M, Naeem U, Moeed A, et al. (2022) Novel drug delivery systems for inflammatory bowel disease. *World J Gastroenterol*. 28:1922–33.
45. Sedghi S, Fields JZ, Klamut M, Urban G, Durkin M, Winship D, et al. (1993) Increased production of luminol enhanced chemiluminescence by the inflamed colonic mucosa in patients with ulcerative colitis. *Gut*. 34:1191–7.
46. Simmonds NJ, Allen RE, Stevens TRJ, Niall R, Van Someren M, Blake DR, et al. (1992) Chemiluminescence assay of mucosal reactive oxygen metabolites in inflammatory bowel disease. *Gastroenterology*. 103:186–96.
47. Vong LB, Tomita T, Yoshitomi T, Matsui H, Nagasaki Y. (2012) An orally administered redox nanoparticle that accumulates in the colonic mucosa and reduces colitis in mice. *Gastroenterology*. 143:1027–1036.
48. Naserifar M, Hosseinzadeh H, Abnous K, Mohammadi M, Taghdisi SM, Ramezani M, et al. (2020) Oral delivery of folate-targeted resveratrol-loaded nanoparticles for inflammatory bowel disease therapy in rats. *Life Sci*. 262:118555.
49. Akbarzadeh A, Rezaei-Sadabady R, Davaran S, Joo SW, Zarghami N, Hanifehpour Y, et al. (2013) Liposome: classification, preparation, and applications. *Nanoscale Res Lett*. 8:102.
50. Trucillo P, Campardelli R, Reverchon E. (2020) Liposomes: From Bangham to Supercritical Fluids. *Processes*. 8:1022.
51. Zhao Y, Yang Y, Zhang J, Wang R, Cheng B, Kalambhe D, et al. (2020) Lactoferrin-mediated macrophage targeting delivery and patchouli alcohol-based therapeutic strategy for inflammatory bowel diseases. *Acta Pharm Sin B*. 10:1966–76.
52. Wang C, Han Z, Wu Y, Lu X, Tang X, Xiao J, et al. (2021) Enhancing stability and anti-inflammatory properties of curcumin in ulcerative colitis therapy using liposomes mediated colon-specific drug delivery system. *Food and Chemical Toxicology*. 151:112123.
53. Yasmin F, Najeeb H, Shaikh S, Hasanain M, Naeem U, Moeed A, et al. (2022) Novel drug delivery systems for inflammatory bowel disease. *World J Gastroenterol*. 28:1922–33.
54. Setia A, Sahu RK, Ray S, Widyowati R, Ekasari W, Saraf S. (2022) Advances in hybrid vesicular-based drug delivery

systems: Improved biocompatibility, targeting, therapeutic efficacy and pharmacokinetics of anticancer drugs. *Curr Drug Metab.* 23:757–80.
55. Zhang L, Li M, Zhang G, Gao C, Wang S, Zhang T, et al. (2021) Micro- and Nanoencapsulated Hybrid Delivery System (MNEHDS): A novel approach for colon-targeted oral delivery of berberine. *Mol Pharm.* 18:1573–81.
56. Manchanda S, Das N, Chandra A, Bandyopadhyay S, Chaurasia S. (2020) Fabrication of advanced parenteral drug-delivery systems. In: *Drug Delivery Systems*. Elsevier, p. 47–84.
57. Lautenschläger C, Schmidt C, Fischer D, Stallmach A. (2014) Drug delivery strategies in the therapy of inflammatory bowel disease. *Adv Drug Deliv Rev.* 71:58–76.
58. Guo Q, Wang Y, Xu D, Nossent J, Pavlos NJ, Xu J. (2018) Rheumatoid arthritis: pathological mechanisms and modern pharmacologic therapies. *Bone Res.* Apr 27;6(1):15.
59. Mourya A, Famta P, Shah S, Srinivasarao DA, Sharma A, Vambhurkar G, et al. (2024) Raloxifene loaded d-α-tocopherol polyethylene glycol 1000 succinate stabilized poly (ε-caprolactone) nanoparticles augmented drug delivery and apoptosis in breast cancer cells. *J Drug Deliv Sci Technol.* 92:105399.
60. Janoušková O. (2018) Synthetic polymer scaffolds for soft tissue engineering. *Physiol Res.* 67: S335–48.
61. Filion MC, Phillips NC. (1997) Anti-inflammatory activity of cationic lipids. *Br J Pharmacol.* 122:551–7.
62. Reston JT, Gould-Fogerite S, Mannino RJ. (1991) Potentiation of DNA mediated gene transfer in NIH3T3 cells by activators of protein kinase C. *Biochimica et Biophysica Acta (BBA) - Gene Structure and Expression.* 1088:270–6.
63. Jia M, Deng C, Luo J, Zhang P, Sun X, Zhang Z, et al. (2018) A novel dexamethasone-loaded liposome alleviates rheumatoid arthritis in rats. *Int J Pharm.* 540:57–64.
64. Wang Q, He L, Fan D, Liang W, Fang J. (2020) Improving the anti-inflammatory efficacy of dexamethasone in the treatment of rheumatoid arthritis with polymerized stealth liposomes as a delivery vehicle. *J Mater Chem B.* 8:1841–51.
65. Hippalgaonkar K, Majumdar S, Kansara V. (2010) Injectable lipid emulsions—Advancements, opportunities and challenges. *AAPS PharmSciTech.* 11:1526–40.
66. de Meijer VE, Gura KM, Le HD, Meisel JA, Puder M. (2009) Fish oil–based lipid emulsions prevent and reverse parenteral nutrition–associated liver disease: The boston experience. *J. Parenter. Enteral. Nutr.* 33:541–7.
67. Vinnars E, Wilmore D. (2003) History of parenteral nutrition. *J. Parenter. Enteral. Nutr.* 27:225–31.
68. Liu H, Yang M, Wu P, Guan J, Men L, Lin H, et al. (2015) Simultaneous determination of triamcinolone acetonide palmitate and triamcinolone acetonide in beagle dog plasma by UPLC-MS/MS and its application to a long-term pharmacokinetic study of triamcinolone acetonide palmitate lipid emulsion injection. *J Pharm Biomed Anal.* 104:105–11.
69. Tiwari AP, Rohiwal SS. (2019) Synthesis and bioconjugation of hybrid nanostructures for biomedical applications. In: *Hybrid Nanostructures for Cancer Theranostics*. Elsevier; p. 17–41.
70. Wang Q, Jiang J, Chen W, Jiang H, Zhang Z, Sun X. (2016) Targeted delivery of low-dose dexamethasone using PCL–PEG micelles for effective treatment of rheumatoid arthritis. *Journal of Controlled Release.* 230:64–72.
71. Saadat E, Shakor N, Gholami M, Dorkoosh FA. (2015) Hyaluronic acid based micelle for articular delivery of triamcinolone, preparation, in vitro and in vivo evaluation. *Int J Pharm.* 489:218–25.
72. Henriques J, Sousa J, Veiga F, Cardoso C, Vitorino C. (2019) Process analytical technologies and injectable drug products: Is there a future? *Int J Pharm.* 554:21–35.
73. Gharbavi M, Manjili HK, Amani J, Sharafi A, Danafar H. (2019) In vivo and in vitro biocompatibility study of novel microemulsion hybridized with bovine serum albumin as nanocarrier for drug delivery. *Heliyon.* 5:e01858.
74. Rahdar A, Taboada P, Hajinezhad MR, Barani M, Beyzaei H. (2019) Effect of tocopherol on the properties of Pluronic F127 microemulsions: Physico-chemical characterization and in vivo toxicity. *J Mol Liq.* 277:624–30.
75. Theochari I, Goulielmaki M, Danino D, Papadimitriou V, Pintzas A, Xenakis A. (2017) Drug nanocarriers for cancer chemotherapy based on microemulsions: The case of Vemurafenib analog PLX4720. *Colloids Surf. B Biointerfaces* 154:350–6.
76. Li Y, Angelova A, Liu J, Garamus VM, Li N, Drechsler M, et al. (2019) In situ phase transition of microemulsions for parenteral injection yielding lyotropic liquid crystalline carriers of the antitumor drug bufalin. *Colloids Surf. B Biointerfaces* 173:217–25.
77. Hsu CY, Sung CT, Aljuffali IA, Chen CH, Hu KY, Fang JY. (2018) Intravenous anti-MRSA phosphatiosomes mediate enhanced affinity to pulmonary surfactants for effective treatment of infectious pneumonia. *Nanomedicine.* 14:215–25.
78. Karim R, Somani S, Al Robaian M, Mullin M, Amor R, McConnell G, et al. (2017) Tumor regression after intravenous administration of targeted vesicles entrapping the vitamin E α-tocotrienol. *J. Control. Release.* 246:79–87.
79. Lamaisakul S, Tantituvanont A, Lipipun V, Ritthidej G. (2020) Development of novel cationic microemulsion as parenteral adjuvant for influenza vaccine. *Asian J Pharm Sci.* 15:591–604.
80. Webster DM, Sundaram P, Byrne ME. (2013) Injectable nanomaterials for drug delivery: Carriers, targeting moieties, and therapeutics. *Eur. J. Pharm. Biopharm.* 84:1–20.
81. Pineros I, Slowing K, Serrano DR, de Pablo E, Ballesteros MP. (2017) Analgesic and anti-inflammatory controlled-released injectable microemulsion: Pseudo-ternary phase diagrams, in vitro, ex vivo and in vivo evaluation. *Eur. J. Pharm. Sci.* 101:220–7.
82. Sulistio A, Mansfeld FM, Reyes-Ortega F, D'Souza AM, Ng SMY, Birkett S, et al. (2019) Intra-articular Treatment of Osteoarthritis with Diclofenac-Conjugated Polymer Reduces Inflammation and Pain. *ACS Appl Bio Mater.* 2:2822–32.
83. Coclitu C, Constantinescu CS, Tanasescu R. (2016) The future of multiple sclerosis treatments. *Expert Rev Neurother.* 16:1341–56.
84. Javan MR, Shahraki S, Safa A, Zamani MR, Salmaninejad A, Aslani S. (2017) An interleukin 12 B single nucleotide polymorphism increases IL-12p40 production and is associated with increased disease susceptibility in patients with relapsing-remitting multiple sclerosis. *Neurol Res.* 39:435–41.
85. Wallin MT, Culpepper WJ, Nichols E, Bhutta ZA, Gebrehiwot TT, Hay SI, et al. (2019) Global, regional, and national burden of multiple sclerosis 1990–2016: A systematic analysis for the Global Burden of Disease Study 2016. *Lancet Neurol.* 18:269–85.
86. Ali R, Nicholas RSJ, Muraro PA. (2013) Drugs in development for relapsing multiple sclerosis. *Drugs.* 73:625–50.
87. Dhib-Jalbut S. (2003) Glatiramer acetate (Copaxone®) therapy for multiple sclerosis. *Pharmacol Ther.* 98:245–55.

88. Jones JL, Coles AJ. (2014) Mode of action and clinical studies with alemtuzumab. *Exp Neurol*. 262:37–43.
89. Cohen JA, Arnold DL, Comi G, Bar-Or A, Gujrathi S, Hartung JP, et al. (2016) Safety and efficacy of the selective sphingosine 1-phosphate receptor modulator ozanimod in relapsing multiple sclerosis (RADIANCE): a randomised, placebo-controlled, phase 2 trial. *Lancet Neurol*. 15:373–81.
90. Coles AJ, Twyman CL, Arnold DL, Cohen JA, Confavreux C, Fox EJ, et al. (2012) Alemtuzumab for patients with relapsing multiple sclerosis after disease-modifying therapy: a randomised controlled phase 3 trial. *The Lancet*. 380:1829–39.
91. Bielekova B, Catalfamo M, Reichert-Scrivner S, Packer A, Cerna M, Waldmann TA, et al. (2006) Regulatory CD56 BRIGHT natural killer cells mediate immunomodulatory effects of IL-2Rα-targeted therapy (daclizumab) in multiple sclerosis. *Proc. Natl. Acad. Sci*. 103:5941–6.
92. Dalmau J, Gleichman AJ, Hughes EG, Rossi JE, Peng X, Lai M, et al. (2008) Anti-NMDA-receptor encephalitis: case series and analysis of the effects of antibodies. *Lancet Neurol*. 7:1091–8.
93. Titulaer MJ, McCracken L, Gabilondo I, Armangué T, Glaser C, Iizuka T, et al. (2013) Treatment and prognostic factors for long-term outcome in patients with anti-NMDA receptor encephalitis: an observational cohort study. *Lancet Neurol*. 12:157–65.
94. Chen-Xu M, Yokose C, Rai SK, Pillinger MH, Choi HK. (2019) Contemporary prevalence of gout and hyperuricemia in the united states and decadal trends: The national health and nutrition examination survey, 2007–2016. *Arthritis Rheum*. 71:991–9.
95. Gerwin N, Hops C, Lucke A. (2006) Intraarticular drug delivery in osteoarthritis☆. *Adv Drug Deliv Rev*. 58:226–42.
96. Aboumanei MH, Fayez H. (2021) Intra-articular formulation of colchicine loaded nanoemulsion systems for enhanced locoregional drug delivery: *in vitro* characterization, ^{99m} Tc coupling and *in vivo* biodistribution studies. *Drug Dev Ind Pharm*. 47:770–7.
97. Peng X, Li X, Xie B, Lai Y, Sosnik A, Boucetta H, et al. (2023) Gout therapeutics and drug delivery. *J. Control. Release* 362:728–54.
98. Sejvar JJ, Baughman AL, Wise M, Morgan OW. (2011) Population Incidence of Guillain-Barré Syndrome: A Systematic Review and Meta-Analysis. *Neuroepidemiology*. 36:123–33.
99. Esposito S, Longo MR. (2017) Guillain–Barré syndrome. *Autoimmun Rev*. 16:96–101.
100. Chalela JA. (2001) Pearls and Pitfalls in the Intensive Care Management of Guillain-Barré Syndrome. *Semin Neurol*. 21:399–406.
101. Gorson KC. (2012) An update on the management of chronic inflammatory demyelinating polyneuropathy. *Ther Adv NeurolDisord*. 5:359–73.
102. Ashman EJ, Gronseth GS. (2012) Level of evidence reviews. *Neurology*. 79:13–4.
103. Börü ÜT, Erdoğan H, Alp R, Taşdemir M, Yıldırım S, Bilgiç A, et al. (2014) Treatment of chronic inflammatory demyelinating polyneuropathy with high dose intravenous methylprednisolone monthly for five years: 10-Year follow up. *Clin Neurol Neurosurg*. 118:89–93.
104. Nobile-Orazio E, Cocito D, Jann S, Uncini A, Beghi E, Messina P, et al. (2012) Intravenous immunoglobulin versus intravenous methylprednisolone for chronic inflammatory demyelinating polyradiculoneuropathy: a randomised controlled trial. *Lancet Neurol*. 11:493–502.
105. Pingle P, Mourya A, Namdeo M, Babu KC, Veerabomma H, Maurya R, et al. (2023) Andrographolide-Soya-L-α-Phosphatidyl Choline Complex Augmented Solubility and Drug Delivery in Leishmania donovani, a Causative Agent for Cutaneous and Visceral Leishmaniasis. *AAPS PharmSciTech*. 24:46.
106. Dewangan P, Mourya A, Singh PK, Chaudhary M, Sharma R, Bajwa N, et al. (2023) Chapter 1 - Drug delivery: The conceptual perspectives and therapeutic applications. In: Madan J, Baldi A, Chaudhary M, Chopra N, editors. *Polymer-Drug Conjugates*. Academic Press, p. 1–38.
107. Katta C Babu, Bahuguna D, Veerabomma H, Gollapalli S, Shaikh AS, Bhale NA, et al. (2024) Naringenin-zinc oxide nanocomposites amalgamated polymeric gel augmented drug delivery and attenuated experimental cutaneous candidiasis in Balb/c Mice: In vitro and in vivo studies. *AAPS PharmSciTech*.25:130.
108. Medzhitov R. (2010) Inflammation 2010: New Adventures of an Old Flame. *Cell*. 140:771–6.
109. Ferrero-Miliani L, Nielsen OH, Andersen PS, Girardin SE. (2007) Chronic inflammation: importance of NOD2 and NALP3 in interleukin-1β generation. *Clin Exp Immunol*. 147:227–35.
110. Zhou Y, Hong Y, Huang H. (2016) Triptolide Attenuates Inflammatory Response in Membranous Glomerulo-Nephritis Rat via Downregulation of NF-κBSignaling Pathway. *Kidney Blood Press Res*. 41:901–10.
111. Takeuchi O, Akira S. (2010) Pattern Recognition Receptors and Inflammation. *Cell*. 140:805–20.
112. Chertov O, Yang D, Howard OMZ, Oppenheim JJ. (2000) Leukocyte granule proteins mobilize innate host defenses and adaptive immune responses. *Immunol Rev*. 177:68–78.
113. Proksch E, Brandner JM, Jensen J. (2008) The skin: an indispensable barrier. *Exp Dermatol*. 17:1063–72.
114. Elias PM, Feingold KR. (1992) Lipids and the epidermal water barrier: metabolism, regulation, and pathophysiology. *Semin Dermatol*. 11:176–82.
115. Cork MJ, Danby SG, Vasilopoulos Y, Hadgraft J, Lane ME, Moustafa M, et al. (2009) Epidermal Barrier Dysfunction in Atopic Dermatitis. *J. Invest. Dermatol*. 129:1892–908.
116. Iyer A, Jyothi VGSS, Agrawal A, Khatri DK, Srivastava S, Singh SB, et al. (2021) Does skin permeation kinetics influence efficacy of topical dermal drug delivery system?: Assessment, prediction, utilization, and integration of chitosan biomacromolecule for augmenting topical dermal drug delivery in skin. *J. Adv. Pharm. Technol Res* 12: 345–55.
117. Roden DM, McLeod HL, Relling M V, Williams MS, Mensah GA, Peterson JF, et al. (2019) Pharmacogenomics. *The Lancet*. 394:521–32.
118. Barthel HR, Axford-Gatley RA. (2010) Topical Nonsteroidal Anti-Inflammatory Drugs for Osteoarthritis. *Postgrad Med*. 122:98–106.
119. Hengge UR, Ruzicka T, Schwartz RA, Cork MJ. (2006) Adverse effects of topical glucocorticosteroids. *J Am Acad Dermatol*. 54:1–15.
120. Cury Martins J, Martins C, Aoki V, Gois AF, Ishii HA, da Silva EM. (2015) Topical tacrolimus for atopic dermatitis. *Cochrane Database Syst. Rev*. 2016 (10): CD009864.
121. Sidbury R, Davis DM, Cohen DE, Cordoro KM, Berger TG, Bergman JN, et al. (2014) Guidelines of care for the management of atopic dermatitis. *J. Am. Acad. Dermatol*. 71:327–49.
122. Wollina U, Hansel G. (2008) The use of topical calcineurin inhibitors in lupus erythematosus: an overview. *J. Eur. Acad. Dermatol. Venereol*. 22:1–6.

123. Paller AS, Tom WL, Lebwohl MG, Blumenthal RL, Boguniewicz M, Call RS, et al. (2016) Efficacy and safety of crisaborole ointment, a novel, nonsteroidal phosphodiesterase 4 (PDE4) inhibitor for the topical treatment of atopic dermatitis (AD) in children and adults. *J Am Acad Dermatol*. 75:494–503.
124. Kolipaka T, Sen S, Mane SS, Bajad GD, Dengale SJ, Ranjan OP. (2023) Development of posaconazole nanosuspension for bioavailability enhancement: Formulation optimization, in vitro characterization, and pharmacokinetic study. *J Drug Deliv Sci Technol*. 83:104434.
125. Steinhoff M, Schmelz M, Szabó IL, Oaklander AL. (2018) Clinical presentation, management, and pathophysiology of neuropathic itch. *Lancet Neurol*. Aug;17(8):709–20.
126. Yosipovitch G, Greaves MW, Schmelz M. (2003) Itch. *The Lancet*.361(9358):690–4.
127. Cevc G, Vierl U. (2010) Nanotechnology and the transdermal routeA state of the art review and critical appraisal. *J. Control Release* 141:277–99.
128. Illum L. (2007) Nanoparticulate Systems for Nasal Delivery of Drugs: A Real Improvement over Simple Systems? *J Pharm Sci*. 96:473–83.
129. Benson H. (2005) Transdermal Drug Delivery: Penetration Enhancement Techniques. *Curr Drug Deliv*. 2:23–33.
130. Mitragotri S. (2006) Current status and future prospects of needle-free liquid jet injectors. *Nat Rev Drug Discov*. 5:543–8.
131. Sainaga Jyothi VGS, Ghouse SM, Khatri DK, Nanduri S, Singh SB, Madan J. (2022) Lipid nanoparticles in topical dermal drug delivery: Does chemistry of lipid persuade skin penetration? *J Drug Deliv Sci Technol*. 69:103176.
132. Katta C, Shaikh AS, Bhale N, Jyothi VGSS, Kaki VR, Dikundwar AG, et al. (2023) Naringenin-Capped Silver Nanoparticles Amalgamated Gel for the Treatment of Cutaneous Candidiasis. *AAPS PharmSciTech*. 24:126.
133. Alsaidan OA, Elmowafy M, Shalaby K, Alzarea SI, Massoud D, Kassem AM, et al. (2023) Hydrocortisone-Loaded Lipid–Polymer Hybrid Nanoparticles for Controlled Topical Delivery: Formulation Design Optimization and In Vitro and In Vivo Appraisal. *ACS Omega*. 8:18714–25.
134. Chahal SK, Sodhi RK, Madan J. (2020) Duloxetine hydrochloride loaded film forming dermal gel enriched with methylcobalamin and geranium oil attenuates paclitaxel-induced peripheral neuropathy in rats. *IBRO Rep*. 9: 85–95.
135. Trivedi D, Goyal A. (2020) Formulation and evaluation of transdermal patches containing dexketoprofen trometamol. *Int. J. Pharm. Chem. Anal*. 7:87–97.
136. Ohn J, Jang M, Kang BM, Yang H, Hong JT, Kim KH, et al. (2021) Dissolving Candlelit Microneedle for Chronic Inflammatory Skin Diseases. *Advanced Science*. 8(14).
137. Changsan N, Srichana T, Atipairin A, Sawatdee S. (2023) Wound healing efficacy of a polymeric spray film solution containing *Centella asiatica* leaf extract on acute wounds. *J Wound Care*. 32:S22–32.
138. Bhattacharjee S, Harithasree V, Mourya A, Babu CK, Bahuguna D, Devangan P, et al. (2024) Hyaluronic Acid Hydrogel in Therapeutics Delivery and Biomedical Applications. In: Jana S, editor. *Biomaterial-based Hydrogels: Therapeutics Carrier and Tissue Regeneration*. Singapore: Springer Nature Singapore. p. 139–77.
139. Kolipaka T, Pandey G, Abraham N, Srinivasarao DA, Raghuvanshi RS, Rajinikanth PS, et al. (2024) Stimuli-responsive polysaccharide-based smart hydrogels for diabetic wound healing: Design aspects, preparation methods and regulatory perspectives. *CarbohydrPolym*. 324:121537.
140. Vishwakarma U, Mathews R, Baby B, Rajarajan S, Das S. (2023) Formulation and Evluation of Hydrogel of Meloxicam for the Management of Rheumatoid Arthritis. *Int. J. Pharm. Res. Appl* 8:575–92.
141. Donthi MR, Munnangi SR, Krishna KV, Saha RN, Singhvi G, Dubey SK. (2023) Nanoemulgel: A Novel Nano Carrier as a Tool for Topical Drug Delivery. *Pharmaceutics*. 15:164.
142. Anand K, Ray S, Rahman M, Shaharyar A, Bhowmik R, Bera R, et al. (2019) Nano-emulgel: Emerging as a Smarter Topical Lipidic Emulsion-based Nanocarrier for Skin Healthcare Applications. *Recent Pat Antiinfect Drug Discov*. 14:16–35.
143. Rao K, Kumari S, Minocha N. (2024) Formulation and Characterization of Glucosamine Sulphate Potassium Chloride (GSPC) Loaded Emulgel for the Treatment of Osteoarthritis. *Curr Rheumatol Rev*. 20: 97–108.
144. Moldoveanu B, Otmishi P, Jani P, Walker J, Sarmiento X, Guardiola J, et al. (2009) Inflammatory mechanisms in the lung. *J Inflamm Res*. 2:1–11.
145. Wang B, Wang L, Yang Q, Zhang Y, Qinglai T, Yang X, et al. (2024) Pulmonary inhalation for disease treatment: Basic research and clinical translations. *Mater Today Bio*. 25:100966.
146. He S, Gui J, Xiong K, Chen M, Gao H, Fu Y. (2022) A roadmap to pulmonary delivery strategies for the treatment of infectious lung diseases. *J Nanobiotechnology*. 20:101.
147. Cheng YS. (2014) Mechanisms of Pharmaceutical Aerosol Deposition in the Respiratory Tract. *AAPS PharmSciTech*. 15:630–40.
148. Darquenne C. (2012) Aerosol Deposition in Health and Disease. *J Aerosol Med Pulm Drug Deliv*. 25:140–7.
149. Hoppentocht M, Hagedoorn P, Frijlink HW, de Boer AH. (2014) Technological and practical challenges of dry powder inhalers and formulations. *Adv Drug Deliv Rev*. 75:18–31.
150. Dhoble S, Kapse A, Ghegade V, Chogale M, Ghodake V, Patravale V, (2024) et al. Design, development, and technical considerations for dry powder inhaler devices. *Drug Discov Today*. 29:103954.
151. Lavorini F, Fontana GA, Usmani OS. (2014) New Inhaler Devices - The Good, the Bad and the Ugly. *Respiration*. 88:3–15.
152. Myrdal PB, Sheth P, Stein SW. (2014) Advances in Metered Dose Inhaler Technology: Formulation Development. *AAPS PharmSciTech*. 15:434–55.
153. Carvalho TC, McConville JT. (2016) The function and performance of aqueous aerosol devices for inhalation therapy. *J. Pharm. Pharmacol*. 68:556–78.
154. Longest W, Spence B, Hindle M. (2019) Devices for Improved Delivery of Nebulized Pharmaceutical Aerosols to the Lungs. *J Aerosol Med Pulm Drug Deliv*. 32:317–39.
155. Lin YW, Wong J, Qu L, Chan HK, Zhou Q. (2015) Powder Production and Particle Engineering for Dry Powder Inhaler Formulations. *Curr Pharm Des*. 21:3902–16.
156. Courrier HM, Butz N, Vandamme ThF. (2002) Pulmonary Drug Delivery Systems: Recent Developments and Prospects; *Crit Rev Ther Drug Carrier Syst*. 19:425–98.
157. Chaurasiya B, Zhao YY. (2020) Dry Powder for Pulmonary Delivery: A Comprehensive Review. *Pharmaceutics*. 13:31.
158. Brogden RN, Pinder RM, Sawyer PR, Speight TM, Avery GS. (1975) Beclomethasone Dipropionate Inhaler. *Drugs*. 10:166–210.
159. El-Gendy N, Bailey MM, Berkland C. (2011) Particle Engineering Technologies for Pulmonary Drug Delivery. In:

Controlled Pulmonary Drug Delivery. New York, Springer New York. p. 283–312.

160. Naikwade SR, Bajaj AN, Gurav P, Gatne MM, Singh Soni P. (2009) Development of Budesonide Microparticles Using Spray-Drying Technology for Pulmonary Administration: Design, Characterization, In Vitro Evaluation, and In Vivo Efficacy Study. *AAPS PharmSciTech*. 10:993–1012.
161. Sakagami M, Kinoshita W, Sakon K, Sato J Makino Y (2002) Mucoadhesive beclomethasone microspheres for powder inhalation: their pharmacokinetics and pharmacodynamics evaluation. *J Control Release* 80:207–18.
162. Buhecha MD, Lansley AB, Somavarapu S, Pannala AS. (2019) Development and characterization of PLA nanoparticles for pulmonary drug delivery: Co-encapsulation of theophylline and budesonide, a hydrophilic and lipophilic drug. *J Drug Deliv Sci Technol*. 53:101128.
163. Abdelaziz HM, Gaber M, Abd-Elwakil MM, Mabrouk MT, Elgohary MM, Kamel NM, et al. (2018) Inhalable particulate drug delivery systems for lung cancer therapy: Nanoparticles, microparticles, nanocomposites and nanoaggregates. *J Control Release* 269:374–92.
164. Mohamed A, Pekoz AY, Ross K, Hutcheon GA, Saleem IY. (2019) Pulmonary delivery of Nanocomposite Microparticles (NCMPs) incorporating miR-146a for treatment of COPD. *Int J Pharm*. 569:118524.
165. Niven RW, Schreier H. (1990) Nebulization of Liposomes. I. *Effects of Lipid Composition. Pharm Res*. 07:1127–33.
166. He C, Yin L, Song Y, Tang C, Yin C. (2015) Optimization of multifunctional chitosan–siRNA nanoparticles for oral delivery applications, targeting TNF-α silencing in rats. *Acta Biomater*. 17:98–106.
167. Yadav S, Gandham SK, Panicucci R, Amiji MM. (2016) Intranasal brain delivery of cationic nanoemulsion-encapsulated TNFα siRNA in prevention of experimental neuroinflammation. *Nanomedicine*. 12:987–1002.
168. Shrivastava AK, Pandey A. (2013) Inflammation and rheumatoid arthritis. *J PhysiolBiochem*. 2013 Jun 31;69(2):335–47.
169. Hua P, Liang R, Yang S, Tu Y, Chen M. (2024) Microneedle-assisted dual delivery of PUMA gene and celastrol for synergistic therapy of rheumatoid arthritis through restoring synovial homeostasis. *Bioact Mater*. 36:83–95.
170. Haase S, Linker RA. (2021) Inflammation in multiple sclerosis. *Ther Adv NeurolDisord*. 14:175628642110076.
171. Makar TK, Bever CT, Singh IS, Royal W, Sahu SN, Sura TP, et al. (2009) Brain-derived neurotrophic factor gene delivery in an animal model of multiple sclerosis using bone marrow stem cells as a vehicle. *J Neuroimmunol*. 210:40–51.
172. Tashiro T, Sawada Y. (2022) Psoriasis and Systemic Inflammatory Disorders. *Int J Mol Sci*. 23:4457.
173. Suzuki IL, de Araujo MM, Bagnato VS, Bentley MVLB. (2021) TNFα siRNA delivery by nanoparticles and photochemical internalization for psoriasis topical therapy. *Journal of Controlled Release*. 338:316–29.
174. Selva-O'Callaghan A, Pinal-Fernandez I, Trallero-Araguás E, Milisenda JC, Grau-Junyent JM, Mammen AL. (2018) Classification and management of adult inflammatory myopathies. *Lancet Neurol*. 17:816–28.
175. Smith TJ, Hegedüs L. Graves' Disease. (2016) *New England Journal of Medicine*. 375:1552–65.
176. Effraimidis G, Wiersinga WM. (2014) Mechanisms in Endocrinology: Autoimmune thyroid disease: old and new players. *Eur J Endocrinol*. 170:R241–52.
177. Brix TH, Kyvik KO, Christensen K, Hegedüs L. (2001) Evidence for a Major Role of Heredity in Graves' Disease: A Population-Based Study of Two Danish Twin Cohorts [1]. *J Clin Endocrinol Metab*. 86:930–4.
178. Hasham A, Tomer Y. (2012) Genetic and epigenetic mechanisms in thyroid autoimmunity. *Immunol Res*. 54:204–13.
179. Aalbers CJ, Bevaart L, Loiler S, de Cortie K, Wright JF, Mingozzi F, et al. (2015) Preclinical Potency and Biodistribution Studies of an AAV 5 Vector Expressing Human Interferon-β (ART-I02) for Local Treatment of Patients with Rheumatoid Arthritis. *PLoS One*. 10:e0130612.
180. Limanskiy V, Vyas A, Chaturvedi LS, Vyas D. (2019) Harnessing the potential of gene editing technology using CRISPR in inflammatory bowel disease. *World J Gastroenterol*. 25:2177–87.

4 Drug Delivery Based on Synthetic Polymers for Lung Diseases

C. Sarath Chandran, Krishnameera Sajayan, and P. K. Anjana
College of Pharmaceutical Sciences, Govt. Medical College Kannur, Kannur, India

Shijina Kappally
Sharjah University, UAE

4.1 INTRODUCTION

The third edition of the *Global Impact of Respiratory Diseases* emphasizes the vital role of our lungs in enabling us to lead fulfilling lives, a fact that has become more apparent in recent history, as the COVID-19 pandemic claimed the lives of 3.4 million individuals. According to a recent global analysis, chronic pulmonary obstructive disorder (COPD) and lower respiratory infections remain a significant cause of mortality among both pediatric and geriatric populations, while the incidences of asthma have increased sharply over the past three decades. It is also noteworthy that lung carcinomas are considered some of the deadliest forms of cancer, with survival rates of only 10%–20%, even in developed countries. Among the various chronic respiratory diseases, including asthma, COPD, interstitial lung diseases, pneumoconiosis, and pulmonary sarcoidosis, the socioeconomic burden they impose on a global scale is substantial (1, 2). Optimal treatment of lung infections necessitates a pharmaceutical formulation capable of depositing drugs in the lower regions of the lung, ensuring optimal bioavailability. To achieve this goal, it is essential to consider the distribution and retention of the administered drug at an appropriate site. Prolonged drug retention in the lungs is crucial for local administration of antibiotics. Factors such as inhaler formulation, breathing technique, and physicochemical stability of the drug, as well as particle characteristics, must all be considered. Innovative drug delivery technologies have garnered increasing interest in the treatment of respiratory disorders. Drug-targeted delivery has recently been investigated as a possible method for delivering medication to particular parts of the lungs.

Advancements in polymer science have contributed significantly to the development of innovative delivery systems for pulmonary drug targeting. Selecting the right polymers for developing drug delivery systems is a challenging task that requires a deep knowledge of their various structures and attributes. In addition to their physicochemical properties, polymers must undergo extensive biochemical characterization and specific preclinical tests to guarantee their safety. Surface properties, such as hydrophilicity, lubricity, smoothness, and surface energy, as well as physical properties, such as durability, permeability, and degradability, play a crucial role in determining biocompatibility with tissues, blood, and water sorption capacity for polymer-based drug delivery systems. The structural properties of polymers are critical for mass transport, which is vital in controlled delivery systems. Polymers can be categorized as either natural or synthetic, with options for biodegradable and nonbiodegradable materials. While natural polymers have several advantages and show promise for drug delivery applications, synthetic polymers are frequently chosen over natural polymers because of their possible disadvantages, which include microbial contamination, uneven quality, and unpredictable hydration rates. Novel drug delivery systems for the treatment of different lung infections are frequently developed using synthetic polymers, such as poly(lactic acid) (PLA), poly(lactic-co-glycolic acid) (PLGA), polyethylene glycol (PEG), polyacrylates, and polyanhydrides(3). Considering their versatile properties and design flexibility, the acceptance of synthetic polymers has increased several-fold over the last few years.

4.2 SYNTHETIC POLYMER-BASED DRUG DELIVERY SYSTEMS

Site-targeted delivery has been investigated as a potential strategy for successful treatment of various lung diseases. This approach has demonstrated great promise for drug delivery to specific sites within the lungs. Drug delivery systems, such as liposomes, nanoparticles, polymeric micelles, microparticles, and dendrimers, are composed of biodegradable and biocompatible synthetic polymers (Figure 4.1). These delivery systems allow the transport of both hydrophilic and hydrophobic medications to specific locations within the lungs. These vehicles enhance the cellular uptake of the targeted drugs and exhibit a high therapeutic response. Furthermore, they can maintain therapeutic concentrations for extended periods, reduce dosing frequency,

DOI: 10.1201/9781003477389-4

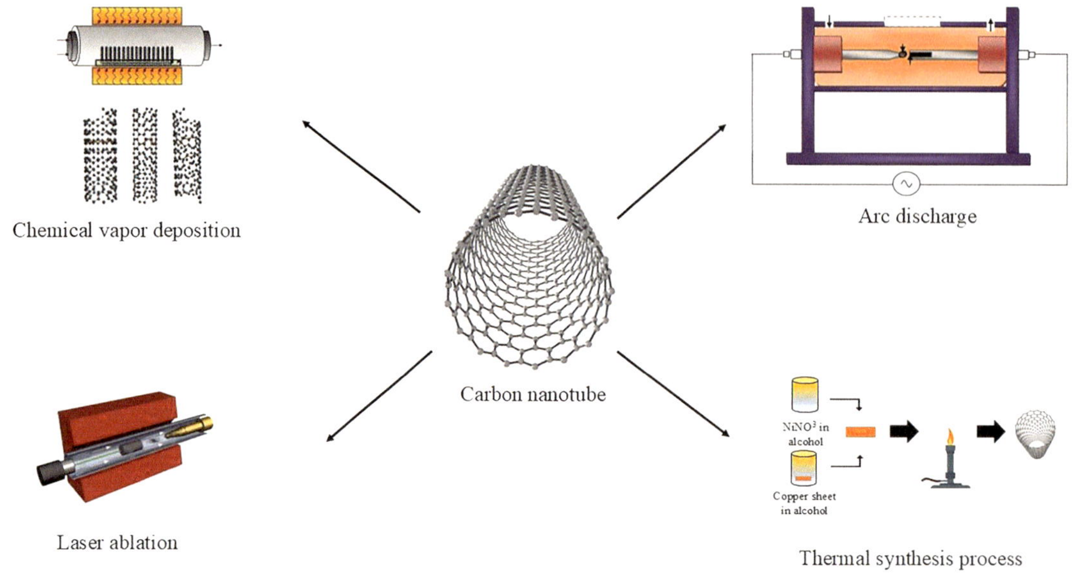

FIGURE 4.1 Various synthetic polymer-based drug delivery approaches for lung targeting.

and improve treatment outcomes for various lung diseases. In addition to increasing solubility, dissolution, and bioavailability, these delivery systems also decrease drug toxicity and extend the duration of drug presence at the site of delivery (3).

4.2.1 Liposomes

Liposomes are made up of one or more phospholipid layers and based on their size and layers, it is generally categorized as small unilamellar, large unilamellar, or multilamellar vesicles. They are used in the management of various lung diseases. Administering drug-loaded liposomes to the lungs can improve the local curative effect while reducing adverse effects, ensuring a prolonged presence of the drug in the lungs. Liposomes are primarily composed of phospholipids and cholesterol; however, synthetic polymers are used to modify their surfaces to improve their stability and prevent drug leakage. The US Food and Drug Administration (USFDA) has approved the application of the synthetic polymer PEG in a variety of products, including food, cosmetics, and pharmaceuticals, and it is well tolerated by the body. The use of PEG to modify liposomes has been reported to be a valuable strategy for enhancing the efficacy of pulmonary drug targeting in the treatment of lung diseases. Biocompatible PEG is utilized for the surface modification of liposomes, which can increase the dwelling time of the drug in the biological environment. Under normal circumstances, PEGylated proteins are considered safe; however, the production of antibodies against PEG may occur under extreme conditions (4).

In conventional methods, PEG is covalently bonded to liposomes through chemical groups, such as phosphates, amides, and esters. The chain length and density of PEG molecules on the liposome surface influence the pharmacokinetics and biodistribution of the formulation. PEG on the liposome surface forms a thick protective film around the liposome, known as a hydration film or fixed aqueous film. The thickness of this film increased with increasing PEG chain length, ultimately enhancing the systemic circulation of liposomes. In a significant finding, liposomes modified with both 2000 and 500 Da PEG-coated 1,2-distearoyl-sn-glycero-3-phosphoethanolamine-PEG (DPSE-PEG) demonstrated a longer circulation time than those with individual PEG modifications (5, 6).

PEGylated liposomes exhibit several advantages over conventional liposomes, including a longer half-life, reduced plasma clearance, and preferential distribution to targeted tissues. This is because of the integration of PEG into the lipid bilayer of liposomes, which generates a hydrated shell that protects them from protein-mediated destruction. Additionally, PEGylated liposomes are less likely to leak drugs into the bloodstream during circulation, and are absorbed by the reticuloendothelial system (RES). PEG on the liposomal surface delays the rate of systemic circulation by inhibiting phagocytosis and fast breakdown *in vivo*. Therapy of lung cancer with triptolide-loaded liposomes containing an anti-carbonic anhydrase IX antibody (CA IX liposomes) was examined in a study by Lin et al. The outcomes proved that direct drug administration to the lungs could achieve a sustained drug release profile at the intended site, leading to greater drug accumulation at the targeted site and fewer systemic adverse effects. Studies have demonstrated that PEGylated liposomes can achieve an extended drug release profile at targeted sites, yielding greater drug accumulation in the lungs and reducing systemic side effects (7).

Abnormal remodeling of lung tissue and consequent scarring of alveolar structures are hallmarks of a lung debilitating disease known as idiopathic pulmonary fibrosis (IPF). Researchers have investigated the use of PEGylated liposome-based aerosolized drug delivery systems to increase therapeutic choices. Both fibrotic and healthy mouse lungs showed the extended distribution of these liposomes. Using MicroSprayer, Togami et al. studied the circulation properties of atomized PEGylated liposomes in the lungs of mice with pulmonary fibrosis. Through the use of in vivo imaging, liposome concentration measurements in bronchoalveolar lavage fluid (BALF) and alveolar macrophages (AMs), and lung tissue slice assessment, intrapulmonary pharmacokinetics were assessed. According to these findings, PEGylated liposomes were less common in AMs but more common in BALF than in unmodified liposomes. The fibrotic areas of the tissue sections had high concentrations of both liposome types. According to these results, aerosolized PEGylated liposomes have potential as IPF medication delivery systems. They also offer important information regarding the distribution properties of these liposomes and their possible therapeutic applications for progressive and chronic lung illnesses (8).

Vancomycin is one of the few antimicrobial drugs licensed by the USFDA for the treatment of Methicillin-resistant Staphylococcus aureus (MRSA). However, vancomycin treatment has a significant clinical failure rate. Studies have indicated that antimicrobials encapsulated in liposomes may have better pharmacokinetics, pharmacodynamics, and lower toxicity than formulations that are currently used. Muppidi et al. reported that surface PEGylated liposome encapsulation delivers the "drug-to-bug" more effectively than typical vancomycin formulations by depositing a larger concentration of the antibiotic into the lung tissue. Conventional liposomes with unmodified surfaces are quickly taken up by hepatosplenic and intravascular phagocytes belonging to the monocyte-macrophage line. PEGylation-modified surfaces, on the other hand, produce "stealth" liposomes that avoid opsonization and postpone hepatosplenic clearance, enabling extended circulation and extravasation into targeted organs. S. aureus is capable of surviving inside neutrophils and AMs, as well as for three to four days in phagolysosomes before escaping into the cytoplasm and causing lysis and dissemination of host cells. It has been demonstrated that liposome encapsulation improves the ability of vancomycin to kill MRSA intracel lularly when compared to traditional formulations. In a murine pharmacokinetic and biodistribution investigation, PEGylated liposomal vancomycin, correlated with standard and non-PEGylated formulations, considerably extended blood circulation time, augmented deposition in the lung, spleen, and liver, and lowered accumulation in kidney tissue. PEG-treated surface modification enabled liposomes to function better, indicating the possibility of using synthetic polymers as excipients for the development of effective drug delivery systems for lung disorder therapy (9). The potential of synthetic polymers to enhance liposomal performance has been demonstrated by continuing research, which holds tremendous promise for the development of effective and targeted therapeutics for a variety of respiratory disorders.

4.2.2 Microparticulate System

Microparticles offer various advantages as drug delivery systems for the treatment of lung diseases, including sustained, controlled, and targeted drug delivery, resulting in improved bioavailability. They exhibit low immunogenicity, high entrapment efficiency, and favorable drug-loading characteristics. For pulmonary delivery, microparticles ideally have a particle size of 1–5 µm to maximize efficiency. Particles larger than this range may be deposited in the larynx or esophagus, whereas smaller particles may be exhaled. Achieving an optimum aerodynamic diameter is crucial for preventing agglomeration owing to London forces. To address this issue, porous microparticles have recently been used. Larger porous microparticles, with increased particle size, show reduced fractional surface area, minimize particle-particle interactions, and lower the tendency to form agglomerates. This decrease in aggregation suggests that very low energy is needed to atomize the particles, highlighting the superiority of large porous microparticles (10).

Synthetic polymers, such as poly (lactic-co-glycolic acid) (PLGA), poly (butyl cyanoacrylate), poly (lactic) acid, and poly (lactic-co-lysine graft lysine) are used in the preparation of microparticulate systems for the introduction of drugs into the respiratory region (11). Among these, PLGA has been extensively explored for the development of porous and nonporous microparticles. A previous study demonstrated that porous PLGA microparticles are better taken up by the lungs than nonporous PLGA microparticles; therefore, they have a higher drug delivery efficiency (12). This could be due to their high homogeneity and porosity, which make them light, enabling them to reach the respiratory regions of the lungs. PLGA is a biodegradable synthetic polymer formed through the random copolymerization of polylactic acid and polyglycolic acid. An advantageous feature of this excipient is that the monomer composition can be carefully manipulated to yield PLGA for specific applications. One such application is the manipulation of the extent of drug release; that is, the deterioration rate of PLGA is altered by the copolymer blend and molecular weight, allowing for the adjustment of drug release over a period ranging from weeks to months. Drug diffusion through the matrix and polymer degradation-controlled drug release have been reported with the PLGA matrix. Diverse processes are involved in the breakdown of PLGA, and research to date has indicated that the main mechanism for drug release is diffusion-based. A biphasic pattern, consisting of an initial "burst," followed by a zero-order release phase, is usually observed in PLGA-based microparticle formulations of drugs. This pattern is likely to match Higuchi's concept (13). The elimination of PLGA from the system can be initiated by hydrolysis of the polymer into its metabolic monomers. The resultant lactic and glycolic acid monomers can be easily eliminated through the Krebs cycle. This characteristic makes PLGA a compelling choice considering its minimal systemic toxicity, making it a prominent ingredient in the formulation of drug delivery

TABLE 4.1
PLGA-Based Microparticulate Systems for the Treatment of Lung Disorders

Microparticulate Delivery System	Drug	Method of Preparation	Advantages of the Delivery System
Injectable microparticles (16)	Ivacaftor	Single emulsion solvent evaporation	The delivery system had an in vitro release profile for six weeks. An in vivo pharmacokinetic study demonstrated a sixfold increase in $AUC_{0\text{-}t}$ compared to conventional ivacaftor injection with steady plasma levels for 28 days. Avoid daily administration of the drug in cystic fibrosis conditions.
Inhalable microspheres (17)	Monoammoniuma Glycyrrhizinate	Electrostatic spray technology	Aerosolized microspheres displayed sustained release for five days and were able to induce therapeutic efficiency in the management of acute lung injury.
Microspheres (18)	Docetaxel and Celecoxib	Emulsion solvent evaporation	The porous PLGA-based microspheres containing the drugs were feasible to produce the desired features. The pulmonary co-administration of the drugs could lead to an effective strategy to treat various lung cancers.
Inhalable large porous microparticles (19)	Oridonin	Emulsion solvent evaporation and freeze-drying	In vitro studies disclosed that the microparticles were accepted by the AMs only after 8 h, avoiding the phagocytosis of the particles for a long time. The formulation resulted in elevated anti-tumor activity of the drug in non-small cell lung cancer rat models.
Injectable microspheres (20)	Sophoridine	Oil-in-oil emulsion solvent evaporation	Developed microspheres had an in vitro sustained drug release for 14 days. The delivery system had significantly high in vivo drug levels in the lung tissue of rats, demonstrating its potential for the therapy of lung cancer.

systems (14, 15). Most PLGA microparticles were crafted using the emulsion solvent evaporation technique, and some of the latest studies using this polymer are discussed below and in Table 4.1.

Pulmonary arterial hypertension (PAH) is a debilitating pulmonary hemodynamic disorder that encompasses various clinical conditions that can be fatal. Sildenafil citrate has been approved for PAH treatment. Studies have been performed on sildenafil formulations such as PLGA nanoparticles, lipid nanocarriers, metal-organic frameworks, and PLGA microparticles. Among these formulations, porous PLGA microparticles designed for intratracheal administration and integration into dry powder inhalers (DPIs) are emerging as strong contenders for PAH treatment. The effectiveness of this approach is supported by in vitro, in vivo, and ex vivo studies in rodent models (21). PLGA stands out for its outstanding tensile strength, which guarantees its structural stability throughout the delivery process. The attributes of PLGA microspheres, including soaring dispersity, porous framework, and tunable size, make them well-suited for pulmonary drug delivery. Additionally, PLGA exhibits optimal solvency in a broad spectrum of organic solvents, facilitating precise formulation of controlled-release microspheres. Artesunate, a semisynthetic traditional Chinese medicine with anticancer properties, has been widely studied as an intervention for lung cancer. However, its limited water solubility poses a challenge to bioavailability. To overcome this limitation, a suitable drug delivery system, specifically microspheres, has been explored for successful delivery. Artesunate was encapsulated in a PLGA polymer to generate porous PLGA microparticles, and the efficacy of this delivery system for drug transport was investigated. The drug was efficiently released from the PLGA microspheres after degradation, with an onset of burst release, resulting in a cumulative release of the drug within eight days (22).

While the success of PEG modifications on liposomes has been established, the following question arises: Is it possible to apply this approach to a microparticulate system? Li et al. addressed this question by investigating the feasibility of employing PEG modifications on PLGA microspheres, aiming to achieve dual benefits such as enhanced lung retention and absorption in lung tissue. In this study, PLGA microspheres were modified with PEG-distearoyl glycerophosphoethanolamine (DSPE). Notably, when PEG-DSPE and PLGA were modified with a 1:1 molar ratio and included a high molecular weight PEG2000, systemic exposure was reduced in rats. Furthermore, the same formulation improves mucosal permeability and decreases uptake by macrophages (23). A similar approach was investigated in the treatment of tuberculosis, where the efficiency of antibiotics was hindered by microbial biofilms and lung mucosal barriers. The mucolytic agent N-acetyl cysteine (NAC) was decorated on the surface of PLGA microparticles containing synthetic host defense peptides with anti-TB and anti-biofilm properties. The NAC film on porous PLGA-MS significantly boosted the transit of particles across the mucus barrier according to the results of the multiple-tracking approach. The specially formulated inhalable formulations distribute the medication to the lungs consistently following pulmonary delivery, do not stick to the lung mucosa, and break up the microbial biofilms (24). Therefore, PLGA can serve as a platform for implementing modifications to the microparticulate system, allowing for fine-tuning of drug delivery to the pulmonary region.

Although synthetic polymer-based microparticulate systems offer numerous advantages, they also have drawbacks, including challenges for large-scale production. A notable example is the Nutropin Depot, which encapsulates growth hormone in PLGA microparticles and was withdrawn from sales due to its high manufacturing expenses. Additionally, microparticles pose difficulties in effectively controlling drug release and carry the risk of drug deterioration during the fabrication process (10). The research and development of synthetic polymer-based microparticulate systems for lung targeting holds the potential to overhaul pulmonary therapy, offering a substantial advantage over alternative drug delivery systems.

4.2.3 Nanoparticulate System

Nanotechnology has been extensively employed to deliver drugs with low solubility profiles to their intended destinations. Owing to their reduced particle size, these systems offer a large surface area available for dissolution, resulting in the faster dissolution of solid medications. The improved particle roundness and introduction of imperfections into the crystal lattice could contribute to the enhanced solubility of nanoparticles compared with that of microparticles. Notably, the steric barriers of extracellular polymeric substances are broken by nanoparticles of size ≤100 nm, which are better at penetrating the right proportions of mucus and biofilms (11). Considering their potential, nanoparticles based on synthetic polymers are widely employed as carriers for delivering drugs to specific sites in the lungs.

PLGA nanoparticles can also be prepared and evaluated for their drug delivery potential. These nanoparticles carry high negative charges on their surfaces, particularly when a common emulsifying agent, polyvinyl alcohol (PVA), is used. Even though negatively charged nanocarriers have demonstrated low levels of local inflammation, they are rapidly cleared by airway defense mechanisms in experimental animals. To address this challenge, PLGA nanoparticles surface-coated with PEG were introduced. This modification endowed the nanoparticles with airway defense and mucus-evading properties, ultimately allowing them to be retained in the lungs. This accumulation in the lungs facilitates the release of molecular cargo (25). Similarly, using PLGA alone might lead to hydrophobicity, potentially causing challenges in loading hydrophilic drugs and in vivo toxicity. This issue can also be addressed by coating the surface with PEG. This approach was employed to develop febuxostat-loaded PLGA nanoparticles coated with PEG for prophylaxis against tumor lysis syndrome in lung cancer. The combination of these two synthetic polymers resulted in improved loading and extended release of the drug (26).

Synthetic polymers have garnered considerable attention in this field because they offer the capability to achieve both sustained drug release and targeted localization with polymeric nanoparticle systems. Hydrophilic polymers built upon poly(2-hydroxyethyl methacrylate) (PHEMA) outshine biomedical engineering applications because of their high water content. Their beneficial tissue compatibility, low toxicity, and ease with which physical properties can be adjusted through formulation chemistry. The presence of hydroxyl and carboxyl groups on each repeat unit in PHEMA renders this polymer water-compatible, whereas the hydrophobic α-methyl groups and backbone contribute to the hydrolytic stability and mechanical strength of the polymer matrix (27).

Molecular imprinted nanoparticles (MIP) can be created using molecular imprinting technology, which involves processing polymers with attachment sites that specifically identify target molecules. These MIPs are selectivity comparable to that of natural receptors and exhibit greater durability than that of natural molecules. In the development of double-imprinted nanoparticles for the drugs fluticasone propionate and salmeterol xinafoate, both used in combination to treat COPD and asthma, the PHEMA polymer was employed. Functional monomers with a chemical affinity for drugs, such as N-methacryloyl-(L)-histidine (MAH) and N-methacryloyl-(L)-alanine (MAAL), were polymerized with PHEMA to create p(HEMA-MAAL-MAH). This study suggests that the double-imprinted p(HEMA-MAAL-MAH) nanoparticles are valuable nanomaterials with the potential to serve as effective drug carriers (28).

A nanocarrier was developed using a lysine-PHEMA polymer-encapsulating formoterol. The investigation covered both in vitro experiments and in vivo responses in ovalbumin (Balb/c mice) asthma models to test toxicity levels. The formulation exhibited a steady in vitro drug release profile with zero cytotoxicity. However, an inflammatory response, which elevated the asthma condition, was induced in the mouse model. Several challenges must be addressed before the implementation of lysine-PHEMA-based nanocarrier therapy before being applied in clinical practice (29).

Polymer-based nanoparticles are also considered to be a safer option for the delivery of genetic materials to treat lung disorders. An excellent example of a synthetic polymer is polyethyleneimine (PEI). PEI has been broadly used for this purpose due to its affinity toward nucleic acids. This polymer has a positive charge and it can electrostatically interact with the negative charges usually seen on the phospholipids of nucleic acid. PEI also releases the bound nucleic acid during the endolysosomal break-free of polymeric nanoparticles. Even though PEI has cytotoxicity effects, it remains to be a valuable synthetic polymer in gene delivery. This is because the cytotoxicity of the polymer can be controlled by carefully adjusting the molecular weight and its configurations to a suitable level. PEI was used in the aerosolized delivery of interleukin-12 into the lungs for the therapy of metastasis lung osteosarcoma. In vivo studies conducted on the mice demonstrated an efficacious expression of the delivered drug. There was an 87% reduction in the size of metastatic lung nodules compared to freely treated interleukin-12 (30).

Nevertheless, there are many obstacles to the practical application of polymeric nanomedicines, despite their adaptability and proficiency. The potential benefits of synthetic polymer-based nanoparticles may also present a dilemma that needs to be carefully considered on an individual basis, especially if they are introduced into the human body without appropriate handling. Their vast surface area, small size, and reactivity are directly linked to a variety of toxicities, and they can interact with biological substances in a manner that enhances the risk of pulmonary toxicity.

4.2.4 Polymeric Micelles

Amphiphilic polymers self-aggregate above the critical micelle concentration (CMC) in water to form micelles. The micelle had a core-shell structure characterized by a size of 10–100 nm. The system carries the advantage of carrying both lipophilic and hydrophilic drug molecules because of the presence of a hydrophobic environment at the central core and a hydrophilic corona (31). Biodegradable and nonbiodegradable polymers are simultaneously used for micelle formation. Synthetic polymers such as PEG, Pluronic, PLGA, Polyethylenimines (PEIs), and α,β-poly(N-2-hydroxyethyl)-D,L-aspartamide (PHEA) fall under this category (32–34). The formulation of micelles depends primarily on the concentration of the polymer used. Solvent evaporation, sonication, and thin-film hydration are methods used for the formulation of micelles (33–35). The stability of these systems was greater than that of micelles formed by surfactants. Along with good stability, the functionalization of traditional micelle surfaces with certain groups or materials helps achieve targeting efficiency in lung diseases (34). The biocompatibility, better stability, diffusing capacity, prolonged circulation time, and high capacity to solubilize the drug make micelles effective in lung disease treatment (32–34). The micelles formed by synthetic polymers can readily penetrate and translocate through respiratory epithelial cells by endocytosis mediated by clathrin, cholesterol, and energy (32).

Drugs that are normally used for tuberculosis treatment can be effectively delivered via the respiratory route using polymeric micelles. The poorly soluble rifampicin used in the triple-line therapy of tuberculosis cannot directly reach macrophages that carry bacteria. Soluplus, chemically known as poly (vinyl caprolactam)-poly (vinyl acetate)-poly (ethylene glycol), is a grafted polymer that increases rifampicin solubility. Soluplus has a low CMC value and thus shows high stability and solubility for the encapsulated drug. It shows a 2.5-fold rise in microbicidal activity than that of pure rifampicin solution, and micelles accumulated for 24 h in the lungs of the experimental rats. Micelles formed by the synthetic polymer Soluplus were found to be safer than the drug solution in cytotoxicity studies (31). In addition, we can encapsulate two drugs instead of one in the case of polymeric micelles synthesized with a polyvinylpyrrolidone-polycaprolactone (PVP-b-PCL) block copolymer. However, rifampicin showed greater binding efficiency with the micelle core than pyrazinamide and isoniazid. This stronger binding of rifampicin to the polymer decreases the release of rifampicin from polymeric micelles and thus takes more time to release the entire drug encapsulated within the micelles (36). Therefore, to increase the binding capacity of the drug to the polymer, the system requires more time for complete delivery at a specific site.

The efficiency of beclomethasone dipropionate was compromised in the treatment of COPD and asthma owing to its hydrophobicity. In addition, this pro-drug undergoes rapid hydrolysis to form metabolites, which decreases bioavailability. Mucociliary clearance also affects the use of drugs in conventional dosage forms. The polymeric micelles lend a hand to the drug by prolonging its retention time and increasing its solubility in water. Water-soluble synthetic PHEA polymers can be utilized for the encapsulation of beclomethasone dipropionate, instead of using the polymer as a starting material in certain preparations. Together with the improved solubility of enclosed drugs, local administration enhances the penetration of drugs through the mucus layer and provides efficiency in treating lung diseases (37). However, when poly- (ethylene oxide)-block-distearoyl phosphatidylethanolamine (mPEG-DSPE) polymer micelles were encapsulated with beclomethasone dipropionate, the polymeric micelles were deposited in small amounts in the throat region. Usually, inhalation therapy with corticosteroids for respiratory diseases results in drug deposition in the throat region and causes candidiasis (35). Hence, polymeric micelles can reduce the side effects of drug deposition during conventional inhalation therapy. Researchers have conducted several studies aimed at overcoming the shortcomings of conventional drug therapies using polymeric micelles. The advantages of these systems are critical for making them a superior alternative to other drug delivery systems. Moreover, these studies have facilitated the commercialization of these systems, opening new avenues for drug discovery and development (Table 4.2).

4.2.5 Dendrimers

Dendrimers are drug delivery systems with highly branched structures that exhibit uniformity in their molecular structures and low polydispersity. Several chains arose from the center core of the dendrimer to form a branched structure. Along with the chains, functionalization of the system helps to target the dendrimers at specific sites. Additionally,

TABLE 4.2
Polymeric Micelles-Based Delivery Systems for the Treatment of Lung Diseases

Drug	Synthetic Polymer	Indication	Reference
Curcumin	PEG–PLGA	Chronic hypoxic hypercapnic and monocrotaline (MCT) induced PAH	(38)
Curcumin	PEG–PLGA	Lung cancer (non-small cell)	(39)
Anti-inflammatory gene	PAMAM	Lung inflammatory disease	(40)
Cisplatin	poly(2-oxazoline)s (POx) poly(2-methyl-2-oxazoline-block-2-butyl-2-oxazoline-block-2-methyl-2-oxazoline) (P(MeOx-b-BuOx-b-MeOx)	Small cell and non-small cell lung cancer	(41)
Plasmid DNA	PAMAM	Acute lung injury	(42)
Cisplatin	PEG-poly (aspartic acid)	Lung cancer	(43)
Hematoporphyrin	Pluronic L122	Lung cancer	(44)
Rifampicin	Poly(l-lactide)-PEGpoly(d-lactide)-PEG	Release characteristics	(45)
Isoniazid and pyrazinamide	PEG-poly(aspartic acid)	Tuberculosis	(46)

Marketed Products

Product	Drug	Polymer	Indication	Company
Nanoxel	Paclitaxel	PLA	Lung cancer (non-small cell)	FRESENIUS KABI INDIA PVT LTD
Genexol	Paclitaxel	PLA	Lung cancer (non-small cell)	LUPIN LTD
Mitotax	Paclitaxel		Lung cancer (non-small cell)	Dr REDDY'S LABORATORIES LTD

they increase the rate of cellular uptake and influence the mechanism associated with uptake and lung epithelial cell permeation (47, 48). The methods of entrapment of the drug in dendrimers can be divided into three categories based on the area where entrapment occurs. The first is entrapment at the core or void space by molecular entrapment. The second one is at the branching points with the help of hydrogen bonds, and finally is on the surface group utilizing charge interactions. Hence, the selection and screening of different dendrimers are necessary for the preparation of drug-dendrimer conjugates (49).

In a series of studies conducted by scientists, polyamidoamine (PAMAM), a synthetic polymer, has been utilized for the development of dendrimers. The core of the PAMAM is made up of ethylene diamine, followed by dendritic branches attached via Michael addition. The branches of the dendrimers formed depended on the generation selected (G0 to G10) (50). The terminal group of the half-generation dendrimers consisted of amine groups and a full generation with carboxylate groups. These terminal groups make them more water-soluble and thus enhance the solubility of entrapped hydrophobic drug molecules, such as beclometasone dipropionate, used in asthma. In addition, sustained drug release is possible with dendrimers, as demonstrated by the study conducted by Nasr et al. (51). Apart from sustained drug delivery, PAMAM dendrimers also exhibit controlled drug release from conjugated drugs (52). Factors such as the pH of the surrounding medium influence drug solubility and, in turn, the release of drugs from dendrimers (53).

In a study consisting of the entrapment of erlotinib in the PAMAM dendrimer (G4 and G5), a drug used in lung cancer, dendrimers showed differences in antiproliferative activity against A549 lung cells. Hence, the generation of dendrimers or dendritic branches affects the activity of the drug at cancer-specific sites. The similar activity of the blank dendrimers in cytotoxic studies revealed the anticancer activity of the PAMAM dendrimers due to surface charge (50). Dendrimers can be utilized with the aid of microparticles by encapsulating a drug-dendrimer combination (G4-12 doxorubicin/mannitol microparticles, PAMAM-rifampicin microspheres) in lung diseases. However, it affects the size of microparticles and enhances the cytotoxicity of doxorubicin in DPI formulations (54). A study involving rifampicin instead of doxorubicin revealed that the pharmacokinetic parameters and bioavailability of drugs depend on the generation of PAMAM dendrimers used for their preparation (55). Systemic administration of lung anticancer drugs causes systemic toxicity and low drug concentrations at specific sites within the lung. However, the administration of doxorubicin-PAMAM dendrimers through the pulmonary route increases the retention of doxorubicin in tumor lung tissues compared to systemic administration (56).

TNF-α plays a crucial role in lung inflammation; therefore, siRNA can be used as a therapeutic agent against TNF-α. By utilizing characteristics such as high functionalization, structure, and size maintenance after conjugation with siRNA, PAMAM dendrimers (G3) are a better treatment strategy for acute inflammatory conditions. The complex named "dendriplexes" (PAMAM dendrimers with siRNA) manifested efficiency in the cellular uptake by macrophages and TNF-α inhibition than the pure siRNA in the study conducted in an in vivo murine model (57). Pressurized metered dose inhalers (pMDIs), a less expensive and user-friendly inhaler, have been reported for the effective delivery of dendriplexes. This is because pMDIs assist the dendriplexes to deposit in deep lung tissues under various disease conditions (58). Thus, the studies draw the inference that PAMAM dendrimers will be a promising synthetic polymer-based drug delivery system for siRNA in lung diseases.

Surface functionalization of dendrimers with PEG promises better targeting and transport efficiency in lung tissues (59). Transport across the lung epithelium can be enhanced by increasing the PEGylation of PAMAM dendrimers (G3) (47). Propellant-based inhalers used for pulmonary administration favor PEGylation because of their effects on drug delivery (60). Dendriplexes (G5) modified with triphenylphosphonium (TPP) show better transport across the epithelium and deep lung deposition, even in the delivery of siRNA by inhalers (61). Polylysine is a synthetic polymer used in dendrimers for lung delivery. Upon PEGylation, the polylysine dendrimer influences the clearance of the drug from the lungs. If the PEGylated dendrimer is small, mucociliary clearance becomes easier for the dendrimer, and hence, it reaches the blood quickly. However, a large, PEGylated dendrimer was deposited in the lung tissues (62). Research in the area of synthetic polymer-based dendrimers is underway, and the choice of specific dendrimer type, size, surface modifications, and drug encapsulation strategy will depend on the targeted lung disease and therapeutic goals. The development of novel synthetic polymers may lead to further progress in dendrimer-based drug delivery to the lungs. Additionally, thorough preclinical and clinical studies are necessary to assess the safety and efficacy of dendrimer-based drug delivery systems in lung diseases.

4.2.6 Miscellaneous Approaches for Synthetic Polymer-Based Applications

In addition to drug delivery systems, such as nanoparticles, microparticles, micelles, and dendrimers made up of synthetic polymers, there are several further applications of synthetic polymers in lung diseases. These include polymeric particles, polyplexes, and tissue engineering in lung transplantation using synthetic polymers (63–65). Polymeric particles are normally prepared in the form of dry powder for local administration of the drug to the lung. PHEA, an amphiphilic synthetic polymer, can be used to formulate polymeric particles. In addition, the activity of the system can be increased by the functionalization of the synthetic polymer with PLA and PEG. The biocompatible nature, controlled or sustained drug release, etc., help polymeric particles to deliver drugs in diseases such as COPD and asthma (63).

Similarly, the application of polyplexes, which are complexes formed from the condensation of a polymer with the nucleic acids of a gene or siRNA, has been attempted for lung delivery. These two parts of the polyplexes interact through their carrying charges. Normally, interactions occur between the negative charges of nucleic acids and the cationic charges of polymers. Synthetic polymers, such as virus-inspired polymers for endosomal release (VIPER) and polyethylenimines (PEIs), are commonly utilized for the composition of polyplexes in pulmonary fibrosis and lung tumors. PEIs incorporated with siRNAs have been reported to inhibit plasminogen activity inhibitor -1 in pulmonary fibrosis. Furthermore, the polymer VIPER is prepared from PEIs, a block polymer that effectively silences disease-associated genes in combination with siRNAs. Successful intake of the therapeutic agent in lung cells is possible with the aid of these complexes, thus reducing the systemic toxicity associated with local treatment (64, 66).

The invention of advanced synthetic polymers has had a significant impact on the research and development of artificial tissues and organs. Artificial lung creation and transplantation are effective strategies for the treatment of lung diseases. The selection of scaffolds for tissue engineering is an important parameter. Scaffolds are usually made up of polymers, and a few synthetic polymers such as poly(vinylidene fluoride-co-trifluoroethylene) [P(VDF TrFE)] and PCL have been used in this field. As a sophisticated organ, the polymer selected for scaffold preparation should possess characteristics, such as elastic properties and biocompatibility, to participate in the engineering process. By incorporating antimicrobials into scaffolds, immunomodulation and antibacterial activity can be achieved in the transplantation process (67, 68). Immunomodulation of the artificial lung by the polymer-DNA complex before transplantation is also possible (69). Therefore, medical professionals can ensure the safety of the developed lung because of its good functioning within the human body before transplantation. Placing an artificial lung in the living environment of the human body may cause coagulation of the blood and, in turn, thrombocytosis. This may lead to lung failure in the human body. It has been reported that the coating of poly-carboxy betaine (pCB) conjugated with 3,4-dihydroxyphenylalanine (DOPA) on the surface of an artificial lung can protect the artificial lung from failure. The material behaves as an anticoagulant in transplanted lungs (65). Instead of complete replacement of a lung in living organisms, it can also treat chronic respiratory diseases with the development of certain nonfunctional or diseased regions of the lung. Scaffolds can be developed using synthetic polymers such as poly L-lactic acid (PLLA) (70). Although several investigations are ongoing in the field of artificial organs and safer transplantation, real-time outcomes in this field are still awaited.

The 3D structure made up of polysaccharides having the capacity to hold water molecules is known as hydrogels. The hydration of the hydrogels depends on the networking of the polymer units. Furthermore, the polymer networks control the entrapment of the drug and its release from the polymer matrix. The unloading of the drug from the hydrogel happens based on polymer characteristics such as pH sensitivity, temperature, and pressure. Hence, the hydrogels can be classified based on response to stimulation, size and shape, polymer type, biodegradability, etc. The method of preparation of hydrogel depends on how the polymer network is formed. There are two types of interactions, i.e., physical and chemical. The chemical interaction includes enzyme interaction, free radical polymerization, ionizing irradiation, etc. However, the physical interactions consist of hydrogen bonds, crystallization, etc. (71). Synthetic polymers such as PLGA, PVA, PEG, Pluronics, etc. were used in lung diseases to develop the model for lung diseases, diagnosis purposes, and treatment. With the availability of synthetic polymers, the popularity of hydrogel as a drug delivery system for the treatment of lung diseases has increased significantly. Synthetic polymers like PEG can successfully be utilized to study tumor-associated angiogenesis is an example of the development of a treatment model for the disease (72). The metals for the inhibition of tumors in cancer display limitations due to their side effects. Therefore, in place of platinum, the copper complexes showed higher in vitro anticancer activity. The polymer pluronic F127 extends the release of drug from the copper complex incorporated hydrogel and reduces tumor growth. The hydrogel formulation with metal complexes also decreases systemic toxicity more than the naked complexes (73). Wu et al. created a dual drug delivery system consisting of hydrogel and polymeric micelles incorporating the drugs cisplatin (DDP) and paclitaxel (PTX) (PEG–PCL–PEG/DDP + MPEG–PCL/PTX) as a treatment strategy for lung tumor in 2014. The tumor in the dual drug delivery system inhibited the angiogenesis in tumor tissue and its proliferation activity (74). The study reveals the effectiveness and scope of two drug delivery systems put together for lung disease treatment. The removal of tumors from the lung and thoracoscopic resection were also found to be adopted for the treatment of lung cancer. The process of thoracoscopic resection needs sealing of the lung tissues after the process. PleuraSeal consists of a synthetic polymer having PEG as one of the ingredients that can be used as an adjunct in the surgery. The water solubility, nontoxic behavior, and biocompatibility of PEG make it an effective part of the PleuraSeal. The hydrogel reduces the virus and bacterial infections and allergic reactions associated with the thoracoscopic resection (75). Also, synthetic polymers in combination with natural polymers were used to prepare the hydrogels. Here the researchers report that the limitations of the natural polymers can be improved by synthetic polymers. The properties like purity, capacity to control degradation, sustained release for a longer duration, etc. of synthetic polymers are some of them. The combination of chitosan with PVA can incorporate the antimicrobial ZnO nanoparticles against the bacteria causing pneumonia (76). The hydrogel helps to further improve the activities of drug delivery systems such as nanoparticles and microparticles in certain therapies.

4.3 CONCLUSION

Various applications of synthetic polymers in the treatment of lung diseases, such as the development of drug delivery systems and organ transplantation, have demonstrated their adaptability and potential for advancing the field. However, the development and commercialization of synthetic polymer-based products present significant challenges that must be addressed. Selecting appropriate encapsulation methods and additives is essential to ensure the optimal characteristics of vesicular and particulate drug carriers. The challenge lies in finding a balance between the small size required for effective respiratory deposition and the large size needed for precise measurements during production. Further research is needed to understand the mechanisms of in vivo clearance and persistence of these products. Although biodegradable synthetic polymers are generally considered nontoxic, comprehensive studies are required to ensure their safety. Moreover, there is a lack of chronic toxicological data highlighting the need for further research in this area.

REFERENCES

1. Agustí A, Celli BR, Criner GJ, Halpin D, Anzueto A, Barnes P, et al. Global Initiative for Chronic Obstructive Lung Disease 2023 Report: GOLD Executive Summary. *American Journal of Respiratory and Critical Care Medicine* 2023 Apr 1;207(7):819–37.
2. Xie M, Liu X, Cao X, Guo M, Li X. Trends in prevalence and incidence of chronic respiratory diseases from 1990 to 2017. *Respiratory Research* 2020 Dec;21(1):49.
3. Emami F, Mostafavi Yazdi SJ, Na DH. Poly (lactic acid)/poly (lactic-co-glycolic acid) particulate carriers for pulmonary drug delivery. *Journal of Pharmaceutical Investigation*. 2019;49:427–42.
4. Harris JM, Chess RB. Effect of pegylation on pharmaceuticals. *Nature Reviews Drug discovery*. 2003;2(3):214–21.
5. Muralidharan P, Mallory E, Malapit M, Hayes Jr D, Mansour HM. Inhalable PEGylated phospholipid nanocarriers and PEGylated therapeutics for respiratory delivery as aerosolized colloidal dispersions and dry powder inhalers. *Pharmaceutics*. 2014;6(2):333–53.
6. Nag OK, Awasthi V. Surface engineering of liposomes for stealth behavior. *Pharmaceutics*. 2013;5(4):542–69.
7. Lin C, Wong BCK, Chen H, Bian Z, Zhang G, Zhang X, et al. Pulmonary delivery of triptolide-loaded liposomes decorated with anti-carbonic anhydrase IX antibody for lung cancer therapy. *Scientific Reports*. 2017;7(1):1097.
8. Togami K, Maruta Y, Nanbu M, Tada H, Chono S. Prolonged distribution of aerosolized PEGylated liposomes in the lungs of mice with bleomycin-induced pulmonary fibrosis. *Drug Development and Industrial Pharmacy*. 2020;46(11):1873–80.
9. Muppidi K, Wang J, Betageri G, Pumerantz AS. PEGylated liposome encapsulation increases the lung tissue concentration of vancomycin. *Antimicrobial Agents and Chemotherapy*. 2011;55(10):4537–42.
10. Pulivendala G, Bale S, Godugu C. Inhalation of sustained release microparticles for the targeted treatment of respiratory diseases. *Drug Delivery and Translational Research*. 2020;10:339–53.

11. Kumar M, Hilles AR, Almurisi SHA, Bhatia A, Mahmood S. Micro and nano-carriers-based pulmonary drug delivery system: Their current updates, challenges, and limitations–A review. *JCIS Open*. 2023; 43: 100095.
12. Oh YJ, Lee J, Seo JY, Rhim T, Kim SH, Yoon HJ, et al. Preparation of budesonide-loaded porous PLGA microparticles and their therapeutic efficacy in a murine asthma model. *Journal of Controlled Release*. 2011;150(1):56–62.
13. Rytting E, Nguyen J, Wang X, Kissel T. Biodegradable polymeric nanocarriers for pulmonary drug delivery. *Expert Opinion on Drug Delivery*. 2008;5(6):629–39.
14. García MC. Drug delivery systems based on nonimmunogenic biopolymers. In: *Engineering of Biomaterials for Drug Delivery Systems*. Elsevier; 2018. p. 317–44.
15. Zhao S, Huang C, Yue X, Li X, Zhou P, Wu A, et al. Application advance of electrosprayed micro/nanoparticles based on natural or synthetic polymers for drug delivery system. *Materials & Design*. 2022;220:110850.
16. Nakhla DS, Mekkawy AI, Naguib YW, Silva AD, Gao D, Kim JA, et al. Injectable long-acting ivacaftor-loaded poly (lactide-co-glycolide) microparticle formulations for the treatment of cystic fibrosis: in vitro characterization and in vivo pharmacokinetics in mice. *International Journal of Pharmaceutics*. 2023;650: 123693.
17. Jin J, Shou X, Shi K, Zan X, Zu Y, Huang X. MAG-encapsulated PLGA microspheres for Lipopolysaccharide-induced acute lung injury. *Journal of Drug Delivery Science and Technology*. 2023; 86: 104632.
18. Ziaei E, Emami J, Rezazadeh M, Kazemi M. Pulmonary delivery of docetaxel and celecoxib by PLGA porous microparticles for their synergistic effects against lung cancer. *Anti-Cancer Agents in Medicinal Chemistry (Formerly Current Medicinal Chemistry-Anti-Cancer Agents)*. 2022;22(5):951–67.
19. Zhu L, Li M, Liu X, Du L, Jin Y. Inhalable oridonin-loaded poly (lactic-co-glycolic) acid large porous microparticles for in situ treatment of primary non-small cell lung cancer. *Acta Pharmaceutica Sinica B*. 2017;7(1):80–90.
20. Wang W, Cai Y, Zhang G, Liu Y, Sui H, Park K, et al. Sophoridine-loaded PLGA microspheres for lung targeting: preparation, in vitro, and in vivo evaluation. *Drug Delivery*. 2016;23(9):3674–80.
21. Lazo REL, de Paula Oliveira B, de Fátima Cobre A, Ferreira LM, Felipe KB, de Oliveira PR, et al. Engineering porous PLGA microparticles for pulmonary delivery of sildenafil citrate. *Powder Technology*. 2023;430:118999.
22. Xiong B, Chen Y, Liu Y, Hu X, Han H, Li Q. Artesunate-loaded porous PLGA microsphere as a pulmonary delivery system for the treatment of non-small cell lung cancer. *Colloids and Surfaces B: Biointerfaces*. 2021;206:111937.
23. Li J, Zheng H, Xu EY, Moehwald M, Chen L, Zhang X, et al. Inhalable PLGA microspheres: Tunable lung retention and systemic exposure via polyethylene glycol modification. *Acta Biomaterialia*. 2021;123:325–34.
24. Sharma A, Vaghasiya K, Gupta P, Singh AK, Gupta UD, Verma RK. Dynamic mucus penetrating microspheres for efficient pulmonary delivery and enhanced efficacy of host defence peptide (HDP) in experimental tuberculosis. *Journal of Controlled Release*. 2020;324:17–33.
25. Vij N. Synthesis and evaluation of airway-targeted PLGA-PEG nanoparticles for drug delivery in obstructive lung diseases. In: *Nanoparticles in Biology and Medicine: Methods and Protocols*. Springer; 2020. p. 147–54.
26. Alfaifi MY, Shati AA, Elbehairi SEI, Fahmy UA, Alhakamy NA, Md S. Anti-tumor effect of PEG-coated PLGA nanoparticles of febuxostat on A549 non-small cell lung cancer cells. *3 Biotech*. 2020;10:1–10.
27. Saini RK, Bajpai J, Bajpai AK. Synthesis of Poly (2-Hydroxyethyl Methacrylate)(PHEMA)-Based Superparamagnetic Nanoparticles for Biomedical and Pharmaceutical Applications. *Nanoparticles in Biology and Medicine: Methods and Protocols*. 2020;21118: 165–74.
28. Feyzioğlu Demir E, Akgöl S. Synthesis and characterization of double molecular imprinted nanoparticles and investigation to adsorption of respiratory drugs. *Polymer-Plastics Technology and Materials*. 2022;61(4):384–99.
29. Bakan B, Jonckheere AC, Decaesteker T, Marain NF, Murugadoss S, Karabay Yavasoglu NU, et al. Impact of a Polymer-Based Nanoparticle with Formoterol Drug as Nanocarrier System In Vitro and in an Experimental Asthmatic Model. *Toxics*. 2023;11(12):974.
30. Ray L. Polymeric nanoparticle-based drug/gene delivery for lung cancer. In: *Nanotechnology-Based Targeted Drug Delivery Systems for Lung Cancer*. Elsevier; 2019. p. 77–93.
31. Grotz E, Tateosian NL, Salgueiro J, Bernabeu E, Gonzalez L, Manca ML, et al. Pulmonary delivery of rifampicin-loaded soluplus micelles against Mycobacterium tuberculosis. *Journal of Drug Delivery Science and Technology*. 2019;53:101170.
32. Hu X, Yang FF, Liu CY, Ehrhardt C, Liao YH. In vitro uptake and transport studies of PEG-PLGA polymeric micelles in respiratory epithelial cells. *European Journal of Pharmaceutics and Biopharmaceutics*. 2017;114:29–37.
33. Pellosi DS, d'Angelo I, Maiolino S, Mitidieri E, di Villa Bianca R d'Emmanuele, Sorrentino R, et al. In vitro/in vivo investigation on the potential of Pluronic® mixed micelles for pulmonary drug delivery. *European Journal of Pharmaceutics and Biopharmaceutics*. 2018;130:30–8.
34. Triolo D, Craparo EF, Porsio B, Fiorica C, Giammona G, Cavallaro G. Polymeric drug delivery micelle-like nanocarriers for pulmonary administration of beclomethasone dipropionate. *Colloids and Surfaces B: Biointerfaces*. 2017; 151:206–14.
35. Gaber NN, Darwis Y, Peh KK, Tan YTF. Characterization of polymeric micelles for pulmonary delivery of beclomethasone dipropionate. *Journal of Nanoscience and Nanotechnology*. 2006;6(9–10):3095–101.
36. Veeren A, Bhaw-Luximon A, Jhurry D. Polyvinylpyrrolidone–polycaprolactone block copolymer micelles as nanocarriers of anti-TB drugs. *European Polymer Journal*. 2013;49(10):3034–45.
37. Triolo D, Craparo EF, Porsio B, Fiorica C, Giammona G, Cavallaro G. Polymeric drug delivery micelle-like nanocarriers for pulmonary administration of beclomethasone dipropionate. *Colloids and Surfaces B: Biointerfaces*. 2017; 151:206–14.
38. Hu X, Yang FF, Quan LH, Liu CY, Liu XM, Ehrhardt C, et al. Pulmonary delivered polymeric micelles–pharmacokinetic evaluation and biodistribution studies. *European Journal of Pharmaceutics and Biopharmaceutics*. 2014; 88(3):1064–75.
39. Zhu WT, Liu SY, Wu L, Xu HL, Wang J, Ni GX, et al. Delivery of curcumin by directed self-assembled micelles enhances therapeutic treatment of non-small-cell lung cancer. *International Journal of Nanomedicine*. 2017;12: 2621–34.
40. Kim G, Piao C, Oh J, Lee M. Self-assembled polymeric micelles for combined delivery of anti-inflammatory gene and drug to the lungs by inhalation. *Nanoscale*. 2018; 10(18):8503–14.

41. Wan X, Min Y, Bludau H, Keith A, Sheiko SS, Jordan R, et al. Drug combination synergy in worm-like polymeric micelles improves treatment outcome for small cell and non-small cell lung cancer. *ACS Nano*. 2018;12(3): 2426–39.
42. Piao C, Zhuang C, Choi M, Ha J, Lee M. A RAGE-antagonist peptide potentiates polymeric micelle-mediated intracellular delivery of plasmid DNA for acute lung injury gene therapy. *Nanoscale*. 2020;12(25):13606–17.
43. Nishiyama N, Kato Y, Sugiyama Y, Kataoka K. Cisplatin-loaded polymer-metal complex micelle with time-modulated decaying property as a novel drug delivery system. *Pharmaceutical Research*. 2001;18:1035–41.
44. Yang YT, Chen CT, Yang JC, Tsai T. Spray-dried microparticles containing polymeric micelles encapsulating hematoporphyrin. *The AAPS Journal*. 2010;12:138–46.
45. Chen L, Xie Z, Hu J, Chen X, Jing X. Enantiomeric PLA–PEG block copolymers and their stereocomplex micelles used as rifampin delivery. *Journal of Nanoparticle Research*. 2007;9:777–85.
46. Silva M, Ricelli NL, El Seoud O, Valentim CS, Ferreira AG, Sato DN, et al. Potential Tuberculostatic Agent: Micelle-forming Pyrazinamide Prodrug. *Archiv der Pharmazie: An International Journal Pharmaceutical and Medicinal Chemistry*. 2006;339(6):283–90.
47. Bharatwaj B, Mohammad AK, Dimovski R, Cassio FL, Bazito RC, Conti D, et al. Dendrimer nanocarriers for transport modulation across models of the pulmonary epithelium. *Molecular Pharmaceutics*. 2015;12(3):826–38.
48. Fatani WK, Aleanizy FS, Alqahtani FY, Alanazi MM, Aldossari AA, Shakeel F, et al. Erlotinib-loaded dendrimer nanocomposites as a targeted lung cancer chemotherapy. *Molecules*. 2023;28(9):3974.
49. Chauhan AS. Dendrimers for drug delivery. *Molecules*. 2018;23(4):938.
50. Fatani WK, Aleanizy FS, Alqahtani FY, Alanazi MM, Aldossari AA, Shakeel F, et al. Erlotinib-loaded dendrimer nanocomposites as a targeted lung cancer chemotherapy. *Molecules*. 2023;28(9):3974.
51. Nasr M, Najlah M, D'Emanuele A, Elhissi A. PAMAM dendrimers as aerosol drug nanocarriers for pulmonary delivery via nebulization. *International Journal of Pharmaceutics*. 2014 Jan;461(1–2):242–50.
52. Morris CJ, Aljayyoussi G, Mansour O, Griffiths P, Gumbleton M. Endocytic Uptake, Transport and Macromolecular Interactions of Anionic PAMAM Dendrimers within Lung Tissue. *Pharmaceutical Research* 2017 Dec;34(12):2517–31.
53. Chawla M, Kaushik RD, Malik MK, Pundir V, Singh J, Rehmaan H. Development and optimization of RofA-PAMAM dendrimer complex materials for sustained drug delivery. *Materials Today Communications*. 2022 Dec;33: 104881.
54. Zhong Q. Co-Spray Dried Mannitol/Poly(amidoamine)-Doxorubicin Dry-Powder Inhaler Formulations for Lung Adenocarcinoma: Morphology, In Vitro Evaluation, and Aerodynamic Performance. *AAPS PharmSciTech*. 2018 Feb;19(2):531–40.
55. Rajabnezhad S, Casettari L, Lam JKW, Nomani A, Torkamani MR, Palmieri GF, et al. Pulmonary delivery of rifampicin microspheres using lower generation polyamidoamine dendrimers as a carrier. *Powder Technology*. 2016 Apr;291:366–74.
56. Zhong Q, Bielski ER, Rodrigues LS, Brown MR, Reineke JJ, Da Rocha SRP. Conjugation to Poly(amidoamine) Dendrimers and Pulmonary Delivery Reduce Cardiac Accumulation and Enhance Antitumor Activity of Doxorubicin in Lung Metastasis. *Mol Pharmaceutics*. 2016 Jul 5;13(7):2363–75.
57. Bohr A. Treatment of acute lung inflammation by pulmonary delivery of anti-TNF-α siRNA with PAMAM dendrimers in a murine model. *European Journal of Pharmaceutics and Biopharmaceutics*. 2020;156: 14–120.
58. Conti DS, Brewer D, Grashik J, Avasarala S, Da Rocha SRP. Poly(amidoamine) Dendrimer Nanocarriers and Their Aerosol Formulations for siRNA Delivery to the Lung Epithelium. *Mol Pharmaceutics*. 2014 Jun 2;11(6):1808–22.
59. Wang H, Zheng L, Peng C, Shen M, Shi X, Zhang G. Folic acid-modified dendrimer-entrapped gold nanoparticles as nanoprobes for targeted CT imaging of human lung adencarcinoma. *Biomaterials*. 2013;34(2):470–80.
60. Zhong Q, Humia BV, Punjabi AR, Padilha FF, Da Rocha SRP. The interaction of dendrimer-doxorubicin conjugates with a model pulmonary epithelium and their cosolvent-free, pseudo-solution formulations in pressurized metered-dose inhalers. *European Journal of Pharmaceutical Sciences*. 2017 Nov;109:86–95.
61. Bielski E, Zhong Q, Mirza H, Brown M, Molla A, Carvajal T, et al. TPP-dendrimer nanocarriers for siRNA delivery to the pulmonary epithelium and their dry powder and metered-dose inhaler formulations. *International Journal of Pharmaceutics*. 2017 Jul;527(1–2):171–83.
62. Ryan GM, Kaminskas LM, Kelly BD, Owen DJ, McIntosh MP, Porter CJH. Pulmonary Administration of PEGylated Polylysine Dendrimers: Absorption from the Lung versus Retention within the Lung Is Highly Size-Dependent. *Mol Pharmaceutics*. 2013 Aug 5;10(8):2986–95.
63. Craparo EF, Drago SE, Costabile G, Ferraro M, Pace E, Scaffaro R, et al. Sustained-Release Powders Based on Polymer Particles for Pulmonary Delivery of Beclomethasone Dipropionate in the Treatment of Lung Inflammation. *Pharmaceutics*. 2023;15(4):1248.
64. Ding L, Zhu C, Yu F, Wu P, Chen G, Ullah A, et al. Pulmonary delivery of polyplexes for combined PAI-1 gene silencing and CXCR4 inhibition to treat lung fibrosis. *Nanomedicine: Nanotechnology, Biology and Medicine*. 2018 Aug;14(6):1765–76.
65. Ukita R, Wu K, Lin X, Carleton NM, Naito N, Lai A, et al. Zwitterionic poly-carboxybetaine coating reduces artificial lung thrombosis in sheep and rabbits. *Acta Biomaterialia*. 2019;92:71–81.
66. Feldmann DP, Cheng Y, Kandil R, Xie Y, Mohammadi M, Harz H, et al. In vitro and in vivo delivery of siRNA via VIPER polymer system to lung cells. *Journal of Controlled Release*. 2018 Apr;276:50–8.
67. Azimi B, Sorayani Bafqi MS, Fusco A, Ricci C, Gallone G, Bagherzadeh R, et al. Electrospun ZnO/Poly(Vinylidene Fluoride-Trifluoroethylene) Scaffolds for Lung Tissue Engineering. *Tissue Engineering Part A*. 2020 Dec 1;26(23–24):1312–31.
68. Kosmala A, Fitzgerald M, Moore E, Stam F. Evaluation of a Gelatin-Modified Poly(ε -Caprolactone) Film as a Scaffold for Lung Disease. *Analytical Letters*. 2017 Jan 2;50(1):219–32.
69. Uduehi AN, Stammberger U, Kubisa B, Gugger M, Buehler TA, Schmid RA. Effects of linear polyethylenimine and polyethylenimine/DNA on lung function after airway instillation to rat lungs. *Molecular Therapy*. 2001;4(1):52–7.
70. Young BM, Shankar K, Allen BP, Pouliot RA, Schneck MB, Mikhaiel NS, et al. Electrospun Decellularized Lung Matrix Scaffold for Airway Smooth Muscle Culture.

ACS Biomaterials Science and Engineering 2017 Dec 11;3(12):3480–92.
71. Ahmed EM. Hydrogel: Preparation, characterization, and applications: A review. *Journal of Advanced Research*. 2015;6(2):105–21.
72. Roudsari LC, Jeffs SE, Witt AS, Gill BJ, West JL. A 3D poly (ethylene glycol)-based tumor angiogenesis model to study the influence of vascular cells on lung tumor cell behavior. *Scientific Reports*. 2016;6(1):32726.
73. Deng J, Peng C, Hou L, Wu Y, Liu W, Fang G, et al. Dithiocarbazate-copper complex loaded thermosensitive hydrogel for lung cancer therapy via tumor in situ sustained-release. *Inorganic Chemistry Frontiers*. 2022;9(23):6190–201.
74. Wu Z, Zou X, Yang L, Lin S, Fan J, Yang B, et al. Thermosensitive hydrogel used in dual drug delivery system with paclitaxel-loaded micelles for in situ treatment of lung cancer. *Colloids and Surfaces B: Biointerfaces*. 2014;122:90–8.
75. Saxena AK. Synthetic biodegradable hydrogel (PleuraSeal) sealant for sealing of lung tissue after thoracoscopic resection. *The Journal of Thoracic and Cardiovascular Surgery*. 2010;139(2):496–7.
76. Rata DM, Cadinoiu AN, Popa M, Atanase LI, Daraba OM, Popescu I, et al. Biocomposite hydrogels for the treatment of bacterial infections: physicochemical characterization and in vitro assessment. *Pharmaceutics*. 2021;13(12):2079.

5 Synthetic Polymeric Material-Based Drug Delivery System for Skin Disorders

Shashwat Pandey, Vivek Bakshi, and Nimisha
Amity University Uttar Pradesh, Noida, India

5.1 INTRODUCTION

Polymers are high molecular weight compounds mainly associated with an immense amount of simple repeating units that are generated from molecules with low molecular weights known as monomers. Moreover, polymers are usually obtained from natural sources (for example, chitosan and albumin) or formulated synthetically (for example, polylactic acid (PLA) and poly-lactic-co-glycolic acid (PLGA)). In today's era, natural polymers are in high demand because of their superior biocompatibility and biodegradability along with low cytotoxicity, but synthetic polymers are presently getting the upper hand because of their versatility with the structure of their chemicals and their versatility of their mechanical features, along with improved release kinetics without accumulating the carrier. By getting so many multiple considerations, it has also been found that from the last two decades, the polymeric drug delivery system has received a lot of attention because of the high therapeutic effectiveness associated with its formulation technique. Additionally, it also has one more advantage in today's generation – i.e., the controlled release of the drug by maintaining its therapeutic level, along with the composition adjustment.

In today's era, polymers are mainly used for targeted drug delivery systems, especially for skin disorders because of their high effectiveness, and they are mainly used because of the adaptability of synthetic polymers with the physiochemical properties of the skin. Furthermore, they may facilitate internalization and intracellular localization to enable active movement through the deeper layers of the skin, demonstrating their therapeutic potential for disorders of the skin, such as psoriasis. Polymers have been constructed for the drug delivery system of skin disorders so that they may effectively transport drug molecules via opposing physiological regions; solvent-activated and diffusion-controlled formulations are investigated. Furthermore, by understanding the underlying processes and nature of behavioral transitions, the polymeric structure of a controlled molecular architecture may be designed to provide the accurate reaction when external factors are added.

5.2 SYNTHETIC AND NATURAL POLYMERS FOR SKIN DISORDERS

Polymeric nanoparticles are important drug carriers because they can deliver drugs in a controlled and sustained basis. They are composed of either natural or synthetic polymers that can be further functionalized for precise targeting [1]. Polymeric NPs, on the other hand, are stiff and thicker, exceeding 50 nm in diameter. They are better suited for topical treatment with the goal of concentrating drugs on the skin's outermost layer or in skin extensions.

5.2.1 Polymers Generated Organically

Natural polymers are recommended since they are considered eco-friendly with reduced pathogenicity [2]. Chitosan, albumin, and hyaluronic acid (HA) are three frequently used organic polymers.

5.2.1.1 Chitosan

Chitosan is an organic polymer derived from chitin that is mainly discovered in the exoskeletons of crustaceans, such as shrimp, crab, and several others. Due to its unique properties, chitosan has gained attention in the field of dermatology for its potential therapeutic applications in managing various skin disorders. Chitosan possesses hemostatic properties and promotes wound healing. It aids in blood clotting and provides a scaffold for tissue regeneration [3]. Chitosan is used in treating various skin disorders due to its unique properties, including biocompatibility, biodegradability, and antimicrobial activity. Chitosan accelerates wound healing through several mechanisms, and it promotes blood clotting, which helps in stopping bleeding and protecting the wound from infection; chitosan has a broad-spectrum antimicrobial effect, inhibiting the growth of bacteria, fungi, and viruses. This helps in preventing infections in wounds and other skin conditions [4]. Chitosan forms a film over the wound, maintaining a moist environment that is conducive to healing. This moist environment helps to facilitate cell migration, proliferation, and tissue regeneration. Chitosan can activate macrophages

DOI: 10.1201/9781003477389-5

and other immune cells, enhancing the immune response and promoting tissue repair. Chitosan can reduce inflammation, which is beneficial in conditions like eczema, psoriasis, and dermatitis. Its anti-inflammatory properties help in soothing the skin, reducing redness, and alleviating itching [5]. Chitosan can enhance the skin's barrier function by forming a protective film over the skin surface. This barrier helps in preventing the loss of moisture and protects the skin from external irritants and pathogens. Chitosan has been shown to have antioxidant properties, which can help in reducing oxidative stress on the skin. This is particularly useful in managing skin disorders caused or exacerbated by oxidative damage, such as aging and certain inflammatory conditions. Chitosan can promote the synthesis of collagen, a vital protein for skin health and regeneration. This can be particularly beneficial in treating scars, wrinkles, and other conditions where collagen degradation or insufficient collagen production is a concern. Chitosan can be used as a carrier for delivering therapeutic agents to the skin [6]. Its biocompatibility and ability to form gels and films make it an excellent medium for controlled drug release, enhancing the effectiveness of treatments for various skin disorders. Chitosan-based dressings or films can be applied to wounds, promoting a moist environment conducive to healing while preventing infection. Chitosan exhibits antimicrobial activity against a wide range of bacteria and fungi. This property is beneficial in managing skin infections associated with certain skin disorders [7]. Chitosan-based formulations can help prevent the proliferation of microorganisms on the skin, aiding in the treatment of conditions like dermatitis or infected wounds. Chitosan has been studied for its anti-inflammatory effects. In skin disorders characterized by inflammation, such as psoriasis or eczema, chitosan may help in reducing inflammation and associated symptoms. Chitosan is often used as a component in drug delivery systems for topical applications [8]. It can encapsulate drugs and facilitate their controlled release onto the skin, improving drug absorption and efficacy in treating various skin conditions. Chitosan-based formulations can help in maintaining skin hydration by forming a barrier on the skin surface, preventing moisture loss. This property is beneficial for managing dry or sensitive skin conditions. Chitosan is composed of repeating units of glucosamine and N-acetylglucosamine. Its structure varies based on the degree of deacetylation (the extent to which acetyl groups have been removed), molecular weight, and the distribution of the two monomers [9]. It has a linear and flexible structure, consisting of long chains of the glucosamine and N-acetylglucosamine units linked by β-(1–4) glycosidic bonds. Chitosan's properties, such as solubility, charge density, and biocompatibility, are influenced by the degree of deacetylation. Higher deacetylation levels lead to increased water solubility and enhanced bioactivity [10]. Chitosan's unique properties arise from its cationic nature due to the presence of amine groups. In an acidic environment, these amino groups become protonated, resulting in the formation of positively charged chitosan molecules. This property allows chitosan to interact with negatively charged molecules. Chitosan's positive charge enables it to bind with anionic cellular components and cellular tissues. This imparts biological activity and also increases the therapeutic efficacy, antimicrobial activity, and tissue regeneration. Chitosan's ability to form gels or films is due to its polyelectrolyte behavior. When chitosan interacts with acidic solutions or other polyanions, it can form gels or films through electrostatic interactions and hydrogen bonding. Chitosan's unique properties allow it to be utilized in drug delivery systems. It can encapsulate drugs within its matrix, enabling controlled release by diffusion or degradation of the chitosan matrix in response to environmental schematic representation cell receptor interaction of chitosan during skin regeneration stimuli [11] and thus, it is depicted in Figure 5.1.

5.2.1.2 Hyaluronic Acid

HA is an organic polysaccharide distributed throughout the human body, particularly in connective tissues, skin, and the eyes. It is essential for keeping tissues hydrated, healing wounds, and cushioning joints. In dermatology, HA has gained attention not only as a skincare ingredient but also as a potential component in drug delivery systems for managing various skin disorders. HA's unique properties, including its ability to retain water and form a gel-like substance, make it suitable as a vehicle for delivering active pharmaceutical ingredients to the skin [12]. When combined with other substances, HA can enhance the penetration of drugs or active compounds into the skin layers, improving their efficacy in treating skin disorders. HA has an extraordinary ability to retain water, holding up to 1,000 times its weight in water. This property makes it an excellent moisturizer, helping to keep the skin hydrated. By retaining moisture, HA strengthens the skin's barrier, preventing transepidermal water loss (TEWL) and keeping the skin plump and smooth. This is particularly beneficial for dry or dehydrated skin conditions. HA helps to plump up the skin, reducing the appearance of fine lines and wrinkles. This is achieved by the hydration it provides, which fills in the gaps between collagen and elastin fibers. Some studies suggest that HA may stimulate the production of collagen, which is crucial for maintaining skin elasticity and firmness [13]. This can help to combat signs of aging. It is involved in cell proliferation and migration, which are essential processes in wound healing. HA has anti-inflammatory properties, which help in reducing inflammation in wounds and promoting a quicker healing process. HA is noncomedogenic (doesn't clog pores), making it suitable for acne-prone skin [14]. It provides hydration without adding excess oil, which can help in managing acne. HA can be used in the treatment of scars, particularly atrophic (depressed) scars, by promoting hydration and skin regeneration, which can improve the appearance of the skin over time. The hydrating and anti-inflammatory properties of HA can soothe and calm irritated skin, making it beneficial for conditions like eczema and dermatitis [15]. By enhancing the skin's

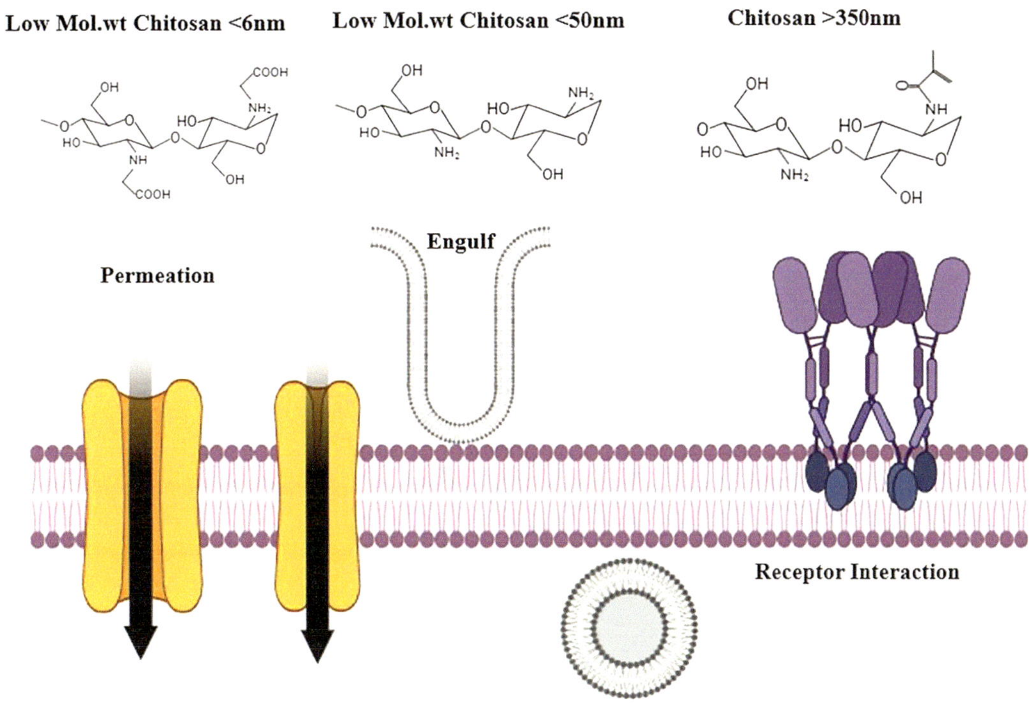

FIGURE 5.1 Schematic representation cell receptor interaction of chitosan during skin regeneration.

barrier function, HA helps to protect against irritants and allergens, which can exacerbate these conditions. HA can help in the recovery of skin from sunburn and UV-induced damage by promoting hydration and repair. It can also be used in formulations that provide a protective barrier against environmental damage [16]. HA is often used in post-procedure skincare (e.g., after laser treatments, chemical peels, or micro needling) to aid in the recovery process by promoting hydration and soothing the skin. HA's ability to hydrate, soothe, and promote skin repair makes it a versatile and effective ingredient in managing various skin disorders [17]. HA is known for its moisturizing properties. In conditions where the skin barrier is compromised, such as in eczema or dermatitis, formulations containing HA can help in hydrating the skin and supporting barrier repair. This aids in reducing symptoms and promoting skin healing. By encapsulating drugs within HA-based formulations, it's possible to achieve a gradual release of the drug, ensuring prolonged therapeutic effects and minimizing potential side effects [18]. Hyaluronic acid is generally well-tolerated by the skin due to its biocompatibility and similarity to the body's natural HA. HA can also be modified to facilitate targeted delivery of drugs to specific skin layers or cells, treatment for certain skin disorders [19], and thus, it is depicted in Figure 5.2.

5.2.1.3 Albumin

Albumin is a globular protein found in blood plasma. It has been investigated for its wide ranging application in drug delivery systems for various medical attributes like biocompatibility and biodegradability, which elevate its ability to carry a diverse range of drugs to targeted sites [20]. Albumin can be used to encapsulate drugs for topical applications. These formulations can improve the solubility and stability of drugs, enhancing their penetration into the skin layers and ensuring targeted delivery to the affected areas. Moist wounds promote cell migration and tissue regeneration, which are crucial for healing. As a carrier protein, albumin can transport essential nutrients and growth factors to the wound site, aiding in tissue repair and regeneration. Albumin has antioxidant properties, which can help protect cells from oxidative stress and damage during the healing process [21]. Albumin can modulate inflammatory responses, helping to reduce inflammation in various skin conditions. This is particularly useful in chronic inflammatory skin disorders such as eczema and psoriasis. Albumin helps to improve skin hydration and elasticity. In cosmetic formulations, it is used for its skin-tightening effects, helping to reduce the appearance of fine lines and wrinkles. While albumin itself is not collagen, it can support the skin's collagen structure by maintaining a hydrated environment, which is crucial for the integrity and function of collagen fibers [22]. Albumin can be used in the management of scars by promoting hydration and supporting the skin's natural healing processes. It may also help in reducing the appearance of scars by improving the overall texture and elasticity of the skin. In burn treatment, albumin-based products can be applied to create a protective layer over the skin, helping to retain moisture and prevent infection. This can aid in the healing of burn wounds and reduce the risk of scarring [23]. Albumin can serve as a carrier for delivering therapeutic agents to

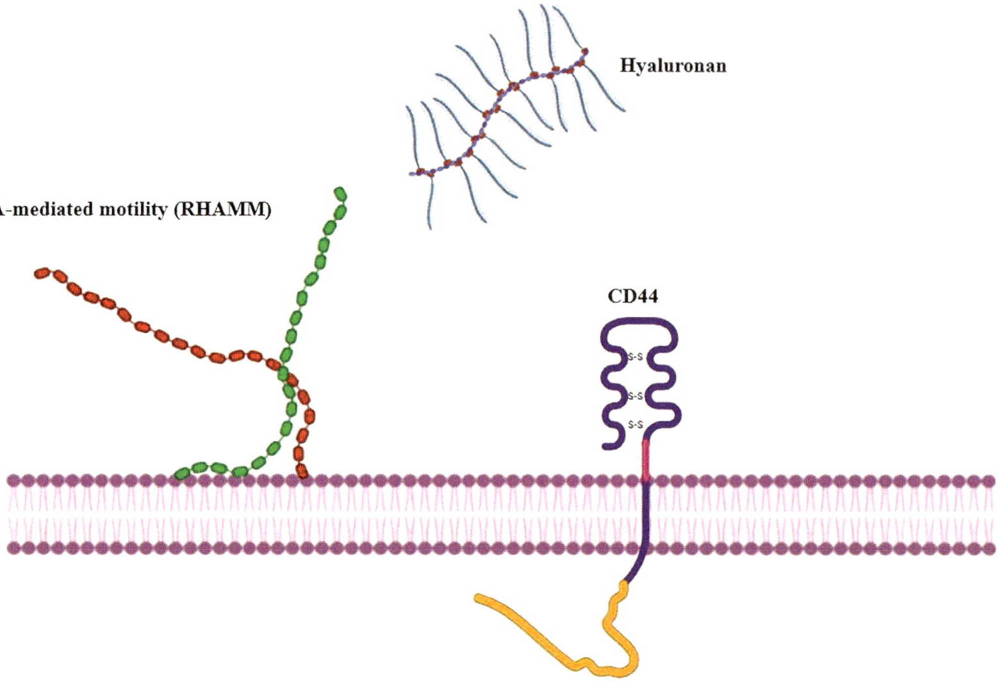

FIGURE 5.2 Diagrammatic representation of cell receptor interaction of HA during skin regeneration.

the skin. Its ability to bind various substances makes it an effective vehicle for targeted drug delivery, enhancing the efficacy of treatments for specific skin conditions. Albumin is often used in face masks and creams for its firming and tightening effects. It can give the skin a temporary lift and smooth appearance, making it popular in anti-aging skin-care products. By modulating immune responses, albumin can help manage autoimmune or inflammatory skin disorders, where the immune system attacks the skin or causes chronic inflammation [24]. Albumin provides a stable environment for drug molecules, protecting them from degradation and enhancing their bioavailability. This can be particularly beneficial for drugs susceptible to degradation when exposed to the external environment or skin enzymes, along with its controlled release profile, can prolong the therapeutic effects of medications, reducing the frequency of application and potentially enhancing patient compliance [25]. Albumin is naturally present in the body, making it well-tolerated and reducing the risk of adverse reactions or immunogenic responses when used in drug delivery systems. By encapsulating various drugs or active compounds in albumin carriers, it's possible to enhance the therapeutic efficacy in treating skin disorders such as psoriasis, eczema, acne, and wound healing. Albumin is a multifunctional globular protein primarily produced by the liver, found in the blood plasma of mammals, including humans. Its structural and functional characteristics play crucial roles in various physiological processes. The structure and mechanism of albumin are key to understanding its diverse functions in the body. Albumin is composed of a single polypeptide chain consisting of 585 amino acids in humans. Its primary structure involves a sequence of amino acids arranged in a specific order. The secondary structure of albumin primarily comprises α-helices and β-sheets, forming distinct domains within the protein. These secondary structural elements contribute to the overall stability of the protein. The tertiary structure of albumin involves the folding and arrangement of the secondary structures into a compact and globular shape. This 3D structure determines albumin's functions and interactions with other molecules [26]. Unlike many proteins, albumin is primarily a monomeric protein, meaning it exists as a single polypeptide chain and does not generally form larger protein complexes or aggregates.

5.2.2 Polymers of synthetic origin

Synthetic polymers have found various applications in the field of medicine, including dermatology and the treatment of skin disorders. They have unique characteristics that may be customized to specific therapeutic specifications, such as controlled administration of drugs, wound healing, tissue engineering, and so on. Synthetic polymers can be used to develop drugs delivery methods for skin conditions such as psoriasis, eczema, and acne [27]. They can encapsulate medications, allowing controlled release at the affected area, reducing side effects and improving therapeutic outcomes. Polymers like polyurethane, polyethylene glycol, and others are used in wound dressings for chronic wounds or burns. These dressings provide a protective barrier, promote healing, maintain a moist environment, and prevent infection. Polymer-based formulations are used in topical creams, ointments, and gels for various skin conditions. They help in enhancing the stability and bioavailability of active pharmaceutical ingredients, ensuring better penetration and efficacy. Synthetic polymers are employed in creating skin substitutes and scaffolds for tissue engineering applications. These substitutes can aid in the regeneration of skin tissue for patients with severe burns or chronic wounds. Certain skin disorders involve compromised skin barriers. Polymers can be utilized to create films or coatings that act as barriers, protecting the skin from irritants, allergens, or microbial invasion in conditions like atopic dermatitis [28] In some cases, implantable devices made from biocompatible synthetic polymers are used for long-term drug delivery or to assist in the treatment of specific skin disorders. synthetic polymers offer advantages in treating skin disorders, there are challenges such as biocompatibility, potential toxicity, and long-term effects that need to be addressed during their development and use. Researchers continually work on refining these materials to improve their safety and efficacy for dermatological applications.

5.2.2.1 PLGA (Poly (lactic-co-glycolic acid))

PLGA is an organic copolymer that is frequently used in biomedical fields because of its biocompatibility and configurable disintegration features. In dermatology, PLGA has shown promise in addressing various skin disorders and conditions due to its versatility and controlled drug release capabilities. PLGA is commonly used to create nanoparticles, microparticles, or nanospheres that encapsulate drugs, allowing for controlled and sustained release of active ingredients. This can enhance the efficacy and reduce the side effects of topical medications used in treating skin disorders such as acne, eczema, psoriasis, and fungal infections. PLGA-based formulations can be designed to target specific skin layers or cell types, improving the precision of drug delivery [29]. This targeted approach can be particularly useful in conditions like melanoma, where precise delivery of therapeutic agents is crucial. PLGA can be used to fabricate scaffolds that support tissue regeneration. PLGA scaffolds provide structural support, aiding in the regeneration of skin tissue in chronic wounds, burns, and ulcers. PLGA can be incorporated with angiogenic factors to promote the formation of new blood vessels, which is essential for the healing of large wounds and improving skin graft integration [30]. PLGA can encapsulate and deliver anti-inflammatory drugs to inflamed skin areas, providing prolonged relief and reducing the need for frequent applications. By encapsulating antimicrobial agents, PLGA systems can provide a sustained release of these agents, helping to manage infections in wounds and other skin conditions. In cosmetic dermatology, PLGA can be used to encapsulate and deliver active ingredients such as retinoids, peptides, or antioxidants. This controlled release can enhance the effectiveness of anti-aging treatments by ensuring a steady supply of active compounds to the skin. By controlling the release rate of potent active ingredients, PLGA can reduce the risk of irritation or other side effects often associated with high concentrations of these agents [31]. PLGA-based formulations can be designed to deliver agents that modulate the healing process, reducing scar formation and improving skin texture. PLGA can help deliver ceramides or other barrier-repairing agents to strengthen the skin's natural barrier, which is often compromised in conditions like atopic dermatitis [32]. PLGA is used in the development of vaccine delivery systems, including those targeting skin-related diseases. It can encapsulate antigens and adjuvants, providing controlled release and enhancing the immune response. PLGA can be used to encapsulate UV-filtering agents, improving their stability and providing controlled release. This can enhance the effectiveness and safety of sunscreen products. For individuals at high risk of skin cancer, PLGA-based systems can deliver chemo-preventive agents that inhibit carcinogenesis [33]. PLGA is commonly employed as a carrier for controlled drug delivery in dermatology. It can encapsulate various drugs used to treat skin disorders such as psoriasis, acne, atopic dermatitis, and skin cancers. In chronic wounds or injuries, PLGA-based dressings or scaffolds can be used to facilitate wound healing. These materials provide a supportive matrix for tissue regeneration, maintain a moist environment, and can be tailored to release growth factors or drug that aid in wound closure and tissue repair. PLGA, often in combination with other polymers or materials, is used to create construct scaffolds for biological engineering applications in dermatology [34]. These scaffolds replicate the skin's extracellular matrix, facilitating cell proliferation and the growth of new tissue. They are used to treat burns that are severe, chronic ulcers, and other skin defects. PLGA can be shaped into various forms, such as micro particles, nanoparticles, or films, suitable for implantable devices. These devices may be utilized for sustained release of therapeutics or as temporary implants for tissue regeneration in dermatological conditions. PLGA-based films or coatings can be designed to act as barriers to protect the skin from external irritants, allergens, or infections. They can help in managing certain dermatological conditions by providing a protective layer over compromised skin barriers. [35]. However, proper tissue response needs to be addressed during the development and application of PLGA-based systems. PLGA undergoes hydrolysis of its ester bonds under physiological conditions (in the presence of water) due to its susceptibility to hydrolytic degradation. This hydrolysis leads to the breakdown of PLGA into its constituent monomers, lactic acid, and glycolic acid. Hydrolysis of PLGA is catalyzed by acidic conditions, such as those found in tissues with a lower pH. The presence of protons accelerates the hydrolytic cleavage of the ester bonds in PLGA. The ester bonds in PLGA chains undergo cleavage through a hydrolysis reaction. Water molecules attack the ester bond, leading to the breakdown of the polymer chain into smaller oligomers and eventually into monomers. This process is known as bulk erosion, where degradation occurs uniformly throughout the polymer matrix [36].

5.2.2.2 Poly(ε-caprolactone) (PCL)

PCL is renewable and bioresorbable synthetic polymer widely investigated for drug delivery applications due to its favorable properties, including slow degradation rates and versatility in formulation. PCL is commonly used to create scaffolds and matrices for tissue engineering. These structures provide a supportive framework for cell attachment and proliferation, facilitating tissue regeneration in wounds, burns, and ulcers. The slow degradation rate of PCL makes it ideal for long-term tissue support. PCL-based wound dressings can be loaded with therapeutic agents such as antibiotics, growth factors, or anti-inflammatory drugs. These dressings provide controlled and sustained release of the drugs, promoting healing and reducing infection risks. PCL is used to fabricate nanoparticles and microspheres that encapsulate drugs for controlled and sustained release. This approach can enhance the efficacy and reduce the side effects of treatments for various skin disorders, including acne, eczema, and psoriasis. PCL-based systems can be engineered for targeted drug delivery, allowing for localized treatment and minimizing systemic exposure. This is particularly useful in managing conditions like localized infections or inflammatory skin disorders [37]. PCL can be used in the fabrication of bioengineered skin grafts. These grafts provide temporary or permanent coverage for large wounds or burns, supporting tissue regeneration and reducing scarring. In reconstructive surgery and aesthetic dermatology, PCL-based materials can be used as dermal fillers or scaffolds for soft tissue augmentation. These applications help restore skin volume and improve the appearance of scars or other skin irregularities [38]. PCL can be combined with antimicrobial agents to create coatings for medical devices or wound dressings. These coatings can help prevent infections in wounds or during surgical procedures. PCL is used in the production of dermal fillers for cosmetic treatments. These fillers help in reducing wrinkles and restoring facial volume. The slow biodegradation of PCL provides long-lasting results compared to other fillers [39]. PCL-based nanoparticles can deliver active ingredients such as retinoids, peptides, or antioxidants in cosmetic formulations. This controlled release enhances the effectiveness of anti-aging and skin-rejuvenation treatments. PCL can be used to develop topical formulations for chronic skin conditions, such as psoriasis or atopic dermatitis. The controlled release of active ingredients can provide sustained therapeutic effects, improving patient compliance and treatment outcomes. PCL can be used to encapsulate UV filters in sunscreen formulations. This encapsulation can enhance the stability of UV filters, improve their skin penetration, and provide a controlled release for prolonged protection [40]. PCL can be used to create formulations for topical drug delivery to treat skin disorders such as psoriasis, eczema, acne, and dermatitis. PCL's ability to encapsulate drugs and slowly release them over time onto the skin can enhance drug penetration, efficacy, and reduce the frequency of application.[41]. PCL is known for its biodegradability and compatibility with bodily tissues. Besides drug delivery, PCL is utilized in tissue engineering and wound healing applications [42].

5.2.2.3 Poly(lactic acid) (PLA)

PLA renewable and bioresorbable synthetic polymer frequently utilized in drug delivery systems and biomedical applications. PLA-based drug delivery systems have shown promise in managing various skin disorders due to their favorable characteristics [43]. The degradation of PLA into lactic acid also contributes to the acidic environment, which can inhibit microbial growth and support wound healing [44] The degradation rate of PLA can be controlled by adjusting its molecular weight and crystallinity. This allows for the design of wound dressings and scaffolds that degrade at appropriate rates, providing sustained support during the healing process. This technology is beneficial for delivering therapeutic agents to treat various skin disorders, including acne, psoriasis, eczema, and fungal infections [45]. PLA-based delivery systems can provide localized treatment, minimizing systemic side effects and improving the efficacy of the drugs. This is particularly useful in managing localized skin conditions and infections. PLA can be used to develop antimicrobial coatings for wound dressings and medical devices. These coatings can be loaded with antimicrobial agents to prevent infections in wounds or post-surgical sites. PLA-based delivery systems can encapsulate and deliver anti-inflammatory agents, providing sustained relief for chronic inflammatory skin conditions [46]. PLA is used in the production of dermal fillers for cosmetic applications. These fillers help restore facial volume, reduce the appearance of wrinkles, and improve skin texture. The slow degradation of PLA allows for long-lasting effects. PLA can encapsulate and deliver active cosmetic ingredients, such as peptides, vitamins, or antioxidants. This controlled release enhances the effectiveness of anti-aging and skin-rejuvenation products [47]. PLA is used to create biodegradable skin substitutes that can be applied to large wounds or burns. These substitutes provide temporary coverage and support for new tissue growth, reducing the risk of infection and scarring. PLA scaffolds can be loaded with growth factors or stem cells to promote skin regeneration in reconstructive surgery and chronic wound management. PLA can be used to encapsulate UV filters in sunscreen formulations, enhancing the stability and controlled release of these agents. This can improve the effectiveness and duration of sun protection. PLA-based products can be used to deliver treatments that improve the appearance of scars [48]. The controlled release of active ingredients can help soften and smooth scar tissue, improving skin texture. PLA can be formulated into nanoparticles, microparticles, or films to encapsulate drugs for topical application. These formulations enable controlled release of medications, improving drug penetration into the skin and enhancing their therapeutic efficacy for conditions like psoriasis, acne, eczema, and dermatitis. PLA's biodegradability allows for gradual drug release over an extended period. PLA degrades into non-toxic by-products, making it suitable for biomedical applications. Its biocompatibility ensures minimal adverse reactions or inflammation when used for drug delivery to the skin[49]. PLA-based carriers will encapsulate an array of drugs, including hydrophobic and hydrophilic molecules. This flexibility enables the incorporation of different therapeutic agents tailored to specific skin conditions. In addition to drug delivery, PLA is utilized in tissue engineering and wound healing applications. PLA-based scaffolds or dressings support tissue regeneration and wound closure, aiding in the treatment of chronic wounds or severe skin injuries [50].

5.3 RELEASE MECHANISM OF POLYMERS

5.3.1 Topical and Transdermal Delivery Across Skin Barrier

Topical and transdermal techniques for drug delivery offer some benefits.

- Deliver the drugs exactly to the targeted cells or regions.
- Lesser amount of drug is needed to for a therapeutic efficacy.
- Drug's peak plasma level is avoided.
- Increased bioavailability brought on by the liver's removal of first-pass metabolism.
- Significantly improved patient compliance by the removal of frequent dosage.

Pathways of cutaneous penetration are as follows:

- Via intercellular and intracellular pathways through the stratum corneum, and then by partitioning and diffusion into the viable epidermis and dermis
- Employing the appendageal pathway [51]

Enhancement technique for passive and active skin penetration:

- The barrier that limits the pace at which drugs can penetrate the skin is the stratum corneum
- The stratum corneum lipid's distinctive constituents, layout, and morphology selectively and efficiently restricts the penetration of chemicals [52].

Enhancement Techniques for Passive Skin Penetration

The diffusion flux J can be explained by equation 2.

$$J = \frac{D_m C_{sm}.C_v}{L.C_{sv}}$$

D_m = diffusion coefficient of the drug in the membrane (skin) C_{sm}.

C_{sm} = solubility of the drug in the membrane.

C_v and C_{sv} = the drug's solubility and concentration in the vehicle.

L = length of the membrane's diffusion route.

Permeation rates were linked with the drug's ability to diffuse into the skin strata and its partitioning between the stratum corneum and topically administered composition according to the solute structure-skin transport model. Furthermore, it relates to the solute size and hydrogen bonding in addition to the octanol–water partition coefficient [53]. By employing passive penetration enhancement techniques, the formulation is optimized and the drug carriers improve the flow. Therefore, we understand from equation 2 that the methods related to passive penetration enhancement involve adding eutectic systems or chemical penetration enhancers to the formulation [54], which causes a temporary disruption the presence of lipids within the stratum corneum's cell framework and declines the skin's obstruction but additionally raises the diffusion coefficient Dm (enhances the diffusivity) and the solubility of drugs in the skin (increases Csm, and, consequently, increases the partition coefficient) [55].

5.3.2 Techniques for Increasing Active Skin Penetration

Approaches for elevating active penetration through the skin have been established and categorized as electrical, mechanical, or other energy-related. The mechanisms for active penetration supplementation tactics primarily involve the following:

- Microneedle array technique, which is mainly employed by bypassing the stratum corneum [31]
- Removal of stratum corneum by inducing stratum corneum [56]
- Lipid packing of stratum corneum's is disrupted by ultrasound waves [57]
- Electroporation to create temporary aqueous pores in the stratum corneum's lipid bilayers [58]

Among these, the advancement of techniques that employ microneedles, which has received substantial attention, Microneedles are considered tiny structures shaped like patches that can pierce skin without causing pain and inject active ingredients. Microneedles made of a variety of materials, including silicon, metal, and biodegradable polymers, have been created. In order to increase the therapeutic benefits or skin delivery for various small molecules and macromolecules, including proteins, peptides, nucleic acids, and virus-like particles, microneedles have been widely used [59].

5.4 ROLE OF SYNTHETIC POLYMERS IN SKIN DISORDERS

5.4.1 Polydimethylsiloxane

Uses:

- Enhance the efficacy of wound healing in a broad range of events, including burns, blisters, abrasions, split-thickness skin grafts, ulcers, and numerous other wounds.
- Although they are comfortable enough to be altered without distressing the delicate wound site or the primary dressing, they are frequently used as replaceable supplementary dressings that soak up wound discharges [60].
- They also have a significant advantageous effect on scar healing (keloids and hypertrophic) [3].
- In a meta-analysis of three clinical investigations, soft silicone dressings were found to be better than other commonly used dressings for wounds that are acute, leading to less discomfort, quicker recovery, and fewer modifications to the dressing [61].

5.4.2 Polyvinyl Alcohol (PVA)

Uses:

- Because of its high hydrophilicity, it is an excellent absorber of wound fluid, lowering the risk of infection and promoting wound healing.
- PVA hydrogels are ideally suited as a material for wound dressings, and studies have concentrated on enhancing their cellular interaction for additional skin biomaterial development applications [62].

5.4.3 *Poly-N-vinylpyrrolidone*

Uses:

- In a 29-day full-thickness wound model on an ovine, this biodegradable temporary matrix was found to heal equivalently to the Integra® dermal regeneration template [63].
- The by-products of degradation of polyurethane scaffolds induced low level of inflammation in a full-thickness wound model, emphasizing that they could have effectiveness in skin tissue reconstruction [64].

5.4.4 *Polyurethane*

Uses:

- Medtronic Corporation explored polyurethane in the early 1980s, as an insulator for pacemaker leads. The degradation of it in the body and the linked inflammatory and fibrotic reaction lead to its continuous use in developing novel medical devices utilising its microfilamentous scaffold.
- The by-products of degradation of polyurethane scaffolds induced low level of inflammation in a full-thickness wound model, emphasizing that they could have effectiveness in skin tissue reconstruction [64].

5.4.5 *Polylactic Acid*

Uses:

- Used in skin tissue engineering
- Used electro spun fiber matrices

5.5 NATURAL POLYMER IN SKIN DISORDER

5.5.1 Polymers with Properties and Uses

TABLE 5.1
Applications of Polymers with Its Characteristics

Polymer Name	Properties	Uses
Alginate	It consists of β-D-mannuronic acid (M-blocks) and 1–4 α-L-guluronic acid	Used in atopic dermatitis Enhance the rate of wound healing
Chitin and Chitosan	Exhibit pH-dependent solubility Extremely Hydrophobic and insoluble in water and other organic solvent	Used in atopic dermatitis
Starch	Exhibit high degree of hydrophilicity	Improve wound dressing properties Improve wound dressing properties
Hyaluronic Acid	It is polyanionic Exhibit N-acetyl-β-D-glucosamine linked to β-D-glucuronic acid by a 1 → 4 glycosidic bonds. The HA chain itself is formed by 1 → 3 glycosidic bonds	Used in chronic wound healing

5.5.2 Advantages and Disadvantages of Natural Polymers in skin Disorders

TABLE 5.2
Advantages and Disadvantages of Organic Polymers Needed for Skin Disorder

Advantage	Disadvantage
Active drugs enable safeguards against early declination [65]	Not appropriate for transdermal applications [69]
Possibilities for follicular deposits and trans-follicular drug delivery [66]	The importance of refining approaches for organic nanoparticles made from polymers
Controlled release of drugs employing modification of polymers method [67]	Organic polymeric nanoparticles are not homogeneous from batch to batch
Biocompatible and biodegradable nature [68]	A slow pace of polymer breakdown elevates the risk of toxicity

5.6 SYNTHETIC BIODEGRADABLE POLYMERS USED IN DRUG DELIVERY SYSTEM FOR SKIN DISORDERS

5.6.1 Chemical Structure of Synthetic Polymers

TABLE 5.3
Chemical Structure of Synthetic Polymers

Synthetic Polymer	Chemical Structure
Polydimethylsiloxane (PDMS)	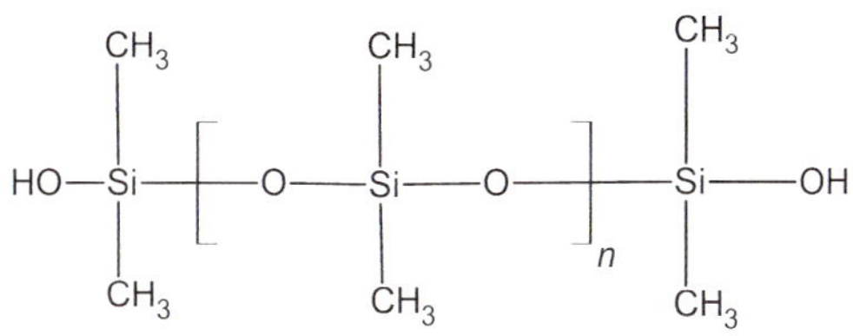
Polyvinyl alcohol (PVA)	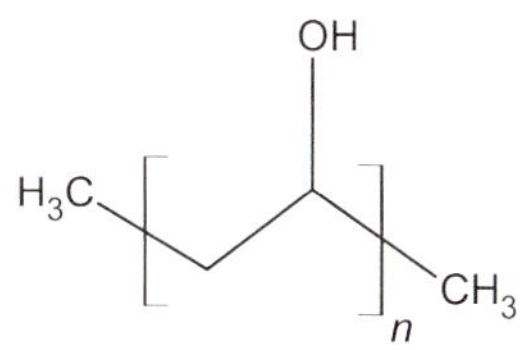
Poly-N-vinylpyrrolidone	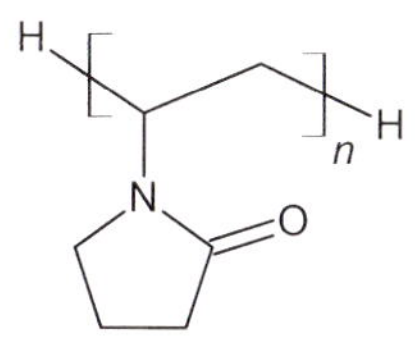
Polyurethane (PU)	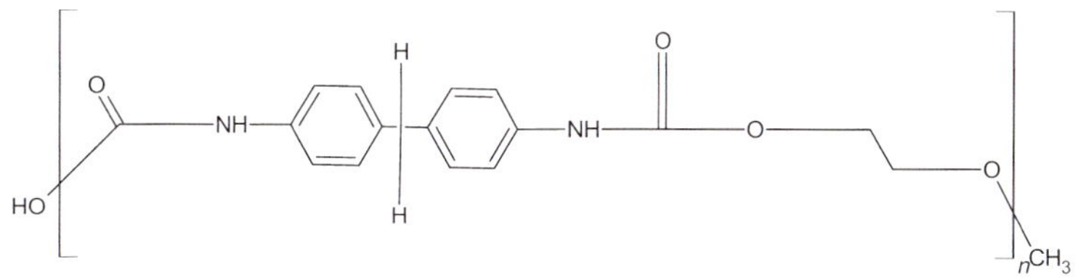
Polylactic acid (PLA)	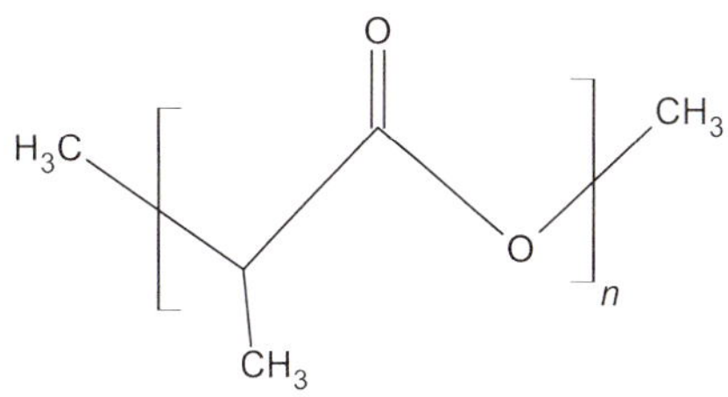

5.6.2 Polymeric-Based Drug Delivery System

It is categorized in three stages:

a. ***Permeation Through Polymer Membrane***: Cell membrane has having phospholipid bilayer structure which has lipophilic activity toward the skin. Moreover, the molecule should have dual properties – i.e., lipid-loving or water-loving attributes. Hydrophobic attributes are responsible for constructing an impedance in the membrane of biological cells so that molecules flow restricted and join the transcellular channel via passive absorption and hydrophilic properties were responsible for dissolving in aqueous media surrounding biological membranes. Endothelial/epithelial cells are present in between the tight junction crossed by an aqueous channel and it will permit the absorption of small water molecules (mol. wt. less than 600 Da) by paracellular pathway [70]. However, the procedure operates on a partially or entirely encased drug reservoir, with a polymeric coating applied to the exterior of the drug emitting region. Therefore, solid dispersing media mainly consist of concentrated drug solution associated with a drug reservoir having this type of drug, and so there is the formation liquid dispersion medium with the dispersion of solid drug particles. The polymeric membrane's semipermeability can either be heterogeneous or homogeneous in nature or the membrane can have nonporous or partially microporous properties [71–73].

b. ***Diffusion-Controlled Polymer Matrix***: A water miscible polymer present in an aqueous solution contains solid and thus forms suspension therefore it is termed a drug reservoir. Moreover, a drug reservoir is formed as a suspension of solid drugs in water miscible solubilizers. Followed by its homogenous dispersion of drug suspension with lipophilic polymer induce mechanical shear force, which results from the formation of thousands of microscopic drug reservoirs. A high dispersion method is employed for a micro-dispersion-controlled system. Micro-reservoir is generally formed by matrix and reservoir dispersion [74].

c. ***Partitioned Controlled Micro-Reservoir***: A water miscible polymer present in an aqueous solution contains solid and thus forms suspension therefore it is termed a "drug reservoir." Moreover, drug reservoir is formed as a suspension of solid drugs in water miscible solubilizers. Followed by its homogenous dispersion of drug suspension with lipophilic polymer induce mechanical shear force which results from the formation of thousands of microscopic drug reservoirs. A high dispersion method is employed for a micro-dispersion-controlled system. A micro-reservoir is generally formed by matrix and reservoir dispersion [74].

5.6.3 Biodegradation of Synthetic Polymers and Their Breakdown Pathways

Abiotic degradation: The weather, aging, sunlight, soil submerging, and water are a few examples of external factors that might hasten the decaying process. As a result, polymers experience mechanical, chemical, thermal, and photodegradation, which turns them into complementary elements that enhance the biodegradation process [75]. Furthermore, temperature effects are involved in thermal factors related to the chemical constitution of polymers, including melting temperature (Tm), the temperature of glass transition (Tg), and the crystalline/amorphous ratio. Agrochemicals in the soil, O_2, H_2O, and atmospheric contaminants can all contribute to chemical breakdown. The most potent agent for abiotic chemical breakdown is O_2, which is used via the assault of covalent bonds to generate free radicals. The polymer materials' exposure to light causes photodegradation, a crucial abiotic factor affecting the biodegradation process. Norrish reactions—photoionization (Norrish I) and chain scission (Norrish II)—are used in their procedure [76].

Biodeterioration and Depolymerization: Microbial activity begins both within and on the material's surface as soon as the polymer is broken apart in Figure 5.3. Microorganisms can stimulate processes chemically, mechanically, and enzymatically. The polymer's chemistry defines its properties' microorganisms (lichenaceae, bacteria, protozoa, algae, fungus, and weather), and the environmental factors (temperature, humidity, and pollution). These microbes function as slime matter and glue layers by adhering to the polymer surface [77]. They infiltrate the porous structure of the polymer, changing the distribution and size of the

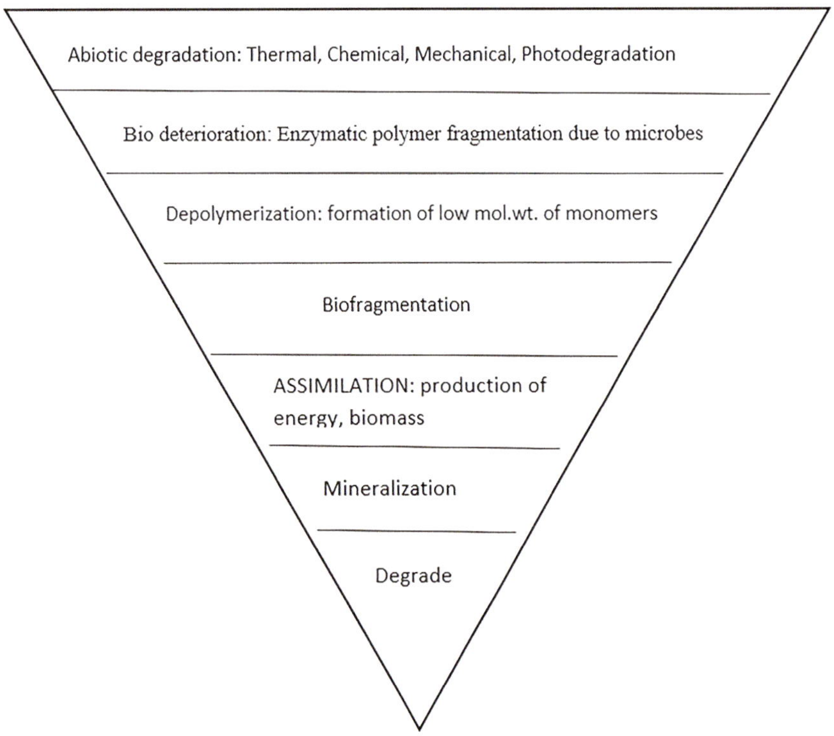

FIGURE 5.3 Breakdown pathways for synthetic polymers.

pores, resulting in mechanical and fracture damage [78]. The polymer components must be broken down into tiny, low-mol. Wt. pieces (oligomers and monomers) to do this. The cytoplasmic membrane and cell wall are incompatible with the high mol. wt. macromolecules, making this process essential. To appropriately rupture the polymer. H_2O_2 is preferred because it's a highly reactive and oxidative chemical produced by some soil decomposing agents, including brown rot fungal organisms. This molecule creates very reactive but nonspecific free radicals OH via the Fenton reaction when it comes into contact with ferrous atoms in an acidic environment. To guard against deterioration, fungi create small molecules that serve as carriers of free radicals. Chain fragmentation results from their diffusion through the polymer matrix [79].

Assimilation and Mineralization: This represents the last phase, and assimilation is coupled with a lack of established guidelines, thus the assimilation mechanisms can be confirmed by mineralization. During mineralization, the polymer fragments interact immediately with the surrounding microbial cells. Disintegration-generated monomers migrate throughout the cellular membrane and are metabolized to adenosine triphosphate (ATP) and other components needed for cell structure growth. The energy required for maintaining cellular activity were readily available through three processes: aerobic respiration, anaerobic respiration, and fermentation [80].

5.6.4 Factors Affecting the Biodegradation of Synthetic Polymers

Light, heat, moisture, chemicals, and biological attacks are some of the factors in the environment which trigger synthetic polymers to degrade. These conditions may also trigger physical and chemical changes in polymers, which can result in molecular chain breakdown [81]. Among the nonbiotic affects to take into account include radiation, degradation by heat, and chemical hydrolysis. By absorbing solar radiation, polymers begin to alter in various ways as a result of photolysis and photo-oxidation, The infrared spectrum (760–2500 nm) induces oxidation from heat, whereas the visible spectrum (400–700 nm) induces thermal destruction. UVA (315–400nm) and UVB (295–315nm) are accountable for the process of photodegradation. In addition to their molecular weight, the polymers have been reported to alter several physical and optical characteristics. These alterations include ductility, embrittlement, chalking, cracking, and color and these are few of the causes that make biodegradation possible.

5.7 ADVANCED SYNTHETIC POLYMERIC DRUG DELIVERY SYSTEM IN SKIN DISORDER

5.7.1 Polymer-Based Smart Drug Delivery Systems for Topical Applications

For topical applications, nanogels or hydrogels which are polymeric in nature are the system discussed widely. Hydrogels are also termed as hydrophilic gels with known pharmaceutical forms used for skin drug delivery so therefore when it is dispersed in a suitable solvent, gelling agent fuse to form a 3D network structure of colloid with limitation like fluid flow responsible for entrapment and mobilization of the solvent molecule as copolymer [82]. The primary structure of hydrogels is a network of polymeric channels, primarily linked by chemical or physical cross-linking of homo polymers or copolymers [83]. These channels expand when placed in an aqueous environment.

Polymeric nano systems present numerous advantages over free drug solutions due to their size range of 1–100 nm and physicochemical composition:

- The drug is protected from premature degradation.
- Premature interaction of the drug with the biological environment is prevented.
- They maximize drug absorption in some tissue, such as the skin.
- They are responsible for regulating pharmacokinetic and the distribution profile of drug tissue.
- They are responsible for solving problems related to formulation [84, 85].

Nanogels are multifaceted hydrogel polymers having a conformation of nanoscale [86]. These smart gels are developed through the integration of stimuli-responsive polymers. The pertaining activation will trigger the gels to alter configuration or morphology. This modification will be accomplished through an array of techniques, including fluctuations in temperature above or below the gel's acidic or basic groups that function on the chains of polymer, as well as a lower critical solution temperature (LCST), and thus, it is depicted in Figure 5.4.

5.7.2 Polymer

5.7.2.1 pH-Responsive Delivery System

pH variation of skin

Basic skin activities (hemostasis, SC integrity/cohesion, and exfoliation) depend primarily on management of lower skin pH (4–6), [87, 88] also the alteration of skin pH leads to disturbances in its microbiota and skin barrier system due to subsequent allergies and swelling, leading to situations like

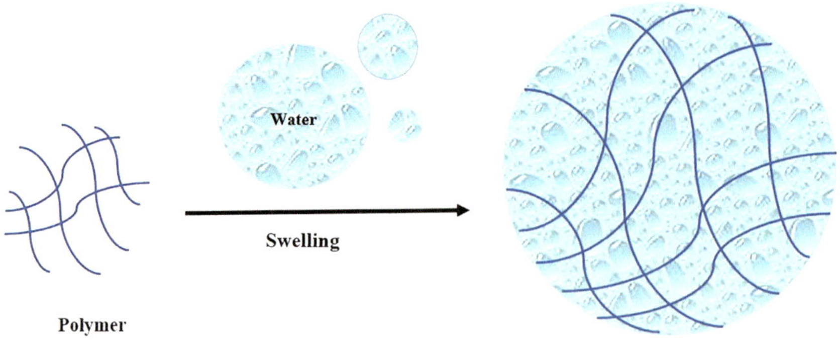

FIGURE 5.4 Diagrammatic representation of hydrophilic gel formation by swelling of polymer.

atopic dermatitis [89, 90–92], irritant contact dermatitis [93], candidiasis [94] and acne [95].

The four stages of the wound healing technique are characterized by variations in pH:

- Hemostasis
- Inflammation
- Proliferation
- Remodeling

As the inflammatory stage is replaced by the proliferative stage, the pH progressively elevates from 5.7–7.6 [96]

Systems centered around pH-responsive polymers have been explored to govern the release of drugs in the therapeutic region, as various skin conditions have been associated with pH disturbances in the skin.

5.7.2.1.1 pH-Responsive Mechanisms in Advanced Polymeric Systems

Functional grp-involved pH-responsive polymers can respond to the alteration in the surrounding ecosystem's pH by emitting and if not then absorbing a proton. pH-sensitive polymers are then categorized into two categories: anionic and cationic polymers [97–99]

The primary functional groups of anionic polymers are carboxylates, or carboxylic acid grps, which ionize at pH values higher than their pKa of around 4, whereas cationic polymers are associated with amine groups that ionize to NH_3^+ and form at pH values below their pKa of approximately 6.5. Additionally, functional groups ionize with the polymer's side chains, causing a shift in the alignment of the polymer as well as inflammation, distortion, or deterioration [100–102]. Rapid system swelling in the case of hydrogels and nanogels is caused by rapid ionization, which results in electrostatic repellent forces throughout the polymer lattice and so system's inflammation causes an increase in pore size, which quickens the rate at which cargo is released at the target site. This phenomenon is illustrated in Figure 5.5.

TABLE 5.4
Difference between Anionic and Cationic Polymer

Anionic Polymer	Cationic Polymer
Possess carboxylates that include carboxylic acid groups	Possess amine groups that include ionized to NH_3^+ form
Ionized at pH above their pKa ≈ 4	Ionized at pH below their pKa ≈ 6.5
Skin diseases are responsible for elevating pH (e.g., wounds, dermatitis) AniOnic pOlymer systems are more pertinent and being examined for cognitive tOpical therapy of skin disOrders in contrast to healthy skin	Cationic polymers (contain amine groups) are used to inhibit the active ingredients from emitting all their constituents in the dosage form immediately when the pH is neutral
Example: Poly (acrylic acid)	Example: Methacrylate chitosan (MAC) hydrogels

5.7.2.2 Thermoresponsive Delivery Systems

Implementation Techniques of a Thermoresponsive Mechanism for Drug Delivery

One of the biomedical techniques that have been examined most extensively is temperature-induced drug release, which has numerous uses in environments where temperature fluctuations occur naturally. There are several approaches available for cutaneous delivery:

- Difference in temperature between skin surface temperature (32°C) and its recipe, which is generally at room temperature [103, 104].
- Drug delivery via the skin's thermal variation (32°C–37°C). Typically, the technique has been modified to function with nanosystems. When the drugs eventually penetrate the deeper regions, the temperature of the SC (37°C) is also beneficial, as its exterior temperature (32°C) restricts immediate emission of drugs [105, 106–108].
- Temperature gradient between wounded and healthy skin enables the active ingredient to be applied to the wounded region, causing the active ingredient to be released [109–114].

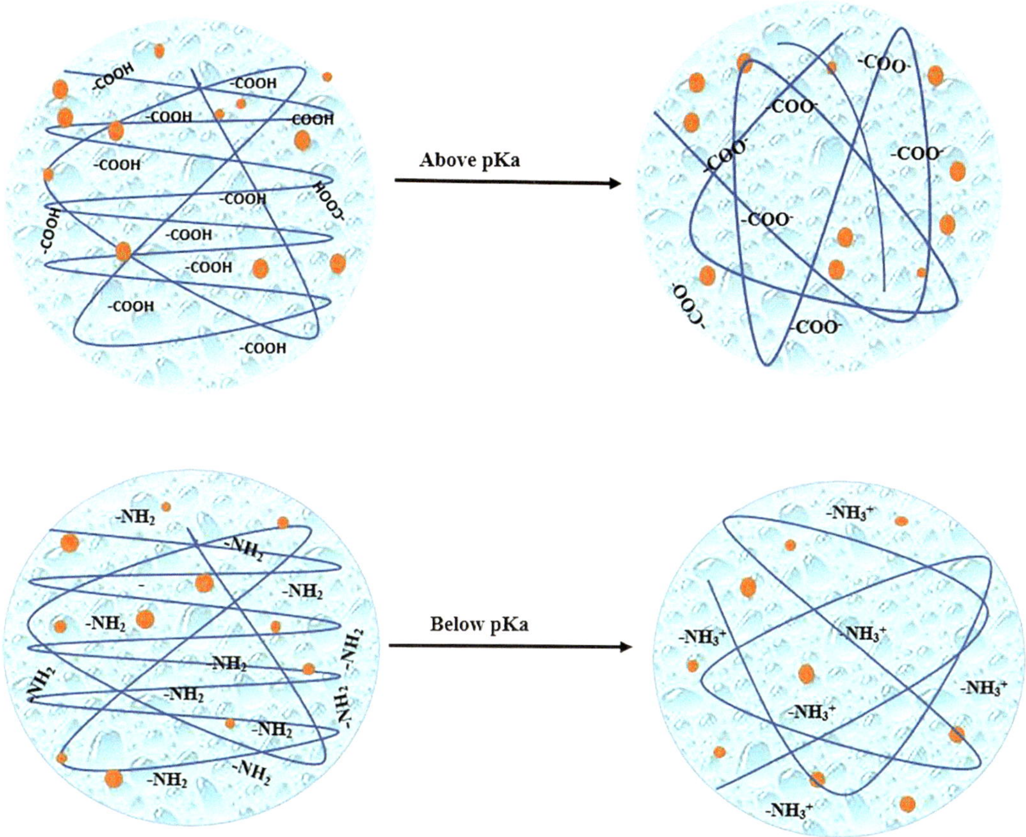

FIGURE 5.5 Diagrammatic representation of how a pH-responsive hydrogel reacts to changes in the pH of its surroundings (above: anionic polymers, below: cationic polymers) Difference between anionic and cationic polymer.

5.7.3 Other Stimuli-Responsive Delivery Systems

a) **Redox-Responsive Systems:** It primarily deals with the detrimental impacts of inflammation, which give rise to oxidants and reactive oxygen species (ROS), which in turn cause an oxidative imbalance and stress. In reaction to ongoing oxidative stress, ROS frequently cause redox adaptation, which increases the levels of glutathione (GSH) and other antioxidant molecules. Disulfide bonds were added to nanocarriers for smart drug delivery into cancer cells, which are shown to have large levels of GSH since they are sensitive to GSH [115–117]. Our group [118] developed redox-responsive nanoparticles for topical applications by including disulfide bonds. In short, three steps were involved in the synthesis of a combination of poly(lactide) (PLA) and redox-responsive poly (ethylene glycol)-block-poly(lactide) (PEGblock-PLA) with a disulfide connection.

b) ***Enzyme-Cleavable Systems*:** Organ-like skin with the activity of the enzyme (e.g., Cytochrome P_{450}). This is essential to take into consideration when formulating drugs that are used topically. Additionally, it regulates the organic ecosystem – which includes the metabolic microenvironment – for the release of active substances [119].

c) ***Electro-Sensitive Systems*:** The external stimulus is considered an electrical field that enables precise regulation of the electrical current's size, Electromagnetic waves and the phase in between the waves. Polyelectrolytes, or polymers with a significantly the core of the chain has an abundance of ionizable groups are used to create delivery systems that take advantage of these external stimuli. Electro-responsive hydrogels often expand or contract in response to an electric field; that feature makes them beneficial for drug delivery systems [120].

5.7.4 New Direction for the Advance of Skin Disorder Treatment

The recent advancement of the role of collagen in skin and wound sites enhances tissue biomechanical durability and pliability, and additionally functions as an organic scaffold for proliferative adhesion and expansion [121–125]. Biofilm-mediated MMP-2 amplification via microRNAs induces a collagenolytic habitat within the injury, substantially decreasing the collagen I/collagen III ratio and impairing the biomechanical features of the healed skin, potentially disposing of tissue susceptible to injury repetition [126].

a) **Role in Inflammation**: The inflammatory phase of the healing process involves hemostasis plus swelling associated with injury-induced clotting, leading to an accumulation of fibrin that blocks the initial hemorrhage. Collagen I and IV fragments serve as mediators of inflammation by recruiting neutrophils, stimulating phagocytosis and immunological feedback, and modifying gene expression [127, 128]. Inflammation is a vital stage in the normal process of healing wounds because it promotes the proliferation of fibroblasts, which generate collagen and ECM [129]. The rapid resolution of inflammation is similarly significant in appropriate wound recovery. Additionally, it is marked by an equilibrium of pro- and anti-inflammatory reactions [126]. Moreover, collagen is being proven to perform a key role in generating an anti-inflammatory, pro-angiogenic wound macrophage phenotype through microRNA signaling [130, 131].
b) **Role in Angiogenesis**: Angiogenesis, an essential aspect of both biological (development, wound healing) and pathological (cancer) procedures and thus it is precisely regulated by the balanced activity of stimulators and inhibitors. ECM remodeling is necessary for vascular formation, and collagens play a significant part in that procedure [132, 133–136]. Collagen I is currently reported to effectively induce angiogenesis in vitro and in vivo through activating certain integrin receptors. Specifically, the C-propeptide component of collagen I attract endothelial cells, potentially triggering angiogenesis in the healing wound [137]. On the other hand, proteolytic collagen components of collagen IV and XVIII (such as endostatin, arresten, canstatin, and tumstatin) have anti-angiogenic properties [128, 129]; likewise, the fragments lower endothelial cell growth and motility along with the activation of apoptosis. These components are highly significant in elevating angiogenesis in numerous clinical circumstances [133, 137, 138].
c) **Role in ECM Remodeling**: Collagens are structural elements of the ECM, which assist retain skin elasticity, anchor growth factors, and modulate cell adhesion and signaling across cells and the ECM [124]. As the tissue surrounding the wound transforms all throughout time, the adult wound heals with the appearance of a 'normal' scar. The density, fiber dimension, and alignment of collagen filaments appear to be the most significant variance among scar and unwounded skin. Tissue with scarring has disproportionately significant amounts of collagen I and III, fibronectin, and laminin proteins [139]. Collagen fiber position in scars (normotrophic, hypertrophic, and keloid) is transverse to the epithelial surface, contrary to normal skin, where the fibers generate a three-dimensional basketweave-like network [140]. These scars fluctuate because of morphology and chemistry. Hypertrophic scars consist of finer collagen bundles than keloid or neurotrophic scars [141–142].

5.8 CONCLUSION

The development and utilization of synthetic material-based drug delivery systems for skin disorders represent a promising avenue in modern pharmaceutical research. These innovative DDS offer several advantages, including controlled drug release, enhanced therapeutic efficacy, reduced side effects, and improved patient compliance. The versatility of synthetic materials allows for tailoring DDS properties, such as biocompatibility, biodegradability, and targeting capabilities, catering to specific requirements of various skin disorders. Furthermore, their tunable characteristics enable the incorporation of different drugs, ranging from small molecules to biologics, facilitating personalized treatment strategies. Despite the remarkable progress made in this field, challenges such as ensuring long-term stability, achieving optimal drug loading and release kinetics, and obtaining regulatory approval remain. Addressing these hurdles through continued research, collaboration between interdisciplinary fields, and innovative technological advancements will further advance the efficacy and safety of synthetic material-based DDS for treating skin disorders. In the future, the continued exploration and refinement of these DDS hold tremendous potential to revolutionize dermatological therapies, offering patients more effective, targeted, and personalized treatment options, ultimately improving the quality of life for individuals affected by various skin conditions.

5.9 FUTURE PERSPECTIVE

The future of synthetic polymeric material-based drug delivery systems for skin disorders holds significant promise due to ongoing advancements in materials science, nanotechnology, and biomedical engineering. Nanotechnology will play a crucial role in the evolution of polymeric drug delivery systems. Nanocarriers, such as nanoparticles, micelles, and liposomes, can encapsulate drugs, protect them from degradation, and facilitate their penetration into deeper skin layers. Future systems will likely focus on achieving more precise targeting and controlled release of therapeutics. By utilizing smart polymers that respond to specific stimuli (e.g., pH, temperature, or enzymes), these systems can deliver drugs directly to the affected areas while minimizing systemic side effects. This precision can enhance treatment efficacy and reduce the required dosage. Synthetic polymeric material-based drug delivery systems represent a promising frontier in the treatment of skin disorders. These systems offer the potential for targeted, controlled, and sustained release of therapeutic agents, minimizing side effects and enhancing treatment efficacy. The integration of advanced materials, nanotechnology, and personalized medicine approaches will likely drive further innovations, making treatments more effective and patient-friendly. As research progresses and these technologies become more refined and widely available, they could significantly improve the management and outcomes of various skin conditions, offering new hope to patients and health-care providers alike.

REFERENCES

1. Takeuchi I, Kobayashi S, Hida Y, Makino K. Estradiol-loaded PLGA nanoparticles for improving low bone mineral density of cancellous bone caused by osteoporosis: Application of enhanced charged nanoparticles with iontophoresis. *Colloids and Surfaces B: Biointerfaces*. 2017 Jul 1; 155:35–40.
2. Jaques JA, Rezer JF, Ruchel JB, Souza VD, Pinheiro KD, Schlemmer KB, Schlemmer JB, Bertoldo TM, Martins NM, Bertoncheli CD, Fontana MC. An experimental model of contact dermatitis: evaluation of the oxidative profile of Wistar rats treated with free and nanoencapsulated clobetasol. *Redox Report*. 2012 Sep 1;17(5):206–13.
3. Jung SM, Yoon GH, Lee HC, Jung MH, Yu SI, Yeon SJ, Min SK, Kwon YS, Hwang JH, Shin HS. Thermodynamic insights and conceptual design of skin-sensitive chitosan coated ceramide/PLGA nanodrug for regeneration of stratum corneum on atopic dermatitis. *Scientific Reports*. 2015 Dec 15;5(1):18089.
4. Auxenfans C, Builles N, Andre V, Lequeux C, Fievet A, Rose S, Braye FM, Fradette J, Janin-Manificat H, Nataf S, Burillon C. Porous matrix and primary-cell culture: a shared concept for skin and cornea tissue engineering. *Pathologie-biologie*. 2008 Jul 3;57(4):290–8.
5. Aranaz I, Mengíbar M, Harris R, Paños I, Miralles B, Acosta N, Galed G, Heras Á. Functional characterization of chitin and chitosan. *Current Chemical Biology*. 2009 May 1;3(2):203–30.
6. Ghormade V, Pathan EK, Deshpande MV. Can fungi compete with marine sources for chitosan production? *International Journal of Biological Macromolecules*. 2017 Nov 1; 104:1415–21.
7. Rinaudc M, Pavlov G, Desbrieres J. Solubilization of chitosan in strong acid medium. *International Journal of Polymer Analysis and Characterization*. 1999 Jun 1;5(3):267–76.
8. Joseph SM, Krishnamoorthy S, Paranthaman R, Moses JA, Anandharamakrishnan C. A review on source-specific chemistry, functionality, and applications of chitin and chitosan. *Carbohydrate Polymer Technologies and Applications*. 2021 Dec 25;2:100036.
9. Milas M, Rinaudo M, Roure I, Al-Assaf S, Phillips GO, Williams PA. Comparative rheological behavior of hyaluronan from bacterial and animal sources with cross-linked hyaluronan (hylan) in aqueous solution. *Biopolymers: Original Research on Biomolecules*. 2001 Oct 5;59(4):191–204.
10. Lapcık Jr L, Lapcık L, De Smedt S, Demeester J, Chabrecek P. Hyaluronan: preparation, structure, properties, and applications. *Chemical Reviews* 1998;98(8):2663–84.
11. Laurent TC (ed): *The Chemistry, Biology and Medical Applications of Hyaluronan and Its Derivates*. Wenner-Gren International Series 72. London, Portland Press, 1998.
12. Peters Jr T. *All about albumin: biochemistry, genetics, and medical applications*. Academic press; 1995 Dec 21.
13. Kouoh F, Gressier B, Luyckx M, Brunet C, Dine T, Cazin M, Cazin JC. Antioxidant properties of albumin: effect on oxidative metabolism of human neutrophil granulocytes. *Il Farmaco*. 1999 Oct 30;54(10):695–9.
14. Quinlan G J, Martin G S, Evans T W. Albumin: biochemical properties and therapeutic potential [J]. *Hepatology*, 2010, 41(6): 1211–9.
15. Lee E, Eom JE, Jeon KH, Kim TH, Kim E, Jhon GJ, Kwon Y. Evaluation of albumin structural modifications through cobalt-albumin binding (CAB) assay. *Journal of Pharmaceutical and Biomedical Analysis*. 2014 Mar 25;91:17–23.
16. Kohli N, Sharma V, Brown SJ, García-Gareta E. Synthetic polymers for skin biomaterials. In *Biomaterials for skin repair and regeneration* 2019 Jan 1 (pp. 125–149). Woodhead Publishing.
17. Grombe R, Gouzy MF, Maitz MF, Freundenberg U, Zschoche S, Simon F, Pompe T, Sperling C, Werner C. Sulfated glycopolymer thin films—Preparation, characterization, and biological activity. *Macromolecular Bioscience*. 2007 Feb 12;7(2):195–200.
18. Gentile P, Chiono V, Carmagnola I, Hatton PV. An overview of poly (lactic-co-glycolic) acid (PLGA)-based biomaterials for bone tissue engineering. *International Journal of Molecular Sciences*. 2014 Feb 28;15(3):3640–59.
19. Gunatillake PA, Adhikari R, Gadegaard N. Biodegradable synthetic polymers for tissue engineering. *European Cells & Materials* 2003 May 20;5(1):1–6.
20. Saghazadeh S, Rinoldi C, Schot M, Kashaf SS, Sharifi F, Jalilian E, Nuutila K, Giatsidis G, Mostafalu P, Derakhshandeh H, Yue K. Drug delivery systems and materials for wound healing applications. *Advanced Drug Delivery Reviews*. 2018 Mar 1;127:138–66.
21. Blasi P. Poly (lactic acid)/poly (lactic-co-glycolic acid)-based microparticles: An overview. *Journal of Pharmaceutical Investigation*. 2019 Jul 1;49:337–46.
22. Carrasco F, Pagès P, Gámez-Pérez J, Santana OO, Maspoch ML. Processing of poly (lactic acid): Characterization of chemical structure, thermal stability and mechanical properties. *Polymer Degradation and Stability*. 2010 Feb 1;95(2):116–25.
23. Sartore L, Inverardi N, Pandini S, Bignotti F, Chiellini F. PLA/PCL-based foams as scaffolds for tissue engineering applications. *Materials Today: Proceedings*. 2019 Jan 1;7:410–7.
24. Hadgraft J, Guy RH. Feasibility assessment in topical and transdermal delivery: mathematical models and in vitro studies. In Transdermal Drug Delivery Systems 2002 Oct 29 (pp. 1–23). CRC Press.
25. Scheuplein R. 1. and Blank, IH, Permeability of the skin. *Physiological Reviews* 1971;51:702–47
26. Zhang Q, Grice JE, Li P, Jepps OG, Wang GJ, Roberts MS. Skin solubility determines maximum transepidermal flux for similar size molecules. *Pharmaceutical Research*. 2009 Aug;26:1974–85.
27. Moser K, Kriwet K, Naik A, Kalia YN, Guy RH. Passive skin penetration enhancement and its quantification in vitro. *European Journal of Pharmaceutics and Biopharmaceutics*. 2001 Sep 1;52(2):103–12.
28. Williams AC, Barry BW. Penetration enhancers. *Advanced Drug Delivery Reviews*. 2012 Dec 1;64:128–37.
29. Cormier M, Johnson B, Ameri M, Nyam K, Libiran L, Zhang DD, Daddona P. Transdermal delivery of desmopressin using a coated microneedle array patch system. *Journal of Controlled Release*. 2004 Jul 7;97(3):503–11.
30. Akomeah FK, Martin GP, Muddle AG, Brown MB. Effect of abrasion induced by a rotating brush on the skin permeation of solutes with varying physicochemical properties. *European Journal of Pharmaceutics and Biopharmaceutics*. 2008 Mar 1;68(3):724–34.

31. Alvarez-Roman R, Merino G, Kalia YN, Naik A, Guy RH. Skin permeability enhancement by low frequency sonophoresis: lipid extraction and transport pathways. *Journal of Pharmaceutical Sciences* 2003, 92:1138–1146.
32. Mori K, Hasegawa T, Sato S, Sugibayashi K. Effect of electric field on the enhanced skin permeation of drugs by electroporation. *Journal of Controlled Release*. 2003 Jun 24;90(2):171–9.
33. Harvey AJ, Kaestner SA, Sutter DE, Harvey NG, Mikszta JA, Pettis RJ. Microneedle-based intradermal delivery enables rapid lymphatic uptake and distribution of protein drugs. *Pharmaceutical Research*. 2011 Jan;28:107–16.
34. Fukushima K, Ise A, Morita H, Hasegawa R, Ito Y, Sugioka N, Takada K. Two-layered dissolving microneedles for percutaneous delivery of peptide/protein drugs in rats. *Pharmaceutical Research* 2011, 28:7–21
35. Coulman SA, Barrow D, Anstey A, Gateley C, Morrissey A, Wilke N, Allender C, Brain K, Birchall JC. Minimally invasive cutaneous delivery of macromolecules and plasmid DNA via microneedles. *Current Drug Delivery*. 2006 Jan 1;3(1):65–75.
36. Quan FS, Kim YC, Compans RW, Prausnitz MR, Kang SM. Dose sparing enabled by skin immunization with influenza virus-like particle vaccine using microneedles. *Journal of Controlled Release*. 2010 Nov 1;147(3):326–32.
37. Weller C, Sussman G. Wound dressings update. *Journal of Pharmacy Practice and Research*. 2006 Dec;36(4):318–24.
38. Weller C, Sussman G. Wound dressings update. *Journal of Pharmacy Practice and Research*. 2006 Dec;36(4):318–24.
39. Mustoe TA, Cooter RD, Gold MH, Hobbs FR, Ramelet AA, Shakespeare PG, Stella M, Téot L, Wood FM, Ziegler UE, International Advisory Panel on Scar Management. *International Clinical Recommendations On Scar Management. Plastic and Reconstructive Surgery*. 2002 Aug 1;110(2):560–71.
40. Rippon M, Davies P, White R, Bosanquet N. Cost implications of using an atraumatic dressing in the treatment of acute wounds. *Journal of Wound Care*. 2008 May;17(5):224–7.
41. Singh B, Pal L. Sterculia crosslinked PVA and PVA-poly (AAm) hydrogel wound dressings for slow drug delivery: mechanical, mucoadhesive, biocompatible and permeability properties. *Journal of the Mechanical Behavior of Biomedical Materials*. 2012 May 1;9:9–21.
42. Liang HY, Lee WK, Hsu JT, Shih JY, Ma TL, Vo TTT, Lee CW, Cheng MT, Lee IT. Polycaprolactone in bone tissue engineering: A comprehensive review of innovations in scaffold fabrication and surface modifications. *Journal of Functional Biomaterials*, 2024 Aug 24;15(9):243.
43. DeStefano V, Khan S, Tabada A, Applications of PLA in modern medicine, *Engineered Regeneration*, 2020, 1:76–87.
44. Zhuo F, Abourehab MA, Hussain Z. Hyaluronic acid decorated tacrolimus-loaded nanoparticles: Efficient approach to maximize dermal targeting and anti-dermatitis efficacy. *Carbohydrate Polymers*. 2018 Oct 1;197:478–89.
45. Razzak MT, Dewi SP, Lely H, Taty E. The characterization of dressing component materials and radiation formation of PVA–PVP hydrogel. *Radiation Physics and Chemistry*. 1999 Jun 11;55(2):153–65.
46. Li A, Dearman BL, Crompton KE, Moore TG, Greenwood JE. Evaluation of a novel biodegradable polymer for the generation of a dermal matrix. *Journal of Burn Care & Research*. 2009 Jul 1;30(4):717–28.
47. Yoon SD, Kwon YS, Lee KS. Biodegradation and biocompatibility of poly L-lactic acid implantable mesh. *International Neurourology Journal*. 2017 Apr;21(Suppl 1):S48.
48. Cui W, Zhu X, Yang Y, Li X, Jin Y. Evaluation of electrospun fibrous scaffolds of poly (dl-lactide) and poly (ethylene glycol) for skin tissue engineering. *Materials Science and Engineering: C*. 2009 Aug 1;29(6):1869–76.
49. Jouybar A, Seyedjafari E, Ardeshirylajimi A, Zandi-Karimi A, Feizi N, Khani MM, Pousti I. Enhanced skin regeneration by herbal extract-coated poly-L-lactic acid nanofibrous scaffold. *Artificial Organs*. 2017 Nov;41(11):E296–307.
50. Perisic T, Zhang Z, Foehr P, Hopfner U, Klutz K, Burgkart RH, Slobodianski A, Goeldner M, Machens HG, Schilling AF. Biodegradable poly (lactic acid-co-glycolic acid) scaffolds as carriers for genetically-modified fibroblasts. *PLoS One*. 2017 Apr 5;12(4):e0174860.
51. Batheja P, Sheihet L, Kohn J, Singer AJ, Michniak-Kohn B. Topical drug delivery by a polymeric nanosphere gel: formulation optimization and in vitro and in vivo skin distribution studies. *Journal of Controlled Release*. 2011 Jan 20;149(2):159–67.
52. Alvarez-Román R, Naik A, Kalia YN, Guy RH, Fessi H. Skin penetration and distribution of polymeric nanoparticles. *Journal of Controlled Release*. 2004 Sep 14;99(1):53–62.
53. Todo H, Kimura E, Yasuno H, Tokudome Y, Hashimoto F, Ikarashi Y, Sugibayashi K. Permeation pathway of macromolecules and nanospheres through skin. *Biological and Pharmaceutical Bulletin*. 2010 Aug 1;33(8):1394–9.
54. Tian F, Conde J, Bao C, Chen Y, Curtin J, Cui D. Gold nanostars for efficient in vitro and in vivo real-time SERS detection and drug delivery via plasmonic-tunable Raman/FTIR imaging. *Biomaterials*. 2016 Nov 1;106:87–97.
55. Guterres SS, Alves MP, Pohlmann AR. Polymeric nanoparticles, nanospheres and nanocapsules, for cutaneous applications. *Drug Target Insights*. 2007 Jan;2:11773 9280700200002.
56. Sheihet L, Chandra P, Batheja P, Devore D, Kohn J, Michniak B. Tyrosine-derived nanospheres for enhanced topical skin penetration. *International Journal of Pharmaceutics*. 2008 Feb 28;350(1–2):312–9.
57. Zhang Z, Tsai PC, Ramezanli T, Michniak-Kohn BB. Polymeric nanoparticles-based topical delivery systems for the treatment of dermatological diseases. *Wiley Interdisciplinary Reviews: Nanomedicine and Nanobiotechnology*. 2013 May;5(3):205–18.
58. Ellis LV, Cain MP, Hutchison V, Flodby P, Crandall ED, Borok Z, Zhou B, Ostrin EJ, Wythe JD, Chen J. Epithelial Vegfa specifies a distinct endothelial population in the mouse lung. *Developmental Cell*. 2020 Mar 9;52(5):617–30.
59. Peterson B, Weyers M, Steenekamp JH, Steyn JD, Gouws C, Hamman JH. Drug bioavailability enhancing agents of natural origin (bioenhancers) that modulate drug membrane permeation and pre-systemic metabolism. *Pharmaceutics*. 2019 Jan 16;11(1):33.
60. Pegoraro M. Modification of polymers for the preparation of semipermeable membranes. In *Chemical Transformations of Polymers* 1972 Jan 1 (pp. 199–215). Butterworth-Heinemann.
61. Sabbagh F, Muhamad II, Nazari Z, Mobini P, Khatir NM. Investigation of acyclovir-loaded, acrylamide-based hydrogels for potential use as vaginal ring. *Materials Today Communications*. 2018 Sep 1;16:274–80.
62. Yahya I, Atif R, Ahmed L, Eldeen TS, Omara A, Eltayeb M. Mathematical modeling of diffusion controlled drug release profiles from nanoparticles. *International Journal of Research and Scientific Innovation (IJRSI)*. 2019 May;6(5):287–91.

63. Nauman EB, Patel K, Karande P. Design of optimized diffusion-controlled transdermal drug delivery systems. *Drug Development and Industrial Pharmacy*. 2011 Jan 1;37(1):93–102.
64. Yang WW, Pierstorff E. Reservoir-based polymer drug delivery systems. *Journal of Laboratory Automation*. 2012 Feb;17(1):50–8.
65. Kim B, Seong KY, You I, Selvaraj V, Yim SG, O'Cearbhaill ED, Jeong U, Yang SY. Touch-actuated transdermal delivery patch for quantitative skin permeation control. *Sensors and Actuators B: Chemical*. 2018 Mar 1;256:18–26.
66. Jakubowicz I, Yarahmadi N, Petersen H. Evaluation of the rate of abiotic degradation of biodegradable polyethylene in various environments. *Polymer Degradation and Stability*. 2006 Jul 1;91(7):1556–62.
67. Siracusa V. Microbial degradation of synthetic biopolymers waste. *Polymers*. 2019 Jun 20;11(6):1066.
68. Briassoulis D. The effects of tensile stress and the agrochemical Vapam on the ageing of low density polyethylene (LDPE) agricultural films. Part I. Mechanical behaviour. *Polymer Degradation and Stability*. 2005 Jun 1;88(3): 489–503.
69. Nakamura H, Nakamura T, Noguchi T, Imagawa K. Photodegradation of PEEK sheets under tensile stress. *Polymer Degradation and Stability*. 2006 Apr 1;91(4):740–6.
70. Hueck HJ. The biodeterioration of materials—an appraisal. *International Biodeterioration & Biodegradation*. 2001 Jan 1;48(1–4):5–11.
71. Walsh JH. Ecological considerations of biodeterioration. *International Biodeterioration & Biodegradation*. 2001 Jan 1;48(1–4):16–25.
72. Cappitelli F, Principi P, Sorlini C. Biodeterioration of modern materials in contemporary collections: can biotechnology help?. *Trends in Biotechnology*. 2006 Aug 1;24(8):350–4.
73. Mohanan N, Montazer Z, Sharma PK, Levin DB. Microbial and enzymatic degradation of synthetic plastics. *Frontiers in Microbiology*. 2020 Nov 26;11:580709.
74. Siracusa V. Microbial degradation of synthetic biopolymers waste. *Polymers*. 2019 Jun 20;11(6):1066.
75. Lugauskas A, Levinskait L, Pečiulyt D. Micromycetes as deterioration agents of polymeric materials. *International Biodeterioration & Biodegradation*. 2003 Dec 1;52(4):233–42.
76. Warscheid T, Braams J. Biodeterioration of stone: a review. *International Biodeterioration & Biodegradation*. 2000 Dec 1;46(4):343–68.
77. Lucas N, Bienaime C, Belloy C, Queneudec M, Silvestre F, Nava-Saucedo JE. Polymer biodegradation: Mechanisms and estimation techniques–A review. *Chemosphere*. 2008 Sep 1;73(4):429–42.
78. van der Zee M. Biodegradability of polymers: mechanisms and evaluation methods. In *Handbook of biodegradable polymers* 2005. Rapra Technology.
79. Sood N, Bhardwaj A, Mehta S, Mehta A. Stimuli-responsive hydrogels in drug delivery and tissue engineering. *Drug delivery*. 2016 Mar 23;23(3):748–70.
80. Mota AH, Rijo P, Molpeceres J, Reis CP. Broad overview of engineering of functional nanosystems for skin delivery. *International Journal of Pharmaceutics*. 2017 Nov 5;532(2):710–28.
81. Sood N, Bhardwaj A, Mehta S, Mehta A. Stimuli-responsive hydrogels in drug delivery and tissue engineering. *Drug Delivery*. 2016 Mar 23;23(3):748–70.
82. Alexis F, Pridgen EM, Langer R, Farokhzad OC. Nanoparticle technologies for cancer therapy. *Drug Delivery*. 2010: 45: 55–86.
83. Soni KS, Desale SS, Bronich TK. Nanogels: An overview of properties, biomedical applications and obstacles to clinical translation. *Journal of Controlled Release*. 2016 Oct 28;240:109–26.
84. Van Gheluwe L, Chourpa I, Gaigne C, Munnier E. Polymer-based smart drug delivery systems for skin application and demonstration of stimuli-responsiveness. *Polymers*. 2021 Apr 15;13(8):1285.
85. Lambers H, Piessens S, Bloem A, Pronk H, Finkel P. Natural skin surface pH is on average below 5, which is beneficial for its resident flora. *International Journal of Cosmetic Science*. 2006 Oct;28(5):359–70.
86. Rippke F, Schreiner V, Doering T, Maibach HI. Stratum corneum pH in atopic dermatitis: impact on skin barrier function and colonization with Staphylococcus aureus. *American Journal of Clinical Dermatology*. 2004 Aug;5:217–23.
87. Van Gheluwe L, Chourpa I, Gaigne C, Munnier E. Polymer-based smart drug delivery systems for skin application and demonstration of stimuli-responsiveness. *Polymers*. 2021 Apr 15;13(8):1285.
88. Eberlein-König B, Schäfer T, Huss-Marp J, Darsow U, Möhrenschlager M, Herbert O, Abeck D, Krämer U, Behrendt H, Ring J. Skin surface pH, stratum corneum hydration, trans-epidermal water loss and skin roughness related to atopic eczema and skin dryness in a population of primary school children. *Acta dermato-venereologica*. 2000 May 1;80(3):188–91.
89. Sparavigna A, Setaro M, Gualandri V. Cutaneous pH in children affected by atopic dermatitis and in healthy children: a multicenter study. *Skin Research and Technology*. 1999 Nov;5(4):221–7.
90. Seidenari S, Francomano M, Mantovani L. Baseline biophysical parameters in subjects with sensitive skin. *Contact Dermatitis*. 1998 Jun;38(6):311–5.
91. Runeman B, Faergemann J, Larkö O. Experimental Candida albicans lesions in healthy humans: dependence on skin pH. *Acta dermato-venereologica*. 2000 Nov 1;80(6):421–4.
92. Surber C, Humbert P, Abels C, Maibach H. The acid mantle: a myth or an essential part of skin health. *pH of the Skin: Issues and Challenges*. 2018;54:1–10.
93. Schneider LA, Korber A, Grabbe S, Dissemond J. Influence of pH on wound-healing: a new perspective for wound-therapy?. *Archives of Dermatological Research*. 2007 Feb;298(9):413–20.
94. Zhu L, Bratlie KM. pH sensitive methacrylated chitosan hydrogels with tunable physical and chemical properties. *Biochemical Engineering Journal*. 2018 Apr 15;132:38–46.
95. Vuković JS, Perić-Grujić AA, Mitić-Ćulafić DS, Božić Nedeljković BD, Tomić SL. Antibacterial activity of pH-sensitive silver (I)/poly (2-hydroxyethyl acrylate/itaconic acid) hydrogels. *Macromolecular Research*. 2020 Apr;28(4):382–9.
96. Dong P, Sahle FF, Lohan SB, Saeidpour S, Albrecht S, Teutloff C, Bodmeier R, Unbehauen M, Wolff C, Haag R, Lademann J. pH-sensitive Eudragit® L 100 nanoparticles promote cutaneous penetration and drug release on the skin. *Journal of Controlled Release*. 2019 Feb 10;295:214–22.
97. Quattrone A, Czajka A, Sibilla S. Thermosensitive hydrogel mask significantly improves skin moisture and skin tone; bilateral clinical trial. *Cosmetics*. 2017 Jun 1;4(2):17.

98. Zhu L, Bratlie KM. pH sensitive methacrylated chitosan hydrogels with tunable physical and chemical properties. *Biochemical Engineering Journal*. 2018 Apr 15;132:38–46.
99. Vuković JS, Perić-Grujić AA, Mitić-Ćulafić DS, Božić Nedeljković BD, Tomić SL. Antibacterial activity of pH-sensitive silver (I)/poly (2-hydroxyethyl acrylate/itaconic acid) hydrogels. *Macromolecular Research*. 2020 Apr;28(4):382–9.
100. Dong P, Sahle FF, Lohan SB, Saeidpour S, Albrecht S, Teutloff C, Bodmeier R, Unbehauen M, Wolff C, Haag R, Lademann J. pH-sensitive Eudragit® L 100 nanoparticles promote cutaneous penetration and drug release on the skin. *Journal of Controlled Release*. 2019 Feb 10;295:214–22.
101. Rozman B, Zvonar A, Falson F, Gasperlin M. Temperature-sensitive microemulsion gel: an effective topical delivery system for simultaneous delivery of vitamins C and E. *AAPS PharmSciTech*. 2009 Mar;10:54–61.
102. Yamazaki N, Sugimoto T, Fukushima M, Teranishi R, Kotaka A, Shinde C, Kumei T, Sumida Y, Munekata Y, Maruyama KI, Yuba E. Dual-stimuli responsive liposomes using pH- and temperature-sensitive polymers for controlled transdermal delivery. *Polymer Chemistry*. 2017;8(9):1507–18.
103. Giulbudagian M, Hönzke S, Bergueiro J, Işık D, Schumacher F, Saeidpour S, Lohan SB, Meinke MC, Teutloff C, Schäfer-Korting M, Yealland G. Enhanced topical delivery of dexamethasone by β-cyclodextrin decorated thermoresponsive nanogels. *Nanoscale*. 2018;10(1):469–79.
104. Kang JH, Chon J, Kim YI, Lee HJ, Oh DW, Lee HG, Han CS, Kim DW, Park CW. Preparation and evaluation of tacrolimus-loaded thermosensitive solid lipid nanoparticles for improved dermal distribution. *International Journal of Nanomedicine*. 2019 Jul 18:5381–96.
105. Witting M, Molina M, Obst K, Plank R, Eckl KM, Hennies HC, Calderón M, Frieß W, Hedtrich S. Thermosensitive dendritic polyglycerol-based nanogels for cutaneous delivery of biomacromolecules. *Nanomedicine: Nanotechnology, Biology and Medicine*. 2015 Jul 1;11(5):1179–87.
106. Asadian-Birjand M, Bergueiro J, Rancan F, Cuggino JC, Mutihac RC, Achazi K, Dernedde J, Blume-Peytayi U, Vogt A, Calderón M. Engineering thermoresponsive polyether-based nanogels for temperature dependent skin penetration. *Polymer Chemistry*. 2015;6(32):5827–31.
107. Ferreira NN, Ferreira LM, Cardoso VM, Boni FI, Souza AL, Gremião MP. Recent advances in smart hydrogels for biomedical applications: From self-assembly to functional approaches. *European Polymer Journal*. 2018 Feb 1;99:117–33.
108. Gandhi A, Paul A, Sen SO, Sen KK. Studies on thermoresponsive polymers: Phase behaviour, drug delivery and biomedical applications. *Asian Journal of Pharmaceutical Sciences*. 2015 Apr 1;10(2):99–107.
109. Mohapatra S, Ranjan S, Dasgupta N, Kumar R, Thomas S, editors. *Applications of targeted nano drugs and delivery systems: nanoscience and nanotechnology in drug delivery*. Elsevier; 2018 Oct 5.
110. Spěváček J. NMR investigations of phase transition in aqueous polymer solutions and gels. *Current Opinion in Colloid & Interface Science*. 2009 Jun 1;14(3):184–91.
111. Spěváček J, Konefał R, Dybal J, Čadová E, Kovářová J. Thermoresponsive behavior of block copolymers of PEO and PNIPAm with different architecture in aqueous solutions: A study by NMR, FTIR, DSC and quantum-chemical calculations. *European Polymer Journal*. 2017 Sep 1;94:471–83.
112. Kuppusamy P, Li H, Ilangovan G, Cardounel AJ, Zweier JL, Yamada K, Krishna MC, Mitchell JB. Noninvasive imaging of tumor redox status and its modification by tissue glutathione levels. *Cancer Research*. 2002 Jan 1;62(1):307–12.
113. Rao NV, Ko H, Lee J, Park JH. Recent progress and advances in stimuli-responsive polymers for cancer therapy. *Frontiers in Bioengineering and Biotechnology*. 2018 Aug 13;6:110.
114. Wen H, Li Y. Redox sensitive nanoparticles with disulfide bond linked sheddable shell for intracellular drug delivery. *Medicinal Chemistry* 2014;4(11):748–55.
115. Van Gheluwe L, Buchy E, Chourpa I, Munnier E. Three-Step Synthesis of a Redox-Responsive Blend of PEG–block–PLA and PLA and Application to the Nanoencapsulation of Retinol. *Polymers*. 2020 Oct 14;12(10):2350.
116. Kim SE, Lee PW, Pokorski JK. Biologically triggered delivery of EGF from polymer fiber patches. *ACS Macro Letters*. 2017 Jun 20;6(6):593–7.
117. Bawa P, Pillay V, Choonara YE, Du Toit LC. Stimuli-responsive polymers and their applications in drug delivery. *Biomedical Materials*. 2009 Mar 5;4(2):022001.
118. Im JS, Bai BC, Lee YS. The effect of carbon nanotubes on drug delivery in an electro-sensitive transdermal drug delivery system. *Biomaterials*. 2010 Feb 1;31(6):1414–9.
119. Schultz G, Chin G, Moldawer L, Diegelmann R. *Ch23 Mechanisms of Vascular Disease*. University of Adelaide Press; 2011.
120. Sorushanova A, Delgado LM, Wu Z, Shologu N, Kshirsagar A, Raghunath R, Mullen AM, Bayon Y, Pandit A, Raghunath M, Zeugolis DI. The collagen suprafamily: from biosynthesis to advanced biomaterial development. *Advanced Materials*. 2019 Jan;31(1):1801651.
121. Guo SA, DiPietro LA. Factors affecting wound healing. *Journal of Dental Research*. 2010 Mar;89(3):219–29.
122. Gurtner GC, Werner S, Barrandon Y, Longaker MT. Wound repair and regeneration. *Nature*. 2008 May 15;453(7193):314–21.
123. Roy S, Santra S, Das A, Dixith S, Sinha M, Ghatak S, Ghosh N, Banerjee P, Khanna S, Mathew-Steiner S, Ghatak PD. Staphylococcus aureus biofilm infection compromises wound healing by causing deficiencies in granulation tissue collagen. *Annals of Surgery*. 2020 Jun;271(6):1174.
124. Ricard-Blum S. The collagen family. *Cold Spring Harbor Perspectives in Biology*. 2011 Jan 1;3(1):a004978.
125. Ricard-Blum S, Ballut L. Matricryptins derived from collagens and proteoglycans. *Frontiers in Bioscience* 2011 Jan 1;16(1):674–97.
126. Rosique RG, Rosique MJ, Farina Junior JA. Curbing inflammation in skin wound healing: a review. *International Journal of Inflammation*. 2015; 2015: 316235.
127. El Masry MS, Chaffee S, Ghatak PD, Mathew-Steiner SS, Das A, Higuita-Castro N, Roy S, Anani RA, Sen CK. Stabilized collagen matrix dressing improves wound macrophage function and epithelialization. *The FASEB Journal*. 2019 Feb;33(2):2144.
128. Elgharably, H.; Roy, S.; Khanna, S.; Abas, M.; Dasghatak, P.; Das, A.; Mohammed, K.; Sen, C.K. A modified collagen gel enhances healing outcome in a preclinical swine model of excisional wounds. *Wound Repair Regen*. 2013, 21, 473–481.
129. Marchand M, Monnot C, Muller L, Germain S. Extracellular matrix scaffolding in angiogenesis and capillary homeostasis. In Seminars in cell & developmental biology 2019 May 1 (Vol. 89, pp. 147–156). Academic Press.

130. Twardowski T, Fertala A, Orgel JP, San Antonio JD. Type I collagen and collagen mimetics as angiogenesis promoting superpolymers. *Current Pharmaceutical Design*. 2007 Dec 1;13(35):3608–21.
131. Feng, X.; Tonnesen, M.G.; Mousa, S.A.; Clark, R.A.F. Fibrin and Collagen Differentially but Synergistically Regulate Sprout Angiogenesis of Human Dermal Microvascular Endothelial Cells in 3-Dimensional Matrix. *International Journal of Cell Biology* 2013, 2013, 231279.
132. Elgharably H, Ganesh K, Dickerson J, Khanna S, Abas M, Ghatak PD, Dixit S, Bergdall V, Roy S, Sen CK. A modified collagen gel dressing promotes angiogenesis in a preclinical swine model of chronic ischemic wounds. *Wound Repair and Regeneration*. 2014 Nov;22(6):720–9.
133. Kisling A, Lust RM, Katwa LC. What is the role of peptide fragments of collagen I and IV in health and disease?. *Life Sciences*. 2019 Jul 1;228:30–4.
134. Wenzel D, Schmidt A, Reimann K, Hescheler J, Pfitzer G, Bloch W, Fleischmann BK. Endostatin, the proteolytic fragment of collagen XVIII, induces vasorelaxation. *Circulation Research*. 2006 May 12;98(9):1203–11.
135. Kareva I, Abou-Slaybi A, Dodd O, Dashevsky O, Klement GL. Normal wound healing and tumor angiogenesis as a game of competitive inhibition. *PLoS One*. 2016 Dec 9;11(12):e0166655.
136. Ricard-Blum S, Ballut L. Matricryptins derived from collagens and proteoglycans. *Frontiers in Bioscience* 2011 Jan 1;16(1):674–97.
137. Profyris C, Tziotzios C, Do Vale I. Cutaneous scarring: Pathophysiology, molecular mechanisms, and scar reduction therapeutics: Part I. The molecular basis of scar formation. *Journal of the American Academy of Dermatology*. 2012 Jan 1;66(1):1–10.
138. Ehrlich HP, Desmoulière A, Diegelmann RF, Cohen IK, Compton CC, Garner WL, Kapanci Y, Gabbiani G. Morphological and immunochemical differences between keloid and hypertrophic scar. *The American Journal of Pathology*. 1994 Jul;145(1):105.
139. Gauglitz GG, Korting HC, Pavicic T, Ruzicka T, Jeschke MG. Hypertrophic scarring and keloids: pathomechanisms and current and emerging treatment strategies. *Molecular Medicine*. 2011 Jan;17(1):113–25.
140. Shih B, Garside E, McGrouther DA, Bayat A. Molecular dissection of abnormal wound healing processes resulting in keloid disease. *Wound Repair and Regeneration*. 2010 Mar;18(2):139–53.
141. Verhaegen PD, Van Zuijlen PP, Pennings NM, Van Marle J, Niessen FB, Van Der Horst CM, Middelkoop E. Differences in collagen architecture between keloid, hypertrophic scar, normotrophic scar, and normal skin: an objective histopathological analysis. *Wound Repair and Regeneration*. 2009 Sep;17(5):649–56.
142. Xue M, Jackson CJ. Extracellular matrix reorganization during wound healing and its impact on abnormal scarring. *Advances in Wound Care*. 2015 Mar 1;4(3):119–36.

6 Applications of Synthetic Polymers for Drug Delivery Systems for the Management of Cancer

Supriya Singh, Sanket Kumar, and Sheikh Shahnawaz Quadir
Mohanlal Sukhadia University, Udaipur, India

Girima Nagda and Chhavi Bhalothia
University College of Science, Mohanlal Sukhadia University, Udaipur, India

Garima Joshi and Deepak Choudhary
Mohanlal Sukhadia University, Udaipur, India

6.1 INTRODUCTION

In 2022, there were about 20 million new cases of cancer, as well as 9.7 million cancer-related fatalities. According to estimates, around one in every five men and women are expected to get cancer at some point in their lives [1]. In 2022, lung cancer was the most often diagnosed cancer, accounting for about 2.5 million new cases, or one in every eight cancers globally (12.4% of all cancers), followed by malignancies of the female breast, colorectum, prostate and stomach with an average estimated percentage of 9.6, 7.3, 4.9, and 11.6, respectively [2]. Presently, clinical treatment options for cancer include radiation therapy, chemotherapy, and surgery. Surgical treatment is challenging for removing infiltrating and spreading malignancies, while chemotherapy can cause serious complications such as organ damage, gastrointestinal distress, leukopenia, and anemia [3]. Additionally, limited buildup of drugs in deep tumors may cause resistance after reiterated treatment. Chemotherapy can have a negative impact on treatment efficacy in contrast to toxicity ratios [4]. Innovations in nanotechnology have led to the discovery of novel treatments for severe diseases like cancer. Nanotechnology breakthroughs can reduce drug adverse effects by using regulated drug delivery systems. Nanomedicine-based treatments offer a promising technique for targeting tumor tissue. These formulations use submicrometer-sized carriers to specifically diagnose and treat tumors using appropriate targeting vectors. This improves the therapeutic index and pharmacokinetic profile of anticancer medicines [5].

Various therapeutic nanoparticle frameworks are being explored for cancer treatment, such as lipid-based systems, inorganic and polymer-drug coupled systems. Over 20 nanotechnology-based therapies have received clinical approval over the last two decades. Liposomal methods and polymer-drug conjugates are two key product categories [6]. Polymers offer effective drug administration because to their drug kinetics, circulation time, biological compatibility, and biodegradability properties [7]. They can be classified into natural, synthetic, and pseudo-synthetic macromolecules. Compared to natural polymers, synthetic polymers are easier to synthesize in large numbers, have low batch-to-batch variability, and may be easily modified to attain desired qualities. These features make polymers a possible solution for managing medication difficulties [8].

Synthetic polymers have unique qualities based on their molecular makeup, functional groups, degree of the polymerization process and production technique [9]. For instance, acrylates, with their film-forming properties, are compatible with mucosal membranes despite their lack of pharmacological activity. Some polymers are inert at physiological pH and can swell, while others are pH responsive and only soluble around pH 6–7 [10]. Polycationic polymers exhibit better mucoadhesive characteristics. They are quintessential vehicles and are ideal for nucleic acids, oligonucleotides, DNA, peptides, and proteins [11].

A Recent study emphasizes the usefulness of stimuli-responsive polymeric nanoparticles for anticancer therapy administration and targeting the "trimethyl-locked" benzoquinone polymer used for encapsulation of paclitaxel nanoparticles. It is redox receptive and biodegradable. It responds to multiple stimuli, which involves pH, temperature, magnetic field, enzyme and guest molecules [12]. Also, poly-d,l-lactide-co-glycolide (PLGA) nanoparticles designed to deliver camptothecin (CPT) to intracranial tumors in immune-competent albino mice has been studied [13].

In this chapter, the abundant synthetic polymers like polyesters, polyamides, polyanhydrides, phosphorous based polymers, silicones, and acrylic polymers, will be

DOI: 10.1201/9781003477389-6

discussed. They are being used along with anticancer drugs for the management of different kinds of cancer.

6.2 CANCER PATHOLOGY

The significance of cancer pathology cannot be overstated when it comes to diagnosing and treating cancer. It encompasses diverse areas, such as anatomical, clinical, and molecular pathology. Anatomical pathology is dedicated to scrutinizing tissue samples for cancer diagnosis and characterization, while clinical pathology entails the analysis of bodily fluids and tissues in the laboratory. Molecular pathology, on the other hand, focuses on identifying and characterizing diseases at the molecular level [14]. They supply vital information for precise diagnosis, personalized treatment strategies, and predicting patient outcomes. Thus, cancer pathology is crucial in steering clinical decision-making and enhancing patient care [15].

Genetic irregularities play a pivotal role in tumor evolution and are linked to the enlistment of immunosuppressive and pro-tumorigenic cells, such as tumor-associated macrophages (TAMs) and myeloid-derived suppressor cells (MDSCs). Moreover, the release of chemokines, growth factors, and cytokines by the stroma of tumor and cancer cells contributes to the mobilization and polarization of immune cells, including neutrophils, fostering enhanced tumor growth and metastatic potential. The intricate interplay between cancer cells and the immune microenvironment stands as a crucial factor in shaping cancer pathophysiology, presenting vital targets for cancer therapy [16].

Genetic aberrations, encompassing mutations in tumor suppressor genes, DNA repair genes, and oncogenes, lead to the loss of genetic stability and tumor formation. Concurrently, tumor-associated inflammation facilitated by immune cells like macrophages, neutrophils, and lymphocytes, significantly impacts tumor growth, angiogenesis, and metastasis [17].

The cancer microenvironment, consisting of cancer cells, stromal cells, extracellular matrix components, and immune cells, further shapes tumor growth, invasion, and metastasis [18, 19]. Understanding this complexity is crucial for developing effective therapeutic strategies, with the microenvironment serving as a potential target for combined treatments [20]. Figure 6.1 gives an overview of elements contributing to inflammation associated with tumors, arising either from the microenvironment of the tumor or directly from cells of cancer.

Cancer pathology encompasses a range of subfields, including clinical pathology, anatomical pathology, and molecular pathology.

6.2.1 Anatomical Pathology

Anatomical pathologists play a pivotal role in the comprehensive process of cancer diagnosis. The scrutiny of tissue samples, obtained through biopsies or surgical resections, is instrumental in discerning the presence of cancer, enabling the determination of the specific type and evaluation of key characteristics such as tumor grade and stage [21]. Additionally, anatomical pathologists are crucial participants in the staging of cancer, a critical process that involves determining the extent of the disease and its potential spread to adjacent tissues or lymph nodes. Biomarker analysis, facilitated through tests like immunohistochemistry and molecular testing, further enhances the role of anatomical pathology [22].

6.2.1.1 Limitations of Anatomical Pathology

Anatomical pathology, essential in cancer diagnosis and treatment, faces limitations. Tissue sampling relies on representative samples, impacting accuracy when tumor heterogeneity is not fully captured [23]. While providing crucial diagnostic information, anatomical pathology has limited predictive capacity, necessitating additional

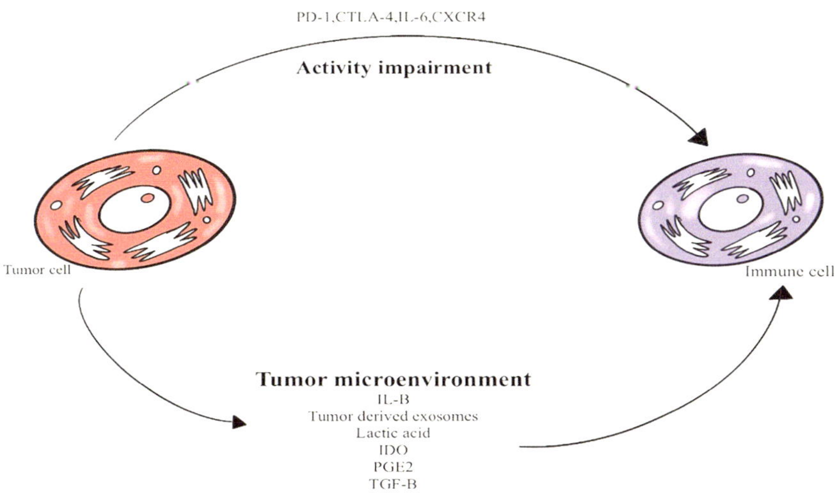

FIGURE 6.1 Summary of factors inducing inflammation in tumors, stemming from either the microenvironment of the tumor or directly from cancer cells [16].

molecular and genetic testing [24]. Turnaround time for sample analysis can delay diagnosis and treatment initiation. Resource intensity, requiring specialized expertise and infrastructure, may limit access in some health-care settings. Addressing these limitations is crucial for optimizing anatomical pathology's role in cancer care [25].

6.2.2 Clinical Pathology

Clinical pathology delves into the intricate interactions between cancer cells and stromal cells, unraveling their impact on tumor growth, invasion, and metastasis. Immune cells like MDSCs, TAMs, and Tregs take the center stage in modulating tumor progression and immune evasion [20]. Utilizing molecular biology techniques, clinical pathology identifies genetics, protein, and other molecular markers associated with cancer development and progression. These biomarkers, ranging from genetic mutations and gene fusions to protein expression and microRNA signatures, aid in early detection, risk stratification, and guiding treatment decisions [26]. A crucial technique employed by clinical pathology, immunohistochemistry (IHC) detects specific proteins like estrogen receptor, HER2/neu, progesterone receptor, and human epidermal growth factor receptor 2 in cancer cells [16]. These insights prove instrumental in assessing the effectiveness of various treatments, including immune checkpoint inhibitors (ICI) and immunotherapies [27].

6.2.2.1 Limitations of Clinical Pathology

Clinical pathology, integral to cancer diagnosis and treatment, faces challenges such as interpretative subjectivity, potentially affecting diagnosis consistency and tumor characteristic identification. While providing essential diagnostic information, clinical pathology has limitations in predicting treatment responses or disease recurrence, often requiring supplementary molecular and genetic testing for precision. Addressing these challenges necessitates strategic considerations in resource allocation and testing procedures [28].

6.2.3 Molecular Pathology

Molecular pathology plays a pivotal role in cancer diagnosis and treatment based on the molecular characteristics of cancer cells through the study of DNA, RNA, and proteins. Molecular pathology aids in identifying genetic abnormalities in cancer cells, crucial for tumor development [29]. It delineates their roles in modulating tumor progression, invasion, and metastasis through the secretion of cytokines and other mechanisms [30]. Molecular pathology contributes to deciphering immune evasion strategies employed by cancer cells, such as the expression of PD-L1 and CTLA-4, leading to T-cell exhaustion. This understanding is fundamental for the development of targeted ICI therapies [31].

6.2.3.1 Limitations of Molecular Pathology

Tumor heterogeneity, showcasing diverse genetic profiles, can hinder the accuracy of molecular testing and targeted therapy selection due to uneven treatment responses within a tumor [32]. Complexities in interpreting molecular test results demand expertise, with testing method variability impacting result reliability. Immunotherapies, though transformative, exhibit response variations among patients, and immune-related adverse events can occur, influenced by the tumor microenvironment and immunosuppressive factors [33]. Cost and accessibility issues in molecular testing and targeted therapies may hinder patient access, emphasizing the need to address financial and accessibility challenges in cancer care [34].

6.3 CLASSIFICATION AND BRIEF DESCRIPTION OF POLYMERS USED FOR CANCER MANAGEMENT

Polymeric carriers for drug delivery must have the ability to dissolve in water, lack of immunogenicity, acceptable pharmacokinetics, and presence of desired functional groups for drug conjugation, targeting, or copolymerization.

Classifying polymers in applications for drug delivery can be challenging due to their diverse topologies. However, classifications are important because they can reveal common features among polymer groups [35]

Polymers used for drug administration are often classified by their origin, chemical nature, backbone stability and water solubility.

Based on their origin, polymers are classified into three main groups/types: natural, semi-synthetic, and synthetic. Natural polymers are derived directly from nature, while semi-synthetic polymers are the modified versions of natural polymers. Synthetic polymers are thoroughly synthesized in laboratories [36].

Natural polymers are plentiful and generally biodegradable. Natural materials like chitosan, dextran, gelatin, alginate, and collagen have been used in drug delivery from a very long period of time. But, the main challenge with natural polymers in pharmaceuticals is developing reproducible production methods and the need for multiple purification steps due to their structural complexity. This can make modification, purification, and GMP-compliant production challenging. Biopreparation by living organisms, such as plants, algae or bacteria causes significant batch-to-batch variability. Furthermore, their utilization is limited due to their substantial immunogenicity [37].

Synthetic polymers like polyethylene glycol (PEG), polyglutamic acid (PGA), polycarprolactone (PCL), N-(2-hydroxypropyl)-methacrylamide copolymer (HPMA), and PLGA, are commonly used to encapsulate drugs for cancer therapy. They offer versatile formulations with customizable characteristics. Table 6.1 contains the detailed classification of polymers.

TABLE 6.1
Classification of Polymers [38]

Classification	Example
NATURAL POLYMERS	
• Polysaccharides	Alginate, dextran, cyclodextrins, agarose, carrageenan, chitosan, heparin, hyaluronic acid
• Protein-based polymers	Albumin, collagen, gelatin, spider silk
• Polyamino acids	Poly (L-aspartic acid), poly (α-L-lysin)
• Cellulose derivatives	Ethyl cellulose, cellulose acetate propionate, carboxymethyl cellulose
SYNTHETIC POLYMERS	
• *Biodegradable*	
Polyesters	Poly (lactic acid), poly (dioxanones), poly (glycolic acid), poly (phosphoesters)
Polyamides	Poly (imino carbonates)
Polyanhydrides	Poly (adipic acid), poly (sebacic acid), poly (terphthalic acid)
Phosphorous-based polymers	Polyphosphonates, polyphosphates
• *Nonbiodegradable*	
Silicones	Colloidal silica, polydimethylsiloxane
Acrylic polymers	Polyhydro(ethyl-methacrylate), polymethacrylate
Others	Poly (vinyl pyrrolidone), poloxamer, polyethylene glycol

6.3.1 Polyesters

Polyesters prove advantageous in medical applications owing to their degradation through nonenzymatic hydrolysis of ester bonds in the polymer backbone. The by-products of degradation can be readily absorbed through metabolic pathways, and the degradation rates can be influenced by tailoring the polymer structures. Within the polyester family, differentiation can be made based on the monomer structure. Poly (a-hydroxy acids), for instance, are linear thermoplastic materials typically produced through the ring-opening polymerization of cyclic dimers of their constituent components [39]. PGA, poly (lactic acid) (PLA), and their copolymers, such as PLGA, exemplify biodegradable polyesters and, more broadly, materials with biodegradable properties [40]. Poly(α-hydroxy acids) have been selected as the predominant material for tissue engineering, a process that entails combining isolated cells with three-dimensional macroporous carriers in vitro. Polymers derived from lactic acid and glycolic acid (GA) continue to be the favored scaffold materials for orthopedic applications, specifically in the context of bone and cartilage [41].

6.3.2 Polyamides

Introduced by DuPont in 1938, polyamide filaments were initially utilized for toothbrushes. By 1950, their widespread adoption in plastic manufacturing became prevalent, attributed to their appealing characteristics such as their stability at extreme temperatures, resistance to chemicals and a very high tensile strength and stiffness. They exhibit exceptional mechanical properties, including impressive resistance to friction and wear. Their mechanical classification includes PA 66 for hardness and robustness, and PA 12 for softness and flexibility [42]. The use of water-soluble polymeric conjugates in drug delivery through carrier-based systems provides advantages such as enhanced pharmacokinetics, decreased toxicity to vital cells, tissues and organs, and heightened drug accumulation at the designated site. In biomedical applications, poly (aspartic acid) finds common use for gene and medication administration, attributed to its barely minimal toxicity, non-antigenic properties, high biocompatibility, and biodegradability [43]. Poly-L-lysine serves as a water-soluble carrier for medications and macromolecules, easily degraded by cells. Comprising about 25–30 L-lysine residues, it forms a positively charged homo-polypeptide with a molecular weight of 4,700 Da. Utilized as a positively charged polymer, poly-(L-lysine) facilitates the delivery of negatively charged compounds through ionic interactions. The cationic character of poly-(L-lysine) allows for selective internalization into tumor cells through electrostatically adsorptive endocytosis [44].

6.3.3 Polyanhydrides

Biodegradable polymers based on anhydrides find extensive use in the controlled release of medications. Polyanhydrides demonstrate elevated reactivity with water, resulting in a comparatively swift hydrolytic breakdown. As a consequence, they undergo surface erosion as opposed to bulk degradation, setting them apart from other recyclable polymers. This particular attribute positions polyanhydrides as the favored option for applications related to controlled release. Effective investigations into polyanhydrides culminated in the endorsement of the Gliadel wafer, a bioresorbable implant crafted for dispensing anticancer medications to cerebral tumor locations [45]. Salem et al. investigated how polyanhydride chemistry affects particle-based cancer vaccinations. The researchers investigated how poly anhydride chemistry affects immune responses [46]. Polyanhydrides are valuable materials in medical, biological, and pharmaceutical applications due to their distinctive disintegration and erosion properties, as well as their structural diversity. Poly (anhydride esters) have been investigated as a promising technique for the regulated release of bioactive compounds. Polyanhydrides can contain bioactive compounds such as anti-inflammatory medicines, analgesics, and antiseptics as monomers or co-monomers [47].

6.3.4 Phosphorous-Based Polymers

Phosphorus-containing materials have diverse technological applications. For example, they were widely employed in industry because of their ability to bind metals. Organophosphonates have unique complexing capabilities and can be utilized as dispersants, corrosion inhibitors, or deposit preventers [48]. New laws have enabled the use of phosphorinated polymers in flame retardancy, wherein phosphorus is believed to be effective [49]. Phosphorus can be incorporated into macromolecules as terminal groups or within the main chain, forming an integral part of the polymer backbone. Polyphosphazenes and polyphosphoesters are two common phosphorus-based polymers with phosphorus in the backbone.

In the field of biomedicine, materials containing phosphorus, including polyphosphates, polyphosphonates, and polyphosphoesters, are becoming increasingly favored. This is attributed to their biodegradability, compatibility with blood, minimal protein adsorption, and robust interactions with dentin, enamel, and the skeletal system [50].

6.3.5 Silicones

Polysiloxanes, commonly referred to as silicones, are polymers that feature a backbone, consisting of silicon-oxygen bonds, with organic groups (generally methyl), which are attached to silicon atoms. The synthesis of silicones, which can exhibit properties ranging from liquid to rigid polymers, involves manipulating factors such as the length of the silicon-oxygen chain, the nature of side chains, and the greater extent of cross-linking [51]. Silicone is synthesized by hydrolyzing chlorosilanes to form straight chain or cyclic siloxane oligomers, which are subsequently polymerized through polycondensation or polymerization. Linear poly(dimethylsiloxane) is the most often used polysiloxane. They are known for their biocompatibility and endurance, are widely used in health care [52]. The materials' great biodurability is due to their ability to repel water, minimal surface tension, and their stability in chemical and thermal environments [53].

6.3.6 Acrylic Polymers

Polymethyl methacrylate (PMMA), which is a nondegradable polyacrylate, is among the most often used nonmetallic implant component in orthopedics. PMMA was initially utilized in dentures before being introduced for orthopedic surgery in the mid-1950s. Health-care equipment like blood pumps and dialyzers use PMMA due to its good biocompatibility and hemocompatibility, as well as its simplicity of manipulation. Its optical qualities make it suitable for implanted ophthalmic lenses as well as firm contact lenses. PMMA has physical and coloring qualities that are useful for denture production [54]. Hydrogels made from poly (2-hydroxyethyl methacrylate) (PHEMA) and copolymers have been thoroughly investigated for controlled drug delivery and biological purposes. PHEMA gels, with limited mechanical qualities, have been employed for reconstructing nasal cartilages, wound dressings and artificial corneas. Porous PHEMA sponges were found to enhance proliferative in growth and neo-vascularization in a subcutaneous rabbit framework, while also being cytocompatible [55].

6.4 SYNTHETIC POLYMERS FOR DRUG DELIVERY SYSTEM IN CANCER

Cancer treatment is not a one-size-fits-all approach due to the diverse nature of the tumor cell population. Local nonmetastatic malignancies are typically treated with radiotherapy and surgery, while metastatic cancers require chemotherapy. Chemotherapy uses anticancer medications to prevent tumor cells from rapidly multiplying. Polymers, synthetic and natural, have several applications in chemotherapy, including stimuli-responsive and targeted drug delivery, drug-eluting implants, and theranostics [56].

Chemotherapy's inefficiency and negative effects stem from its formulation and biodistribution. Conventional anticancer medications delivered intravenously circulate throughout the body and harm both cancerous and healthy cells. Tumor cell resistance to treatment is a significant issue. Tumor tissue's interstitium has significant hydrostatic pressure, resulting in outward convective flow that can remove drugs from the tumors [57]. To address these issues, a suitable anticancer drug delivery system should deliver a significant amount of the administered medicine to the intended site at the appropriate timing and duration. These qualities can be acquired by creating nanoscale-controlled formulations. Polymeric nanocarriers with enclosed or chemically conjugated medicines offer a promising approach to enhance tumor targeting and release. Nanocarriers circulate in the blood and accumulate in tumors via increased enhanced permeability and retention (EPR) [58].

Fast-growing malignancies require rapid vascularization to give oxygen and nutrients, resulting in a faulty vascular system in tumor tissues. This reduces lymphatic outflow and makes bloodstream vessels at the tumor location susceptible to nanoparticle compositions [59]. Targeting ligands help deliver drug-encapsulated nanoparticles to tumor locations, decreasing systemic side effects. Specific interactions between ligands on nanocarrier surfaces and receptors expressed on tumor cells can also aid nanoparticle internalization by activating receptor-mediated endocytosis. Polymers, including nanoparticles, micelles, polyplexes, polymer-drug mixtures, and dendrimers, can deliver targeted drugs to tumor locations [60].

Each medication has a unique method of action that can have varying effects on various sorts of cancer cells. For such reasons, chemotherapy for cancer may consist of multiple medications administered in combination. As a result, fresh drugs and drug delivery technologies for the treatment of cancer are required to lower overall toxicity and boost therapeutic efficacy [61].

Localized administration of chemotherapeutics is gaining popularity in cancer therapy, despite previous research into other methods. This method balances low systemic medication levels with therapeutic doses at tumor locations.

TABLE 6.2
Various Approaches for Delivering Cancer Chemotherapy [62]

Type	Route	Platforms for Drug Delivery	Clinically Tested Systems
Systemic	IV SC/IM	Liposomes, micelles, nanoparticles	Abraxane, Doxil
		Polymer implants/ injectables	Eligard, Suprefact, Zoladex
Local	Oral	Sustained release system	Tarceva, Xeloda
	Intraperitoneal	Micro/nanoparticle	Paclimer
	Intratumoral	Polymer implants/ injectables	Gliadel

This method can significantly improve medication therapy for solid tumors, which make up 85% of the instances of cancer [61]. Table 6.2 provides the various approaches for delivering cancer chemotherapy.

There are numerous possibilities for designing polymeric drug delivery technologies. The versatile nature of polymers makes them a popular choice in drug delivery, allowing for the development of an extensive spectrum of systems, including surgical implants, micro/nanoparticles, hydrogels, and pastes [63].

6.4.1 Polymer-Based Drug Distribution Strategies for Cancer Therapy

Nanoparticulate drug delivery systems (NDDSs) offer substantial benefits over free drugs, making them a potential alternative for cancer therapy. Nanoparticles (NPs) can be made from various materials, including polymers, fatty acids, and metals [64]. They can include a variety of cargoes, including chemotherapeutics, DNA, protein, diagnostic agents, and fluorophores. Amphiphilic block copolymers are gaining popularity for their ability to distribute and release payloads with precision [65]. Lipid-mediated targeting allows for site-specific cargo delivery by using ligand at the exterior of nanoparticles to recognize specific receptors overexpressed in sick cells. In addition to tumor targeting, nanoparticles can release payloads in response to either internal or external stimuli. Polymeric NDDSs' site-specific administration and stimuli-triggered drug release enhance therapeutic efficacy while reducing negative effects [65].

6.4.2 Stimuli-Responsive Drug Delivery System

To enable on-demand drug administration, several techniques have been devised by incorporating stimuli responsiveness in polymers. Polymers exhibit noticeable changes in physiochemical properties, including polarity, conformity, chemical conformation, and phase structure, in response to environmental changes. Changes in polymer properties affect nanocarriers shape and chemistry, resulting in the release of payloads in specific biological environments [66].

6.4.2.1 pH-Responsive Drug Delivery System

pH fluctuations at the tissue, systemic, cellular, and organelle stages have been extensively studied as an internal stimulus for developing environment-sensitive NDDS. Cancer cells undergo greater aerobic glycolysis, leading to a slightly acidic external microenvironment (pH 6.5–6.8) as opposed to regular tissue and systemic blood (pH 7.4). Furthermore, compartments within cells, including premature eondosomes (pH 5–6) and mature lysosomes (pH 4-5), have much lower pH values [67]. Figure 6.2 shows the release of drugs in the pH-responsive system.

There are various methods to produce pH-sensitive NDDSs. One strategy involves covalently attaching

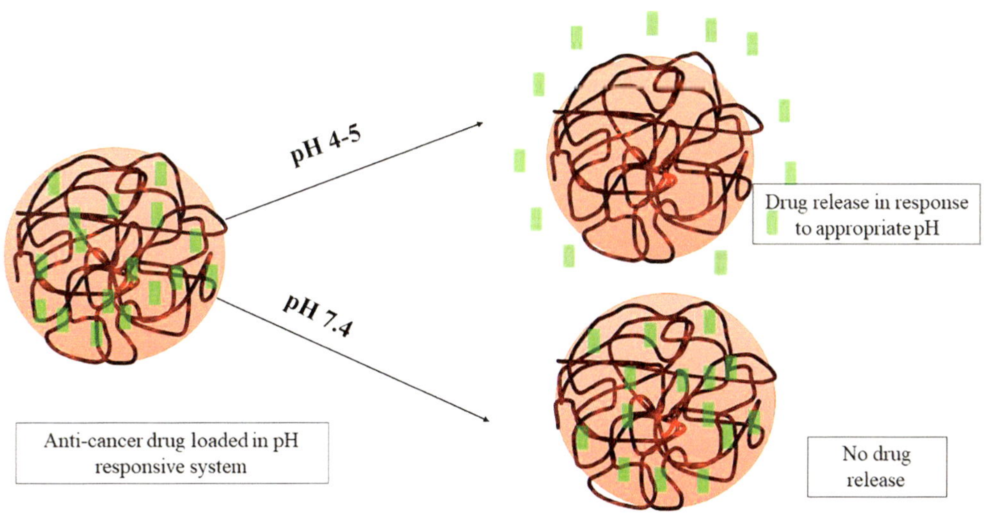

FIGURE 6.2 Release of drugs in the pH-responsive system.

medicines to the polymer framework via acid-labile bonds. Cleaving these connections at acidic tumor conditions allows for fast drug release and dissolution of nanocarriers. To assemble NDDSs, using polymers with charged groups, such as amino and carboxy groups should be considered. Polymers can flip between aqueous and phobic states by protonating and deprotonating their labile functional groups at different pH levels [68].

6.4.2.2 pH-Sensitive Polymer-Based Drug Delivery System

Polymers can be classed as cationic or anionic based on their charge at the physiological pH. Cationic polymers typically contain amine groups, which receive protons at an acidic pH and release protons at an alkaline pH. NDDSs made from cationic polymers disassemble when exposed to acid due to enhanced hydrophilicity caused by amine group protonation [69]. Polymers containing amine groups can effectively neutralize acidic endo/lysosomal compartments through protonation. Increased endo/lysosomal pH causes ionic transport into the lumen, causing the compartment to inflate and burst, releasing nanoparticles. This is known as the proton sponge effect. Numerous β-carboxylic amides-based polymers like2,3-dimethylmaleic anhydride, polyamidoamine, poly(allyl ethylene phosphate) have been synthesized in the last decade [70].

Deprotonation (ionization) occurs in anionic polymers, making them hydrophilic when above their pKa, while cationic polymers are protonated (nonionized) and hydrophobic when below their pKa. Alterations in polymer solubility result in the deformation of NPs and subsequent drug release. Examples of anionic polymers comprise polyacids such as aspartic acid, glutamic acid, and acrylic acid [71].

6.4.2.3 Reduction Responsive Drug Delivery System

The external microenvironment has a higher redox potential than the intracellular compartment, which is more reducing. The inner cytosol and subcellular regions contain larger levels of glutathione tripeptide (GSH) than the extracellular compartments, resulting in this gradient. The high intracellular redox potential is ideal for creating NDDSs that facilitate endo/lysosomal escape and therapeutic drug release. PEG, polyanhydrides and PCL, reduction-triggered shell-sheddable nano drug delivery with a PEG exterior and a core of additional polymers, including as PLA, PBLA, and polyphenylene sulfide (PPS), have been developed. Polyethyleneimine (PEI), which is yet another aquaphilic polymer, has been used to develop a drug delivery that has a reduction-sensitive sheddable shell [72].

6.4.2.4 Thermoresponsive Drug Delivery System

Tumor tissue differs significantly from normal tissue due to its distinct temperature gradient. Tumors' hyperthermia and sensitivity to high temperatures may contribute to the development of stimuli-sensitive NDDSs. Thermosensitive polymer-based NDDSs have been extensively researched for their ability to combine hyperthermia and chemotherapy in a single system. The polymers that are thermo responsive are classed as lower critical solution temperature (LCST) or upper critical solution temperature (UCST) based on how their hydrophilic-hydrophobic balance changes when heated [73]. Poly (ethyleneoxide) (PEO), poly (propyleneglycol) (PPG), and poly(methyl vinyl ether) (PMVE) are the most commonly used LCST polymers, whereas poly(6-(acryloyloxymethyl) uracil) (PAU) and poly(methacrylamide) (PMAAm) are commonly used UCST polymers [74].

6.4.2.5 Light-Responsive Drug Delivery System

Light distinguishes itself from other stimuli by its ability to be localized in time as well as space. Light-triggered processes typically involve chemical bond formation or breakage, isomer interconversion, charge conversion, and reaction rearrangement. Photochromic compounds, such as salicylideneaniline, azobenzene and spiropyran, are commonly employed to modify the characteristics of light-responsive polymers. Light-triggered processes can either be reversible or irreversible based on where the chromophore is linked to the polymer (backbone or pendant groups). Polymers with photocleavable units exhibit irreversible light-responsive behavior, resulting in the cleavage of a light-responsive constituent into an even more polar species following irradiation. O-nitrobenzyl ester (ONB) is a key photocleavable molecule with an irreversible light response [75].

6.4.3 Targeted Drug Delivery System

Nanocarriers ability to target cancer cells improves treatment efficacy while preventing harm to normal cells. Targeting mechanisms are classified into two discrete heads: passive and active. Passive targeting takes advantage of the differences between tumors and normal tissue. Passive targeting involves successfully delivering medications to the target place to achieve therapeutic effects [76]. Cancer cell growth induces neovascularization, and large puncture in the vascular wall diminish cancer arteries permselectivity when in comparison with normal vessels. Active targeting targets cancer cells by directly interacting with ligands and receptors. The ligands on the outside of nanoparticles target overexpressed molecules on cancer cells, distinguishing them from healthy ones. Ligands on nanoparticles bind with receptors on cancer cells, leading to receptor-mediated endocytosis and effective drug release. Active targeting is ideal for macromolecular medication delivery, including proteins and siRNAs. The other examples of targeting constituents comprise monoclonal antibodies, carbohydrates, amino acids and vitamins [77].

Transferrin is a suitable ligand for tumor targeting as its receptors are overexpressed in malignancies and correlate with the degree of malignancy. Transferrin receptors are commonly found on malignant cells, making them a promising target for cancer treatment. A gene delivery system was developed by Bellocq and colleagues using a cyclodextrin polymer modified with transferring [78]. This system involves nanoparticles with polymer/DNA surfaces modified with PEG, enabling targeted delivery to cancer cells through transferrin. Sahoo and Labhasetwar recently developed nanoparticles loaded with paclitaxel, featuring biodegradable polymer shells composed of PLGA linked to transferrin through epoxy links. The transferrin-conjugated nanoparticles exhibited enhanced cellular absorption and reduced exocytosis, leading to more potent antiproliferative effects and prolonged duration compared to free drugs or unconjugated nanoparticles [79].

The luteinizing hormone-releasing hormone (LHRH) receptor, discovered on the finest human cells, is overexpressed in uterine and other cancer cells, making it a potential target [80]. Dharap et al. designed an anticancer drug delivery method using LHRH-PEG-camptothecin, which targets cancer cell receptors. PEG is employed as a carrier to increase blood circulation, while camptothecin serves as an anticancer medication [81].

6.4.4 Gene Polymer Conjugation–Based Drug Delivery

The Human Genome Project has greatly improved understanding of inherited genetic illnesses like hemophilia and cystic fibrosis. Inherited genes may contribute to cancer development. Gene therapy is a novel method to cancer treatment that targets the aberrant genes responsible for the disease. Gene therapy involves two key components: a successful therapeutic gene that can be produced in a target cell and a prudent and structured delivery mechanism that can transport the gene to the desired tissue or organ [82].

Two categories of carriers are employed for transporting genes to specific locations: viral and nonviral avenues. Viral vectors, including adenoviruses, retroviruses, and adeno-related viruses, are adept at introducing the genetic materials into host cells, contributing to elevated rates of gene transfection. Nonviral vectors, particularly polymers, offer a promising alternative with lower safety hazards and the ability to adjust to individual therapeutic demands through simple manufacturing, purification, and chemical modifications [83]. Nonviral vectors are a cost-effective and simple way to transmit genetic material of different sizes.

Nonviral vectors, which are polymer based, are typically cationic and interact with anionic DNA to form polyplexes, which are nanosized ionic complexes. Polyplexes with a positive net charge from cationic polymers outperform DNA in terms of transfection efficiency. PEI outperforms other nonviral gene carriers in terms of transfection efficiency and consistency across multiple cell types. In order to diminish cytotoxicity and prolong the half-life of PEI/DNA complexes, scientists covalently linked PEG to the PEI backbone [84].

Poly(L-lysine), known for developing polyelectrolyte complexes with DNA, can act as a nonviral cistron carrier. It is a straight chain polypeptide composed of replicated lysine remnants, and its main α-amino groups become protonated in physiological conditions. Poly(L-lysine)'s (PLL) cationic composition allows for an electrostatic kind of interaction with the anionic DNA, resulting in nanoparticulate polyplexes with varying characteristics based on molecular weight [85].

Apart from this, polymeric micelles have proved to be a boon as drug delivery agents. Ringsdorf's invention of polymeric micelles involves the formation of amphiphilic copolymers in an aqueous state. Polymeric micelles can boost the solubility of water insoluble drugs due to their distinctive structural framework, which consists of a hydrophobic inner core supported by a hydrophilic outer shell. Polymeric micelles are nanosized containers that can hold pharmaceuticals through physical, chemical or electrostatic interactions [86].

6.5 CONCLUSION AND FUTURE PERSPECTIVE

Systems for delivery of drugs are a promising technique for cancer treatment, enabling sophisicated and targeted treatments. These systems can improve anticancer drug solubility and permeability, enhance retention, extend circulation half-life, improve biological distribution, and decrease toxicity. The clinical and preclinical data support the rationale for combining chemotherapy and immunotherapy. Combining immunotherapy and chemotherapy medications holds potential for long-term cancer maintenance therapy, maximizing synergistic antitumor effects. NDDS are effective platforms for enhancing chemo immunotherapy efficacy, with several multifunctional designs currently available. These NDDS for chemo immunotherapy demonstrated numerous benefits, including elevated solubility and bioavailability of chemotherapy and immunotherapy drugs, longer circulation time in vivo, targeted accumulation of therapeutic substances in tumor sites, and improved pharmacokinetics in vivo. These improvements significantly improved the therapeutic effects of low-dose chemotherapy agents and reduced side effects. Various stimuli-responsive polymers are commonly used to develop drug delivery systems for the treatment of cancer. Modern polymer synthesis methods have yielded several polymers in response to changes in pH, temperature and reductant, demonstrating their success. Nanocarriers derived from amphiphilic copolymers are attractive due to their small dimensions, biocompatibility, biological degradation, nontoxic nature, and improved kinetics and pharmacodynamics.

Personalized cancer nanomedicine has significant potential for enhancing treatment techniques for cancer patients. However, it is vital to highlight that the majority of these investigations are still in the early phases, and their clinical effects need to be confirmed.

REFERENCES

[1] Woodring R. N., Gurysh E. G., Bachelder E. M. et al (2023) Drug delivery systems for localized cancer combination therapy. *ACS Applied Bio Materials* 6(3): 934–950.

[2] Bray F., Laversanne M., Sung H. et al (2024) Global cancer statistics 2022: GLOBOCAN estimates of incidence and mortality worldwide for 36 cancers in 185 countries. *CA: A Cancer Journal for Clinician* 74(3): 229–263.

[3] Liu L., Fei Z., Chen M. et al (2018) Induction chemotherapy plus concurrent chemoradiotherapy versus induction chemotherapy plus volumetric modulated arc therapy alone in the treatment of stage II-IVB nasopharyngeal carcinoma patients: a retrospective controlled study. *Radiation Oncology* 13: 1–9.

[4] Cashin P. H., Mahteme H., Spang N. et al (2016) Cytoreductive surgery and intraperitoneal chemotherapy versus systemic chemotherapy for colorectal peritoneal metastases: a randomised trial. *European Journal of Cancer* 53: 155–162.

[5] Oerlemans C., Bult W., Bos M. et al (2010) Polymeric micelles in anticancer therapy: targeting, imaging and triggered release. *Pharmaceutical Research* 27: 2569–2589.

[6] Wicki A., Witzigmann D., Balasubramanian V. et al (2015) Nanomedicine in cancer therapy: challenges, opportunities, and clinical applications. *Journal of Controlled Release* 200: 138–157.

[7] Salari N., Faraji F., Torghabeh F. M. et al (2022) Polymer-based drug delivery systems for anticancer drugs: A systematic review. *Cancer Treatment and Research Communications* 32: 100605.

[8] Englert C., Brendel J. C., Majdanski T. C. et al (2018) Pharmapolymers in the 21st century: Synthetic polymers in drug delivery applications. *Progress in Polymer Science* 87: 107–164.

[9] Vijay R., Angayarkanny S., Reddy B. S. R. et al (2010). Adsorption and emulsification properties of amphiphilic poly (styrene-co-octadecyl maleamic acid salt) with comb-like architecture. *Journal of Colloid and Interface Science* 346(1): 143–152.

[10] Marschütz M. K., Bernkop-Schnürch A. (2002) Thiolated polymers: self-crosslinking properties of thiolated 450 kDa poly (acrylic acid) and their influence on mucoadhesion. *European Journal of Pharmaceutical Sciences* 15(4): 387–394.

[11] Srividhya M., Preethi S., Gnanamani A et al (2006) Sustained release of protein from poly (ethylene glycol) incorporated amphiphilic comb like polymers. *International Journal of Pharmaceutics* 326(1–2): 119–127.

[12] Colson Y. L., Grinstaff M. W. (2012) Biologically responsive polymeric nanoparticles for drug delivery. *Advanced Materials* 24(28): 3878–3886.

[13] Householder K. T., DiPerna D. M., Chung E. P. et al (2015). Intravenous delivery of camptothecin-loaded PLGA nanoparticles for the treatment of intracranial glioma. *International Journal of Pharmaceutics* 479(2): 374–380.

[14] Ozretić L., Heukamp L. C., Odenthal M. et al (2012) The role of molecular diagnostics in cancer diagnosis and treatment. *Oncology Research and Treatment* 35(Suppl. 1): 8–12.

[15] Connolly J. L., Schnitt S. J., Wang H. H. et al (2003) Role of the Surgical Pathologist in the Diagnosis and Management of the Cancer Patient. In: Kufe DW, Pollock RE, Weichselbaum RR, et al, (ed) *Holland-Frei Cancer Medicine*. 6th edn. BC Decker

[16] Pęczek P., Gajda M., Rutkowski K. et al (2023) Cancer-associated inflammation: pathophysiology and clinical significance. *Journal of Cancer Research and Clinical Oncology* 149(6): 2657–2672.

[17] Wellenstein M. D., Visser K. E. (2018) Cancer-cell-intrinsic mechanisms shaping the tumor immune landscape. *Immunity* 48(3): 399–416.

[18] Hinshaw D. C., Shevde L. A. (2019) The tumor microenvironment innately modulates cancer progression. *Cancer Research* 79(18): 4557–4566

[19] Zhang R., Qi F., Zhao F. et al (2019) Cancer-associated fibroblasts enhance tumor-associated macrophages enrichment and suppress NK cells function in colorectal cancer. *Cell Death & Disease* 10(4): 273.

[20] Ferrari S. M., Fallahi P., Galdiero M. R. et al (2019) Immune and inflammatory cells in thyroid cancer microenvironment. *International Journal of Molecular Sciences* 20(18): 4413.

[21] Yu G., Gail M. H., Consonni D. et al (2016) Characterizing human lung tissue microbiota and its relationship to epidemiological and clinical features. *Genome biology* 17: 1–12.

[22] Rotte A. (2019) Combination of CTLA-4 and PD-1 blockers for treatment of cancer. *Journal of Experimental & Clinical Cancer Research* 38(1): 255.

[23] Pisapia P., L'Imperio V., Galuppini F. et al (2022) The evolving landscape of anatomic pathology. *Critical Reviews in Oncology/Hematology* 178: 103776..

[24] Gil J., Wu H. S. (2003) Applications of image analysis to anatomic pathology: realities and promises. *Cancer Investigation*, 21(6): 950–959.

[25] Salmon A. S., Salgado J. B., Morell A. R. (2006) PET in abdominal pathology: advantages and limitations. *Abdominal Imaging* 31(2): 174–181.

[26] Echle A., Rindtorff N. T., Brinker T. J. et al (2021) Deep learning in cancer pathology: a new generation of clinical biomarkers. *British Journal of Cancer* 124(4): 686–696.

[27] Vito A., El-Sayes N., Mossman K. (2020) Hypoxia-driven immune escape in the tumor microenvironment. *Cells* 9(4): 992.

[28] Jahn S. W., Plass M., Moinfar F. (2020) Digital pathology: advantages, limitations and emerging perspectives. *Journal of Clinical Medicine* 9(11): 3697.

[29] Fu L. Q., Du W. L., Cai M. H. et al (2020) The roles of tumor-associated macrophages in tumor angiogenesis and metastasis. *Cellular Immunology* 353: 104119.

[30] Zhu H, Gu Y, Xue Y et al (2017) CXCR2+ MDSCs promote breast cancer progression by inducing EMT and activated T cell exhaustion. *Oncotarget* 8(70): 114554.

[31] Tu L, Guan R, Yang H et al (2020) Assessment of the expression of the immune checkpoint molecules PD-1, CTLA4, TIM-3 and LAG-3 across different cancers in relation to treatment response, tumor-infiltrating immune cells and survival. *International Journal of Cancer* 147(2): 423–439.

[32] Li F, Yang Q, He AT et al (2021) Circular RNAs in cancer: limitations in functional studies and diagnostic potential. In: *Seminars in Cancer Biology*, Vol. 75, Academic Press, p 49–61.

[33] Cerretelli G, Ager A, Arends M J et al (2020) Molecular pathology of Lynch syndrome. *The Journal of Pathology* 250(5): 518–531.

[34] Harada S, Morlote D (2020) Molecular Pathology of Colorectal Cancer. *Advances in Anatomic Pathology* 27(1): 20–26.

[35] Elbert DL, Hubbell JA (1996) Surface treatments of polymers for biocompatibility. *Annual Review of Materials Research* 26(1): 365–394.

[36] Deb PK, Kokaz SF, Abed SN et al (2018) Pharmaceutical and biomedical applications of polymers. In: *Basic Fundamentals of Drug Delivery*, Academic Press, p. 203–267.

[37] Angelova N, Hunkeler D (1999) Rationalizing the design of polymeric biomaterials. *Trends in biotechnology* 17(10): 409–421.

[38] Pillai O, Panchagnula R (2001) Polymers in drug delivery. *Current Opinion in Chemical Biology* 5(4): 447–451.
[39] Hu Y, Daoud W. A., Cheuk K. K. L. et al (2016). Newly developed techniques on polycondensation, ring-opening polymerization and polymer modification: Focus on poly (lactic acid). *Materials*, 9(3): 133.
[40] Cutright D. E., Beasley III J. D., Perez B (1971) Histologic comparison of polylactic and polyglycolic acid sutures. *Oral Surgery, Oral Medicine, Oral Pathology*. 32(1): 165–173
[41] Agrawal C. M., Athanasiou K. A. (1997) Technique to control pH in vicinity of biodegrading PLA-PGA implants. *Journal of Biomedical Materials Research* 38(2): 105–114.
[42] Boddu S. H., Bhagav P, Karla P. K. et al (2021) Polyamide/poly (amino acid) polymers for drug delivery. *Journal of Functional Biomaterials* 12(4): 58.
[43] Shah S. S., Casanova N, Antuono G et al (2020) Polyamide backbone modified cell targeting and penetrating peptides in cancer detection and treatment. *Frontiers in Chemistry* 8: 218.
[44] Eivazzadeh-Keihan R, Radinekiyan F, Maleki A et al (2020) A new generation of star polymer: magnetic aromatic polyamides with unique microscopic flower morphology and in vitro hyperthermia of cancer therapy. *Journal of Materials Science* 55(1): 319–336.
[45] Basu A, Domb A. J. (2018) Recent advances in polyanhydride based biomaterials. *Advanced Materials* 30(41): 1706815.
[46] Wafa E. I., Geary, S. M., Goodman J. T. et al (2017) The effect of polyanhydride chemistry in particle-based cancer vaccines on the magnitude of the anti-tumor immune response. *Acta biomaterialia* 50: 417–427.
[47] Snyder B. L., Mohammed H. S., Samways D. S. et al (2020) Drug delivery and drug efficacy from amorphous poly (thioether anhydrides). *Macromolecular Bioscience* 20(5): 1900377.
[48] Wang X, Shao J, Abd El Raouf M et al (2018) Near-infrared light-triggered drug delivery system based on black phosphorus for in vivo bone regeneration. *Biomaterials* 179: 164–174.
[49] Monge S, Canniccioni B, Graillot A et al (2011) Phosphorus-containing polymers: a great opportunity for the biomedical field. *Biomacromolecules* 12(6): 1973–1982.
[50] Henke H, Brüggemann O, Teasdale I (2017) Branched Macromolecular Architectures for Degradable, *Multifunctional Phosphorus-Based Polymers. Macromolecular rapid communications* 38(4): 1600644
[51] Curtis J, Steichen S. D (2020) Silicones. In: Wagner WR, Sakiyama-Elbert SE, Zhang G, Yaszemski MJ (ed) *Biomaterials Science*, Cambridge: Academic Press, p. 109–123.
[52] Bai L, Yan H, Bai T et al (2019) High fluorescent hyperbranched polysiloxane containing β-cyclodextrin for cell imaging and drug delivery. *Biomacromolecules* 20(11): 4230–4240.
[53] Abu-Thabit NY, Makhlouf ASH (2018) Historical development of drug delivery systems: From conventional macroscale to controlled, targeted, and responsive nanoscale systems. In: Hamdy Makhlouf AS, Abu-Thabit NY (ed) *Stimuli Responsive Polymeric Nanocarriers for Drug Delivery Applications*, Volume 1. Woodhead Publishing, Sawston, p 3–41.
[54] Feldman D (2019) Polymers and polymer nanocomposites for cancer therapy. *Applied Sciences*, 9(18): 3899.
[55] Liu H, Chen H, Cao F et al (2019). Amphiphilic block copolymer poly (acrylic acid)-b-polycaprolactone as a novel pH-sensitive nanocarrier for anti-cancer drugs delivery: In-Vitro and in-vivo evaluation. *Polymers* 11(5): 820.
[56] Park JH, Lee S, Kim JH et al (2008). Polymeric nanomedicine for cancer therapy. *Progress in Polymer Science* 33(1): 113–137.
[57] Fonseca AC, Serra AC, Coelho JF (2015) Bioabsorbable polymers in cancer therapy: latest developments. *EPMA Journal*, 6: 1–18.
[58] Doppalapudi S, Jain A, Domb AJ et al (2016) Biodegradable polymers for targeted delivery of anti-cancer drugs. *Expert Opinion on Drug Delivery*, 13(6): 891–909.
[59] Selvolini G, Marrazza G (2017) MIP-based sensors: promising new tools for cancer biomarker determination. *Sensors* 17(4): 718.
[60] Lee Y, Thompson D. H (2017) Stimuli-responsive liposomes for drug delivery. *Wiley Interdisciplinary Reviews: Nanomedicine and Nanobiotechnology* 9(5): e1450.
[61] Zhang L, Gu FX, Chan JM et al (2008) Nanoparticles in medicine: therapeutic applications and developments. *Clinical Pharmacology & Therapeutics* 83(5): 761–769.
[62] De Souza R, Zahedi P, Allen CJ et al (2010) Polymeric drug delivery systems for localized cancer chemotherapy. *Drug Delivery* 17(6): 365–375.
[63] Jagur-Grodzinski J (2009) Polymers for targeted and/or sustained drug delivery. *Polymers for Advanced Technologies* 20(7): 595–606.
[64] Anselmo AC, Mitragotri S (2016) Nanoparticles in the clinic. *Bioengineering & Translational Medicine* 1(1): 10–29.
[65] Elsabahy M, Wooley KL (2012) Design of polymeric nanoparticles for biomedical delivery applications. *Chemical Society Reviews* 41(7): 2545–2561.
[66] Mura S, Nicolas J, Couvreur P (2013) Stimuli-responsive nanocarriers for drug delivery. *Nature Materials* 12(11): 991–1003.
[67] Fleige E, Quadir MA, Haag R (2012) Stimuli-responsive polymeric nanocarriers for the controlled transport of active compounds: concepts and applications. *Advanced Drug Delivery Reviews*, 64(9): 866–884.
[68] Bahcecioglu G, Basara G, Ellis BW, Ren X, Zorlutuna P (2020) Breast cancer models: Engineering the tumor microenvironment. *Acta biomaterialia* 106: 1–21
[69] Veselov VV, Nosyrev AE, Jicsinszky L et al (2022) Targeted delivery methods for anticancer drugs. *Cancers* 14(3): 622.
[70] Liu G, Yang L, Chen G et al (2021) A review on drug delivery system for tumor therapy. *Frontiers in Pharmacology* 12: 735446.
[71] Edis Z, Wang J, Waqas MK Ijaz, et al (2021) Nanocarriers-mediated drug delivery systems for anticancer agents: an overview and perspectives. *Int J Nanomedicine*, 17(16): 1313–1330.
[72] Yao Y, Zhou Y, Liu L et al (2020) Nanoparticle-based drug delivery in cancer therapy and its role in overcoming drug resistance. *Frontiers in Molecular Biosciences* 7: 193.
[73] López Mendoza CM, Alcántara Quintana LE (2022) Smart drug delivery strategies for cancer therapy. *Frontiers in Nanotechnology* 3: 753766.
[74] Liyanage PY, Hettiarachchi SD, Zhou Y et al (2019) Nanoparticle-mediated targeted drug delivery for breast cancer treatment. Biochimica et Biophysica Acta (BBA)-Reviews on *Cancer* 1871(2): 419–433.
[75] Zarrintaj P, Jouyandeh M, Ganjali MR et al (2019) Thermo-sensitive polymers in medicine: A review. *European Polymer Journal* 117: 402–423.

[76] Guo X, Wang L, Wei X et al (2016) Polymer-based drug delivery systems for cancer treatment. *Journal of Polymer Science Part A: Polymer Chemistry*, 54(22): 3525–3550.

[77] Hacker MC, Krieghoff J, Mikos AG (2019) Synthetic polymers. In: *Principles of regenerative medicine*, Third edn, Academic Press p. 559–590.

[78] Bellocq NC, Pun SH, Jensen, GS et al (2003) Transferrin-containing, cyclodextrin polymer-based particles for tumor-targeted gene delivery. *Bioconjugate Chemistry* 14(6): 1122–1132

[79] Sahoo SK, Ma W, Labhasetwar V (2004) Efficacy of transferrin-conjugated paclitaxel-loaded nanoparticles in a murine model of prostate cancer. *Int J Cancer* 112(2): 335–340

[80] Mu W, Chu Q, Liu Y et al (2020) A review on nano-based drug delivery system for cancer chemoimmunotherapy. *Nano-Micro Letters* 12: 1–24.

[81] Dharap SS, Wang Y, Chandna P Khandare et al (2005) Tumor-specific targeting of an anticancer drug delivery system by LHRH peptide. *Proceedings of the National Academy of Sciences* 102(36): 12962–12967.

[82] Sharma P, Negi P, Mahindroo N (2018) Recent advances in polymeric drug delivery carrier systems. *Adv. Polym. Biomed. Appl* 10: 369–388.

[83] Yu J, Qiu H, Yin S et al (2021) Polymeric drug delivery system based on pluronics for cancer treatment. *Molecules* 26(12): 3610.

[84] Avramović N, Mandić B, Savić-Radojević A et al (2020) Polymeric Nanocarriers of Drug Delivery Systems in Cancer Therapy. *Pharmaceutics* 12(4): 298

[85] Mogoşanu GD, Grumezescu AM, Bejenaru LE et al (2016) Natural and synthetic polymers for drug delivery and targeting. In: Grumezescu AM, (ed) *Nanobiomaterials in Drug Delivery*. William Andrew Publishing, Amsterdam, The Netherlands, pp. 229–284.

[86] Kulthe SS, Choudhari YM, Inamdar NN et al (2012) Polymeric micelles: authoritative aspects for drug delivery. *Designed Monomers and Polymers* 15(5): 465–521.

7 Drug Delivery Based on Synthetic Polymers for CNS and Neurodegenerative Disorders

Ruchika Kabra and Atul Kabra
Chandigarh University, Mohali, India

Deepak Gupta
Shoolini University, Solan, India

Rajbala
RIMT University, Mandi Gobindgarh, India

Rajwant Kaur
Chandigarh University, Mohali, India

Mohamed El-Shazly
Ain Shams University, Cario, Egypt

7.1 INTRODUCTION

7.1.1 Neurodegenerative Disorders

Neurons play an essential role in enabling communication and are necessary for the human brain to function efficiently. Although neurons originate mostly in the brain, they are found across the entirety of the body [1]. Neural stem cells produce most of the neurons in childhood, but after this age, they produce fewer neurons overall. Although neurons have a finite lifespan, the gradual loss of neurons, changes in their structure, and impaired functions, collectively recognized as neurodegeneration, are central to the development of various brain ailments [2]. Neurodegeneration is characterized by improperly changed versions of proteins accumulating in the brain, as well as failure of neuronal networks and synapses [3]. Diseases characterized by neurodegeneration, termed neurodegenerative diseases (NDs), pose significant health concerns. Spinocerebellar ataxia (SCA), amyotrophic lateral sclerosis (ALS), Alzheimer's disease, Parkinson's disease, Huntington's disease, and spinal muscular atrophy (SMA) are frequently encountered neurodegenerative disorders [4, 5].

Numerous factors influence the initiation and progression of neurodegenerative disorders and central nervous system (CNS) diseases depicted in Figure 7.1. Genetic predispositions, environmental exposures, and the aging process are key elements influencing the development of these conditions. In disorders like Alzheimer's and Parkinson's, specific genetic mutations have been pinpointed as influential risk factors, impacting susceptibility. Environmental elements, encompassing exposure to toxins, pollutants, and certain lifestyle choices, also contribute to the initiation or worsening of neurodegenerative processes. Moreover, the natural aging of the body is often associated with an elevated vulnerability to CNS diseases, as age-related alterations in cellular function and molecular pathways may contribute to their emergence.

Owing to the brain's ability to regulate a wide range of physiological processes, neurodegenerative illnesses have a widespread impact and impair vital functions, such as speech, movement, balance, and stability. They also impact more complex functions like cognitive capacity, bladder and bowel control [6–8]. Many NDs follow a progressive course without remission, although some treatments focus on symptom management, pain relief, and enhancing balance and mobility [9, 10].

7.1.1.1 Alzheimer's Disease

The rapid and permanent loss of cognitive function is the hallmark of Alzheimer's disease (AD), a neurodegenerative disorder. Tau protein tangles and beta amyloid plaques accumulate in the brain, impeding communication between neurons and eventually leading to the death of cells. Neuronal communication is an important pathogenic

DOI: 10.1201/9781003477389-7

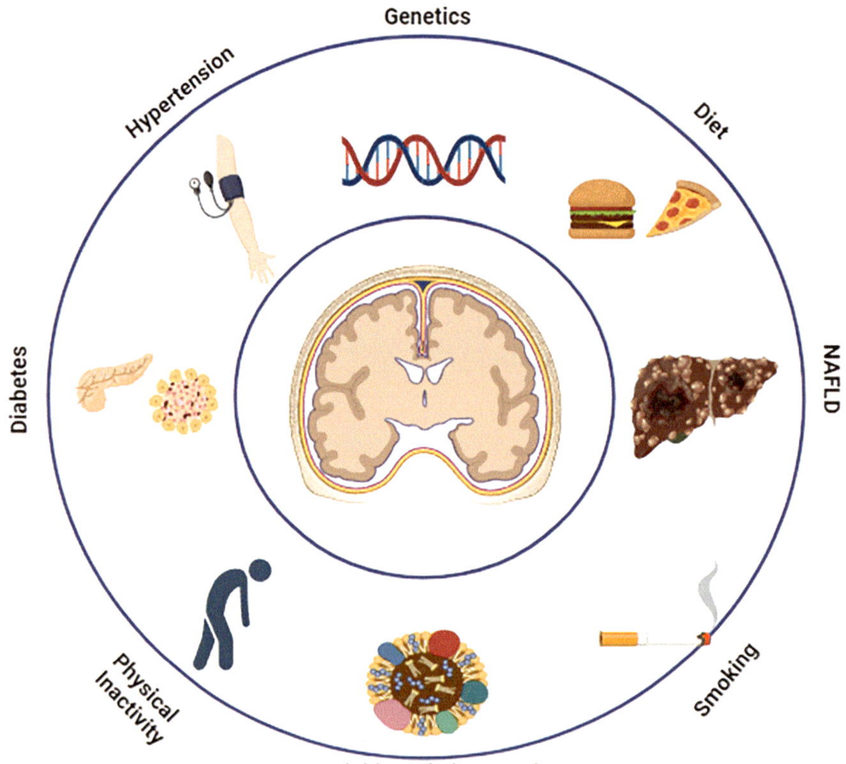

FIGURE 7.1 Causative factors associated with neurodegenerative and CNS diseases.

hallmark. With its primary impact on memory, thinking, and behavior, AD represents a significant public health challenge, affecting both individuals and their families profoundly. As the disease progresses, individuals may encounter challenges in daily activities, undergo personality changes, and face difficulties in language comprehension. While advancing aging is a notable risk feature, genetic elements, including the APOE gene, also contribute to susceptibility [3, 11].

7.1.1.2 Parkinson's Disease

A chronic neurodegenerative illness called Parkinson's disease (PD) is distinguished as the gradual breakdown of dopamine producing neurons particularly in substantia nigra region of the midbrain. Central to its defining characteristics are the development of Lewy bodies, irregular protein deposits that result in a deficiency of dopamine – a crucial neurotransmitter for motor control. Consequently, individuals with Parkinson's commonly endure motor manifestations such as stiffness, tremors, muscle movement, and unstable posture. Alongside these motor impairments, non-motor manifestations like mood swings, autonomic dysfunction and cognitive decline can have significant detrimental impact on overall standard of living. However, PD's precise etiology still remains an uncertain, and a combination of environmental and genetic variables are thought to play a role in disease progression [7, 12, 13].

7.1.1.3 Amyotrophic Lateral Sclerosis (ALS)

The motor neurons that reside in the CNS and the spinal cord are impacted by the chronic neurodegenerative disease known as ALS. Characterized by the gradual degeneration and loss of these motor neurons, The hallmark of ALS is gradual weakening and atrophy of voluntary muscle as a result of gradual loss and degeneration of these motor neurons. As the disease progresses, people with ALS may have trouble breathing, speaking, and, eventually, swallowing. Despite continuous study, the true cause of ALS is still unknown. In numerous instances, it is thought that a combination of genetic and environmental variables leads to its development [14–16] (Table 7.1).

7.2 DIFFICULTIES IN THE DRUG DELIVERY TO THE BRAIN

Current therapies for NDs emphasize managing disease progression instead of addressing the root causes [25]. Nevertheless, a significant challenge lies in overcoming the barrier between blood and the brain, which prevents adequate distribution of medicinal substances to the brain. Most formulations do not successfully permeate the barrier, limiting their efficacy in treating NDs [26, 27]. The complexity of the BBB, as well as the restricted entry of many medicines, contributes to the scarcity of viable therapy options for NDs.

TABLE 7.1
Current Pharmacological Approaches NDs

Disease	Class	Mechanism of Action	Drugs	References
AD	Cholinesterase inhibitors	Protect against the reduction of acetylcholine	Rivastigmine, Donepezil, Galantamine	[17, 18]
	Glutamate regulators	Block N-methyl-D-aspartate (NMDA) receptors	Memantine	[17]
	Amyloid β-antibody	Eliminating amyloid-β plaques	Aducanumab	[17, 19]
PD	Dopamine supplements	Restore diminished dopamine levels	L-dopa	[20, 21]
	Decarboxylase inhibitors	Inhibit the peripheral degradation of levodopa	Carbidopa	[20–22]
	Dopamine agonist	Induces effects resembling dopamine	Pramipexole Dihydrochloride, Ropinirole Apomorphine Hydrochloride, Pergolide, Hydrochloride, Rotigotine	[20–22]
ALS	Glutamate-receptor antagonist	Blocks the activity of glutamate receptors	Riluzole	[23]
	Free-radical scavenger	Eliminates or neutralizes free radicals	Edaravone	[24]

7.2.1 Blood-Brain Barrier

The blood-brain barrier (BBB) is a crucial physiological boundary that closely controls the flow of constituents among the systemic circulation and the extracellular fluid of the brain [28]. Comprising specialized endothelial cells, tight junctions, and supporting cells, such as astrocytes, it serves as a barrier to keep potentially dangerous chemicals out of the brain. The close connection across endothelial cells creates a barrier that prevents the diffusion of large or hydrophilic molecules, maintaining a stable and controlled environment for optimal neural function. Additionally, the BBB actively pumps out toxins and regulates the entry of specific substances through various transport mechanisms [29, 30].

While the BBB serves as a formidable defense mechanism, it also presents a difficulty for the brain's delivery system. Numerous healing mediators struggle to traverse this barrier, limiting their effectiveness in treating neurological disorders [31]. Investigators are looking into innovative drug delivery approaches, like nanoparticles and specialized carriers, to enhance the drug transportation via BBB, providing new avenues for the advance of targeted treatments for diseases affecting the CNS. Understanding the intricacies of the BBB is fundamental for advancing medication distribution technologies and improving the treatment of neurological conditions [32, 33].

7.2.2 Principles of Pharmacokinetics and Their Impact on Medication Delivery to the Brain

The impact of pharmacokinetic characteristics predominantly shapes the efficacy of drugs administered systemically. Yet, the voyage from administration to the intended destination, especially the brain, encounters considerable hurdles that frequently impede the conveyance of therapeutic substances. A notable impediment involves the existence of diverse plasma proteins to which certain drugs form robust attachments, diminishing the quantity of unbound drugs accessible for conveyance to the brain. Additionally, specific drugs face swift elimination by significant organs responsible for clearance, leading to a further decrease in their bloodstream concentration [34].

An additional element influencing drug delivery involves the interplay between the drug and target cells, potentially hindering drug absorption. For example, drug molecules may engage with cells, causing blockage of channels, shifts in membrane potential, or modifications in cell conformation. These temporary effects have the potential to restrict the cell's reaction to the administered drug molecule and its absorption. In general, drug molecules that are small and lipophilic are better suited for transportation to the brain [35].

7.2.3 Challenges Associated with a Targeted Approach

The effectiveness of drug delivery systems relies on their ability to precisely target specific sites for therapeutic effects [36]. However, numerous factors affect their efficacy, making it challenging to achieve desired therapeutic outcomes. Challenges include a lack of suitable targets and the need to align targets with therapeutic requirements. Additionally, selecting appropriate target epitopes is crucial for facilitating drug uptake by cells, necessitating consideration of various physiological variables such as blood flow, disease status, and tissue architecture. Designing targeting systems involves addressing physicochemical constraints like nanocarrier structure and drug carrier composition. Despite their potential, targeted delivery systems face hurdles such as low drug bioavailability at target sites, susceptibility to opsonization leading to unwanted uptake by the reticuloendothelial system, and potential leakage from movement through vascular gaps or tumor angiogenesis-related defects [37].

7.3 ADVANTAGES ASSOCIATED WITH A TARGETED APPROACH

In addition to the challenges they face, targeted drug delivery systems offer several benefits. These systems help maintain optimal drug levels in the bloodstream and tissues, minimizing potential harm to healthy cells [38]. Targeted delivery systems improve drug absorption at diseased sites and are more specific and desirable compared to conventional drugs [39]. Overall, they enhance the therapeutic profile of medications.

7.4 SYNTHETIC POLYMERS FOR DRUG DELIVERY SYSTEMS

7.4.1 Poly(lactic acid)-poly(ethylene glycol) [PLA-PEG]

PLA is a polymer that is both biodegradable and bioactive, and it comes from natural sources like sugarcane or maze starch. Widely applied in biomedical domains like pharmaceutical distribution systems. PLA's popularity stems from its inherent biocompatibility and biodegradability. PEG, a synthetic water-soluble polymer recognized for its hydrophilic and flexible characteristics, is often working to improve the water solubility of hydrophobic substances and improve material biocompatibility. When PLA and PEG are combined in a copolymer, such as PLA-PEG, a versatile material emerges with desirable properties. The presence of PEG enhances water solubility and overall biocompatibility, making it well-suited for diverse biomedical applications. Retaining PLA's biodegradable nature, PLA-PEG is advantageous in situations where controlled degradation is crucial. In drug delivery systems, the copolymer serves as an effective carrier, with hydrophobic PLA segments encapsulating drugs, while hydrophilic PEG segments enhance stability and circulation time in the bloodstream. PLA-PEG finds application in tissue engineering scaffolds, promoting tissue regeneration through its biocompatibility and controlled degradation. Additionally, the PEG segments on the copolymer's surface prevent nonspecific protein adsorption, dropping the probability of immune responses and improving overall biomaterial performance. The combination of PLA and PEG in PLA-PEG copolymers thus offers a multifaceted solution for biomedical and related fields [40–42].

7.4.2 Poly(ethylene glycol)-block-poly(caprolactone) (PEG-PCL)

PEG-PCL is a block copolymer that strategically integrates the distinct attributes of its two constituent segments, poly(ethylene glycol) (PEG) and poly(caprolactone) (PCL), for a variety of uses, with a primary concentrate on medication distribution methods. PEG, a synthetic water-soluble polymer renowned for its hydrophilic and flexible nature, finds extensive use in biomedical applications by enhancing the solubility of hydrophobic substances, improving biocompatibility, and minimizing nonspecific protein adsorption. On the other hand, PCL, a biodegradable polyester known for providing mechanical strength and stability, boasts a slower degradation rate compared to other biodegradable polymers, making it particularly suited for controlled release applications. The synergistic combination of PEG and PCL in the block copolymer PEG-PCL yields a material with several advantages. The PEG segment imparts hydrophilicity and biocompatibility, rendering the copolymer suitable for biological applications. In drug delivery systems, PEG-PCL serves as a proficient carrier; the hydrophobic PCL block encapsulates medicines that are hydrophobic, but the hydrophilic PEG block contributes to longer circulation time and improved stability and reduced immune response. The incorporation of PCL allows for controlled and sustained drug release, leveraging its slow degradation rate. Furthermore, the PEG-PCL copolymer strikes a balance between the stability of PCL and the solubilizing and biocompatible properties of PEG, making it a well-suited material for a several uses in the biomedical industry [43–45].

7.4.3 Poly(lactic-co-glycolic acid) (PLGA)

PLGA stands out as a versatile copolymer essential in pharmaceutical and biological realms because of its biocompatibility, biodegradability, and adjustable degradation rates. Comprising lactic acid and glycolic acid monomers, PLGA's synthesis allows for tailoring its characteristics, such as deterioration rate and mechanical strength, by manipulating the ratio of these constituents. Renowned for its excellent biocompatibility, PLGA finds use in drug delivery systems, devices for health care, and tissue engineering, where its breakdown products, glycolic acid and lactic acid, naturally metabolize in the body. Its biodegradability is a pivotal advantage, facilitating gradual replacement by natural tissues in various applications. In drug delivery systems, PLGA takes various forms like microspheres and nanoparticles, releasing encapsulated drugs steadily over time. Employed in tissue engineering, PLGA scaffolds support cell growth and tissue regeneration, leveraging their porous structure for cell infiltration and integration. Additionally, PLGA contributes to medical implants, such as sutures and screws, offering temporary mechanical support during healing while gradually degrading as surrounding tissues regain strength. The tunable nature of PLGA, encompassing degradation rate, molecular weight, and mechanical strength, underscores its versatility, making it indispensable across a wide spectrum of biomedical applications [46, 47].

7.4.4 Poly(methacrylate) (PMA)

PMA is a polymer formed from monomers derived from methacrylic acid, a variant of acrylic acid featuring a methyl group. The polymerization of derivatives of methacrylic acid gives rise to polymethacrylates, exhibiting a variety of properties based on their unique structures and compositions. These polymers, including the notable example polymethyl methacrylate (PMMA), play a pivotal role across diverse industries. PMMA, recognized by trade names like Plexiglas and Acrylic, serves as a transparent and rigid plastic material with applications in optical lenses, display screens, medical devices, and other products where qualities like transparency, impact resistance, and ease of processing are essential. The versatility of polymethacrylates stems from the capacity to tailor them for specific purposes by adjusting the chemical structure and molecular weight during the polymerization process. This enables the creation of materials with diverse characteristics, including optical clarity, impact resistance, and biocompatibility. In essence, polymethacrylates serve as a dynamic class of polymers with a broad spectrum of applications, owing to their tunable properties and adaptability to various industrial needs [48–50].

7.4.5 Poly(D,L-lactide) (PLA)

PLA stands as a thermoplastic polymer within the polyester family, renowned for its biodegradable and bioactive properties. Originating from lactic acid, a naturally occurring compound often sourced from renewable materials such as corn starch or sugarcane, PLA exhibits distinct advantages in terms of biocompatibility, biodegradability, and versatility. These attributes have propelled their widespread utilization across diverse applications, including medical implants, drug delivery systems, and various industrial contexts. Due to its thermoplastic nature being able to be shaped into many forms, PLA serves as an environmentally friendly and adaptable material with a range of uses, further emphasizing its significance in both biomedical and general manufacturing fields [51, 52].

7.4.6 Poly(ethylene glycol)-poly(lactic-co-glycolic acid) (PEG-PLGA)

PEG-PLGA is a strategically designed copolymer amalgamating the advantageous traits of PEG and PLGA segments, primarily applied in drug delivery systems. PEG, recognized for its water-soluble and flexible nature, is widely employed in biomedical applications, enhancing solubility for hydrophobic substances, improving biocompatibility, and mitigating nonspecific protein adsorption. PLGA contributes biodegradability, particularly in controlled drug release applications. PEG-PLGA manifests biocompatibility through the hydrophilic PEG segment, rendering it suitable for diverse biomedical applications. Its role as a carrier in drug delivery systems is substantial, utilizing the hydrophobic PLGA to encapsulate hydrophobic drugs, while the hydrophilic PEG enhances stability, extends circulation time, and diminishes immune responses. The incorporation of PLGA permits sustained and controlled drug release, a crucial feature in medication delivery applications. Furthermore, the versatility of PEG-PLGA is exemplified by the adjustability of its properties, such as the ratio of PEG to PLGA and molecular weight, allowing tailored customization for specific biomedical needs. Overall, PEG-PLGA stands as a versatile copolymer, strategically blending PEG and PLGA attributes to address challenges in drug delivery and enhance the efficacy of therapeutic formulations [53, 54].

7.4.7 Polyethylene Glycol

PEG stands as a polyether compound crafted from repeating units of ethylene glycol through the polymerization of ethylene oxide. Acknowledged for its synthetic nature, water solubility, and broad applicability, PEG's key attributes and diverse applications are notable. Its pronounced hydrophilicity, characterized by a robust affinity for water, renders it fitting for formulations necessitating water solubility. PEG's inherent biocompatibility positions it prominently in biomedical realms, notably in drug delivery systems, medical devices, and pharmaceutical formulations. The polymer's capacity to diminish nonspecific protein adsorption proves invaluable in scenarios where interactions with proteins or cells must be minimized. Comprising a polymer backbone of ethylene glycol units, PEG exhibits flexibility and a relatively inert structure. In drug delivery systems, PEG plays a pivotal role by enhancing the solubility and bioavailability of drugs, either through conjugation or encapsulation in nanoparticles, micelles, or other drug carriers. The tunability of PEG's properties, achieved by adjusting its molecular weight, further underscores its versatility across a spectrum of applications [55].

7.4.8 Polyethylenimine (PEI)

PEI stands out as a synthetic polymer featuring a linear or branched structure composed of repeating ethylenimine units, showcasing remarkable versatility with applications across diverse fields. Notably, PEI possesses a distinctive cationic (positively charged) nature attributed to the presence of amino groups in its molecular structure, rendering it particularly adept in interactions with negatively charged substances, such as nucleic acids. In the realm of biomedicine, PEI demonstrates a degree of biocompatibility, making it suitable for specific applications; however, its use in biological systems may be constrained by factors such as cytotoxicity. Consequently, endeavors are frequently undertaken to modify or optimize PEI, aiming for enhanced biocompatibility to expand its utility in biomedical contexts [56, 57].

7.4.9 Poly(ethylene oxide) (PEO)

PEO is a lower-molecular-weight polymer that stands out as a water-soluble polyether polymer composed of repeating ethylene oxide units, offering a high degree of flexibility and widespread applicability. Its exceptional water solubility is instrumental in formulating aqueous solutions, catering to diverse industries requiring water compatibility. PEO's notable biocompatibility renders it well-suited for biomedical and pharmaceutical applications, particularly in drug delivery systems, where its water-soluble and biocompatible nature enhances drug solubility and bioavailability. The polymer's backbone, consisting of repeating ethylene oxide units, provides both flexibility and a relatively inert structure, advantageous in applications desiring a pliable material. Recognized for its high hydrophilicity and strong affinity for water, PEO finds utility in formulations requiring enhanced wetting or dispersibility. In the realm of pharmaceuticals, PEO plays a crucial role in drug delivery systems such as nanoparticles and micelles, offering solutions to improve the solubility of hydrophobic drugs and enhance overall drug stability [58, 59].

7.4.10 Poly-butyl-cyanoacrylate (PBCA)

PBCA stands as a polymer originating from butyl cyanoacrylate, a renowned type of cyanoacrylate adhesive known for its rapid and potent adhesive qualities, forming robust bonds upon polymerization. As a distinct variant of cyanoacrylate, PBCA undergoes polymerization to achieve an extended chain structure. Noteworthy features of PBCA encompass its adhesive prowess, akin to other cyanoacrylate adhesives, facilitating swift and durable connections on diverse surfaces. Explored for potential biocompatibility, PBCA has garnered attention in medical applications, particularly in tissue adhesives and wound closure, capitalizing on its capacity to form strong bonds in biological environments. The polymer may also exhibit biodegradable characteristics, contingent on its formulation, presenting advantages in scenarios where gradual degradation is desired, such as in medical devices or tissue engineering applications. With established utility in medical fields, especially in surgical procedures requiring prompt closure and healing, PBCA has ventured into drug delivery applications. Research endeavors have delved into its potential for encapsulating and releasing drugs in a controlled manner, positioning PBCA as a prospective candidate for therapeutic applications [60, 61].

7.4.11 Poly(deca-4,6-diynedioic acid) (PDDA)

PDDA is a polymer consisting of repetitive units derived from deca-4,6-diynedioic acid. Classified within the polydiacetylene family, PDDA is distinguished by its polymer backbone featuring conjugated carbon-carbon triple bonds (diyne groups). This structural characteristic contributes to the unique electronic and optical properties exhibited by PDDA [62].

7.4.12 Methoxy-Polyethylene Glycol (mPEG)

mPEG is a variant of PEG distinguished by the attachment of methoxy (CH_3O-) groups to its structure. PEG, a synthetic and water-soluble polymer renowned for its hydrophilic and flexible characteristics, finds diverse applications in the biomedical and pharmaceutical domains. The introduction of methoxy groups transforms PEG into mPEG, maintaining its highly hydrophilic nature crucial for water solubility. Notably, mPEG shares PEG's intrinsic biocompatibility, rendering it suitable for numerous biomedical uses. Its common deployment involves surface modification of nanoparticles, drugs, or biomaterials to enhance biocompatibility and mitigate interactions with the immune system. Particularly prominent is mPEG's role in drug delivery systems, where it is incorporated or conjugated with drug carriers to amplify solubility and extend drug circulation time in the bloodstream, ultimately contributing to enhanced therapeutic efficacy [63, 64].

7.4.13 α,β-Poly(N-2-hydroxyethyl)-d,l-aspartamide (PHEA)

PHEA is a polymer featuring a backbone consisting of repeating units of N-2-hydroxyethyl-d,l-aspartamide, classifying it within the poly(α-amino acids) category. Notable for its specific chemical structure, PHEA incorporates hydroxyethyl groups into its chain, contributing to distinctive properties and functionality. Designed for biodegradability, PHEA undergoes breakdown over time into components well-tolerated by the body, making it advantageous for various biomedical applications. The hydroxyethyl groups enhance the polymer's hydrophilicity, rendering it suitable for scenarios where water compatibility is essential. PHEA exhibits crucial biocompatibility, a pivotal characteristic for materials intended for use in biological systems, opening avenues for applications in drug delivery, tissue engineering, and other biomedical fields. Particularly noteworthy is PHEA's exploration in drug delivery systems, leveraging its biocompatibility, tunable degradation rate, and controlled release capacity for the targeted and sustained delivery of therapeutic agents.

The unique chemical structure of PHEA positions it as a versatile material in the biomedical realm, with ongoing research focusing on optimizing its applications for improved health-care solutions [65].

7.4.14 1,2-Distearoyl-sn-glycero-3-phosphoethanolamine-N-(methoxy(polyethyleneglycol)2000): (DSPE-PEG2000)

DSPE-PEG2000 is a pivotal phospholipid derivative extensively utilized in the construction of liposomal delivery systems. Comprising the phospholipid DSPE and a PEG chain having atypical molecular mass of 2000 Daltons, DSPE-PEG2000 brings forth several crucial features. Its integration into liposomes significantly enhances their stability and prolongs circulation time in the bloodstream. The PEGylation process, facilitated by the attachment of the PEG chain, augments liposomal hydrophilicity, mitigating issues like rapid clearance and immune system recognition. In drug delivery systems, DSPE-PEG2000 plays a pivotal role, as the PEGylation helps overcome limitations inherent in traditional liposomes, contributing to enhanced drug delivery efficiency. The inherent biocompatibility of PEG further ensures the compatibility of DSPE-PEG2000 with biological systems, a critical consideration in medical applications. Notably, the PEGylation-induced "stealth effect" reduces opsonization and phagocytosis, ultimately leading to prolonged circulation in the bloodstream – an especially advantageous trait for targeted drug delivery strategies. The versatile properties of DSPE-PEG2000 position it as a cornerstone in the realm of nanomedicine, shaping the landscape of liposomal formulations for improved pharmacokinetics and therapeutic outcomes [66–68].

7.4.15 Poly-(lactide-co-glycolide) [PLGA-b-PEG]

PLGA-b-PEG is a block copolymer integrating PLGA and PEG segments, synergizing the desirable traits of both materials for biomedical applications. PLGA, derived from lactic acid and glycolic acid, boasts biocompatibility, biodegradability, and adjustable degradation rates, making it a staple in controlled release systems and tissue engineering. On the other hand, PEG, a synthetic, water-soluble polymer, contributes hydrophilicity and flexibility, enhancing material biocompatibility. This copolymer, designed for its unique blend of properties, presents numerous advantages. The incorporation of PEG augments water solubility and overall biocompatibility, crucial in diverse biomedical applications. Additionally, PLGA-b-PEG maintains the biodegradable nature of PLGA, ensuring controlled degradation over time. The copolymer works well in medication delivery systems because the hydrophilic PEG segments increase stability and lengthen the period that hydrophobic medication are in the bloodstream, while the hydrophobic PGLA segments effectively encapsulate hydrophobic drugs [69].

7.4.16 Poly(2-methacryloyloxyethyl Phosphorylcholine (MPC)–co-n-butyl methacrylate (BMA)–co-p-nitrophenyl-oxycarbonyl Polyethylene Glycol Methacrylate (MEONP)) (PBMN)

PBMN is a sophisticated copolymer comprising three distinct monomer components. The incorporation of 2-methacryloyloxyethyl phosphorylcholine (MPC) bestows the polymer with a phosphorylcholine moiety, enhancing its biocompatibility and antifouling properties. The addition of n-butyl methacrylate (BMA) imparts hydrophobic characteristics, diversifying the copolymer's attributes. Lastly, the introduction of p-nitrophenyl-oxycarbonyl polyethylene glycol methacrylate (MEONP) incorporates a PEG segment featuring a p-nitrophenyl group, offering versatility for functionalities such as drug conjugation or bioconjugation reactions. This intricate PMBN copolymer exhibits potential applications in the biomedical field, particularly in the advance of biomaterials, drug delivery systems, or coatings where the combination of biocompatibility, antifouling properties [70].

7.5 ADVANTAGES OF SYNTHETIC POLYMERS IN DRUG DELIVERY TO CNS

Synthetic polymers present a host of benefits in streamlining drug delivery to the CNS, rendering them highly appealing for overcoming the limitations encountered with traditional drug delivery approaches [71, 72]. There are several advantages of synthetic polymers in drug delivery to CNS:

Tunable Properties: Through precise engineering, synthetic polymers can be manipulated to display particular physical characteristics like molecular weight, size, surface charge, hydrophobicity/hydrophilicity, and degradation kinetics. This level of control enables researchers to tailor polymer characteristics according to the particular demands of medication distribution to the CNS. For instance, polymers can be customized to mirror the dimension and surface charge of natural molecules, thereby aiding their passage across the BBB or improving their affinity for target cells within the CNS [73].

Biocompatibility: Numerous synthetic polymers demonstrate exceptional biocompatibility, indicating their ability to be well-received by biological systems without triggering significant immune responses or tissue toxicity. This characteristic is vital for drug delivery purposes, as it decreases the probability of adverse reactions and confirms compatibility with the physiological environments of the CNS. Common examples of biocompatible polymers utilized in CNS drug delivery encompass PLGA, PEG, and poly(vinyl alcohol) (PVA) [74].

Protection of Encapsulated Drugs: Synthetic polymers can act as protective vehicles for enclosed drugs, safeguarding them from degradation, enzymatic breakdown, and swift elimination from the bloodstream. This shielding mechanism extends the duration that drugs remain in circulation and boosts their stability, a critical factor when delivering delicate therapeutic agents to the CNS. Through encapsulation within polymeric nanoparticles or hydrogels, researchers can ensure the preservation of drug integrity and bioactivity while navigating them to their designated sites within the brain or spinal cord [74, 75].

Controlled Release Kinetics: The release kinetics of encapsulated pharmaceuticals can be precisely controlled using synthetic polymers, allowing for targeted and extended drug delivery inside the CNS. Through adjustments in polymer composition, structure, and formulation parameters, researchers can craft drug delivery systems with customized release patterns, including continuous, pulsatile, or stimulus-triggered release. This controlled release mechanism serves to enhance drug effectiveness, mitigate systemic adverse effects, and extend therapeutic benefits, particularly beneficial for managing chronic neurological disorders necessitating extended drug therapy [74, 76].

Versatility in Formulation: Synthetic polymers offer a wide range of possibilities in designing formulations, enabling the creation of diverse drug delivery systems tailored to specific therapeutic requirements. These polymers can be shaped into various forms, such as nanoparticles, microparticles, hydrogels, films, fibers, or implants, providing flexibility in how medications are administered (e.g., intravenous, intrathecal, intracerebral) and presented (e.g., sustained-release depots, injectable gels, implantable devices). This adaptability allows researchers to explore numerous approaches for delivering diverse kinds of therapeutics, encompassing tiny molecules, peptides, proteins, nucleic acids, and cells, to targeted sites within the CNS [74, 77].

Surface Modification and Targeting: Synthetic polymers possess the capability to undergo straightforward modification or functionalization with targeting ligands like antibodies, peptides, or aptamers, aiming to heighten their specificity for affected tissues and facilitate penetration via the BBB. This surface alteration enables the active targeting of drug delivery systems toward particular cell types, receptors, or biomarkers inside the CNS, thereby improving the effectiveness of treatment while reducing off-target effects. Moreover, polymeric nanoparticles or hydrogels can be engineered with stealth coatings, such as PEGylation, to avoid immune recognition and extend the period of time that they spend in the bloodstream, ultimately enhancing their capacity to reach the CNS [78, 79].

7.6 POLYMERIC DRUG DELIVERY SYSTEM

A polymeric drug delivery system pertains to a formulation or apparatus crafted to release therapeutic substances into the body in a regulated manner. This strategy seeks to improve the safety and efficiency of drugs by managing the pace, timing, and location of their release. Despite notable advancements in drug delivery in recent decades, attaining precise control over drug penetration into specific body regions, notably the brain, continues to pose challenges [80]. Nevertheless, ongoing research advancements in comprehending carrier-mediated transport mechanisms through the BBB are providing valuable insights into strategies for controlling drug dispersion to the brain. These transport systems involve uptake transporters responsible for facilitating the passage of natural nutrients like amino acids, peptides, hexoses, monocarboxylates, and even stem cells across the BBB [81].

7.7 NANOMEDICAL DRUG DELIVERY FOR NEURODEGENERATIVE DISEASE

The manipulation and application of materials at the nanoscale between 1 to 100 nanometers, is known as nanotechnology [82]. This field encompasses the design, characterization, formulation, and utilization of nanosystems, achieved through both top-down lithographic and nonlithographic fabrication techniques. The idea of extreme miniaturization and precise manipulation at a small scale was first introduced by Richard P. Feynman in his talk titled "There's Plenty of Room at the Bottom." Since Feynman's groundbreaking insights, nanotechnology has witnessed significant growth across various domains, including medicine and engineering. In the realm of nanomedicine, nanoparticles play a crucial role in diagnosing, preventing, and treating diverse diseases. Nanoparticles possess a high surface area-to-volume ratio, resulting in distinctive optical, magnetic, and biological properties within the human body [83].

Nanoparticles can be broadly categorized as organic (e.g., Liposomes) or inorganic (e.g., gold nanoparticles). Polymeric nanoparticles, a specific type, are crafted from biocompatible and biodegradable polymers. These polymers not only ensure compatibility with the biological system but also provide a protective shield, safeguarding drugs from premature degradation [84].

Nanoparticles play a pivotal role in addressing the challenges associated with delivering water-insoluble drugs characterized by low bioavailability. One illustrative example is micelles, nanoscale colloidal dispersions created from amphiphilic polymers. These micelles feature a hydrophobic tail and a hydrophilic head, enabling the hydrophobic core to effectively transport drugs in the body, particularly those with low solubility in water. Additionally, nanocarriers, formulated with polymers, offer a controlled and sustained-release effect, which proves advantageous in various medical conditions. Notably, the Food and Drug Administration (FDA) has granted approval for numerous nanoparticles for human use. These nanoparticles serve as effective nanocarriers for therapeutic applications or function as contrast agents for diagnostic imaging [85].

Nanoparticles can be further tailored by modifying them with polymers and ligands, thereby enhancing their binding affinities with genes and improving their targeting capabilities. Coating nanoparticles also contributes to refining their targeting capability. Their unique size and physicochemical properties empower nanoparticles to penetrate anatomical barriers and provide sustained-release actions. In the context of NDs, nanotechnology primarily focuses on targeting the brain and facilitating the local release of drugs after traversing the BBB [86].

7.7.1 Polymeric Nanoparticles

Polymeric nanoparticles (PNPs) can be derived from various natural and synthetic biomaterials, encompassing chitosan, albumin, gelatin, collagen, PLA, PLGA, PEG, and PCL.

The characteristics of PNPs encompass parameters such as particle size, shape, hydrophobicity, surface charges, and the composition of the coating material, all of which exert influence on their interaction with targeted cells or tissues. These properties also play a crucial role in determining the particles' ability to penetrate cell membranes and navigate through biological barriers.

Particle size is a fundamental factor governing the delivery of nanoparticles in the bloodstream and to specific target sites. Spherical nanoparticles are increasingly preferred due to their simplicity in synthesis and application. Additionally, the surface charges of nanoparticles have significant implications for their distribution in circulation, impacting their stability. Positively charged nanoparticles, as suggested by prior research, may exhibit enhanced targeting capabilities toward specific cells [87–89].

7.7.2 Polymeric Micelles

Polymeric micelles represent supramolecular nanostructures formed through the self-assembly of amphiphilic block copolymers in aqueous solutions, typically water [90]. These copolymers consist of two distinct components: a hydrophilic segment with an affinity for water and a hydrophobic segment that repels water. The spontaneous self-assembly occurs when these copolymers are introduced into a solvent, driven by the goal of minimizing the system's free energy. In the context of polymeric micelles, the hydrophobic segments come together at the core of the structure to create a hydrophobic inner compartment, while the hydrophilic segments extend outward to establish a hydrophilic outer shell. This core-shell arrangement is essential for the micelle's stability in aqueous environments, providing a conducive environment for encapsulating poorly water-soluble substances like hydrophobic drugs or imaging agents. Polymeric micelles offer numerous advantages in drug delivery and biomedical applications. They effectively solubilize hydrophobic compounds, enhancing their bioavailability and therapeutic effectiveness. With a small size typically ranging from 10 to 100 nanometers, polymeric micelles capitalize on the enhanced permeability and retention (EPR) effect, allowing for passive targeting and accumulation in diseased tissues with compromised vasculature, such as tumors. Furthermore, the outer shell of polymeric micelles can be customized with targeting ligands, such as antibodies or peptides, facilitating active targeting and improving the specificity of drug delivery to particular cells or tissues. Controlled release of the encapsulated payload can be achieved through modifications in the copolymer structure or the incorporation of stimuli-responsive elements, enabling triggered drug release in response to environmental cues like pH, temperature, or enzymatic activity. Polymeric micelles exhibit significant potential in enhancing the therapeutic index of drugs, minimizing side effects, and optimizing the efficiency of drug delivery systems. Their versatility, straightforward fabrication, and capacity to encapsulate a wide array of therapeutic agents position them as a promising platform for various biomedical applications [91–93].

7.7.3 Nanogels

Nanogels are submicrometer hydrogel particles characterized by their three-dimensional, crosslinked polymer structures. Typically falling within the nanometer scale, ranging from 1 to 100 nanometers, these structures are designed to absorb and retain water, establishing a distinct platform with diverse applications, especially in the biomedical sector. Comprising hydrophilic polymers, nanogels can be produced using various techniques such as emulsion polymerization or inverse microemulsion, enabling precise control over their size, structure, and surface attributes. A primary application of nanogels is in drug delivery, leveraging their porous configuration for efficient encapsulation of therapeutic agents. The nanogels serve to shield these agents from degradation and facilitate controlled release, responding to environmental cues like pH, temperature, or enzyme concentration for targeted delivery. Furthermore, nanogels demonstrate biocompatibility, rendering them suitable for biomedical use without causing substantial adverse effects. Biodegradable variations, in particular, offer advantages, as they can break down into nontoxic components after fulfilling their therapeutic role [94, 95].

Expanding beyond drug delivery, nanogels prove valuable in imaging and diagnostics. Through the incorporation of imaging agents like fluorescent dyes or contrast agents, nanogels function as versatile carriers, facilitating both therapeutic delivery and real-time monitoring using various imaging techniques. Additionally, nanogels play a pivotal role in tissue engineering, serving as scaffolds that replicate the extracellular matrix. Their nanoscale characteristics enhance critical aspects such as cell adhesion, proliferation, and differentiation, making them instrumental in the field

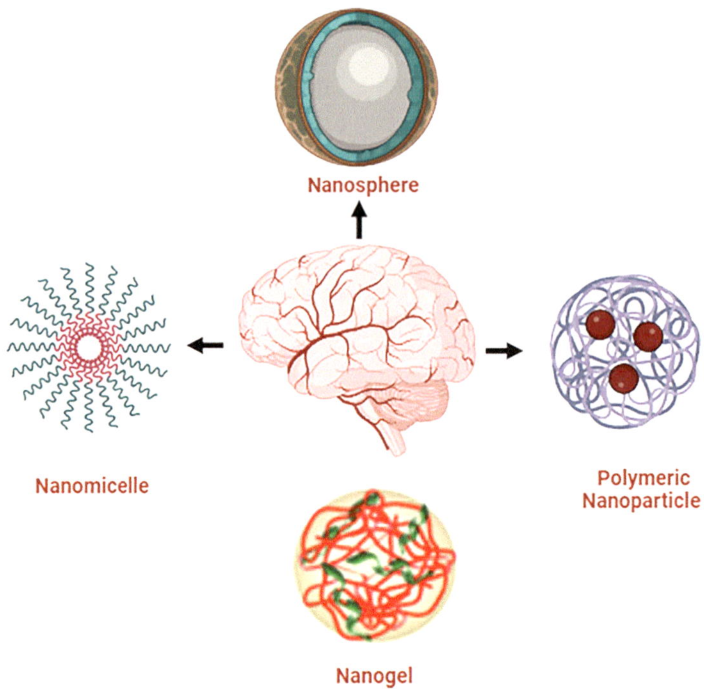

FIGURE 7.2 Types of nanoparticles used in treating neurological disorders.

of regenerative medicine. In summary, nanogels emerge as a versatile and promising category of nanomaterials, providing customized solutions for targeted drug delivery, diagnostics, and tissue engineering to drive advancements in biomedical research and health care [96–97].

7.7.4 Nanosphere

Nanospheres constitute a subset of polymeric nanoparticles, representing matrix-type structures within the size range of 10–200 nm. Widely employed as carriers in clinical drug delivery systems, nanospheres play a crucial role in enhancing the effectiveness of pharmaceutical applications. In essence, drugs are either dissolved, encapsulated, entrapped, or attached to the polymer matrix, resulting in a uniformly dispersed and homogeneous structure. Nanospheres can exhibit either amorphous or crystalline characteristics, offering protection to the drug against enzymatic and chemical degradation [98–100].

These nanospheres can be categorized as either biodegradable or nondegradable. Biodegradable variants include modified starch nanospheres, albumin nanospheres, gelatin nanospheres, polypropylene dextran nanospheres, and polylactic acid nanospheres. The versatility of nanospheres in drug administration is noteworthy, allowing for ingestion or injection. Moreover, nanospheres facilitate organ-targeted drug release, showcasing their potential in formulating diverse systems that extend the patent life of known effective pharmacophores. The inherent advantages of nanospheres make them a promising avenue for advancing drug delivery technologies in clinical applications [101] (Figure 7.2 & Table 7.2).

7.8 CONCLUSION AND FUTURE PROSPECTS

Neurodegenerative diseases including Alzheimer's, Parkinson's, and ALS, pose a significant global health challenge due to the loss of neuron function. Existing treatments predominantly focus on managing symptoms rather than tackling the underlying causes, exacerbated by the BBB hindering the entry of therapeutic agents. This review emphasizes the crucial role of polymers in advancing drug delivery technology, providing controlled release and the capacity to transport both hydrophilic and hydrophobic drugs. Nanoparticles, recognized for their targeted delivery capabilities, offer promising solutions to overcome BBB restrictions. The discussion on the pathophysiology and current management approaches for these disorders underscores the urgent need for innovative therapeutic strategies. The obstacles linked to brain-drug delivery highlight the pressing requirement for effective and targeted interventions. The examination of synthetic polymers in drug delivery, as outlined in this review, sheds light on the potential of diverse polymeric formulations to address the intricacies of NDs. Looking ahead, a collaborative push in research and development becomes imperative to translate these encouraging advancements into practical solutions, targeting the root causes of these debilitating conditions and paving the way for more impactful therapeutic interventions.

TABLE 7.2
Polymeric Nano Formulations of AD

S. No.	Delivery System	Polymer Used	Therapeutic Agent	Study *(In Vitro/In Vivo)*	Outcome	References
1	Nanoparticle	PLGA-b-PEG	Methylene blue	Invitro SHSY-5Y cells	Decrease endogenous and over expressed tau protein levels	[102]
2	Nanoparticle	PLGA	Memantine	Invitro bEnd.3 and astrocytes and *in vivo* transgenic APPswe/PS1dE9 mice	Decrease memory impairment	[103]
3	Nanoparticle	PMBN	Frankincense	Rats	Decrease Stress and depression	[104]
4	Nanoparticle	PLGA	Curcumin	Transgenic mice (5XFAD)	Inhibition of Aβ aggregation	[105]
5	Nanoparticle	PLGA and PBCA	Rivastigmine tartrate	Mice Scopolamine-induced amnesia	Antagonized the scopolamine-induced amnesia	[106]
6	Nanoparticle	PLGA-PEG	OX26	SH-SY5Y cells	Increase brain delivery of OX26	[107]
7	Nanoparticle	PLGA	Zinc	APP23 mouse model	Decrease Aβ aggregation	[108]
8	Nanoparticle	PLGA	Donepezil	Sprague-Dawley rats	Increase level of donepezil in the brain	[109]
9	Nanoparticle	PLGA	Curcumin	GI-1 cells	Decrease Aβ aggregates, antioxidant potential	[110]
10	Nanoparticle	PLGA	Naringin	Scopolamine-induced AD in rats	Increase Neuroprotection and ↓ neuro-inflammation	[111]
11	Nanoparticle	PLGA-PEG	Curcumin	HT22 cells and APP/PS1 Al transgenic mice	Increase spatial learning and memory capability	[112]
12	Nanoparticle	PLGA	Quercetin	SH-SY5Y cells in vitro APP/PS1 mice	Increase Viability of neuron cells and ↓ cognition and memory impairments	[113]
13	Nanoparticle	PLGA-PEG	Pioglitazone	APP/PS1 transgenic mice	Decrease cognition and memory impairments	[114]
14	Nanoparticle	PLGA	Vitamin D-binding protein (DBP)	5XFAD mice	Decrease Aβ aggregation	[115]
15	Nanoparticle	PLGA-PEG	Epigallocatechin gallate EGCG	APP/PS1 mice	Decrease cognition and memory impairments; ↓ Aβ aggregation	[116]
16	Nanoparticle	PLGA	Withaferin-A	DPPH assay	Antioxidant and neuroprotection potential	[117]
17	Polymeric micelles	Pluronic P127 and Pluronic F123	Morin	Wistar rats	Increase Memory	[118]
18	Polymeric micelles	DSPE-PEG(2000)	Curcumin	APPswe/PS1ΔE9 transgenic mice	Decrease Aβ aggregation,	[119]
19	Polymeric micelles	PHEA	rivastigmine	Mouse neuroblastoma cells (Neuro2a)	Decrease Aβ aggregation	[120]
20	Polymeric micelles	mPEG	Resveratrol	PC12 cells	Decrease Aβ aggregation	[121]
21	Nanogel	PDDA	Shiitake-derived lentinan	APP/PS1 transgenic mice	Decrease Aβ aggregation	[122]

TABLE 7.3
Polymeric Nano Formulations of PD

S. No.	Delivery System	Polymer Used	Therapeutic Agent	Study (*In Vitro/In Vivo*)	Outcome	References
1	Nanosuspension	PLGA and Pluronic F68	Pramipexol dihydrochloride	Invitro SH-SY5Y cell line	Increase cell viability	[123]
2	Nanoparticle	PEG-PCL	Ginkgolide B	Sprague-Dawley (SD) rats and C57BL/6 mice	Decrease behavioral deficits, ↑DA, DOPAC, and HVA levels	[124]
3	Nanoparticle	PEI	Retinoic acid	Male C57BL6 mice	Increase DA	[125]
4	Nanoparticle	PEG-PLGA	Lactoferrin (LF)	SD Rats	Lf-NP decrease striatum lesion	[126]
5	Nanoparticle	PLGA	Levodopa	Wistar rats	Increase bioavailability of levodopa	[127]
6	Nanoparticle	PEO and PCL	Levodopa and curcumin	Invitro PC12 cells	Increase viability of cells	[128]
7	Nanoparticle	PEG-PLGA and PLGA	microRNA(miR)-124-loaded Rabies virus glycoprotein (RVG)29	male C57BL/6 mice	Inhibit pro-inflammatory signaling, increase neuroprotection	[129]
8	Nanoparticle	PLGA	Selegiline	SD rats	Increase brain bioavailability	[130]
9	Nanoparticle	PEG-PLGA	Odorranalectin (OL)	SD rats	Increase OL brain delivery	[131]
10	Nanoparticle	PBCA	Dopamine	Rat	Decrease α-synucleinopathy	[132]
11	Nanoparticle	PEG	Dopamine	In vitro SH-SY5Y cellsIn vivo mice	Increase neuroprotective effects	[133]
12	Nanoparticle	PLGA	Dopamine	Mice	Increase motor symptom improvement	[134]
13	Nanoparticle	PEG-PLGA	Dopamine	SD rats	Increase dopamine	[135]

TABLE 7.4
Polymeric Nano Formulations of ALS

S. No.	Delivery System	Polymer Used	Therapeutic Agent	Study type (*In Vitro/In Vivo)*	Outcome	References
1	Nanospheres	PMA and PLA	Amantadine	In vitro NSC-34 cell	Increase cell viability	[136]
2	Nanomicelles	PEG-PCL	*N*-acetyl-L-cysteine	G93A mice	Increase distribution to the brain and spinal cord	[137]
3	Nanoparticle	PLA-PEG	Adapalene	SOD1^{G93A} mouse	Increase motor performance, ↑ neuroprotection	[138]

ABBREVIATIONS

APOE Apolipoprotein E
Aβ Amyloid beta
APP Amyloid precursor protein (APP)
PS1dE9 Presenilin 1 (PS1) with the deletion of exon 9
PC12 cell line Pheochromocytoma 12 cell line
PLA-PEG Poly(lactic acid)-poly(ethylene glycol)
PEG-PCL Poly(ethylene glycol)-block-poly (caprolactone)
PLGA Poly(lactic-co-glycolic acid)
PMA Polymethacrylate
PLA Poly(D,L-lactide)
PEG-PLGA Poly(ethylene glycol)–poly(lactic-co-glycolic acid)
PEG Polyethylene glycol
PEI Polyethylenimine
PEO Poly (ethylene oxide)
PBCA Poly-butyl-cyanoacrylate
PCL Poly (ε-caprolactone)
PDDA Poly(deca-4,6-diynedioic acid)
mPEG Methoxy-polyethyleneglycol

PHEA	α,β-Poly(N-2-hydroxyethyl)-d, l-aspartamide
DSPE-PEG2000	1,2-distearoyl-sn-glycero-3-phosphoethanolamine-N- [methoxy (polyethylene glycol)-2000]
PLGA-b-PEG	Poly-(lactide-co-glycolide)
PMBN	Poly (2- methacryloyloxyethyl phosphorylcholine (MPC)–co-n-butyl methacrylate (BMA)–co-p-nitrophenyloxycrabonyl polyethylene glycolmethacrylate (MEONP))
SD rat	Sprague-Dawley rat
BCS	Biopharmaceutics classification system

REFERENCES

1. Lerner TN, Ye L, Deisseroth K et al. (2016) Communication in neural circuits: tools, opportunities, and challenges. *Cell* 164:1136–1150.
2. Gage FH, Temple S (2013) Neural stem cells: generating and regenerating the brain. *Neuron* 80:588–601.
3. Sharma R, Kabra A, Rao MM et al. (2018) Herbal and holistic solutions for neurodegenerative and depressive disorders: leads from Ayurveda. *Curr Pharm Des* 24:2597–2608.
4. Ayaz M, Mosa OF, Nawaz A et al. (2024) Neuroprotective potentials of lead phytochemicals against Alzheimer's disease with focus on oxidative stress-mediated signaling pathways: pharmacokinetic challenges, target specificity, clinical trials and future perspectives. *Phytomedicine* 124:155272.
5. Kabra A, Sharma R, Kabra R et al. (2018) Emerging and alternative therapies for Parkinson disease: an updated review. *Curr Pharm Des* 24:2573–2582.
6. Hou Y, Dan X, Babbar M et al. (2019) Ageing as a risk factor for neurodegenerative disease. *Nat Rev Neurol* 15:565–581.
7. Lamptey RN, Chaulagain B, Trivedi R et al. (2022) A review of the common neurodegenerative disorders: current therapeutic approaches and the potential role of nanotherapeutics. *Int J Mol Sci* 23:1851.
8. Khan AU, Akram M, Daniyal M et al. (2019) Awareness and current knowledge of Parkinson's disease: a neurodegenerative disorder. *Int J Neurosci* 129:55–93.
9. Deeb O, Nabulsi M (2020) Exploring multiple sclerosis (MS) and amyotrophic lateral sclerosis (ALS) as neurodegenerative diseases and their treatments: a review study. *Curr Top Med Chem* 20:2391–2403.
10. Dal Bello-Haas V (2002) A framework for rehabilitation of neurodegenerative diseases: planning care and maximizing quality of life. *J Neurol Phys Ther* 26:115–129.
11. Sharma R, Kuca K, Nepovimova E et al. (2019) Traditional Ayurvedic and herbal remedies for Alzheimer's disease: from bench to bedside. *Expert Rev Neurother* 19:359–374.
12. Adeloye OO, Olatunji MA, Olawuwo S et al. (2019) Effect of therapeutic exercise on motor symptoms of Parkinson disease: a mini review. *Mediterr J Basic Appl Sci (MJBAS)* 3:214–217.
13. Sahoo S, Padhy AA, Kumari V et al. (2022) Role of ubiquitin–proteasome and autophagy-lysosome pathways in α-synuclein aggregate clearance. *Mol Neurobiol* 59:5379–5407.
14. Morris J et al. (2015) Amyotrophic lateral sclerosis (ALS) and related motor neuron diseases: an overview. *The Neurodiagnostic J* 55:180–194.
15. Alsultan AA, Waller R, Heath PR et al. (2016) The genetics of amyotrophic lateral sclerosis: current insights. *Degener Neurol Neuromuscul Dis* 49–64.
16. Redler RL, Dokholyan NV et al. (2012) The complex molecular biology of amyotrophic lateral sclerosis (ALS). *Prog Mol Biol Transl Sci* 107:215–262.
17. Tayeb HO, Yang HD, Price BH et al. (2012) Pharmacotherapies for Alzheimer's disease: beyond cholinesterase inhibitors. *Pharmacol Ther* 134:8–25.
18. Munoz-Torrero D et al. (2008) Acetylcholinesterase inhibitors as disease-modifying therapies for Alzheimer's disease. *Curr Med Chem* 15:2433–2455.
19. Lozupone M, Solfrizzi V, D'Urso F et al. (2020) Anti-amyloid-β protein agents for the treatment of Alzheimer's disease: an update on emerging drugs. *Expert Opin Emerging Drugs* 25:319–335.
20. Khanam S, Siddique YH et al. (2018) Dopamine: agonists and neurodegenerative disorders. *Curr Drug Targets* 19:1599–1611.
21. Latt MD, Lewis S, Zekry O et al. (2019) Factors to consider in the selection of dopamine agonists for older persons with Parkinson's disease. *Drugs Aging* 36:189–202.
22. Müller T et al. (2021) Experimental dopamine reuptake inhibitors in Parkinson's disease: a review of the evidence. *J Exp Pharmacol* 397–408.
23. Heath PR, Shaw PJ et al. (2002) Update on the glutamatergic neurotransmitter system and the role of excitotoxicity in amyotrophic lateral sclerosis. *Muscle Nerve* 26:438–458.
24. Morren JA, Galvez-Jimenez N et al. (2012) Current and prospective disease-modifying therapies for amyotrophic lateral sclerosis. *Expert Opin Investig Drugs* 21:297–320.
25. Lamptey RN, Chaulagain B, Trivedi R et al. (2022) A review of the common neurodegenerative disorders: current therapeutic approaches and the potential role of nanotherapeutics. *Int J Mol Sci* 23:1851.
26. Teixeira MI, Lopes CM, Amaral MH et al. (2023) Surface-modified lipid nanocarriers for crossing the blood-brain barrier (BBB): a current overview of active targeting in brain diseases. *Colloids Surf B Biointerfaces* 221:112999.
27. Tiwari V, Tiwari A, Sharma A et al. (2023) An optimistic approach to nanotechnology in Alzheimer's disease management: an overview. *J Drug Deliv Sci Technol* 104722.
28. Benz F, Liebner S (2020) Structure and function of the blood–brain barrier (BBB). In: *Physiology, pharmacology and pathology of the blood-brain barrier*. Springer International Publishing, Cham, pp 3–31.
29. Ding S, Khan AI, Cai X et al. (2020) Overcoming blood–brain barrier transport: advances in nanoparticle-based drug delivery strategies. *Mater Today* 37:112–125.
30. Fu BM et al. (2018) Transport across the blood-brain barrier. In: *Molecular, Cellular, and Tissue Engineering of the Vascular System*, pp 235–259. Springer.
31. Lochhead JJ, Yang J, Ronaldson PT et al. (2020) Structure, function, and regulation of the blood-brain barrier tight junction in central nervous system disorders. *Front Physiol* 11:914.
32. Zhao Y, Gan L, Ren L et al. (2022) Factors influencing the blood-brain barrier permeability. *Brain Res* 1788:147937.
33. Moretti R, Pansiot J, Bettati D et al. (2015) Blood-brain barrier dysfunction in disorders of the developing brain. *Front Neurosci* 9:40.
34. Rajpoot K, Tekade RK, Sharma MC et al. (2021) Pharmacokinetics modeling in drug delivery. In: *Biopharmaceutics and Pharmacokinetics Considerations*, pp 279–334. Academic Press.

35. Bader RA (2013) Fundamentals of drug delivery. In: *Engineering polymer systems for improved drug delivery*, pp 1–28.
36. Liu D, Yang F, Xiong F et al. (2016) The smart drug delivery system and its clinical potential. *Theranostics* 6:1306.
37. De R, Mahata MK, Kim KT et al. (2022) Structure-based varieties of polymeric nanocarriers and influences of their physicochemical properties on drug delivery profiles. *Adv Sci* 9:2105373.
38. Jain KK et al. (2020) An overview of drug delivery systems. *Drug Deliv Syst* 1–54.
39. Sahu T, Ratre YK, Chauhan SK et al. (2021) Nanotechnology based drug delivery system: current strategies and emerging therapeutic potential for medical science. *J Drug Deliv Sci Technol* 63:102487.
40. Ilyas RA, Zuhri MYM, Aisyah HA et al. (2022) Natural fiber-reinforced polylactic acid, polylactic acid blends and their composites for advanced applications. *Polymers* 14:202.
41. DeStefano V, Khan S, Tabada A et al. (2020) Applications of PLA in modern medicine. *Eng Regeneration* 1:76–87.
42. Trivedi AK, Gupta MK, Singh H et al. (2023) PLA based biocomposites for sustainable products: a review. *Adv Ind Eng Polym Res* 6:382–395.
43. Dash TK, Konkimalla VB et al. (2012) Polymeric modification and its implication in drug delivery: poly-ε-caprolactone (PCL) as a model polymer. *Mol Pharm* 9:2365–2379.
44. Bhadran A, Shah T, Babanyinah GK et al. (2023) Recent advances in polycaprolactones for anticancer drug delivery. *Pharmaceutics* 15:1977.
45. Kuznetsova EV, Sedush NG, Puchkova YA et al. (2023) Highly stable docetaxel-loaded nanoparticles based on poly (D, L-lactide)-b-poly (ethylene glycol) for cancer treatment: preparation, characterization, and in vitro cytotoxicity studies. *Polymers* 15:2296.
46. Laubach J, Joseph M, Brenza T et al. (2021) Exopolysaccharide and biopolymer-derived films as tools for transdermal drug delivery. *J Control Release* 329:971–987.
47. Nathani K, Damani M, Mangrulkar S et al. (2024) Polymeric nanofibers derived therapeutic approach for skeletal muscle regeneration: the promising future of muscle repair. *J Drug Deliv Sci Technol* 105503.
48. Corsaro C, Neri G, Santoro A et al. (2021) Acrylate and methacrylate polymers' applications: second life with inexpensive and sustainable recycling approaches. *Materials* 15:282.
49. Pereira P, Serra AC, Coelho JF et al. (2021) Vinyl polymer-based technologies towards the efficient delivery of chemotherapeutic drugs. *Prog Polym Sci* 121:101432.
50. Lansalot M, Rieger J, D'Agosto F et al. (2016) Polymerization-induced self-assembly: the contribution of controlled radical polymerization to the formation of self-stabilized polymer particles of various morphologies. *Macromol Self-Assembly* 33–82.
51. Hamad K, Kaseem M, Yang HW et al. (2015) Properties and medical applications of polylactic acid: a review. *Express Polym Lett* 9:435–455.
52. Ruiz-Rubio L, Sánchez-Bodón J, Moreno I et al. (2020) Polyester-based biodegradable polymers for commodities. *Sustainab Polym Mater* 45: 135–172.
53. Tabatabaei Mirakabad FS, Nejati-Koshki K, Akbarzadeh A et al. (2014) PLGA-based nanoparticles as cancer drug delivery systems. *Asian Pac J Cancer Prev* 15:517–535.
54. Wu JW, Li GH, Cui SQ et al. (2016) The current research status of PLGA as drug and gene carrier. *J Nano Res* 42:80–91.
55. Prete S, Dattilo M, Patitucci F et al. (2023) Natural and synthetic polymeric biomaterials for application in wound management. *J Funct Biomater* 14:455.
56. Rai R, Alwani S, Badea I et al. (2019) Polymeric nanoparticles in gene therapy: new avenues of design and optimization for delivery applications. *Polymers* 11:745.
57. Kumar R, Santa Chalarca CF, Bockman MR et al. (2021) Polymeric delivery of therapeutic nucleic acids. *Chem Rev* 121:11527–11652.
58. Bercea M, Wolf BA et al. (2021) Detection of polymer compatibility by means of self-organization: poly (ethylene oxide) and poly (sodium 4-styrenesulfonate). *Soft Matter* 17:5214–5220.
59. Chattopadhyay DK, Raju KVSN et al. (2007) Structural engineering of polyurethane coatings for high performance applications. *Prog Polym Sci* 32:352–418.
60. Hyldbakk A, Mørch Y, Snipstad S et al. (2022) Identification of novel cyanoacrylate monomers for use in nanoparticle drug delivery systems prepared by miniemulsion polymerisation: a multistep screening approach. *Int J Pharm X* 4:100124.
61. Zhang W, Mehta A, Tong Z et al. (2021) Development of polymeric nanoparticles for blood–brain barrier transfer—strategies and challenges. *Adv Sci* 8:2003937.
62. Fang F, Meng F, Luo L et al. (2020) Recent advances on polydiacetylene-based smart materials for biomedical applications. *Mater Chem Front* 4:1089–1104.
63. Hasan A, Pandey LM et al. (2015) Polymers, surface-modified polymers, and self assembled monolayers as surface-modifying agents for biomaterials. *Polymer-Plast Technol Eng* 54:1358–1378.
64. Xiao Y, Yuan M, Zhang J et al. (2014) Functional Poly (ε-caprolactone) based materials: preparation, self-assembly and application in drug delivery. *Curr Top Med Chem* 14:781–818.
65. Kim JH, Son CM, Jeon YS et al. (2011) Synthesis and characterization of poly (aspartic acid) derivatives conjugated with various amino acids. *J Polym Res* 18:881–890.
66. Kowalska M, Broniatowski M, Mach M et al. (2021) The effect of the polyethylene glycol chain length of a lipopolymer (DSPE-PEGn) on the properties of DPPC monolayers and bilayers. *J Mol Liquids* 335:116529.
67. Darwitan A, Wong YS, Nguyen LT et al. (2020) Liposomal nanotherapy for treatment of atherosclerosis. *Adv Healthc Mater* 9:2000465.
68. Beztsinna N, Tsvetkova Y, Bartneck M et al. (2016) Amphiphilic phospholipid-based riboflavin derivatives for tumor targeting nanomedicines. *Bioconjug Chem* 27:2048–2061.
69. Qian H, Wohl AR, Crow JT et al. (2011) A strategy for control of "random" copolymerization of lactide and glycolide: application to synthesis of PEG-b-PLGA block polymers having narrow dispersity. *Macromolecules* 44:7132–7140.
70. Fath PM, Yazdian F, Jamjah R et al. (2017) Synthesis and characterization of PMBN as a biocompatible nanopolymer for bio-applications. *Cell J (Yakhteh)* 19:269.
71. Naahidi S, Jafari M, Edalat F et al. (2013) Biocompatibility of engineered nanoparticles for drug delivery. *J Control Release* 166:182–194.
72. Sahoo SR, Mothilal M, Priyadharshini B et al. (2021) Thermosensitive hydrogels—A potential carrier for the delivery of drugs and macromolecules. *Int J Appl Pharm* 3:102–109.
73. Kamaly N, Yameen B, Wu J et al. (2016) Degradable controlled-release polymers and polymeric nanoparticles: mechanisms of controlling drug release. *Chem Rev* 116:2602–2663.

74. Englert C, Brendel JC, Majdanski TC et al. (2018) Pharmapolymers in the 21st century: Synthetic polymers in drug delivery applications. *Prog Polym Sci* 87:107–164.
75. Waheed S, Li Z, Zhang F et al. (2022) Engineering nano-drug biointerface to overcome biological barriers toward precision drug delivery. *J Nanobiotechnol* 20:395.
76. Gagliardi M, Borri C et al. (2017) Polymer nanoparticles as smart carriers for the enhanced release of therapeutic agents to the CNS. *Curr Pharm Des* 23:393–410.
77. Vilar G, Tulla-Puche J, Albericio F et al. (2012) Polymers and drug delivery systems. *Curr Drug Deliv* 9:367–394.
78. Beygi M, Oroojalian F, Azizi-Arani S et al. (2024) Multifunctional Nanotheranostics for Overcoming the Blood–Brain Barrier. *Adv Funct Mater* 2310881.
79. Ribovski L, Hamelmann NM, Paulusse JM et al. (2021) Polymeric nanoparticles properties and brain delivery. *Pharmaceutics* 13:2045.
80. Liechty WB, Kryscio DR, Slaughter BV et al. (2010) Polymers for drug delivery systems. *Annu Rev Chem Biomol Eng* 1:149–173
81. Song J, Lu C, Leszek J, Zhang J et al. (2021) Design and development of nanomaterial-based drug carriers to overcome the blood–brain barrier by using different transport mechanisms. *Int J Mol Sci* 22:10118.
82. Yadav AK, Shukla R, Flora SJ S (Eds.) (2022) *Nanomedical Drug Delivery for Neurodegenerative Diseases*. Academic Press.
83. Omran BA, Omran BA (2020) Fundamentals of nanotechnology and nanobiotechnology. In: *Nanobiotechnology: A Multidisciplinary Field of Science*, 1–36.
84. Bai C, Liu M (2012) Implantation of nanomaterials and nanostructures on surface and their applications. *Nano Today* 7:258–281.
85. Kashapov R, Gaynanova G, Gabdrakhmanov D et al. (2020) Self-assembly of amphiphilic compounds as a versatile tool for construction of nanoscale drug carriers. *Int J Mol Sci* 21:6961.
86. Kamaly N, Xiao Z, Valencia PM et al. (2012) Targeted polymeric therapeutic nanoparticles: design, development and clinical translation. *Chem Soc Rev* 41:2971–3010.
87. Bhardwaj H, Jangde RK (2023) Current updated review on preparation of polymeric nanoparticles for drug delivery and biomedical applications. *Next Nanotechnology* 2:100013.
88. Anwer AH, Ahtesham A, Shoeb M et al. (2023) State-of-the-art advances in nanocomposite and bio-nanocomposite polymeric materials: A comprehensive review. *Adv Colloid Interface Sci* 102955.
89. Anagnostou K, Stylianakis M, Michaleas S, Skouras A et al. (2020) Biodegradable nanomaterials. In: *Nanomaterials for clinical applications*, pp 123–157. Elsevier.
90. Ge Z, Liu S (2009) Supramolecular self-assembly of nonlinear amphiphilic and double hydrophilic block copolymers in aqueous solutions. *Macromol Rapid Commun* 30:1523–1532.
91. Xu W, Ling P, Zhang T (2013) Polymeric micelles, a promising drug delivery system to enhance bioavailability of poorly water-soluble drugs. *J Drug Deliv* 2013.
92. Gothwal A, Khan I, Gupta U (2016) Polymeric micelles: recent advancements in the delivery of anticancer drugs. *Pharm Res* 33:18–39.
93. Junnuthula V, Kolimi P, Nyavanandi D et al. (2022) Polymeric micelles for breast cancer therapy: recent updates, clinical translation and regulatory considerations. *Pharmaceutics* 14:1860.
94. Karg M, Pich A, Hellweg T (2019) Nanogels and microgels: from model colloids to applications, recent developments, and future trends. *Langmuir* 35:6231–6255.
95. Mauri E, Giannitelli SM, Trombetta M et al. (2021) Synthesis of nanogels: current trends and future outlook. *Gels* 7:36.
96. Manimaran V, Nivetha RP, Tamilanban T et al. (2023) Nanogels as novel drug nanocarriers for CNS drug delivery. *Front Mol Biosci* 10.
97. Pinelli F, Ortolà ÓF, Makvandi P et al. (2020) In vivo drug delivery applications of nanogels: a review. *Nanomedicine* 15:2707–2727.
98. Guterres SS, Alves MP, Pohlmann AR (2007) Polymeric nanoparticles, nanospheres and nanocapsules, for cutaneous applications. *Drug Target Insights* 2:117739280700200002.
99. De R, Mahata MK, Kim KT (2022) Structure-based varieties of polymeric nanocarriers and influences of their physicochemical properties on drug delivery profiles. *Adv Sci* 9:2105373.
100. Kuperkar K, Atanase LI, Bahadur A et al. (2024) Degradable polymeric bio (nano) materials and their biomedical applications: a comprehensive overview and recent updates. *Polymers* 16:206.
101. Curcio M, Altimari I, Spizzirri UG et al. (2013) Biodegradable gelatin-based nanospheres as pH-responsive drug delivery systems. *J Nanopart Res* 15:1–11
102. Jinwal UK, Groshev A, Zhang J et al. (2014) Preparation and characterization of methylene blue nanoparticles for Alzheimer's disease and other tauopathies. *Curr Drug Deliv* 11:541–550.
103. Sánchez-López E, Ettcheto M, Egea MA et al. (2018) Memantine loaded PLGA PEGylated nanoparticles for Alzheimer's disease: in vitro and in vivo characterization. *J Nanobiotechnol* 16:1–16.
104. Moazzam F, Hatamian-Zarmi A, Hosseinzadeh BE et al. (2024) Preparation and characterization of brain-targeted polymeric nanocarriers (Frankincense-PMBN-lactoferrin) and in-vivo evaluation on an Alzheimer's disease-like rat model induced by scopolamine. *Brain Res* 1822:148622.
105. Huo X, Zhang Y, Jin X et al. (2019) A novel synthesis of selenium nanoparticles encapsulated PLGA nanospheres with curcumin molecules for the inhibition of amyloid β aggregation in Alzheimer's disease. *J Photochem Photobiol B* 190:98–102.
106. Joshi SA, Chavhan SS, Sawant KK (2010) Rivastigmine-loaded PLGA and PBCA nanoparticles: preparation, optimization, characterization, in vitro and pharmacodynamic studies. *Eur J Pharm Biopharm* 76:189–199
107. Cui N, Lu H, Li M (2018) Magnetic nanoparticles associated PEG/PLGA block copolymer targeted with anti-transferrin receptor antibodies for Alzheimer's disease. *J Biomed Nanotech* 14:1017–1024.
108. Vilella A, Belletti D, Sauer AK et al. (2018) Reduced plaque size and inflammation in the APP23 mouse model for Alzheimer's disease after chronic application of polymeric nanoparticles for CNS targeted zinc delivery. *J Trace Elem Med Biol* 49:210–221.
109. Bhavna Md S, Ali M, Baboota S et al. (2014) Preparation, characterization, in vivo biodistribution and pharmacokinetic studies of donepezil-loaded PLGA nanoparticles for brain targeting. *Drug Dev Ind Pharm* 40:278–287.
110. Mathew A, Fukuda T, Nagaoka Y et al. (2012) Curcumin loaded-PLGA nanoparticles conjugated with Tet-1 peptide for potential use in Alzheimer's disease. *PLoS One* 7

111. Dashputre NL, Laddha UD, Darekar PP et al. (2023) Potential therapeutic effects of naringin loaded PLGA nanoparticles for the management of Alzheimer's disease: in vitro, ex vivo and in vivo investigation. *Heliyon* 9: e20367.
112. Fan S, Zheng Y, Liu X et al. (2018) Curcumin-loaded PLGA-PEG nanoparticles conjugated with B6 peptide for potential use in Alzheimer's disease. *Drug Deliv* 25:1091–1102.
113. Sun D, Li N, Zhang W et al. (2016) Design of PLGA-functionalized quercetin nanoparticles for potential use in Alzheimer's disease. *Colloids Surf B Biointerfaces* 148:116–129.
114. Silva-Abreu M, Calpena AC, Andrés-Benito P et al. (2018) PPARγ agonist-loaded PLGA-PEG nanocarriers as a potential treatment for Alzheimer's disease: in vitro and in vivo studies. *Int J Nanomed* 13:5577–5590.
115. Jeon SG, Cha MY, Kim JI et al. (2019) Vitamin D-binding protein-loaded PLGA nanoparticles suppress Alzheimer's disease-related pathology in 5XFAD mice. *Nanomedicine* 17:297–307.
116. Cano A, Ettcheto M, Chang JH et al. (2019) Dual-drug loaded nanoparticles of Epigallocatechin-3-gallate (EGCG)/Ascorbic acid enhance therapeutic efficacy of EGCG in a APPswe/PS1dE9 Alzheimer's disease mice model. *J Control Release* 301:62–75.\
117. Madhu S, Komala M, Pandian P (2021) Formulation development and characterization of withaferin-A loaded polymeric nanoparticles for Alzheimer's disease. *BioNanoScience* 11:559–566.
118. Singh M, Thakur V, Deshmukh R et al. (2018) Development and characterization of morin hydrate-loaded micellar nanocarriers for the effective management of Alzheimer's disease. *J Microencapsul* 35:137–148.
119. Chibhabha F, Yang Y, Ying K et al. (2020) Non-invasive optical imaging of retinal Aβ plaques using curcumin loaded polymeric micelles in APP swe/PS1 ΔE9 transgenic mice for the diagnosis of Alzheimer's disease. *J Mater Chem B* 8:7438–7452.
120. Scialabba C, Rocco F, Licciardi M et al. (2012) Amphiphilic polyaspartamide copolymer-based micelles for rivastigmine delivery to neuronal cells. *Drug Deliv* 19:307–316.
121. Lu X, Ji C, Xu H et al. (2009) Resveratrol-loaded polymeric micelles protect cells from Aβ-induced oxidative stress. *Int J Pharm* 375:89–96.
122. Xin Y, Fang F, Yue Q et al. (2024) Microenvironment modulating nanogels by Shiitake-derived lentinan and a reactive oxygen species scavenging conjugated polymer for the treatment of Alzheimer's disease. *Nano Today* 55:102178.
123. Somasundaram I, Yadav BV, Kumar SS (2016) Formulation of PLGA polymeric nanosuspension containing Pramipexole Dihydrochloride for improved treatment of Parkinson's disease. *Res J Pharm Tech* 9:810–816.
124. Zhao Y, Xiong S, Liu P et al. (2020) Polymeric nanoparticles-based brain delivery with improved therapeutic efficacy of ginkgolide B in Parkinson's disease. *Int J Nanomed* 15:10453–10467.
125. Esteves M, Cristóvão AC, Saraiva T et al. (2015) Retinoic acid-loaded polymeric nanoparticles induce neuroprotection in a mouse model for Parkinson's disease. *Front Aging Neurosci* 7:20.
126. Hu K, Shi Y, Jiang W et al. (2011) Lactoferrin conjugated PEG-PLGA nanoparticles for brain delivery: preparation, characterization and efficacy in Parkinson's disease. *Int J Pharm* 415:273–283.
127. Ahmad MZ, Sabri AHB, Anjani QK et al. (2022) Design and development of levodopa loaded polymeric nanoparticles for intranasal delivery. *Pharmaceuticals* 15:370.
128. Mogharbel BF, Cardoso MA, Irioda AC et al. (2022) Biodegradable nanoparticles loaded with levodopa and curcumin for treatment of Parkinson's disease. *Molecules* 27:2811.
129. Gan L, Li Z, Lv Q, Huang W (2019) Rabies virus glycoprotein (RVG29)-linked microRNA-124-loaded polymeric nanoparticles inhibit neuroinflammation in a Parkinson's disease model. *Int J Pharm* 567:118449.
130. Raman S, Khan AA, Mahmood S (2022) Nose to brain delivery of selegiline loaded PLGA/lipid nanoparticles: synthesis, characterization and brain pharmacokinetics evaluation. *J Drug Deliv Sci Technol* 77:103923.
131. Wen Z, Yan Z, Hu K et al. (2011) Odorranalectin-conjugated nanoparticles: preparation, brain delivery and pharmacodynamic study on Parkinson's disease following intranasal administration. *J Control Release* 151:131–138.
132. Jahansooz F, Hosseinzade BE, Zarmi AH et al. (2020) Dopamine-loaded poly (butyl cyanoacrylate) nanoparticles reverse behavioral deficits in Parkinson's animal models. *Ther Deliv* 11:387–399.
133. Gao Y, Cheng Y, Chen J et al. (2022) NIR-assisted MgO-based polydopamine nanoparticles for targeted treatment of Parkinson's disease through the blood–brain barrier. *Adv Healthcare Mater* 11:2201655.
134. Monge-Fuentes V, Biolchi Mayer A, Lima MR et al. (2021) Dopamine-loaded nanoparticle systems circumvent the blood–brain barrier restoring motor function in mouse model for Parkinson's disease. *Sci Rep* 11:15185.
135. Tang S, Wang A, Yan X et al. (2019) Brain-targeted intranasal delivery of dopamine with borneol and lactoferrin co-modified nanoparticles for treating Parkinson's disease. *Drug Deliv* 26:700–707.
136. Mazibuko Z, Indermun S, Govender M et al. (2018) Targeted delivery of amantadine-loaded methacrylate nanosphere-ligands for the potential treatment of amyotrophic lateral sclerosis. *J Pharm Pharm Sci* 21:94–109.
137. Kurano T, Kanazawa T, Iioka S et al. (2022) Intranasal administration of n-acetyl-l-cysteine combined with cell-penetrating peptide-modified polymer nanomicelles as a potential therapeutic approach for amyotrophic lateral sclerosis. *Pharmaceutics* 14:2590.
138. Medina DX, Chung EP, Teague CD et al. (2020) Intravenously administered, retinoid activating nanoparticles increase lifespan and reduce neurodegeneration in the SOD1G93A mouse model of ALS. *Front Bioeng Biotechnol* 8:224.

8 Synthetic Polymer-Derived Drug Delivery Systems in Gastrointestinal Disorders

Manisha Verma, Bina Gidwani, Ravindra Kumar Pandey, and Shiv Shankar Shukla
Columbia Institute of Pharmacy, Raipur, India

8.1 INTRODUCTION

Digestive illnesses are among the most common problems in the tropics. The primary signs of gastrointestinal (GI) disorders are diarrhea, weight loss, malabsorption, abdominal pain, and GI bleeding. Lower digestive tract disease is a broad category that includes many different disorders [1]. Duodenal ulcer disease is one of these, characterized by a mucosal rupture that reaches into the submucosa or deeper. The base of the ulcer is composed of pus and fibrin-filled necrotic tissue. The most common digestive problem is inflammatory bowel disease (IBD). The body's immune system targets the digestive system's components, categorizing IBD as an autoimmune disease. We have divided IBD into two subtypes: ulcerative colitis and Crohn's disease [2, 3].

8.2 TYPES OF GI DISORDERS

GI disorders are divided into two main types – namely, functional and structural. These are further subdivided into various types and discussed in brief in the following sections. The classification of GI disorders is shown in Figure 8.1.

8.2.1 Functional Disorder

The disorders in the human body due to various body parts like the esophagus, oral cavity, stomach, small and large intestine are based on their functions are categorized under functional disorders. These are explained in Figure 8.2.

I. **Irritable bowel syndrome (IBS)**: It's a functional disorder also known as spastic colon. It is an intestinal condition that results in altered stool appearance, constipation, diarrhea, and stomach pain. IBS comes in three different forms:
   - IBS with constipation as the main symptom
   - IBS with diarrhea predominance
   - Mixed-type IBS [4]

II. **Acid reflux (GERD or gastroesophageal reflux)**: This ailment arises from the reflux of stomach acid into the esophagus. There are two possible causes for this:
   - When it shouldn't, the lower aperture that divides the stomach and esophagus relaxes.
   - The aperture becomes insufficient to hold the stomach acid in [5].

III. **Functional dyspepsia**: It is typified by persistent or recurrent symptoms that mimic dyspepsia but don't have an underlying reason. It is believed to be caused by upper GIT inflammation that follows an infection or allergic reaction [6].

8.2.2 Structural Disorders

I. **Hemorrhoids**: Hemorrhoids are enlarged or irritated blood vessels or veins in or near the anus. When the veins are under pressure, they form. They may manifest from the inside or the outside [7].

II. **Diverticulosis**: Diverticula, or protruding pouches, are the cause of diverticulosis. Those over 40 are more likely to experience it [8].

III. **Colitis**: The term "colitis" refers to inflammation of the large intestine (colon). The inflammation linked to this illness may affect the large intestine whole or in part [9]. Under a microscope, the mucosa may seem granular if the disease is mild. In fulminant cases, areas of bleeding, ulceration, and sloughing may coexist with mucosal stripping [10]. On the other hand, inflammation is the hallmark of Crohn's disease, which affects either the small or large intestine. Every layer of the gut is affected by inflammation, including the mouth and anus [11].

There exist five subtypes of colitis:

a) IBD: A type of inflammatory bowel illness called inflammatory.

DOI: 10.1201/9781003477389-8

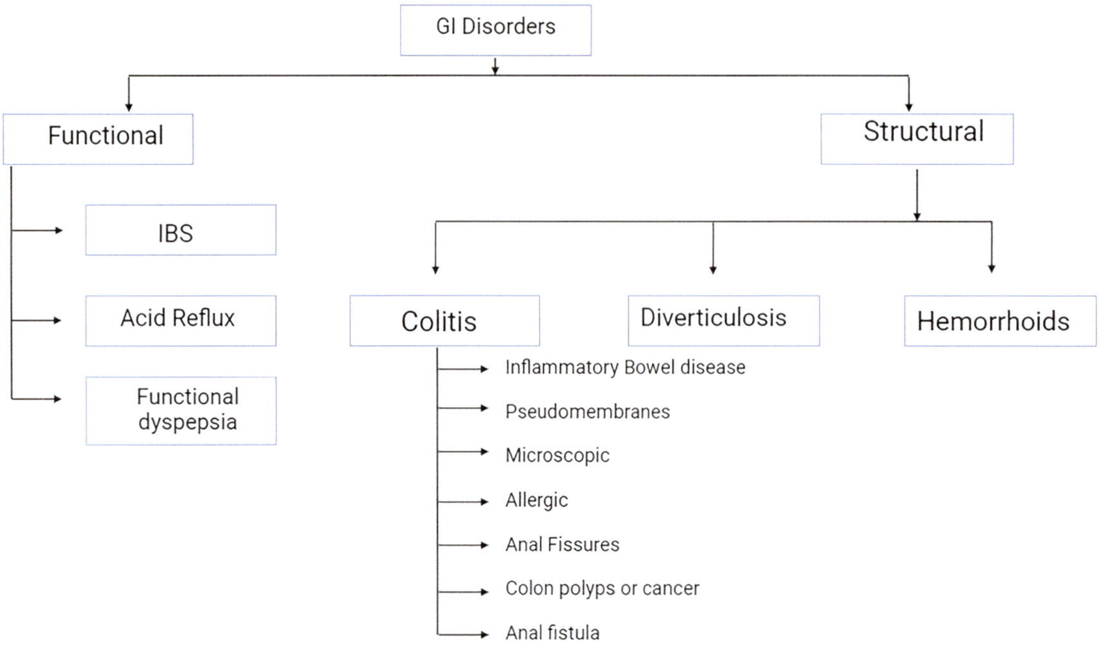

FIGURE 8.1 Classification of GI disorders.

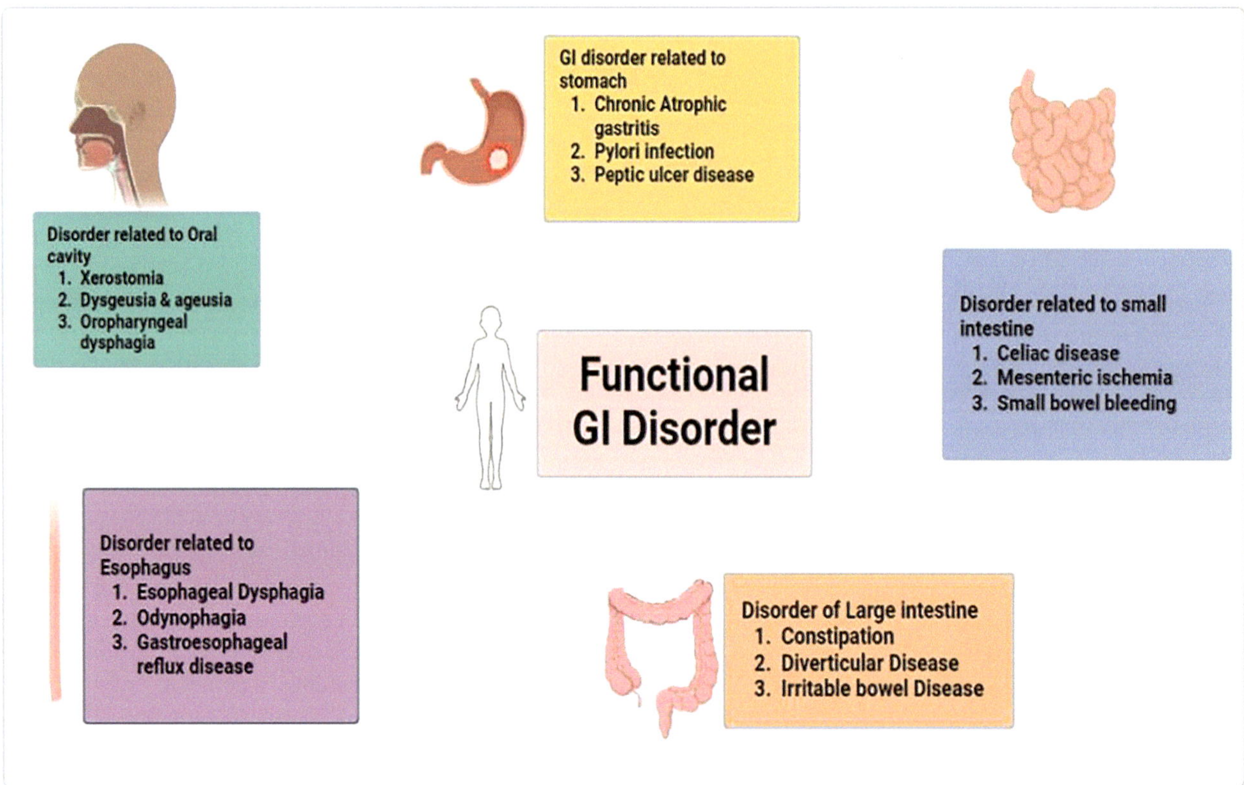

FIGURE 8.2 Functional disorders.

b) Microscopic: This type of inflammation arises from microscopic tissue inflammation in the colon [12].
c) Anal Fissures: These are microscopic rips in the mucosa, the thin, moist tissue lining the anus [13].
d) Anal Fistula: Patients with severe underlying peri-anal Crohn's disease or an infection in the anal glands may develop this form of tunnel between the skin and the anus.
e) Colon Polyps or Cancer: These are masses inside the colon that are made up of clusters of cells. While the majority are not harmful, some can develop into colon cancer [14]. The majority of GI problems may be effectively controlled; however, they may manifest with severe symptoms and negatively impact one's quality of life.

8.3 APPLICATIONS

One drug that is used to manage and treat constipation is polyethylene glycol. It belongs to the laxative medication class. This exercise elucidates the utilization of polyethylene glycol as an efficacious remedy for constipation, encompassing its indications, mechanism of action, and contraindications. This exercise will concentrate on the mechanism of action, adverse event profile, and other crucial features that members of the interprofessional team should be aware of when managing patients with constipation and related diseases. The elements encompass off-label usage, dosage, pharmacodynamics, pharmacokinetics, monitoring, and relevant interactions [15]. The formulations, which can be best suited with polyethylene glycol (PEG) includes polymeric micelles, nanoparticles, and polymeric conjugates. Some of the key formulations of PEG for GI disorder are listed in Table 8.1. Patients with appendicitis, inflammatory bowel illness, perforated intestine, known or suspected intestinal blockage, and hypersensitivity to any formulation ingredient, including polyethylene glycol, should not take PEG [16–18].

8.4 OTHERS

The pH-dependent coating polymers used to enable enteric or colon targeting of drugs often include ethylene glycerides, among other polymers. A recent study has shown that the pH of the human GIT decreases dramatically. However, the basic principle of these systems remains the same: the pH gradually increases from the colon (pH 6.5–7.0) to the small intestine (pH 2.0–3.0) and stomach (pH 7.0–8.0), with the maximum pH observed near the ileocecal junction [19, 20]. We propose integrating pH-responsive polymers with time-dependent release polymers to achieve controlled drug release from the coated system. The use of various polymers, along with their formulations and applications in GI disorders, are listed in Table 8.2.

8.4.1 Eudragit

First introduced in the 1950s, Eudragits is a brand of Rohm GmbH & Co. HG. Darmstadt, Germany. Eudragits made from acrylic and methacrylic acid polymerization, or the esters thereof. Dimethylaminnoethyl ester and butyl ester are two examples [21].

8.4.2 Types of Eudragits Polymers

8.4.2.1 Soluble Poly (meth) Acrylates

Poly (meth)acrylates that dissolve easily in digestive juices are those that produce salts.

Example: Eudragit L, S, FS, and E polymers. The active components can be released in a pH-dependent manner thanks to these polymers having acidic or alkaline groups.

Application ranges from basic taste masking to regulated drug release in all intestinal regions and gastric resistance [22].

8.4.2.2 Insoluble Poly (meth) Acrylates

These include the Eudragit RL and RS polymers, which have alkaline groups and are insoluble, as well as the Eudragit NE polymers, which are neutral and allow for swelling and controlled release of the active component in digestive juices regardless of pH.

Application: Continuous and delayed release of medication.

Under the brand name Eudragit, Eudragit Polymers-Poly(meth)acrylates are well-known throughout the industry.

8.4.2.3 Eudragit Polymers

These polymers facilitate the functionality of the active component in the solid dose form as it goes through the body, facilitating the adaptability of human organism. The ability to combine different polymers allows for the precise control

TABLE 8.1
PEG Formulations for GI Disorders

S.no.	Material	Product	Application
1	Lecithin and PEG	Nanoparticles	IBD
2	PEG	Nanoparticles	IBD
3	Pegylated pterostilbene	Micelles	IBD
4	PEG PLGA	Micelles	Ulcerative colitis
5	Azo-PEG-OMe	Conjugate	Oral delivery

TABLE 8.2
Formulations Using Eudragit Polymer

S.No.	Material	Product	Application
1	Eudragit (R)S100	Nanoparticles	Colon cancer
2	Polyether	Micelles	Ulcerative colitis
3	Acid inuli	Nanoparticles	Colon cancer
4	Ethylene glycoldimethacrylate	Hydrogel	Targeted drug release at colonic site

of drug release, ensuring that treatment is released at the appropriate time, location, and duration as needed. Shielding against external influences is another crucial duty [23]. The purpose of moisture is to disguise tastes or odors in order to enhance patient compliance. The diverse range of products in the portfolio provides optimal performance for enteric, protective, or sustained release characteristics, allowing for complete adaptability to achieve unique drug release profiles [24]. The various grades of Eudragit polymers along with their properties and applications are shown in Table 8.3.

8.4.2.4 Formulations of Eudragit

1. **Gastro resistance and GI targeting**: Eudragit L and S polymers are the recommended option of coating polymers in order to shield the active component from the gastric fluid and to increase therapeutic efficaciousness. They make it possible to target particular intestinal regions.
2. **Drug distribution via the digestive tract**: Extensive efforts have been made in academia and industry to create drug delivery methods that are gastro-retentive due to the requirement for such modes of administration.
3. **Dispensing medication orally**: Through the use of Eudragit polymer coating, sustained intestinal administration of medications was created, allowing the drug to be released into the intestine for extended periods of time while avoiding the stomach [25].

8.5 SYNTHETIC POLYMER IN GI DISORDERS

1. Yuji et al. reviewed medication delivery methods generated from polymers for treating IBD. Serious adverse effects from systemic exposure, long-term inefficiency, and increased risks of infection and malignancy due to immunosuppression hinder the

TABLE 8.3
Grades of Eudragit with Property Solubility and Application

S.no.	Trade Name	Properties	Type of Formulation	Permeability	Solubility	Application
1	Eudragit L100-55	Anionic copolymer consisting of methacrylic acid and methyl methacrylate	Enteric coated	Soluble in intestinal fluid from pH 6	Acetone, Alcohol	Small intestine, gastroesophageal reflux
2	Eudragit S 100	an anionic methacrylate copolymer	Enteric coated	Soluble in intestinal fluid from pH-7	Acetone, Alcohol	Colorectal cancer
3	Eudragit S 12.5	Anionic white free- flowing powder	Enteric coated	Soluble in intestinal fluid from pH 7	Acetone, Alcohol	Colon-specific drug delivery system
4	Eudragit S 12.5P	Anionic, white free-flowing powders	Enteric coated	Soluble in intestinal fluid from pH - 7	Acetone, Alcohol	Colon-specific drug delivery system
5	Eudragit RL 100	Cationic nonbiodegradable	Sustained release	High permeability	Acetone, Alcohol	Rheumatoid arthritis and osteoarthritis
6	Eudragit RL PO	Cationic nonbiodegradable	Sustained release	High permeability	Acetone, Alcohol	upper GI system
7	Eudragit RL 30D	Cationic nonbiodegradable	Sustained release	High permeability	Water	GIT
8	Eudragit RL 12.5	Cationic nonbiodegradable	Sustained release	High permeability	Acetone, Alcohol	Colonic drug delivery
9	Eudragit RS 100	Cationic nonbiodegradable	Sustained release	Low permeability	Acetone, Alcohol	Psoriasis and fungal infections
10	Eudragit RS PO	Cationic nonbiodegradable	Sustained release	Low permeability	Acetone, Alcohol	Small intestine, gastroesophageal reflux
11	Eudragit RS 30 D	Cationic nonbiodegradable	Sustained release	Low permeability	Water	Small intestine, gastroesophageal reflux
12	Eudragit RS 12.5	Cationic nonbiodegradable	Sustained release	Low permeability	Acetone, Alcohol	Small intestine, gastroesophageal reflux
13	Eudragit NE 30 D	Cationic, Yellow in color	Sustained release	Swellable, permeable	Water	IBDs
14	Eudragit E 100	Cationic, Yellow in color	Sustained release	Soluble in gastric fluid to pH 5	Acetone, Alcohol	IBDs
15	Eudragit NE 40D	Cationic, Yellow in color	Sustained release	Swellable, permeable	Water	Intestine

current clinical therapy. Antioxidants, gut microbiota modulators, and naturally bioactive small molecules are some of the names for reactive oxygen species scavengers that have become promising options for treating IBD [26].

2. Yang et al. examined and discussed the therapeutic potential and pharmacological importance of a variety of synthetic and natural biomaterials for effective medication targeting to the GIT through the use of the self-assembly approach. Drug delivery methods that target the GIT are becoming more and more popular as possible means of treating a variety of illnesses, particularly regional colonic illnesses. Because they are able to protect a range of medications that are prone to break down by acidity and enzymes in the upper digestive system, they result in reduced negative effects and enhanced oral administration efficiency. The unique and creative design of self-assembled nanostructures like liposomes, hydrogels, and micelles allows for easy modification in response to external triggers. This makes them well-suited for targeted drug delivery therapies and specialized medicinal needs [27].
3. In 2020, Guarino et al. conducted a review on the mechanism of action of prebiotics and their effects on GI diseases in adults. This review examined the mechanisms by which prebiotics function, as well as their impact on primary GI disorders in adults. In particular, the effects of lactulose, galacto-oligosaccharides, fructo-oligosaccharides, and newly discovered substances that are prebiotic, like xilo oligosaccharides, soybean oligosaccharides, isomalt oligosaccharides, lactobionic acid, resistant starch, and polyphenols were discussed [28].
4. Muhammad et al. briefly discussed the two main categories of polymers – natural and synthetic – that have undergone thorough investigation for their ability to effectively transport a variety of proteinous medications. The delivery of therapeutic proteins, demonstrating a wide range of natural and/or synthetic polymers as effective and safe drug carrier systems were covered [29].
5. A summary of recent developments in oral macromolecular drug administration and the advantages of polymer conjugation are provided by Fuhrmann et al. The advantages of polymer conjugation and recent advancements in oral macromolecule delivery are highlighted in this review. The recent advancements in macromolecule engineering were discussed [30].
6. Anisha et al. provide a review of the clinical status, commercial use, and state-of-the-art utilization of PEG in therapeutic applications. PEG is the preferred polymer for use in medication delivery systems. This Food and Drug Administration–approved polymer's adjustable qualities and established safety profile make it popular; these are important factors to consider when selecting any excipient during the formulation development process [31].
7. Hamed et al. examined current developments in GI diseases and nanotechnology. The fast-developing field of nanomedicine is developing targeted therapeutic techniques based on the pathophysiology of GI diseases. The many uses of the most recently created nanomaterials, which offer enormous promise for the diagnosis and treatment of GI disorders, are the main emphasis of the review [32].
8. Rajpurohit et al. investigated a polymer that targets the colon for medication delivery. The increasing importance of oral, colon-targeted drug delivery systems for both local and systemic administration of various therapeutic medicines was discussed. Oral delivery of medications to the colon protects them from breakdown or release in the stomach and small intestine. The various types of polymers used in colon-focused medication delivery were also discussed [33].
9. Liechty et al. predicted that modern breakthroughs in drug delivery will stem from rational polymer design, specifically tailored to perform diverse biological activities. The physiological impediments to medication administration emphasizing the basic drug delivery systems and their mathematical underpinnings were addressed. The history and uses of stimuli-responsive polymer systems as well as polymer treatments, including conjugates of polymers with drugs and proteins were discussed [34].
10. Dahlal et al. explored the recent progress in the rational development and production of appropriate functional polymers. These developments have yielded a variety of potential polymer-based therapies and biomaterials, some of which are currently available for commercial sale. By utilizing the advantageous characteristics of synthetic polymers, such as controlled variation in size, molecular structure, and large molecular size, a novel class of medicinal substances can be discovered. For example, the presence of large bioactive polymers to the GIT can be limited. In this location, they can recognize, attach to, and remove particular molecules that induce illness in the body. These polymeric medications don't have the systemic toxicity that is usually linked to small-molecule systemic medications because they don't affect the digestive tract and are stable in biological settings [35].
11. Chaurasiya et al. examined the treatment of colonic illnesses, highlighting the substantial variations based on the specific type and severity of the condition. Surgery, medication, and food may all be components of the treatment in specific situations. Currently, there is a significant emphasis on the use of synthetic polymers and their formulations in the treatment of colonic diseases. It enables the administration of medication at specific locations and protects it from degradation. Several synthetic polymers, either individually or in combination,

can provide the desired properties for the colon drug delivery system. Additionally, they can affect the drug's rate of release, absorption, and ability to withstand harsh environments. The treatment of colonic disorders employs various polymers with different mechanisms of action. Some of these polymers, like pectin, chitosan, and cyclodextrin, can be activated by bacteria. Others, like Eudragit and cellulose acetate phthalate (CAP), change when the pH level changes. There are also polymers that do not dissolve in water, such as ethyl cellulose and polyvinyl acetate phthalate (PVAP). Additionally, polymers such as hydroxypropyl methyl cellulose (HPMC), hydroxypropyl cellulose (HPC), and various biodegradable polymers like PLGA and diblock polymers can wear down or swell [36].

12. Wang et al. presented a comprehensive analysis of biomaterials used as therapeutic drug carriers for the purpose of treating IBD. A variety of hydrogels, microparticles, and nanoparticles using inorganic materials, lipids, synthetic polymers, and natural polymers have been developed. These diverse biomaterials can serve as effective drug delivery systems for IBD therapy, ensuring targeted transport to the colon and consistent drug release [37].
13. Kumar et al. investigated the application of polymeric nanoparticles for the management of IBD. Nanotechnology intervention holds potential because it can serve as a carrier for reduced dosages, toxicity, and frequency, thereby facilitating a faster rate of recovery. Furthermore, the field of nanomedicine and nano-diagnostic technologies will offer novel treatment opportunities by improving our understanding of pathophysiology and diagnosis, which are currently deficient [38].
14. Chaubey et al. investigated the importance of ligand-anchored therapeutic breakthroughs in the progress of targeted nanocarriers. There is strong demand for research on a pharmaceutical delivery method that is highly effective in ensuring the drug's successful release specifically to the colon. The targeted drug delivery techniques can be employed to treat several bowel illnesses, such as Crohn's disease, ulcerative colitis, and colon cancers. Ligand-anchored therapy is an effective and accomplished approach for delivering medications to particular target cells, serving both therapeutic and diagnostic objectives. When compared to traditional drugs, ligand-anchored treatment has the advantage of reducing toxicity and undesirable effects. Furthermore, the development of medication resistance is a significant obstacle to treatment effectiveness and a primary factor contributing to chemotherapy failure. Nanocarriers can be designed to encompass a wide range of functions, such as imaging and the controlled release of drugs within cells [39].
15. Mehmood et al. examined the colon-targeted sulfasalazine tablets to treat ulcerative colitis using *in vitro* methods. The sulfasalazine matrix tablets using a wet granulation compression technique with microsomal enzyme-dependent polymers were fabricated [40].
16. Sohail and colleagues presented a comprehensive analysis of intelligent biomaterials derived from both natural and synthetic polymers, specifically designed for the management of ulcerative colitis. A wide range of flexible polymeric biomaterials that can improve the concentration of drugs has been studied. Natural and synthetic polymer-based hydrogels, along with other biomaterials, have shown a lot of promise for specific uses because they can respond to changes in pH, have controlled and prolonged release properties, and release medication components through microbial breakdown in the colon [41].
17. Ray explored the advanced techniques for delivering medication specifically to the colon to treat localized diseases. In recent times, there has been a growing emphasis on the significance of delivering medication specifically to the colon for the treatment of various diseases, such as Crohn's disease, ulcerative colitis, colorectal cancer, and protein and peptide delivery. Many methods to specifically deliver medications to the colon, such as prodrugs, pH-sensitive polymers, pressure-controlled drug delivery systems, microbially triggered osmotic systems, and time-dependent systems have been explored. Due to their readily availability, harmful effects, and secure nature, natural polysaccharides have recently gained significant interest in the advancement of medication delivery systems that specifically target the colon. The majority of natural polymers prevent bioactive from reaching the upper region of the stomach. Nevertheless, the bacteria in the colon degrade glycosidic linkages, allowing bioactives to enter the colon. Pectin, guar gum, amylase, inulin, dextran, chitosan, and chondroitin sulfate are a few instances of polysaccharides [42].
18. Thakur et al. investigated the application of natural and synthetic polymers in delivering medication specifically to the colon. The stomach and small intestine specifically target the colon to break down peptides and proteins through pharmaceutical administration. Targeting a medication specifically to the colon is also beneficial for disease-modifying chemotherapy. Protecting a drug from absorption by the upper GIT is crucial for its efficient transport to the colon. One important strategy for targeting the colon is coating the pharmaceutical core with a polymer or a combination of polymers. Encapsulating the drug core with a pH-sensitive polymer or one that the

colon's bacteria can break down is an extensively studied approach for administering drugs to the colon [43].

19. Many natural remedies for colon ailments, such as ulcerative colitis, diverticulitis, and colon cancer, rely on the use of natural polymers such as guar gum, pectin, dextran, chitosan, inulin, amylose, and others. Kumar et al. conducted a study on various therapeutic approaches. The increasing interest in these biodegradable polymers stems from their low cost, safety, and chemical compatibility with other ingredients in the formulation [44].
20. Kaur et al. examined the specific delivery of drugs to the colon, also known as the lower GI system. Colonic medicine transport plays a crucial role in enhancing the distribution of proteins and peptides in the body and promoting targeted activity in specific regions. The cutting-edge technology was employed to create medications specifically tailored for absorption in the colonic area. Formulations based on prodrugs and polymers sensitive to pH changes have successfully transported drugs to the colon [45].

8.6 CONCLUSION

Synthetic polymers have revolutionized drug delivery systems, especially in managing GI disorders. Their adaptability in designing targeted, controlled-release systems has enhanced therapeutic efficacy and patient compliance. By leveraging the unique properties of synthetic polymers, such as their ability to form complex structures and respond to specific stimuli, researchers have been able to develop drug delivery systems that improve drug stability, control release rates, and target specific areas within the GIT. In conclusion, the integration of synthetic polymers into drug delivery systems for GI disorders represents a significant advancement in pharmaceutical science. These systems not only provide a means to overcome the limitations of conventional therapies but also offer the potential for personalized treatment approaches. As research continues to evolve, further innovations in polymer science and engineering are likely to enhance the effectiveness and precision of these delivery systems, ultimately leading to better patient outcomes and more efficient management of GI disorders.

8.7 FUTURE PROSPECTUS

The future prospects of synthetic polymer-derived drug delivery systems in GI disorders hold immense promise due to ongoing advancements in polymer science and technology. Several key areas are likely to drive the next generation of these systems:

1. **Targeted medication administration**: Enhancing the specificity of medication administration to particular areas inside the GIT will be a major emphasis of future systems. This involves creating polymers that react to external cues, such pH or GI-specific enzymes, to release medications exactly where they are required.
2. **Personalized Medicine**: The development of highly customizable medication delivery systems will be made easier by advancements in polymer technology. In order to improve therapeutic efficacy and reduce adverse effects, medication release patterns can be customized to each patient's needs and genetic profile. This can result in more individualized and efficient therapies for GI illnesses.
3. **Biodegradable and Biocompatible Polymers**: Reducing long-term side effects and enhancing patient safety will need the development of innovative biodegradable and biocompatible synthetic polymers. The likelihood of buildup and possible toxicity can be reduced by engineering these polymers to break down into nontoxic byproducts.
4. **Multifunctional Systems**: It is anticipated that in the future, drug delivery systems may include several functions in a single platform, such as therapy and diagnostics. Polymers might be developed to incorporate sensors or imaging agents that track the evolution of a disease and modify medicine delivery as necessary.
5. **Regenerative Medicine**: Tissue engineering and repair are two regenerative methods for GI problems in which synthetic polymers may be useful. Researchers hope to aid in the regeneration of injured GI tract tissues or organs by employing polymers as scaffolds.
6. **Advanced Manufacturing Techniques**: New developments in manufacturing, such as nanotechnology and 3D printing, will make it possible to construct polymer-based delivery systems that are more intricate and accurate. These methods will make it possible to produce personalized medication delivery systems with improved efficacy and usefulness.

ABBREVIATION

IBD	Inflammatory bowel disease
GERD	Gastroesophageal reflux disease
GIT	Gastrointestinal tract
LDPE	Low-density polyethylene
HDPE	High-density polyethylene
PP	Polypropylene
PVC	Polyvinyl chloride
PS	Polystyrene
TPU	Thermoplastic polyurathanes
PEG	Polyethylene glycol
IBS	Inflammatory bowel syndrome
NP	Nanoparticles

ACKNOWLEDGMENT

The authors are also grateful to the Department of Science and Technology (DST-FIST) Letter no- SR/FIST/COLLEGE/2018/418, New Delhi, for providing financial assistance.

REFERENCES

1. Walker, C. L. F., Rudan, I., Liu, L., Nair, H., Theodoratou, E., Bhutta, Z. A., … & Black, R. E. (2013). Global burden of childhood pneumonia and diarrhoea. *The Lancet*, *381*(9875), 1405–1416.
2. Ng, S. C., Bernstein, C. N., Vatn, M. H., Lakatos, P. L., Loftus, E. V., Tysk, C., … & Colombel, J. F. (2013). Geographical variability and environmental risk factors in inflammatory bowel disease. *Gut*, *62*(4), 630–649.
3. McCallum, R. W. (2019). Gastrointestinal Disorders—An Open Access Journal on Gastroenterology. *Gastrointestinal Disorders*, *1*(1), 1–2.
4. Vahedi, H., Ansari, R., Mir-Nasseri, M. M., & Jafari, E. (2010). Irritable bowel syndrome: a review article. *Middle East J. Dig. Dis*., *2*(2), 66.
5. Clarrett, D. M., & Hachem, C. (2018). Gastroesophageal reflux disease (GERD). *Missouri Medicine*, *115*(3), 214.
6. Ghoshal, U. C., & Singh, R. (2012). Functional dyspepsia: the Indian scenario. *J Assoc Physicians India*, *60*(Suppl), 6–8.
7. Lohsiriwat. (2012) Hemorrhoids: From basic pathophysiology to clinical management; *World J. Gastroenteral*; Vol. 18, No.17.
8. Mallapeta N.S., Farooq U., Patel K.; (2023) *Diverticulosis*; Statpearls.
9. Segal J.P., Leblanc J.F., Hart A.L.; (2021) Ulcerative colitis: An update; *Researchgate*; Vol. 21, No. 2:135–139.
10. Gisbert J.P., Chaparro M.; (2019) Clinical usefulness of proteomics in inflammatory bowel disease: A comprehensive Review; *J. Crohns Colitis*; Vol. 13, No. 3; 374–384.
11. Li, S.; Xie, A.; Li, H.; Zou, X.; Zhang, Q. (2019) A self-assembled, ROS-Responsive Janus-Prodrug for Targeted Therapy of Inflammatory Bowel Disease. *J. Control. Release*; Vol. 316; 66–78.
12. McDowell C., Farooq U., Haseeb M.;(2023) Inflammatory bowel disease; *Statpearls*. PMID: 29262182
13. Jahnny B., Ashurst J.V.; (2022) Anal fissures: Statpearls.
14. Huck M.B., Bohl J.L.; (2016) Colon Polyps: Diagnosis and surveillance. *Clin Colon Rectal Surg* Vol. 29; No. 4: 296–305.
15. D'souza A.A., Shegokar R.; (2016) Polyethylene glycol (PEG)- A versatile polymer for pharmaceutical application; *Researchgate*; Vol. 13, No. 9; 1257–1275.
16. Wang, X.; Yan, J.; Wang, L.; Pan, D.; Xu, Y.; Wang, F.; Sheng, J.; Li, X.; Yang, M.; (2020) Oral delivery of anti-TNF antibody shielded by natural polyphenol-mediated supramolecular assembly for inflammatory bowel disease therapy. *Theranostics* ; Vol. 10; 10808–10822.
17. Zhong, H.; Zhang, H.; Hu, Y.; He, S.; Qu, H.; He, J.; Chen, J.; Chen, J.; Zhang, L.; Deng, C.;(2023) Self-assembled micelle derived from pterostilbene ameliorate acute inflammatory bowel disease. *Int. J. Pharm*.; Vol. 630; 122420.
18. Zhang, C.; Li, J.; Xiao, M.; Wang, D.; Qu, Y.; Zou, L.; Zheng, C.; Zhang, J.; (2022) Oral colon-targeted mucoadhesive micelles with enzyme-responsive control. Release of curcumin for ulcerative colitis therapy; chin. *Chem. Lett*.; Vol. 33; 4924–4929.
19. Sarangi, M.K.; Rao, M.E.B.; Parcha, V.;(2021) Smart polymers for colon targeted drug delivery systems: A review; *Int. J. Polym. Mater. Polym. Biomater*; Vol. 70; 1130–1166.
20. Sunoqrot, S.; Abujamous, L.; (2019) PH-Sensitive polymeric nanoparticles of quercetin as a potential colon cancer-targeted nanomedicine; *J. Drug Deliv. Sci. Technol*.; Vol. 52; 670–676.
21. Hou, Y.; Jin, J.; Duan, H.; Liu, C.; Chen, L.; Huang, W.; Gao, Z.; Jin, M.;(2022) Targeted therapeutic effects of oral inulin-modified double-layered nanoparticles containing chemotherapeutics on orthotopic colon cancer; *Biomaterials*; Vol. 283; 121440.
22. Zhao, X.-X.; Ma, S.-B.; Wen, J.-P.; Hu, D.-T.; Gao, J.-S.; Peng, Q.; Zhang, Y.-C.; Wang, J.-Q.; Wang, K.; Shi, X.-P.;(2022) Reactive oxygen species-responsive polyether micelle nanomaterials for targeted treatment of ulcerative colitis.; *J. Biomed. Nanotechnol*.; Vol. 18; 120–131.
23. Abdullah, O.; Usman Minhas, M.; Ahmad, M.; Ahmad, S.; Ahmad, A.;(2019) Synthesis of hydrogels for combinatorial delivery of 5-fluorouracil and leucovorin calcium in colon cancer: Optimization, in vitro characterization and its toxicological evaluation; *Polym. Bull*.; Vol. 76; 3017–3037.
24. Mohammadi G., Mirzaeei S., Taghe S., Mohammadi P.;(2019) Preparation and evaluation of eudragit L-100 Nanoparticles loaded impregnated with KT tromethamine loaded PVA-HEC insertions for ophthalmic drug delivery. *Adv Pharm Bull*; Vol. 9, No. 4; 593–600.
25. Santos J.J., Silva G. S. D., Velho M.C., Beck R.C.;(2021) Eudragit- A versatile family of polymers for hot melt extrusion and 3D printing processes in pharmaceuticals ;*MDPI*; Vol. 19, No. 9;1424.
26. Yuji P., Fan X., Zhang Z., Guo Z., Pan Q., Gao W. et al.;(2023) Harnessing polymer derived drug delivery system for combating inflammatory bowel disease; *J. Control Release*; Vol. 354; 1–18.
27. Yang X., Yang Y., Yu H., Zhou Y.;(2023) Self assembles polymers for gastrointestinal tract targeted delivery through the oral route: An Update; *MDPI*; Vol. 15 No. 17; 3538.
28. Guarino M.P.L., Altomare A., Emerenziani S., Rosa D.D., Ribolsi M., Balestrieri P. et al.; (2020) Mechanism of action of prebiotics and their effects on gastro-intestinal disorders in adults; *MDPI*; Vol.12; 1037.
29. Muhammad S.H.A., Rehman K., Chen S.; (2017) Natural and synthetic polymers as drug carriers for delivery of therapeutic proteins; *Polym. Rev*.; Vol. 55; 371–406.
30. Fuhrmann K. Fuhrmann G.;(2017) Recent advances in oral delivery of macromolecular drugs and benefits of polymer conjugation; *Curr. Opin. Colloid Interface Sci*.; Vol. 31; 67–74.
31. D'Souza A.A., Shegokar R.;(2016) Polyethylene glycol (PEG): A Versatile polymer for pharmaceutical applications; *Expert Opin. Drug. Deliv*.; Vol. 19, No. 9; 1257–1275.
32. Laroui H., Wilson D. S., Dalmasso G., Salaita K., Murthy N., Sitaraman S.V., Merlin D.;(2011) Nanomedicine in GI; *Am. J. PhysiolGastointest Liver Physiol*.; Vol. 300, No. 3; 371–383.
33. Rajpurohit H., Sharma P., Sharma S., Bhandari A.;(2010) Polymers for colon targeted drug delivery; *Indian J. Pharm Sci*.; Vol 72, No. 6; 689–696.
34. Liechty W.B., Kryscio D. R., Slaughter B.V., Peppas N. A.; (2010) Polymers for drug delivery; *Annu. Rev. Chem. Biomol. Eng*.; Vol. 1; 149–173.
35. Dhal P.K., Polomoscanik S.C., Avila L.Z., Farley R.H., Miller R.J.;(2009) Functional polymers as therapeutic agents: Concept to market place; *Adv. Drug Deliv. Rev*.; Vol. 61, No. 13; 1121–1130.

36. Chaurasia, S., & Mishra, B. (2024). Synthetic polymers as biomaterials for the treatment of colon diseases. In *Advanced Drug Delivery Systems for Colonic Disorders* (pp. 95–130). Academic Press.
37. Wang, C. P. J., Byun, M. J., Kim, S. N., Park, W., Park, H. H., Kim, T. H., … & Park, C. G. (2022). Biomaterials as therapeutic drug carriers for inflammatory bowel disease treatment. *J. Control Release*, *345*, 1–19.
38. Minakshi, P., Kumar, R., Ghosh, M., Brar, B., Barnela, M., & Lakhani, P. (2020). Application of polymeric nano-materials in management of inflammatory bowel disease. *Curr. Top. Med. Chem.*, *20*(11), 982–1008.
39. Chaubey, P., Momin, M., & Sawarkar, S. (2020). Significance of ligand-anchored polymers for drug targeting in the treatment of colonic disorders. *Front. Pharmacol.*, *10*, 1628.
40. Mehmood, Y., Hammad, Y., & Umer, F. (2019). Formulation development using different natural and semi synthetic polymers, in vitro evaluation of colon targeted Sulfasalazine tablets for ulcerative colitis. *Int J Biosci*, *15*(1), 42–55.
41. Sohail, M., Mudassir, Minhas M. U., Khan, S., Hussain, Z., De Matas, M., … & Ullah, K. (2019). Natural and synthetic polymer-based smart biomaterials for management of ulcerative colitis: a review of recent developments and future prospects. *Drug. Deliv. Transl. Res.*, *9*, 595–614.
42. Ray, S. (2019). Advanced colon-specific delivery systems for treating local disorders. In *Polysaccharide carriers for drug delivery* (pp. 737–762). Woodhead Publishing.
43. Thakur, B., Pandit, V., Ashawat, M. S., & Kumar, P. (2016). Natural and synthetic polymers for colon targeted drug delivery. *AJPTech*, *6*(1), 35–44.
44. Kumar, D., Sharma, M., Verma, S., & Saroha, K. (2016). Natural polymers and herbal medicine based therapy for colonic diseases. *Int. J. Herb. Med.*, *4*(3), 49–56.
45. Kaur, A., Kaur, A. P. Kaur, V., Kaur, M., & Murthy, R. S. R. (2014). Polymeric drug delivery approaches for colon targeting: A review. *Drug Deliv. Lett.*, *4*(1), 38–48.

9 Applications of Synthetic Polymers for Drug Delivery Systems for the Management of Eye Disorders

Pooja Jain, Sradhanjali Mohapatra, Mohd. Aamir Mirza, and Zeenat Iqbal
Department of Pharmaceutics, SPER, New Delhi, India

9.1 INTRODUCTION

The eye is an organ that detects light and communicates with the brain via the optic nerve. Perhaps one of the most complicated organs in the body, the eye is made up of various sections, each of which contributes to our capacity to see [1, 2]. Being a complex organ, it is immune-privileged and impedes circulation owing to strong anatomical and physiological barriers. The anterior segment and posterior segment are the two basic segments of the eye. The cornea, conjunctiva, aqueous fluid, iris, ciliary body, and lens are all parts of the anterior segment. The choroid, retina, sclera, optic nerve, and vitreous humor make up the posterior section.

Even though each eye condition has its own specific causes and symptoms, they all eventually lead to total blindness if untreated. Ocular pharmaceutical delivery faces challenges due to the eye's intricate anatomy, unique characteristics, and isolation from other bodily parts [3].

The eye's complex physiology and structure create special barriers that act as a defense against germs and environmental pollutants. In the field of ocular research, tremendous efforts are being made to develop novel drug delivery systems (DDSs) that are patient-acceptable and safe [2, 4, 5]. Topical eye drops and ointments are examples of traditional medication delivery techniques that are marred with poor bioavailability and few therapeutic benefits. This calls for the development of cutting-edge techniques to enhance medication delivery to the intended ocular tissues and boost therapeutic efficacy. Since synthetic biodegradable polymers release drugs precisely, steadily, and with controlled behavior at the target site, they have emerged as the most promising drug candidates for ocular drug delivery. Also, their inherent properties such as biocompatibility, adaptability, and versatile nature in formulation development coincide with the requirements of eye physiology. The utilization of synthetic polymer-based DDSs presents several benefits, including increased drug stability, prolonged residence time, improved bioavailability, and decreased dosing frequency. This approach helps to close the gap between the unfulfilled therapeutic need and the ocular barriers. Overall, this chapter aims to provide a thorough understanding of the role of synthetic polymers in ocular drug delivery, showcasing their potential to enhance the management of eye disorders. By bridging the gap between material science and ophthalmology, synthetic polymers represent a promising frontier in the quest for more effective and patient-friendly ocular therapeutics. The insights and developments presented in this chapter will be invaluable for researchers, clinicians, and regulatory professionals involved in the field of ocular drug delivery.

Although studies are still going on to improvise the drug delivery to eyes but their translation from laboratory to clinical practice is still questionable and warrants clinical studies. This chapter deals with the various types of synthetic polymer used for ocular drug delivery and synthetic polymers-based DDSs to cater to anterior and posterior segment diseases. Also, an attempt has been made to enlighten the readers about the regulatory challenges imposed on ocular DDSs.

9.2 SYNTHETIC POLYMERS FOR OCULAR DRUG DELIVERY SYSTEMS

The growing interest in synthetic polymers can be attributed to their inherent versatility. Recent work has shown the utilization of these artificial materials in a vast array of forms, thanks to their exceptional properties. Synthetic polymers are a broad class of macromolecules that have been produced artificially [6]. Synthetic polymers provide unmatched control over their structure and function, in contrast to their natural counterparts. This is because there is a large range of monomers available, the building blocks from which these polymers are constructed, and the vast array of fabrication methodologies that can be employed. This tailored design allows synthetic polymers to be precisely engineered to meet the specific demands of a particular application, making them especially attractive for use in the biomedical domain [7].

The human eye is a delicate organ with complex tear dynamics that present challenges for delivering therapeutic drugs. Traditional eye drops often suffer from poor bioavailability due to rapid tear turnover and drainage. Synthetic

DOI: 10.1201/9781003477389-9

polymers offer a promising solution by enabling the development of novel drug delivery systems (NDDS) for the eye. These systems have the potential to improve drug release and delivery, patient experience, and treatment compliance by providing targeted, controlled, and sustained drug delivery to the anterior or posterior segment of the eye. They can also overcome the drawbacks of traditional therapies.

Notable artificial polymers that have been authorized by the Food and Drug Administration (FDA) for use in ocular applications and current clinical use are poly(vinyl alcohol) (PVA), poly(ethylene glycol) (PEG), poly(glycolic acid) (PGA), poly(lactic-co-glycolic acid) (PLGA), poly[2-(dimethylamino) ethyl methacrylate) (DMAEM), poly(caprolactone) (PCL), poly(acrylic acid) (PAA), and poly(amidoamine) (PAMAM). Additionally, numerous other polymers are accessible for experimental use or have been permitted for applications in areas other than the eye.

The potential of a broad variety of synthetic polymers in ocular DDS is currently under investigation. Each class possesses unique characteristics that make them suitable for different drug delivery strategies. Here, we will discuss some of the most prominent classes.

9.2.1 Polyesters

This is a significant class of synthetic polymers that is frequently applied to the delivery of drugs. PGA, polylactic acid (PLA), and PLGA are FDA-approved for use in ocular applications and serve as excellent examples. Because these hydrophobic polyesters allow for controlled in vivo biodegradation based on the monomer ratio present in the polymer, they have been used as drug carriers for proteins, genes, and small molecules. They exhibit excellent biocompatibility, making them safe for use in the eye [8].

9.2.2 PLA/poly(lactic alcohol) and PGA/poly(glycolic alcohol)

Natural monomers of lactic acid and glycolic acid are the building blocks of both PLA and PGA; they are both produced via ring-opening polymerization. For up to two years, PLA can biodegrade in vivo; however, PGA degrades more quickly because of its hydrolysis susceptibility. Both polymers' biodegradation yields nontoxic by-products. PLA can be formed in several ways, including injection molding and extrusion, and it is very thermally stable. Although PGA can't be processed as a stand-alone polymer since it's brittle and insoluble in most organic solvents, it's frequently combined with other polymers [9]. Phase 2 clinical trials for Brimo DDS®, an intravitreal PLA implant for the slow release of brimonidine to treat geographic atrophy, were successfully finished. It makes use of PLA's slow rate of biodegradation [10].

9.2.3 PLGA/poly(lactic-co-glycolic acid)

PLGA is a biodegradable polyester made of glycolic and lactic acid copolymerization. The ratio of these monomers determines the degradation rate of the polymer, allowing for tunable release profiles. Lactic acid and glycolic acid, the breakdown products of PLGA, are naturally occurring metabolites that the body can easily eliminate. This biodegradability is particularly advantageous for sustained-release applications, as the polymer carrier degrades after releasing the encapsulated drug. Additionally, PLGA can be formulated to encapsulate both hydrophobic and hydrophilic drugs, offering versatility in drug delivery [11]. When applied topically, a hydrophobic polymer can enable selective penetration through mucous membranes; when given intravitreally, it can reduce the amount of therapy diffused away from the target area and slow release. Owing to these degrading characteristics, one of the most often utilized polymers in authorized polymeric ocular DDSs is PLA. However, because of the consequences of acidic degradation, certain PLGA-based systems have had difficulties. Ozurdex® is a PLGA-based intravitreal implant that has received FDA approval for the extended release of dexamethasone in the treatment of retinal vein occlusion, noninfectious posterior uveitis, and diabetic macular edema (DME). Dexamethasone is coextruded with the implant, permitting four to six months of controlled release before its total in vivo degradation [12]. Durysta® is a recently approved PLGA implant for controlling intraocular pressure (IOP) in open-angle glaucoma patients. For four to six months, the implant lowers IOP by delivering prostaglandin analogue bimatoprost under controlled release [13].

Apart from these biodegradable implants, research is currently being conducted on the use of PLGA in hydrogel, microparticle, and nanoparticle delivery systems. A number of good preclinical studies have evaluated an intravitreal biodegradable system that uses hydrogel-embedded PLGA microspheres to deliver biologics for the management of wet age-related macular degeneration (AMD) [14].

9.2.4 PCL/poly(caprolactone)

Induced ring-opening polymerization is the process that yields PCL, which is composed of ε-caprolactone monomers. Depending on the location and size of the implant, as well as the molecular weight of the polymer, this semicrystalline, hydrophobic polyester polymer deteriorates gradually over months to years. PCL is widely used for experimental drug delivery vehicles due to its low cost, processability, ease of modification, and ease of copolymerization. Because of its crystalline characteristics, PCL is an excellent material for cell delivery and thin films since it can sustain structural integrity even in the final stages of deterioration [15]. In their experimental works, Samy et al [16]have demonstrated that PCL thin films can release drugs for longer periods while causing less intraocular inflammation. Additionally, recent experiments focusing on injectable in situ hydrogels, contact lens embedded nanoparticles, nanoemulsions and nanosuspensions, microparticles, and microcapsules for the management of conditions like wet AMD and glaucoma have shown promise for PCL as an intraocular drug delivery vehicle. It has not yet received FDA approval for use in the eye, despite having approval for other applications, such as sutures [17–21].

9.2.5 Poly(ortho ester) (POE)

POE is another notable synthetic polyester polymer that has been used in ocular medication delivery devices. POE is a promising candidate that relies on surface erosion for therapeutic release, despite not having FDA approval for ocular drug delivery yet. The FDA has given POE approval to deliver drugs in various applications. For ocular use, a thermosetting PEG-polyacetal–POE hydrogel DDS has been created and patented [22]. The use of POE for medicinal delivery has considerable unexplored potential.

9.2.6 Polymethacrylates

Biocompatible thermoplastics with excellent mechanical strength, water transmissibility, and thermal stability are poly(methyl methacrylate) (PMMA) and its derivatives, methacrylate-based polymers [23] FDA-approved for intraocular usage, they were initially utilized in contact lenses in 1936 [24]. PMMA's non-degradable nature makes it suitable for applications requiring long-term implants, such as intraocular lenses (IOLs). Its biocompatibility ensures minimal tissue irritation. Additionally, PMMA possesses good optical clarity, crucial for maintaining vision through implanted lenses. Distinct methacrylate compounds of significance include poly(2-hydroxyethylmethacrylate) (HEMA) and DMAEM. I-vation®: In the context of DME patients, titanium-based intravitreal implants employ polymethylmethacrylate (PMMA) as a long-lasting, non-degradable polymer covering, delivering a continuous release of triamcinolone acetonide [25]. Phase I clinical trials for the implant were successful, but phase IIb was suspended. After adding a vitamin E release barrier, Acuvue® HEMA-based soft contacts have shown promise as drug delivery vehicles by enabling longer release of ofloxacin [26]. Drug-Eluting Contact Lenses: Due to the enormous number of contact wearers, medication-eluting HEMA-based contact lenses are becoming increasingly popular for ocular drug delivery. Pereira-da-Mota et al. used HEMA and other methacrylates to develop atorvastatin-eluting lenses for treating diabetes-related ocular disorders [27]. Methacrylate-based polymer experiments include the use of DMAEM nano gel-based drug carriers to improve muco-adhesion in topically administered delivery vehicles, HEMA-based lenses to elute olopatadine for allergic conjunctivitis, and HEMA contact lenses with temperature-triggered release for the treatment of glaucoma [28].

9.2.7 Polyethylene Oxides/PEG

PEG is a transparent, colorless hydrophilic polymer that improves therapeutic solubility, bioavailability, and biocompatibility. It is derived from ethylene oxide monomers. PEG is available in a wide range of molecular weights, topologies (linear or multiarmed), liquid and solid forms, and activities (bioinert, tetrafunctional). It has been approved by the FDA for several uses, including ophthalmic use, and is generally regarded as safe (GRAS). PEG is highly biocompatible and possesses low protein binding capacity. This reduces the interaction between the polymer and tear proteins, potentially leading to improved drug bioavailability. Additionally, PEG can be conjugated to other polymers or nanoparticles to modify their surface properties, enhancing drug loading and targeting specific ocular tissues. Macugen®: A PEGylated oligoribonucleotide approved in 2004 for late-stage AMD. It is an injectable treatment that binds to VEGFs, or vascular endothelial growth factors [29]. Dextenza® is a PEG-based cylindrical implant that uses hydrolytic erosion to slow down and thus provides controlled release of dexamethasone to reduce inflammation and manage pain after surgery. It was given FDA approval in 2018 and is currently in phase 3 clinical trials for allergic conjunctivitis. It has a one-month release period [30] Ocular Therapeutic Products: Tyrosine kinase inhibitor (OTX-TKI) and dexamethasone intracanalicular ophthalmic insert (OTX-DED) are two examples of these. Phase 2 clinical trials are being conducted for dry eye disease (DED) using OTX-DED, a medication with a lower dose and shorter duration than Dextenza®. In phase 1 clinical trials, OTX-TKI, an injectable PEG hydrogel, offers sustained release of axitinib for the treatment of wet AMD [31]. PEG Hydrogels and Nanostructured Lipid Carriers: Evidence from studies has demonstrated that PEGylation prolongs the half-life of intravitreal injections and increases the bioavailability and corneal absorption of steroids applied topically [32]. Amphotericin B Delivery: Incorporating PEG into nanostructured lipid carriers improves solubility and effectiveness in treating ocular fungal infections [33].

9.2.8 Polyanhydrides

This class of synthetic polymers degrades through a process called hydrolysis, where water breaks down the polymer chains. Polyanhydrides (PAs) offer a unique advantage in terms of their rapid degradation rates compared to other polymers like PLGA. Shorter durations of controlled release of water-soluble medications are made possible by this. Additionally, the degradation products of poly anhydrides are nontoxic and readily eliminated by the body [34]. Each of their subclasses is distinguished by the kind of monomer unit that connects them via an anhydride bond. Aromatic PAs are traditional PA categories. Examples of PAs with aromatic rings include poly(isophthalic anhydride) and poly(terephthalic anhydride), which have higher hydrolytic, mechanical, and thermal stability and degrade more slowly. Aliphatic chains in aliphatic PAs cause decreased stability but quicker degradation, such as in poly(sebacic anhydride). Unsaturated PAs are crystalline and insoluble in typical organic solvents. They are derived from aliphatic or aromatic monomers with unsaturated bonds. In order to balance stability and degradation rates, new subclasses of PAs have been developed, such as aromatic-aliphatic PAs like poly(carboxyphenoxy propane-sebacic acid) [35]. PAs hold promise for the controlled release of water-soluble drugs to the anterior segment of the eye. Their rapid degradation rates make them suitable for situations requiring a shorter duration of drug delivery [36].

9.2.9 Polyvinyl Alcohols

PVA is a water-soluble, biodegradable polymer used for solubilizing water-insoluble drugs, ensuring chemical resistivity, and easy processing. It is available in various hydration states and molecular weights and is on the FDA's GRAS list. PVA's polymer structure provides tunable permeability for controlled-release applications [37]. Vitrasert® is the first FDA-approved PVA and ethylene-vinyl acetate (EVA) copolymer implant for intravitreal treatment of cytomegalovirus retinitis, providing six to eight months of therapeutic dosing [25]. Retisert®: A multilayered implant for treatment of noninfectious posterior uveitis, that possesses an outer layer of PVA for controlled release of fluocinolone acetonide for up to 2.5 years [38]. Iluvien®: Recently FDA-approved for DME, it provides controlled release of fluocinolone acetonide for up to 36 months [39]. Yutiq®: A polyimide/PVA implant for controlled release of fluocinolone acetonide, improving ease of administration and treatment effectiveness of previous implants [40]. PVA is evaluated for topical ocular drug delivery through wafers and other formats as well [41].

9.2.10 Polyolefins

Carbopol®, a synthetic polymer with high water solubility and viscosity-enhancing qualities, is the brand name for PAA. The resulting acrylic acid by-products can induce inflammation, but it is biodegradable and FDA-approved for numerous topical uses, including ocular use [42]. PAA's charged state allows for good mucoadhesive characteristics, making it useful in hydrogel applications for anterior ocular administration. Restasis®: A Carbopol-based drug-eluting ophthalmic emulsion for topical administration of cyclosporine to treat DED. In order to achieve controlled release of epinephrine for glaucoma, PAA is utilized in a number of hydrogel applications for anterior ocular delivery, such as copolymerization with pNIPAAm and thermally sensitive in situ hydrogel formation [43].

9.2.11 Dendrimers

Depending on the functionalization of the branches and the characteristics of the chain, dendrimers (branching chain polymers) have different physical and chemical properties. They are highly biocompatible, water-soluble, and capable of drug entrapment. Because of their adaptability, ability to hold multiple surface charge groups, and simplicity in drug entrapment, dendrimers are the ideal polymer system for drug delivery [44]. PAMAM: PAMAM dendrimers are investigated for ocular drug delivery despite toxicity concerns [45]. Combinations with medications like dexamethasone for diabetic retinopathy (DR) and AMD for the treatment of retinal inflammation in AMD and retinitis pigmentosa and fluocinolone acetonide have demonstrated potential in trials involving prolonged medication release [46]. PAMAM-PEG hydrogel: one research presented a crosslinked PAMAM-PEG hydrogel for controlled delivery of the anti-glaucoma medication brimonidine tartrate [47].

9.2.12 Biodegradable Polyurethanes

Traditional polyurethanes (PUs), synthesized from petroleum-based precursors, are known for their durability and resistance to environmental degradation but pose significant environmental challenges due to their nondegradability [48]. Conversely, biodegradable polyurethanes (BPUs) contain biodegradable segments that are typically derived from natural resources like vegetable oils and are intended to break down under certain biological or environmental conditions [49]. BPUs offer chemical modification flexibility, can be precisely tuned for elasticity and softness, and may be more cost-effective than PLA polymers. Their biodegradability is achieved by integrating hydrolysable oligomers, enhancing their biocompatibility and utility in biomedical applications [50]. In an exploratory study, Kim et al. employed a polyurethane film as a sustained-release vehicle for dexamethasone during strabismus surgeries. This effectively extended the postoperative adjustment period in rabbit eyes to six weeks, removing the need for frequent topical steroid applications [51]. Injectable Vitreous Substitutes: Additionally, Xue et al. used polyurethane thermogels based on biodegradable poly[(R)-3-hydroxybutyrate-(R)-3-hydroxyhexanoate] as possible replacements for injectable transparent vitreous material [52] The aforementioned advancements underscore the auspicious potential of BPUs in forthcoming ocular therapies and interventions [53].

These are just a few examples, and research is ongoing to explore the potential of other synthetic polymers for ocular drug delivery. The key lies in tailoring the polymer properties to achieve the desired drug-release profile and functionality within the specific needs of the targeted ocular disease.

9.3 SYNTHETIC POLYMERS-BASED DRUG DELIVERY SYSTEMS FOR EYE DISORDERS

Local delivery is required in the eye during various ocular difficulties because of the presence of a blood-retinal barrier. However, it needs frequent administration for direct use, which only offers temporary relief by using formulations such as eye drops, intravitreal injections, or subconjunctival injections leading to treatment failure. The development of formulations with prolonged drug release, good retention at the site of action, enhanced bioavailability, and adequate pharmacological action at the ocular site are the current goals of advancements in ocular DDSs. It not only increases patient compliance but also decreases systemic absorption besides adverse effects. This is possible owing to the implementation of materials and technologies in ocular DDSs. FDA authorized both synthetic and natural polymers for manufacturing ocular devices because of their physical and chemical properties and good biocompatibility profile. Some of the approved synthetic polymers used for the ocular DDSs include PEG, metacrylate (MC), PAA, PGA, PLA, PLGA, PVA, DMAEM, PCL, and PAMAM.

Although, both biodegradable and nonbiodegradable polymers are employed for ocular delivery systems, biodegradable polymers have more advantages over nonbiodegradable polymers. The building blocks of biodegradable polymers are monomers joined by various chemical bonds, which can be gone down into monomers through enzymatic activity/hydrolytic processes. Further, these monomers are metabolized or excreted from the body, hence decreasing associated complications. Synthetic biodegradable polymers comprise a broad range of man-made macromolecules designed for specific applications, predominantly in the biomedical arena (Table 9.1). Owing to the presence of an extensive spectrum of available monomers, synthetic polymers provide a wide range of structural and functional diversity, which is one of their advantages, and can be exploited in the biomedical field. Formulating ocular DDSs is made easier by the intrinsic properties of these synthetic polymers, which include biodegradability, biocompatibility, low degradation rates, significant drug/gene loading potential, and readily adjustable structural and functional characteristics [2].

9.4 SYNTHETIC POLYMER-BASED DRUG DELIVERY SYSTEMS FOR INFLAMMATORY EYE DISORDER

Ocular inflammatory diseases fall into two categories: pathological alterations/degenerative disorders with notable inflammatory components, and disease caused by inflammation.

1. **Inflammation-related ocular degenerations** (pathological alterations/degenerative disorders)
   A. Dry eye syndrome (DES)
   B. Glaucoma
   C. DR
   D. AMD

TABLE 9.1
Different Formulations Approach for Management of Ocular Complications Using Synthetic Polymers

Formulation	Polymer (Synthetic)	Advantages	Reference
Nanoparticles (nanosphere and nanocapsules)	PEG and PVP	Nanosphere: ↑ drug retention, ↑ drug loading, ↑ stability and ↓ dosing frequency, ↓ toxicity, promising candidates for treating ocular surface diseases Nanocapsules: ↑ targeted delivery and preserving drug integrity	[54, 55]
Polymeric Micelles	PLGA and PVA	Control drug release, ↑translocation across lipophilic cells in the corneal epithelium and endothelium; augmented interactions with the ocular surface; ↑ stability; form transparent aqueous solutions, and easy application through the eye	[56]
Nanosuspensions	PCL and PLGA	↑ drug-release profile and ↓ the quick release in aqueous solutions	[57]
Hydrogels	PLA, PGA, and PLGA	↑ water content, ↑ hydration properties, vital prospects for bioelectronic connections (owing to electrical conduction capabilities and mechanical attributes) and analogous mechanical properties closely align with the characteristics of the extracellular matrix and soft tissues, making them suitable for ocular drug delivery	[58]
In Situ Gels/smart hydrogels	Carbopol, hydroxy propyl methyl cellulose (HPMC), and polyacrylic acid	Bioadhesive network, ↑ retention time, assisting sustained release, ↓ dosing frequency, cost-effective, ↑ patient compliance	[59]
Biodegradable Implants	PLGA andPLA	Controlled/sustained the release pattern that offers an appealing alternative to recurrent, invasive intravitreal injections	[60]
Non-Biodegradable Implants	Poly (dimethylsiloxane) (PDMS), silicone, PVA, and poly(ethylene-co-vinyl acetate) (PEVA)	↑ precision in controlling drug release and ↑ release durations	[61]
Biodegradable Nanosheets	Natural polymer:chitosan and sodium alginate	↑ flexibility, nanometer-level thickness,↑ transparency, and ↑ adhesive properties, biocompatible and biodegradabledelivery mechanism	[61]
Biodegradable Microneedles	PLA, PLGA, and chitosan	Significant ↓ in ocular tissue damage (due to ↓ invasiveness), and controlled drug release	[62]

2. **Inflammation-driven diseases**
   A. Keratitis
   B. Conjunctivitis
   C. Uveitis

Based on the anatomical location ocular inflammations are classified as.

1. Extraocular inflammations (involving conjunctiva, cornea, and sclera)
2. Intraocular inflammations (involving the uvea and the retina)

9.4.1 Inflammation-Related Ocular Degenerations (Pathological Alterations/Degenerative Disorders)

9.4.1.1 Dry Eye Syndrome

DED is a very common eye problem also called keratoconjunctivitis sicca. The Tear Film and Ocular Surface Society Dry Eye Workshop II defines it as "a multifactorial ocular surface disease wherein the etiology includes tear film instability, hyperosmolarity, inflammation and damage to the ocular surface, and neurosensory abnormalities." The disease is characterized by a loss of tear film homeostasis and is accompanied by ocular symptoms [63]. Sometimes it is also connected with systemic conditions. All of them result in tear film instability, ocular discomfort and vision problems and can potentially damage the ocular surface. The primary management strategy for this disorder involves replenishing or maintaining the tear volume in the eye. This strategy includes the use of lubricants like sodium hyaluronate ocular drops and synthetic tears, immune system suppressants like cyclosporine A (CsA), anti-inflammatory drugs like corticosteroids and nonsteroidal anti-inflammatory drugs (NSAIDs), and other medications like secretagogues and autologous serum eye drops. Because of their well-established strong anti-inflammatory qualities, corticosteroids are widely used to treat inflammation of the ocular surface. Although topical formulations are effective for surface application on the eye, they have many limitations, such as lower drug bioavailability, which leads to frequent administration by decreasing patient compliance. Hence, several studies successfully delivered the desired medications through different delivery systems such as polymeric nanoparticles, micelles, and hydrogels employing a variety of synthetic polymers.

Nanoparticles of CsA were prepared using PLGA and Eudragit®RL and were coated with Carbopol by *Pelin Aksungur* and colleagues. It has been observed that nanoparticles possess small size and positive surface charge owing to the presence of selected polymers that make them appropriate for ocular use. Changing a nanoparticle's surface properties is a practical tactic to increase medication availability and ocular retention when treating DES [63]. In another research Yu at al. [64] formulated synthetic soft hydrogel made up of thiolated polymers incorporating hyaluronic acid for the treatment of DES. The formulation was found to be biocompatible topically in dogs' eyes. Further, it improved the clinical signs by$\geq$ 65% in animals which were unresponsive to cyclosporine treatment. Wagh and Apar [65] reported PLGA-Eudragit RL100 nanoparticles of CsA for the treatment of DES. The system interacts with the surface of the eye, confirming optimal contact between the mucosa and formulation. Small particle size and positive surface charge were present in this formulation, which offers prolonged drug release in the cul-de-sac or the tissues surrounding the eyes following absorption by epithelial cells. In another investigation, K. Mondon et al. prepared MPEG-hexPLA micelles incorporating CsA for DES management. With no compromise to the stability of the ocular surface, it delivers CsA to the cornea selectively while avoiding systemic absorption. Research conducted in vitro and in vivo has confirmed that this formulation is highly effective and patient-compliant, and it can be applied topically [66].

9.4.1.2 Glaucoma

Glaucoma is a multifactorial eye disease that has complex pathogenesis. It mostly happens because of an increase in IOP brought on by a compromised outflow of aqueous humor. It is the leading cause of permanent blindness globally, with progressive optic nerve damage. Chronic mechanical and vascular stress as well as age-related oxidative stress in the retina could be additional causes. The factors that contribute to the advancement of glaucoma are neuroinflammation, mitochondrial dysfunction, cell senescence, and excitotoxicity, which are intimately associated with oxidative stress. The most evident and visible consequence of inflammation in glaucoma is topical treatment-induced ocular surface disease. Since inflammatory changes can occur anywhere along the eye's structure, from the ocular surface to the posterior segment and the visual pathway, glaucoma appears to be a remarkable eye disease [67]. Furthermore, the prevalence of glaucoma is about 20% in patients with uveitis. Combining steroidal and nonsteroidal anti-inflammatory drugs with anti-glaucoma medications is the treatment for combined iridocyclitis and glaucoma. Management of glaucoma includes drugs like beta blockers, alpha agonists, prostaglandin analogues, and carbonic anhydrase inhibitors, etc. Surgery and laser therapy are other options to deal with this disease. However, all these management techniques are associated with limitations. The challenges highlight the importance of developing innovative, efficient, and durable treatments for glaucoma. To overcome the limitations of existing therapies, recent scientific research has shifted toward creating biodegradable polymer-based DDSs. A few of them are highlighted here.

In recent research by Hamed A. Salama et al. [68] PLGA nanoparticles were designed to be administered subconjunctivally to treat glaucoma. The emulsion-diffusion-evaporation method was used to create the brinzolamide-loaded PLGA nanoparticles, which were compatible with the eye's tissue and able to lower the IOP in the Albino rabbit model for up to ten days. Further, Brinzolamide was released from the formulation in a biphasic manner. Additionally, it has been found that nanoparticles with smaller sizes were able to lower the IOP for a longer amount of time than larger particle sizes. In another study, ocular inserts of Dorzolamide were formulated for sustained action using chitosan/hydroxyethyl cellulose by solvent/casting method. The effectiveness of inserts was evaluated in glaucomatous rats along with the pharmacokinetics studies. It has been observed that a single administration of the formulation that exerted a significant hypotensive effect for two weeks also prevented retinal ganglion cell death and proved to be effective for the management of glaucoma [69]. Furthermore, nanoparticles of triblocked oligomer poly (L-lactide)-b-poly(ethylene glycol)-b-poly(L-lactide) (PLEL) and pranoprofen (PF)-loaded PLEL were formulated with the particle size of 151.7 nm for ophthalmic irritation. Tests for ocular toxicity in vitro and in vivo were carried out and proved to be promising for eye inflammation [70].

9.4.1.3 Diabetic Retinopathy

The most common microvascular problem linked to diabetes mellitus in adults of working age is DR. A unique part of DR is played by metabolic dysfunction and the activation of chronic, low-quality inflammatory signaling brought on by high glucose [71]. It causes enlargement of the blood vessels of the retina and leaks blood and fluids. Clinical symptoms of DR include hemorrhages, macular edema, microaneurysms, capillary blockage, and ultimately, neovascularization. The duration of the illness, elevated blood sugar, and hypertension are significant risk factors. These elements set off the physiological and biochemical alterations that result in retinal dysfunction and microvascular damage leads to DR. In patients with DR, DME is the most frequent cause of vision loss. It is a significant DR subtype that is linked to both proliferative and nonproliferative DR severities. Currently, very less management techniques are available for the treatment of DR. Nowadays, nanocarriers-based DDSs using a variety of polymers have been employed to transport many active moieties to the targeted site by increasing the penetration or by extending the duration of action.

In an investigation, sustained released dexamethasone intravitreal implant was formulated for the management of posterior segment diseases management. An extended period of small concentration was observed after a high initial drug concentration in the pharmacokinetic study, which was in line with clinical findings that supported the application of the dexamethasone implant for the long-term management of posterior segment diseases [72]. In another study, Binapani Mahaling et al. formulated a noninvasive nanoparticle to administer triamcinolone acetonide. The core-shell delivery system entails a Pluronic® F68 shell with a hydrophobic polycaprolactone core laden with triamcinolone acetonide active drug. They evaluated the efficacy of the system in a rat model and witnessed a lessening in retinal inflammation (by reduction in the expression of NF-κB, ICAM-1, and TNFα) and vascular abnormalities; hence, it can be used as a topical formulation for managing DR [73]. Further, Zeng L. et al. [74] undertook a study for developing PLGA nanoparticles incorporating Interleukin-12 by the double-emulsion method. This formulation exhibited a sustained drug-release pattern, having better inhibitory efficacy against vascular endothelial growth factor-A and Matrix metalloproteinase-9 expression in rat endothelial cells. This study concluded that the formulation effectively treated the DR in mice model by reducing retinal damage, as evidenced by an increase in thickness and a decrease in neovascularization of the retinas.

9.4.1.4 Age-Related Macular Degeneration

One of the main reasons of vision loss in developed nations is AMD. This is an acquired degeneration of the retina that results in a marked reduction in central vision and has been connected to issues like geographic atrophy and choroidal neovascularization. Focal regions of retinal pigment epithelium loss, subretinal or sub-RPE hemorrhage, serous fluid, and subretinal fibrosis are all possible signs of advanced disease. The risk of AMD rises with age. Patients over 75 years of age have a risk that is more than three times higher than that of patients between the ages of 65 and 74. Bevacizumab, ranibizumab, aflibercept, brolucizumab, and faricimab are examples of anti-vascular endothelial growth factor (VEGF) agents that are frequently used in the clinical management of AMD. However, they are associated with certain limitations. Evidence has been found that targeting VEGF-A alone may not enough hence agents that target several pathways (e.g., aflibercept, faricimab, and others in development) can be preferred for better efficiency.

Currently, an investigation was conducted to formulate albuminate-PLGA-nanoparticles containing bevacizumab by double-emulsion method and evaluated in rabbits by intravitreal injection. Further distribution of the nanoparticles was carried out in the posterior segment of the eye by employing coumarin-6 loaded albuminated-PLGA-nanoparticles. The result of the pharmacokinetic study confirms sufficient drug concentration along with the persistence of nanoparticles in ocular tissues for 56 days. Therefore, they came to the conclusion that the prepared formulation might be more effective in treating posterior segment neovascularization of the eye [75]. In another study, novel dexamethasone nanoparticle comprised of poly (D, L-lactide-co-glycolide) and polyethylenimine was prepared followed by adsorption of bevacizumab by means of electrostatic interactions, onto the surfaces of nanoparticles. Results showed a smaller size of nanoparticles with narrow diameter distribution and neutral surface charge. Furthermore, *in vivo*, chick embryo chorioallantoic membrane assay displayed a sufficient reduction of blood vessels with a decrease in leakage area of choroidal neovascularization in animal models. Hence, it presents a promising new therapeutic option against angiogenesis for potential intravitreal uses, including AMD [76]. Rudeen K M et al. [77] developed a simultaneous release thermoresponsive delivery system having PEG-co-(L-lactic acid) diacrylate-N-isopropylacrylamide hydrogel incorporating dexamethasone-loaded nanoparticles and aflibercept-loaded microparticles by involving modified single- and double-emulsion techniques. They observe a continuous release of the drug up to 224 days with proper preservation of bioactivity of aflibercept in the Combo-system. Conclusively this system eliminates the requirement for distinct dosing regimens of anti-VEGF and corticosteroids with good release profiles for management of wet AMD.

9.4.2 Inflammation-Driven Diseases

9.4.2.1 Keratitis

Keratitis, commonly referred to as a corneal ulcer, is a painful inflammation of the cornea that is marked by ciliary congestion, swelling, and inflammatory cell infiltration. There are two types of keratitis: infectious keratitis, which is caused by viruses, bacteria, fungi, and parasites, and non-infectious keratitis, which can result from dry eye, injury or due to wear of contact lenses for long time. This condition can affect either the surface of the eye or the entire eye. Treatments for keratitis include antivirals, antifungals, antimicrobials, and anti-inflammatory drugs that can target individual cause factors. However, partial absorption of the drug through lipophilic cornea owing to the decreased residence time on the ocular surface, rapid nasolacrimal drainage, and non-productive absorption of the drug through the conjunctiva pose significant challenges. Research has been done to formulate different novel delivery systems to overcome these issues. One of the novel carrier systems is the polymeric system, which can extend the drug release by increasing the residence time to improve therapeutic efficacy.

Shortly, slow-release ketoconazole and ketorolac mucoadhesive gels were created by Hala Mohamed El-Mofty et al. to treat fungal keratitis and associated inflammation. They characterized the system containing different polymeric concentrations and evaluated the system in a rabbit model. They noticed that gels with high polymer concentrations had delayed drug release along with high viscosity and mucoadhesion characteristics. Further, more corneal healing and reduced inflammation were detected in the infected eyes treated with the slow-release formulation comprising both drugs in contrast to the eyes treated with the fast-release formulation [78].

In another investigation, Mudgil and Pawar [79], pioneered the design of Moxifloxacin-loaded PLGA nanosuspensions to increase the residence time of moxifloxacin in the ocular cavity. The formulation was prepared by the solvent evaporation technique, characterized and evaluated (antibacterial activities) against *S. aureus* and *P. aeruginosa*. Additionally, their antibacterial activities are compared with marketed formulations. They found that the optimized formulation had an extended drug-release profile, small and uniform particle size, and good entrapment efficiency. It was also found that, through isolated goat cornea, the in vitro corneal permeability was comparable to that of the commercial formulation. Further, it was turned to be more active against *S. aureus* and *P. aeruginosa* in contrast to the commercially available eye drops. In another work, the development of oxytetracycline-loaded gelatin-polyacrylic acid nanoparticles loaded in situ gelling solution, proposed to sustain release of antibiotic for the management of keratitis was carried out by Muhammad Naseer Abbas et al. [80]. Formulation was prepared and characterized and the in vitro antibacterial activity against *Pseudomonas aeruginosa* was carried out. By means of histological investigations and the Draize test, the system's in vivo biocompatibility was also established. With similar antibacterial activity in the commercial product, the formulation demonstrated a long-lasting effect against keratitis. In other research, Alaa H. Salama et al. [81] fabricated nanoparticles based on polycaprolactone, Kolliphor P188, and chitosan hydrochloride encapsulating ofloxacin, were fabricated using the emulsion solvent evaporation technique. The optimized formulation was characterized and assessed for penetration depth through confocal laser scanning microscopy and antimicrobial activity using rabbits having infections in their eyes with *Escherichia coli*. Further, the optimized formulation was compared with the marketed formulation for antimicrobial activity. Additionally, histopathological examination of eye biopsies was also carried out. Finally, this study demonstrated a novel formulation with superior antibacterial activity in the form of an ocular gel that forms in situ and is mucoadhesive and non-irritating.

9.4.2.2 Conjunctivitis

Conjunctivitis, also referred to as pink eye, is an infection or inflammation of the conjunctiva of the eye. It is one of the most prevalent causes of eye redness though the occurrence varies depending on the underlying cause such as season of the year, causative agent, age and condition of the patient etc. This is characterized by swelling of the conjunctival vessels, which causes edema and hyperemia in the conjunctiva and is typically associated with discharge. It is most often caused by a viral infection (viral conjunctivitis) followed by bacterial conjunctivitis. This is the etiology of infectious conjunctivitis. Among noninfectious etiologies, the most common etiologies are allergic and toxin-induced conjunctivitis.

Mounika et al. [82] designed an in situ gelling system of acyclovir to overcome the limitations of poor bioavailability and therapeutic response exhibited by conventional formulations for the management of conjunctivitis. They used Carbopol 940, HPMCK100M, Polyvinyl alcohol. The prepared formulations were characterized and evaluated for antimicrobial efficacy and ocular irritancy. The resulting formulation has an enhanced drug release over 6 h, improved ocular bioavailability with better patient compliance. Therefore, it can be used successfully to treat conjunctivitis. Presently, Patil et al. [83] formulated and optimized Norfloxacin in situ gels for the treatment of conjunctivitis.

Carbopol-940, HPMC-E50LV, HPMC E4M, and HPMC K4M were used as pH-triggered gelling systems for ocular delivery. Conclusively, a stable, non-irritating formulation that delivered a sustained release of the medication for up to eight hours was formulated successfully for conjunctivitis. Narayana and Ahmed [84] pioneered the formulation and evaluation of ofloxacin and dexamethasone-based in situ gel for the management of conjunctivitis. Here, Carbopol 934 was used as the gelling agent in combination with HPMC as pH pH-triggered gelling system. The formulation was optimized, characterized, and evaluated. The result showed sterile, nonirritant gel released up to 85% at the end of 13 h, which prolonged the residence time in the ocular cavity. Furthermore, this new formulation offers a good substitute for traditional eye drops because of its extended precorneal residence time, which increases bioavailability. Sipos et al. [85] designed erythromycin-loaded polymeric micelles which were entrapped inside Carbopol 934P gel base. They formulated, optimized, characterized, and evaluated the formulation. The optimized formulation showed a high mucoadhesion, which increase the residence time in the ocular cavity. It displayed an extended drug-release pattern through in vitro study. It had an increased permeability through the aqueous humor across the cornea while holding its antimicrobial activity; hence, it can be employed as an innovative formulation for curing bacterial infections of the eye.

9.4.2.3 Uveitis

Inflammation of the uveal tract (the iris, ciliary body, and choroid) is referred to as uveitis. Sometimes it also involves the retina and fluid within the anterior chamber and vitreous. Causes include infection, injury, trauma, autoimmune disorders, etc., or it can occasionally be idiopathic. Symptoms of the disease can be described as redness, ocular ache, decreased vision, photophobia, and floaters. There are four types of uveitis: anterior, intermediate, posterior, and panuveitis, based on the uveal region that is impacted. Uveitis is divided into four categories: anterior, intermediate, posterior, and panuveitis, depending on which area of the uvea is affected. It may be categorized as acute, recurrent, or chronic, and it can happen at any age. Management of the condition often includes the administration of corticosteroids, anti-inflammatory drugs, immunosuppressive agents, etc. There are plenty of systemic and topical anti-inflammatory medications available in the market to treat the condition. However, they are associated with several limitations such as the need for frequent dosing, poor patient compliance, local adverse effects such as cataracts and glaucoma, and many others. Out of several novel formulations polymeric formulation is one which may increase the patient compliance by increasing the residence time and bioavailability. A few are highlighted below.

Sujay Shah et al. [86] engineered an aqueous nano micellar formulation with a higher molecular weight of PEG (2000) (2K), TPGS (1K), modified TPGS (a surfactant polymer), and fluocinolone acetonide (FA) using vitamin E. The formulation was optimized, characterized, and assessed on retinal and corneal cells for cytotoxicity activity. Outcomes revealed 26 times increment in solubility of FA with newly synthesized 2K polymer. With both polymers, nanomicelles showed a narrow size distribution and an extremely small size (<20 nm). It was discovered that the entrapment efficiency was higher than 90% and the viability of the cells was the same as that of the control. Consequently, it may be considered a therapeutic option for posterior uveitis. Wu et al. conducted a study wherein they designed monomethoxy poly(ethylene glycol)-poly(ε-caprolactone) micelles conjugated with rapamycin and assessed their effectiveness in an experimental autoimmune uveitis model in rats. It has been observed that there was a 1000-fold increase in water solubility of rapamycin by forming a micellar formulation. Moreover, intravitreal injection of micelles did not cause vitreous hemorrhage and, for at least 14 days after injection, the retinal pigment epithelium exhibited a high concentration of the micelles. Furthermore, fluorescence microscopy revealed that the retinal tissue had a higher drug concentration of the formulation than unconjugated rapamycin. This showed a local therapeutic effect in the eye by downregulating the Th1 and Th17 response. The findings indicate that it is a potentially effective treatment for reducing sterile intraocular inflammation in cases of uveitis [87]. Further, Gonzalez-Pizarro et al. [88] investigated thermosensitive gels incorporating fluorometholone-loaded PLGA nanoparticles to boost the effectiveness of the ocular anti-inflammatory drug. The formulation was optimized, characterized and compared with the marketed preparation. The outcome showed that there was no ocular cavity irritation, while the nanoparticle maintain its typical size and spherical shape inside the gels. The residence time of the formulation was increased by augmenting the ocular bioavailability. Hence, this topical system is a useful management tool for both acute and ongoing treatment of inflammatory conditions affecting the eyes.

9.5 REGULATORY STATUS OF OCULAR DRUG DELIVERY SYSTEMS

To deliver an active drug directly into the eye at therapeutic concentration is one of the major challenges in ocular drug delivery, and researchers have invested years in designing the safe and effective DDSs for eyes. However, recent advancements in material science and technology have led to the development of various DDSs that have extended the drug release, bioavailability and therapeutic action of drug in the eyes, moreover they have curtailed the systemic drug absorption and side effects thereby enhances the patient comfort. To have a check on quality, physical and chemical properties and biocompatibility of these materials, the FDA authorizes natural and synthetic polymers for ocular drug delivery such as hyaluronic acid, chitosan, gelatin, PEGDA, PLA, PLGA, and PVA.

The fabrication of ocular DDSs based on synthetic polymers faces several regulatory challenges to achieve a successful clinical translation (Table 9.2). The development

TABLE 9.2
Marketed Intraocular Formulations of Various Synthetic Polymers as Matrix Excipients

Trade name	Drug	Polymer system	Dosage form	Indication/ use	Manufacturer
Pilopine HS®	Pilocarpine hydrochloride	Carbopol 940	Gel	Control intraocular pressure	Alcon Laboratories
Zirgan®	Ganciclovir	Carbomer	Gel	Acute herpetic keratitis (dendritic ulcers)	Bausch and Lomb
RESTASIS®	Cyclosporine	Carbomer copolymer type A	Emulsion	Chronic dry eye	Allergan
AzaSite®	Azithromycin	Poloxamer 407/polycarbophil (DuraSite®)	Solution	Bacterial conjunctivitis	Akorn
ALREX®	loteprednol etabonate	Povidone	Suspension	Seasonal allergic conjunctivitis	Bausch and Lomb
Refresh Redness Relief®	Phenylephrine hydrochloride	PVA	Solution	Eye redness and dryness	Allergan

process is made more difficult by international regulatory harmonization, comprehensive in vivo toxicity research, standardized formulation techniques, and regulatory criteria for safety and efficacy.

9.6 CONCLUSION AND FUTURE PROSPECT

The application of synthetic polymers in ocular DDSs represents a significant advancement in the management of eye disorders. These materials offer unparalleled versatility, enabling the design of delivery systems that can address the unique anatomical and physiological challenges presented by the eye. This chapter has comprehensively reviewed the types of synthetic polymers employed in ocular drug delivery, their mechanisms of action, and their therapeutic applications in managing both anterior and posterior segment diseases.

Synthetic polymers such as polyesters (PLA, PVA, PLGA, PCL, POE), polymethacrylates, polyethylene oxides (PEG), polyanhydrides, polyvinyl alcohols, polyolefins, dendrimers, and biodegradable polyurethanes have been explored for their potential in ocular drug delivery. These materials are characterized by their biodegradability, biocompatibility, and ability to be engineered into various forms, including nanoparticles, hydrogels, and implants. Their versatility allows for controlled and sustained drug release, targeting specific regions within the eye, thus improving therapeutic outcomes and patient compliance. The regulatory landscape for ocular DDSs is complex, reflecting the need for stringent safety and efficacy standards. The development and approval of synthetic polymer-based ocular therapeutics require a thorough understanding of the regulatory framework. This chapter has highlighted the critical regulatory considerations, including the approval process, challenges, and strategies to navigate the regulatory pathway. Researchers and developers need to engage with regulatory bodies early in the development process to ensure compliance and facilitate the successful commercialization of these innovative therapies. Looking ahead, the future of synthetic polymer-based ocular DDSs is promising.

BIBLIOGRAPHY

1. Zhou R, Caspi RR (2010) Ocular immune privilege. *F1000 Biology Reports* 2:3.
2. Tsung T-H, Tsai Y-C, Lee H-P, Chen Y-H, Lu D-W (2023) Biodegradable polymer-based drug-delivery Systems for Ocular Diseases. *International Journal of Molecular Sciences* 24:12976.
3. Allyn MM, Luo RH, Hellwarth EB, Swindle-Reilly KE (2022) Considerations for polymers used in ocular drug delivery. *Frontiers in Medicine* 8:787644.
4. Mannermaa E, Vellonen K-S, Urtti A (2006) Drug transport in corneal epithelium and blood–retina barrier: emerging role of transporters in ocular pharmacokinetics. *Advanced Drug Delivery Reviews* 58:1136–1163.
5. Runkle EA, Antonetti DA (2011) The blood-retinal barrier: structure and functional significance. *The Blood-Brain and Other Neural Barriers: Reviews and Protocols* 686:133–148.
6. Englert C, Brendel JC, Majdanski TC, Yildirim T, Schubert S, Gottschaldt M, Windhab N, Schubert US (2018) Pharmapolymers in the 21st century: Synthetic polymers in drug delivery applications. *Progress in Polymer Science* 87:107–164.
7. Wei Q, Deng N-N, Guo J, Deng J (2018) Synthetic polymers for biomedical applications. *International Journal of Biomaterials* 2018:7158621.
8. Makadia HK, Siegel SJ (2011) Poly lactic-co-glycolic acid (PLGA) as biodegradable controlled drug delivery carrier. *Polymers* 3:1377–1397.
9. Farah S, Anderson DG, Langer R (2016) Physical and mechanical properties of PLA, and their functions in widespread applications—A comprehensive review. *Advanced Drug Delivery Reviews* 107:367–392.

10. Kuppermann BD, Patel SS, Boyer DS, Augustin AJ, Freeman WR, Kerr KJ, Guo Q, Schneider S, López FJ (2021) Phase 2 study of the safety and efficacy of brimonidine drug delivery system (Brimo DDS) generation 1 in patients with geographic atrophy secondary to age-related macular degeneration. *Retina* 41:144–155.
11. Baino F, Kargozar S (2020) Regulation of the ocular cell/tissue response by implantable biomaterials and drug delivery systems. *Bioengineering* 7:65.
12. Chan A, Leung L-S, Blumenkranz MS (2011) Critical appraisal of the clinical utility of the dexamethasone intravitreal implant (Ozurdex®) for the treatment of macular edema related to branch retinal vein occlusion or central retinal vein occlusion. *Clinical Ophthalmology* 5:1043–1049.
13. Shirley M (2020) Bimatoprost implant: first approval. *Drugs & Aging* 37:457–462.
14. Allyn MM, Luo RH, Hellwarth EB, Swindle-Reilly KE (2022) Considerations for polymers used in ocular drug delivery. *Frontiers in Medicine* 8:787644.
15. Mondal D, Griffith M, Venkatraman SS (2016) Polycaprolactone-based biomaterials for tissue engineering and drug delivery: Current scenario and challenges. *International Journal of Polymeric Materials and Polymeric Biomaterials* 65:255–265.
16. Samy KE, Cao Y, Kim J, Konichi da Silva NR, Phone A, Bloomer MM, Bhisitkul RB, Desai TA (2019) Co-delivery of timolol and brimonidine with a polymer thin-film intraocular device. *Journal of Ocular Pharmacology and Therapeutics* 35:124–131.
17. Jiang P, Chaparro FJ, Cuddington CT, Palmer AF, Ohr MP, Lannutti JJ, Swindle-Reilly KE (2020) Injectable biodegradable bi-layered capsule for sustained delivery of bevacizumab in treating wet age-related macular degeneration. *Journal of Controlled Release* 320:442–456.
18. Jiang P, Jacobs KM, Ohr MP, Swindle-Reilly KE (2020) Chitosan–polycaprolactone core–shell microparticles for sustained delivery of bevacizumab. *Molecular Pharmaceutics* 17:2570–2584.
19. Kim J, Kudisch M, Mudumba S, Asada H, Aya-Shibuya E, Bhisitkul RB, Desai TA (2016) Biocompatibility and pharmacokinetic analysis of an intracameral polycaprolactone drug delivery implant for glaucoma. *Investigative Ophthalmology & Visual Science* 57:4341–4346.
20. Shahab MS, Rizwanullah M, Alshehri S, Imam SS (2020) Optimization to development of chitosan decorated polycaprolactone nanoparticles for improved ocular delivery of dorzolamide: In vitro, ex vivo and toxicity assessments. *International Journal of Biological Macromolecules* 163:2392–2404.
21. Zhang Z, He Z, Liang R, Ma Y, Huang W, Jiang R, Shi S, Chen H, Li X (2016) Fabrication of a micellar supramolecular hydrogel for ocular drug delivery. *Biomacromolecules* 17:798–807.
22. Souto EB, Dias-Ferreira J, López-Machado A, Ettcheto M, Cano A, Camins Espuny A, Espina M, Garcia ML, Sánchez-López E (2019) Advanced formulation approaches for ocular drug delivery: state-of-the-art and recent patents. *Pharmaceutics* 11:460.
23. Sovadinova I, Palermo EF, Urban M, Mpiga P, Caputo GA, Kuroda K (2011) Activity and mechanism of antimicrobial peptide-mimetic amphiphilic polymethacrylate derivatives. *Polymers* 3:1512–1532.
24. Fan X, Torres-Luna C, Azadi M, Domszy R, Hu N, Yang A, David AE (2020) Evaluation of commercial soft contact lenses for ocular drug delivery: A review. *Acta Biomaterialia* 115:60–74.
25. Kuno N, Fujii S (2010) Biodegradable intraocular therapies for retinal disorders: progress to date. *Drugs & Aging* 27:117–134.
26. Ubani-Ukoma U, Gibson D, Schultz G, Silva BO, Chauhan A (2019) Evaluating the potential of drug eluting contact lenses for treatment of bacterial keratitis using an ex vivo corneal model. *International Journal of Pharmaceutics* 565:499–508.
27. Pereira-da-Mota AF, Vivero-Lopez M, Topete A, Serro AP, Concheiro A, Alvarez-Lorenzo C (2021) Atorvastatin-eluting contact lenses: effects of molecular imprinting and sterilization on drug loading and release. *Pharmaceutics* 13:606.
28. Jung HJ, Chauhan A (2012) Temperature sensitive contact lenses for triggered ophthalmic drug delivery. *Biomaterials* 33:2289–2300.
29. Basile AS, Hutmacher MM, Kowalski KG, Gandelman KY, Nickens DJ (2015) Population pharmacokinetics of pegaptanib sodium (Macugen®) in patients with diabetic macular edema. *Clinical Ophthalmology* 9:323–335.
30. Abdelkader H, Fathalla Z, Seyfoddin A, Farahani M, Thrimawithana T, Allahham A, Alani AW, Al-Kinani AA, Alany RG (2021) Polymeric long-acting drug delivery systems (LADDS) for treatment of chronic diseases: Inserts, patches, wafers, and implants. *Advanced Drug Delivery Reviews* 177:113957.
31. Blizzard CD, Desai A, Driscoll A, Cheung M, Goldstein MH (2021) Ocular pharmacokinetics of OTX-DED, a sustained-release intracanalicular insert delivering dexamethasone, in a canine model. *Investigative Ophthalmology & Visual Science* 62:1323–1323.
32. Foroutan S, Watson D (1999) The in vitro evaluation of polyethylene glycol esters of hydrocortisone 21-succinate as ocular prodrugs. *International Journal of Pharmaceutics* 182:79–92.
33. Lakhani P, Patil A, Wu K-W, Sweeney C, Tripathi S, Avula B, Taskar P, Khan S, Majumdar S (2019) Optimization, stabilization, and characterization of amphotericin B loaded nanostructured lipid carriers for ocular drug delivery. *International Journal of Pharmaceutics* 572:118771.
34. Kumar N, Langer RS, Domb AJ (2002) Polyanhydrides: an overview. *Advanced Drug Delivery Reviews* 54:889–910
35. Reddy PG, Domb AJ (2022) Polyanhydride chemistry. *Biomacromolecules* 23:4959–4984.
36. Martins RDC, Gamazo C, Sánchez-Martínez M, Barberán M, Peñuelas I, Irache JM (2012) Conjunctival vaccination against Brucella ovis in mice with mannosylated nanoparticles. *Journal of Controlled Release* 162:553–560.
37. Teodorescu M, Bercea M, Morariu S (2019) Biomaterials of PVA and PVP in medical and pharmaceutical applications: Perspectives and challenges. *Biotechnology Advances* 37:109–131.
38. Arcinue CA, Cerón OM, Foster CS (2013) A comparison between the fluocinolone acetonide (Retisert) and dexamethasone (Ozurdex) intravitreal implants in uveitis. *Journal of Ocular Pharmacology and Therapeutics* 29:501–507.
39. Habib MS (2018) ILUVIEN® technology in the treatment of center-involving diabetic macular edema: a review of the literature. *Therapeutic Delivery* 9:547–556.
40. García-Estrada P, García-Bon MA, López-Naranjo EJ, Basaldúa-Pérez DN, Santos A, Navarro-Partida J (2021) Polymeric implants for the treatment of intraocular eye diseases: trends in biodegradable and non-biodegradable materials. *Pharmaceutics* 13:701.
41. Shin CS, Marcano DC, Park K, Acharya G (2017) Application of hydrogel template strategy in ocular drug

delivery. *Biomedical Nanotechnology: Methods and Protocols* 1570:279–285.
42. Nguyen DD, Lai J-Y (2020) Advancing the stimuli response of polymer-based drug delivery systems for ocular disease treatment. *Polymer Chemistry* 11:6988–7008.
43. Prasannan A, Tsai H-C, Chen Y-S, Hsiue G-H (2014) A thermally triggered in situ hydrogel from poly (acrylic acid-co-N-isopropylacrylamide) for controlled release of anti-glaucoma drugs. *Journal of Materials Chemistry B* 2:1988–1997.
44. Lancina III MG, Yang H (2017) Dendrimers for ocular drug delivery. *Canadian Journal of Chemistry* 95:897–902.
45. Madaan K, Kumar S, Poonia N, Lather V, Pandita D (2014) Dendrimers in drug delivery and targeting: Drug-dendrimer interactions and toxicity issues. *Journal of Pharmacy and Bioallied Sciences* 6:139–150.
46. Iezzi R, Guru BR, Glybina IV, Mishra MK, Kennedy A, Kannan RM (2012) Dendrimer-based targeted intravitreal therapy for sustained attenuation of neuroinflammation in retinal degeneration. *Biomaterials* 33:979–988.
47. Wang J, Williamson GS, Lancina III MG, Yang H (2017) Mildly cross-linked dendrimer hydrogel prepared via aza-Michael addition reaction for topical brimonidine delivery. *Journal of Biomedical Nanotechnology* 13:1089–1096.
48. Maurya AK, de Souza FM, Gupta RK (2023) P olyurethane and Its Composites: Synthesis to Application. In: *Polyurethanes: Preparation, Properties, and Applications* Volume 1: Fundamentals. ACS Publications, pp 1–20,
49. Kaur R, Singh P, Tanwar S, Varshney G, Yadav S (2022) Assessment of bio-based polyurethanes: Perspective on applications and bio-degradation. *Macromol* 2:284–314,
50. Xie Z, Lu C, Chen X, Chen L, Hu X, Shi Q, Jing X (2007) A facile approach to biodegradable poly (ε-caprolactone)-poly (ethylene glycol)-based polyurethanes containing pendant amino groups. *European Polymer Journal* 43:2080–2087.
51. Kim J, Jeong S, Jung M, Hwang JM (2004) Use of polyurethane with sustained release dexamethasone in delayed adjustable strabismus surgery. *British Journal of Ophthalmology* 88:1450–1454.
52. Xue K, Liu Z, Jiang L, Kai D, Li Z, Su X, Loh XJ (2020) A new highly transparent injectable PHA-based thermogelling vitreous substitute. *Biomaterials Science* 8:926–936.
53. da Silva GR, da Silva-Cunha A Jr, Behar-Cohen F, Ayres E, Orefice RL (2011) Biodegradable polyurethane nanocomposites containing dexamethasone for ocular route. *Materials Science and Engineering: C* 31:414–422.
54. Khiev D, Mohamed ZA, Vichare R, Paulson R, Bhatia S, Mohapatra S, Lobo GP, Valapala M, Kerur N, Passaglia CL (2021) Emerging nano-formulations and nanomedicines applications for ocular drug delivery. *Nanomaterials* 11:173.
55. Carreiró F, Oliveira AM, Neves A, Pires B, Nagasamy Venkatesh D, Durazzo A, Lucarini M, Eder P, Silva AM, Santini A (2020) Polymeric nanoparticles: Production, characterization, toxicology and ecotoxicology. *Molecules* 25:16.
56. Carstens MG, Rijcken CJ, van Nostrum CF, Hennink WE (2008) Pharmaceutical micelles: combining longevity, stability, and stimuli sensitivity. *Multifunctional Pharmaceutical Nanocarriers* 4:263–308.
57. Patel VR, Agrawal Y (2011) Nanosuspension: An approach to enhance solubility of drugs. *Journal of Advanced Pharmaceutical Technology & Research* 2:81–87.
58. Fathi M, Barar J, Aghanejad A, Omidi Y (2015) Hydrogels for ocular drug delivery and tissue engineering. *BioImpacts: BI* 5:159.
59. Agarwal R, Iezhitsa I, Agarwal P, Abdul Nasir NA, Razali N, Alyautdin R, Ismail NM (2016) Liposomes in topical ophthalmic drug delivery: an update. *Drug Delivery* 23:1075–1091.
60. Lee SS, Hughes P, Ross AD, Robinson MR (2010) Biodegradable implants for sustained drug release in the eye. *Pharmaceutical Research* 27:2043–2053.
61. Agrawal AK, Das M, Jain S (2012) In situ gel systems as 'smart'carriers for sustained ocular drug delivery. *Expert Opinion on Drug Delivery* 9:383–402.
62. Thakur RRS, Fallows SJ, McMillan HL, Donnelly RF, Jones DS (2014) Microneedle-mediated intrascleral delivery of in situ forming thermoresponsive implants for sustained ocular drug delivery. *Journal of Pharmacy and Pharmacology* 66:584–595.
63. Aksungur P, Demirbilek M, Denkbaş EB, Vandervoort J, Ludwig A, Ünlü N (2011) Development and characterization of Cyclosporine A loaded nanoparticles for ocular drug delivery: Cellular toxicity, uptake, and kinetic studies. *Journal of Controlled Release* 151:286–294.
64. Yu Y, Chow DWY, Lau CML, Zhou G, Back W, Xu J, Carim S, Chau Y (2021) A bioinspired synthetic soft hydrogel for the treatment of dry eye. *Bioengineering & Translational Medicine* 6:e10227.
65. Wagh VD, Apar DU (2014) Cyclosporine a loaded PLGA nanoparticles for dry eye disease: in vitro characterization studies. *Journal of Nanotechnology* 2014:683153.
66. Mondon K, Zeisser-Labouèbe M, Gurny R, Möller M (2011) Novel Cyclosporin A formulations using MPEG–hexyl-substituted polylactide micelles: A suitability study. *European Journal of Pharmaceutics and Biopharmaceutics* 77:56–65.
67. Baudouin C, Kolko M, Melik-Parsadaniantz S, Messmer EM (2021) Inflammation in Glaucoma: From the back to the front of the eye, and beyond. *Progress in Retinal and Eye Research* 83:100916.
68. Salama HA, Ghorab M, Mahmoud AA, Abdel Hady M (2017) PLGA nanoparticles as subconjunctival injection for management of glaucoma. *Aaps Pharmscitech* 18:2517–2528.
69. Franca JR, Foureaux G, Fuscaldi LL, Ribeiro TG, Castilho RO, Yoshida MI, Cardoso VN, Fernandes SO, Cronemberger S, Nogueira JC (2019) Chitosan/hydroxyethyl cellulose inserts for sustained-release of dorzolamide for glaucoma treatment: In vitro and in vivo evaluation. *International Journal of Pharmaceutics* 570:118662.
70. Luo Y, Yang L, Feng P, Qiu H, Wu X, Lu S, Zhou M, Xu L, Zhu Y (2020) Pranoprofen nanoparticles with poly (L-Lactide)-b-poly (Ethylene Glycol)-b-poly (L-Lactide) as the matrix toward improving ocular anti-inflammation. *Frontiers in Bioengineering and Biotechnology* 8:581621.
71. Das S, Kharbanda K, Suchetha M, Raman R, Dhas E (2021) Deep learning architecture based on segmented fundus image features for classification of diabetic retinopathy. *Biomedical Signal Processing and Control* 68:102600.
72. Chang-Lin J-E, Attar M, Acheampong AA, Robinson MR, Whitcup SM, Kuppermann BD, Welty D (2011) Pharmacokinetics and pharmacodynamics of a sustained-release dexamethasone intravitreal implant. *Investigative Ophthalmology & Visual Science* 52:80–86.
73. Mahaling B, Srinivasarao DA, Raghu G, Kasam RK, Reddy GB, Katti DS (2018) A non-invasive nanoparticle mediated delivery of triamcinolone acetonide ameliorates diabetic retinopathy in rats. *Nanoscale* 10:16485–16498.

74. Zeng L, Ma W, Shi L, Chen X, Wu R, Zhang Y, Chen H, Chen H (2019) Poly (lactic-co-glycolic acid) nanoparticle-mediated interleukin-12 delivery for the treatment of diabetic retinopathy. *International Journal of Nanomedicine* 14:6357–6369.
75. Varshochian R, Riazi-Esfahani M, Jeddi-Tehrani M, Mahmoudi A, Aghazadeh S, Mahbod M, Movassat M, Atyabi F, Sabzevari A, Dinarvand R (2015) Albuminated PLGA nanoparticles containing bevacizumab intended for ocular neovascularization treatment. *Journal of Biomedical Materials Research Part A* 103:3148–3156.
76. Liu J, Zhang X, Li G, Xu F, Li S, Teng L, Li Y, Sun F (2019) Anti-angiogenic activity of bevacizumab-bearing dexamethasone-loaded PLGA nanoparticles for potential intravitreal applications. *International Journal of Nanomedicine* 14:8819–8834.
77. Rudeen KM, Liu W, Mieler WF, Kang-Mieler JJ (2022) Simultaneous release of aflibercept and dexamethasone from an ocular drug delivery system. *Current Eye Research* 47:1034–1042.
78. El-Mofty HM, Abdelhakim MASE, El-Miligi MF, El-Nabarawi MA, Khalil IAH (2014) A new combination formula for treatment of fungal keratitis: an experimental study. *Journal of Ophthalmology* 2014:173298.
79. Mudgil M, Pawar PK (2013) Preparation and in vitro/ex vivo evaluation of moxifloxacin-loaded PLGA nanosuspensions for ophthalmic application. *Scientia pharmaceutica* 81:591–606.
80. Abbas MN, Khan SA, Sadozai SK, Khalil IA, Anter A, Fouly ME, Osman AH, Kazi M (2022) Nanoparticles loaded thermoresponsive in situ gel for ocular antibiotic delivery against bacterial keratitis. *Polymers* 14:1135.
81. Salama AH, AbouSamra MM, Awad GE, Mansy SS (2021) Promising bioadhesive ofloxacin-loaded polymeric nanoparticles for the treatment of ocular inflammation: Formulation and in vivo evaluation. *Drug Delivery and Translational Research* 11:1943–1957.
82. Mounika M, Priya NS, Aruna MS, Rao NR (2017) Formulation and Evaluation of Ocular Insitu Hydrogels of Acyclovir, *International Journal of Research in Pharmacy and Chemistry*, 7(2), 162–170.
83. Patil S, Kadam A, Bandgar S, Patil S (2015) Formulation and evaluation of an in situ gel for ocular drug delivery of anticonjunctival drug. *Cellulose Chemistry and Technology* 49:35–40.
84. Narayana S, Ahmed MG (2022) Design and evaluation of ocular hydrogel containing combination of ofloxacin and dexamethasone for the treatment of conjunctivitis. *Brazilian Journal of Pharmaceutical Sciences* 58:e20180.
85. Sipos B, Budai-Szűcs M, Kókai D, Orosz L, Burián K, Csorba A, Nagy ZZ, Balogh GT, Csóka I, Katona G (2022) Erythromycin-loaded polymeric micelles: In situ gel development, in vitro and ex vivo ocular investigations. *European Journal of Pharmaceutics and Biopharmaceutics* 180:81–90.
86. Shah S, Patel A, Patel S, Pal D, Mitra A (2013) Novel Aqueous Nanomicellar Formulations Containing Fluocinolone Acetonide for the Treatment of Posterior Uveitis. *Investigative Ophthalmology & Visual Science* 54:1076–1076.
87. Wu W, He Z, Zhang Z, Yu X, Song Z, Li X (2016) Intravitreal injection of rapamycin-loaded polymeric micelles for inhibition of ocular inflammation in rat model. *International Journal of Pharmaceutics* 513:238–246.
88. Gonzalez-Pizarro R, Carvajal-Vidal P, Bellowa LH, Calpena AC, Espina M, García ML (2019) In-situ forming gels containing fluorometholone-loaded polymeric nanoparticles for ocular inflammatory conditions. *Colloids and Surfaces B: Biointerfaces* 175:365–374.

10 Synthetic Polymer-Derived Drug Delivery Systems for Diseases Like Rheumatoid Arthritis, Hepatitis, Nephropathy, and Pancreatitis

Vinita Giri, Pooja Pandita, and Sanchit Arora
Delhi Institute of Pharmaceutical Sciences and Research (DIPSAR), Delhi Pharmaceutical Sciences and Research University (DPSRU), New Delhi, India

Viney Lather
Amity University Uttar Pradesh, Noida, India

Dalapathi Gugulothu and Deepti Pandita
Delhi Institute of Pharmaceutical Sciences and Research (DIPSAR), Delhi Pharmaceutical Sciences and Research University (DPSRU), New Delhi, India

10.1 INTRODUCTION

An innovative and hopeful development in the fight against complex and chronic diseases such as pancreatitis, hepatitis, rheumatoid arthritis (RA), and nephropathy has been the investigation and incorporation of synthetic polymer-derived drug delivery systems [1]. Due to the complex pathophysiological mechanisms of these disorders, patients must often deal with substantial burdens that require lifelong care measures [2]. While it can help with symptom alleviation and disease control to some level, traditional pharmacotherapy has many drawbacks. Therapeutic outcomes are severely hampered by problems such as poor medication absorption, the need for high dosages, noticeable side effects, the emergence of disease resistance, and the absence of targeted drug delivery methods [3]. Therefore, there is an urgent need for more advanced, effective, and customized medication delivery systems that can get around these challenges and improve patient care.

The development of drug delivery methods built from synthetic polymers is a major step in the right direction toward tackling these pressing issues [4]. Modern medication delivery systems can be built on a solid foundation thanks to synthetic polymers' chemical flexibility and technological adaptability [5]. It is possible to carefully construct polymers like polyethylene glycol (PEG), polylactic acid (PLA), and poly-lactic-co-glycolic acid (PLGA) to encapsulate medicinal chemicals [6]. This will shield the agents from enzymatic degradation and guarantee their controlled release at the pathological site. Reducing systemic exposure not only increases the bioavailability and efficacy of medications but also dramatically reduces any negative side effects [7]. The use of synthetic polymers in drug administration goes beyond the simple physical encasing of pharmaceuticals. Moreover, these materials can be designed to have special physicochemical characteristics, which are especially useful for targeted drug administration in conditions with distinct clinical indicators [8]. Examples of these features include stimuli responsiveness to pH, temperature, or enzyme activity. For example, polymers that react to particular inflammatory stimuli can make sure that medications are released exactly where they are needed most in the inflammatory environment of pancreatitis or RA [9]. This minimizes systemic negative effects while optimizing treatment efficacy.

Furthermore, a major advancement in the field of nanomedicine has been made with the creation of polymeric nanoparticles (NPs) as drug delivery vehicles [10]. When it comes to their capacity to target particular cell types or tissues, traverse biological barriers, and release their medication payloads gradually, these NPs are unmatched in their benefits [11]. Additionally, these NPs offer a promising way to attain such specificity for diseases like hepatitis and nephropathy, where focused distribution to the liver or kidneys, respectively, is essential for effective treatment [12].

Table 10.1 summarizes the key aspects of synthetic polymer-derived drug delivery systems, including their advantages, special features, benefits, and current issues. Though drug delivery systems created from synthetic polymers have the potential to transform the treatment of

DOI: 10.1201/9781003477389-10

TABLE 10.1
Overview of Synthetic Polymer-Derived Drug Delivery Systems

Aspect	Description	Examples
Targeted diseases	Chronic and complex diseases requiring advanced treatment strategies	Pancreatitis, hepatitis, RA, nephropathy
Challenges with traditional pharmacotherapy	Poor medication absorption, high dosages required, noticeable side effects, diseases resistance, lack of targeted delivery methods	
Advantages of synthetic polymer systems	Controlled release of medications, reduced systemic exposure, increased bioavailability and efficacy, targeted drug delivery	PEG, PLA, PLGA
Special features of polymers	Chemical flexibility, stimuli responsiveness (e.g., pH, temperature, enzyme activity), customized physicochemical properties	
Benefits of polymeric nanoparticles	Target specific cell types or tissues, traverse biological barriers, gradual release of medication payloads	Hepatitis, nephropathy
Current issues	Regulatory challenges, long-term biocompatibility and safety, optimization of synthesis process, ethical considerations	
Future prospects	Improved drug delivery efficacy, development of innovative systems, overcoming regulatory and safety hurdles	

diseases such as nephropathy, RA, pancreatitis, and hepatitis, there are still a number of unanswered questions [13]. These include figuring out how to navigate the complex regulatory landscape in order to have novel drug delivery systems approved, explaining the materials' long-term biocompatibility and safety characteristics, and improving polymer synthesis procedures to ensure scalable and repeatable manufacturing [14]. Furthermore, it is imperative to give due thought to the ethical problems surrounding the use of cutting-edge drug delivery systems, including those about patient consent, privacy, and availability of these life-saving treatments.

To better understand the current status of synthetic polymer-derived drug delivery systems, this chapter will go over their uses, benefits, and drawbacks concerning illnesses such as pancreatitis, hepatitis, RA, and nephropathy. Additionally, we aim to clarify the methods by which these systems improve drug delivery efficacy by thoroughly examining the different kinds of synthetic polymers used in drug administration, their structural characteristics, and the creative formulation techniques used. Furthermore, the prospects for drug delivery using synthetic polymers in the future are examined, along with an overview of the present research, potential advancements, and the numerous challenges that must be solved to fully realize the potential of these innovative therapeutic systems.

10.2 SYNTHETIC POLYMERS IN THE CONTEXT OF DRUG DELIVERY FOR DISEASES

Under physiological settings, natural polymers exhibit rapid breakdown and have weak mechanical characteristics. Thus, a variety of biodegradable and biocompatible synthetic polymers (Figure 10.1), such as poly PEG, poly(vinyl acetate) (PVA), and PLGA, were created to repair cartilage [15]. Covalent bond hydrolysis is often capable of dissolving these polymers, and the rate of hydrolysis is affected by several factors, including steric hindrance at the cleavage site [16]. Furthermore, they were easily adaptable to comply with mechanical, biological, and physical requirements. A good foundation for an implantable device that allows the medication to be gradually released and less frequently administered is provided by polymeric delivery systems [17]. Because polyesters have a high degree of biodegradation and safety, there is no need to remove the device after the medicine has been released. Moreover, the benefits of converting the biodegradable polymer into microspheres allow the device to be injected directly into the illness site, negating the need for a surgical implant [18].

Several polymeric microparticles and NP systems have been tried. PLA and PLGA are two of the several polymers that have been investigated [19]. The covalent bonding between the drug and polymer may prevent drug leakage or burst release in an acidic environment while permitting sustained release in a physiological setting. Polymers, which are extended chains of residues from amino acids, can have a wide variety of chemical and physical properties [20]. Synthetic biodegradable polymers can take the place of

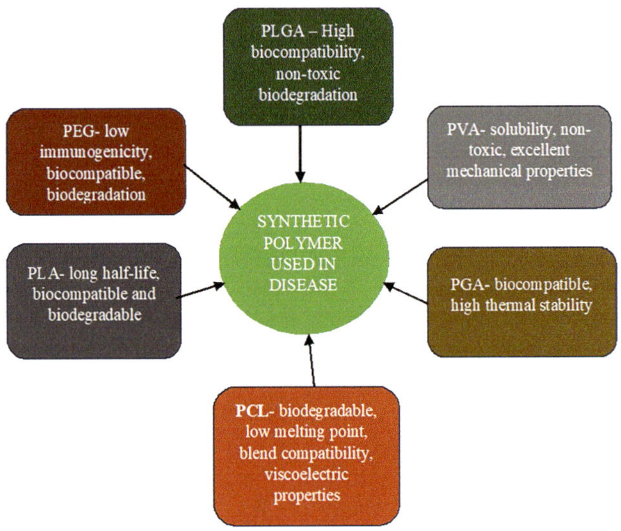

FIGURE 10.1 Majorly used synthetic polymers in diseases such as RA, hepatitis, nephropathy, and pancreatitis.

oligopeptides, which are linear sequences of amino acids, in biomedical applications [21]. Applications and research typically use copolymers of poly(glycoside), poly(lactide), and poly(caprolactone) [22]. Numerous synthetic polymers based on poly(amino ester) (PAE) are suggested as viable options for gene and drug delivery due to their pH responsiveness. For an in-depth understanding of the key aspects and applications of various synthetic polymers in drug delivery, refer to Table 10.2.

10.2.1 Poly(ethylenimine) (PEI)

PEI is soluble in ethanol, chloroform, and boiling water at low pH values. The chemical is not soluble in acetone, benzene, ethyl ether, or cold water. Branched PEI (BPEI) is a polymer consisting of ethylene diamine groups arranged in repeating units [23]. BPEI was formed using aziridine ring-opening polymerization. BPEI is a cationic polymer that has primary, secondary, and tertiary amino groups [24]. Because it can form complexes with negatively charged molecules like DNA and RNA, it has a high amine density that makes it suitable for gene transport [25–29]. PEI also promotes endosomal escape through the "proton sponge" effect, in which its high buffering capacity causes the endosomal membrane to break and result in osmotic swelling, ensuring that its cargo is successfully released into the cytoplasm [30]. This characteristic highlights PEI's potential in gene therapy applications, where efficient intracellular transport and expression of therapeutic genes are critical. It also demonstrates high transfection efficiency across a variety of cell types [31].

10.2.2 PVA

PVA is well-known for its use in consumer, industrial, and adhesive products because of its superior flexibility, clarity, and adhesive qualities [32, 33]. PVA is important for drug delivery since it may be used as a coating material and to formulate systems for the controlled release of drugs [34]. It is a desirable alternative for encasing medications and shielding them until they reach the intended location within the body due to its natural qualities, which include biocompatibility and the capacity to form films [35]. Modifying the characteristics of its composites and the polymer's

TABLE 10.2
Key Aspects of Various Synthetic Polymers in Drug Delivery

Synthetic Polymer Type	Applications in Drug Delivery	Advantages	Limitations	Ref.
PEI	Gene delivery, DNA/RNA transfection	High transfection efficiency, strong nucleic acid binding	Cytotoxicity	[23]
PVA	Hydrogel formulations, ocular drug delivery	Biocompatible, nontoxic	Swelling in water	[32]
Polyesters	Biodegradable implants, microparticles, NPs	Biodegradable versatile	Variable degradation rates	[37]
PEG	Stealth coatings for NPs, drug conjugates	Increases solubility and circulation time	Potential immunogenicity	[45]
PGA	Sutures, tissue engineering scaffolds	High mechanical strength, biodegradable	Acidic degradation products	[50]
PLA	Controlled-release drug delivery system	Biocompatible, biodegradable	Requires catalyst for polymerization	[55]
PLGA	Vaccines, systemic delivery of drugs	Tunable degradation rate, FDA-approved	Cost, drug-polymer interactions	[60]
Polyolefins	Packing materials	Chemical resistance, stability	Nonbiodegradable	[67]
PTFE	Implantable devices, vascular grafts	Chemically inert, biocompatible	Nonbiodegradable high cost	[77]
PVC	Medical tubing and containers	Inexpensive flexible	Plasticizers may leach	[82]
Silicone	Controlled-release systems, implant coatings	Biocompatible flexible	Oil bleeding, cost	[89]
Polyurethanes	Catheters, wound dressings	Elasticity, biostability	Degradation by esterases	[97]
Polyamides	Sutures, microspheres	High mechanical strength, biocompatible	Absorbs moisture, degradation	[102]
Methacrylate	Contact lenses, intraocular lenses	Transparency, tunable hydrophilicity	Requires careful polymerization control	[108]
PPDO	Biodegradable sutures, drug-eluting stents	Excellent biodegradability, good mechanical properties	Limited commercial availability	[116]

formulation can change the rate of medication release in PVA. This allows for the achievement of delayed or sustained drug administration [36].

10.2.3 Polyesters

Polyesters that are biostable and biodegradable are employed in biomedicine. Polycarbonates and poly (ethylene terephthalate) are biostable polyesters with aromatic groups [37]. Membranes, filaments, and meshes are examples of how they are employed. The most often utilized degradable polymers are poly(glycollic acid) (PGA), poly(L-lactic acid) (PLLA), and poly(D lactic acid) (PDLA), which are derived from polyesters having tiny aliphatic glycollic acid or lactic acid. [38]. A biodegradable polyester composed of ether-ester repeating units is called polydioxanone. Nonenzymatic hydrolysis is the main method of breaking down these polymers, and the breakdown products promote more deterioration [39]. Numerous parameters, including as molecular weight, crystallinity, fiber structure, and replacement groups, control the breakdown rates. Clinical application of PGA, PLLA, and PDLA occurred swiftly due to the physiological metabolites that are their breakdown products and monomers. Sometimes concerns about the acidity of these breakdown products result in upper limits on the amount that is permitted [40–42]. In addition to this, polyesters can be manufactured into a variety of shapes, such as scaffolds, microparticles, and NPs, allowing for the targeted delivery and prolonged release of medications [43, 44].

10.2.4 PEG

Molecular weight has a major role in determining the features of PEG, an amphiphilic polymer composed of oxide ethylene monomer units [45–47]. PEGylation is the process of joining PEG chains to molecules or surfaces. It reduces immunogenicity and the reticuloendothelial system's (RES) clearance of therapeutic agents while also improving their solubility and stability against enzymatic degradation [48]. This alteration increases the bioavailability and efficacy of treatments while also extending the duration that medications and biomolecules spend in the bloodstream, enabling less frequent dosage and better patient compliance. PEG is also a perfect stealth coating for liposomes and NPs because it can minimize undesired interactions with biological components and avoid immune system reactions, allowing for focused medication administration with low side effects [49]. PEG's versatility in drug delivery and formulation systems highlights how important it is to the development of more focused and efficient therapeutic interventions.

10.2.5 PGA

PGA is the most fundamental linear aliphatic polyester and is often referred to as polyglycolide. It is a biodegradable thermoplastic polymer [50]. It can be made by ring-opening polymerization or glycolic acid condensation. Since 1954, PGA has been known to be a polymer that may produce strong fibers, but its hydrolytic instability has restricted its use [51, 52]. Because of its hydrolytic breakdown into glycolic acid, a metabolite that is found in the body naturally, PGA is a great material to use in drug delivery systems that reduce long-term toxicity and do not require surgical removal when their therapeutic effect is finished. Owing to PGA's exceptional mechanical strength, robust thermal stability, and adaptability in forming films, fibers, and microparticles, a range of drug delivery systems that are precisely engineered to release medications at a regulated pace can be produced [53, 54].

10.2.6 PLA

For many biomedical applications, PLA is the optimum choice because of its hydrophobic qualities, biodegradability, ease of manufacture, and biocompatibility with host tissue [55, 56]. PLA, a synthetic polymer that is both biodegradable and biocompatible, has become more and more popular in the drug delivery industry [57]. PLA is a major component in the creation of controlled-release systems and has the benefit of being naturally metabolized by the body into lactic acid, which reduces toxicity concerns. Because of its adaptability, a variety of drug delivery platforms, such as scaffolds, microparticles, and NPs, can be created that are specifically designed to release therapeutic chemicals gradually or selectively [58]. Because of its potential and its safety record, PLA has gained favor for a variety of uses, including systemic drug administration and localized cancer therapy, it presents a viable path toward improving treatment efficacy and patient compliance [59].

10.2.7 PLGA

Because of its exceptional mechanical properties, regulated rate of disintegration, excellent cell adhesion, and safety, PLGA is a biodegradable polymer with great promise for usage in a range of applications [60]. The random ring-opening copolymerization of PGA and PLA results in PLGA. Thus, the rate at which PLGA products deteriorate can be controlled by varying the ratio of these two polymers [61]. Consequently, PLGA is favored over PGA and exhibits potential for usage in numerous biological applications, including cancer medication delivery systems and sutures. When implants are composed of PLGA, the rate of bone repair and growth is greatly accelerated [62–65]. The Food and Drug Administration's (FDA) approval of PLA for use in a range of pharmaceutical applications highlights the material's excellent drug delivery capabilities and safety profile. The unique properties of PLGA, such as controlled drug release, biodegradability, and biocompatibility, make it an important material for the development of enhanced drug delivery systems. These qualities are very promising for a range of medical uses [66].

10.2.8 Polyolefins

Polyethylene (PE) and polypropylene (PP) are inert and hydrophobic polyolefins that do not break down *in vivo* [67–72]. Advances in polymer science have made it possible to functionalize and modify polyolefins, which have not historically been linked to biomedical applications, to improve their compatibility with biological environments and medicinal agents [73]. Because of their inertness, resilience to deterioration, and capacity to be shaped into a wide range of forms and structures – from fibers to microparticles – they are well suited for controlled-release applications, which allow for the steady and slow distribution of medications. Furthermore, polyolefins' versatility makes it possible to incorporate targeting moieties or responsive elements, which facilitates stimuli-responsive release and targeted drug delivery [74, 75].

10.2.9 Poly(tetrafluoroethylene) (PTFE)

PTFE (Teflon) has an ethylene backbone with four fluorine molecules that are covalently bonded [76]. It is known as ePTFE (Gore-Tex) in its enlarged, porous form with interconnected fibrils. It is a nonbiodegradable, very hydrophobic substance [77, 78]. Furthermore, because of its exceptional thermal stability, chemical resistance, and biocompatibility, it can be used in specific drug delivery scenarios, like implanted devices and localized drug delivery systems [79–81].

10.2.10 Poly(vinyl chloride) (PVC)

One covalently linked chlorine atom is present in the ethylene backbone of PVC [82–85]. In contrast to other polymers, it is less commonly associated with drug delivery systems, even though it is extensively utilized in numerous industrial and medical applications, including tubing and medical apparatus [86]. Due to its inherent qualities, which include chemical stability, durability, and ease of processing, PVC is the material of choice for producing medical containers that must be sterile and safe, such as IV bags and blood bags. PVC is frequently used to improve its flexibility, but this has generated concerns about its release of plasticizers like phthalates, which have been linked to health and safety issues [87, 88].

10.2.11 Silicone

The -Si-O- backbone of silicones has variable chain lengths and crosslinks that control mechanical properties through a gel structure, from liquid oil to rubber elastomer [89]. Side chains can be changed, although the most common are methyl groups in poly(dimethyl siloxane) (PDMS) [90]. Silicones are hydrophobic, biostable elastomers that do not require plasticizers; depending on the use, the biological reaction changes. Synovitis is a late hazard in intra-articular implants, also there is high tolerance in ophthalmologic applications, and fibrous capsule development occurs with breast implants. In relation with hematologic diseases, it is hypothesized that silicon oil residues cause malignancies and connective tissue problems [91–96].

10.2.12 Polyurethanes

Polyurethanes used in medicine come in polyester, and polycarbonate with aromatic or aliphatic constituents; aromatic formulations have better biostability [97, 98]. Thermoplastic versions combine hard and soft segments, maintaining elasticity without plasticizers. While PVC and polycarbonate-based polyurethanes are oxidation-resistant, bisphenol A concerns exist. Aliphatic polyether-based types soften the body for comfort [99]. Additionally, polyurethanes are distinguished by a unique molecular structure that allows for the modification of their mechanical characteristics and rates of degradation [100, 101].

10.2.13 Polyamides

Polyamides, such as nylon, are used in medical applications due to their strength and stability [102, 103]. They are employed in sutures and angioplasty ballon catheters. Polyamides can be engineered for specific drug release profiles, improving treatment efficacy [104, 105]. Their ability to be customized and biodegradable into safe byproducts makes them valuable for drug delivery systems [106, 107].

10.2.14 Methacrylate

Methacrylate polymers, including polymethyl methacrylate (PMMA) and poly(2-hydroxyethyl methacrylate (PHEMA), are used in dentistry and orthopedics for their rigidity and biocompatibility [108–110]. They can be modified to adjust hydrophilicity and biodegradability for various drug delivery needs, forming hydrogels, nanoparticles, and micelles for controlled release [111–113]. Their low toxicity and adaptability make them suitable for diverse medicinal applications [114, 115].

10.2.15 Poly(p-dioxanone) (PPDO)

PPDO, an FDA-approved material, is valued for its hydrophilicity, biocompatibility, and plasticity [116, 117]. Used in stents and sutures, PPDO degrades into safe components, making it ideal for controlled drug release [118, 119]. Its ability to provide sustained release enhances its use in implants and injectable formulations [120–123].

10.3 Utilizing Polymeric NPs for Targeted Drug Delivery System in Disease Management

10.3.1 RA

RA is a chronic autoimmune disease mostly affecting the diarthrodial joints of the hands and feet. It is typified by inflammation of the synovial joints, which damages the articular cartilage, bone, tendon, and ligament and results in function loss. Comorbid conditions linked to RA include depression, osteoporosis, cancer, lung disease, infection, and cardiovascular disease caused by atherosclerosis [124]. Infections and inflammatory mediators are prominent triggers; however, the underlying disease mechanisms remain unknown.

10.3.1.1 PEI

Zhong et al.'s study looked at the difficulties of using high-dose methotrexate (MTX) to treat RA for a long time. The purpose of the study was to develop a tailored drug delivery system that would target the joint's inflammatory area selectively. The researchers created MTX-PEI@HA NPs by using HA as the hydrophilic negative electrical shell and MTX-linked branching polyethyleneimine (MTX-PEI) NPs as the core. The aqueous phase of MTX-PEI@HA NPs was prepared in a single pot. The purpose of the MTX-PEI@HA NPs is to benefit from the preferential absorption by activated macrophages of endocytosis mediated by the CD44 receptor. In the inflammatory response linked to RA, macrophages are essential. Moreover, using an *in vivo* mouse model, MTX-PEI@HA NPs treatment showed improvement in inflammatory arthritis. Crucially, this focused strategy made it possible to employ high dosages of MTX without compromising safety. The research emphasized the benefits of using HA-coated aqueous-synthesized NPs for targeted delivery that is arthritis-specific. According to the results, MTX-PEI@HA NPs may be viable nanoplatforms for the treatment of RA, providing a more focused and efficient method of MTX administration with fewer systemic adverse effects [125]. The objective of Yin et al.'s second study was to create a temperature-sensitive hydrogel (D-N Gel) with NPs (D-NPs) that could simultaneously transmit MTX and indomethacin (IND). The creation of D-NPs via many noncovalent interactions between PEI-SS (polyethyleneimine with disulfide bonds) and the hydrophobic small molecule drugs MTX and IND was essential to achieving this. After that, these D-NPs were added to a hydrogel matrix that was temperature-sensitive. The temperature-sensitive hydrogel matrix, which contained 10% F68 and 27% F127, showed a sol-to-gel transition temperature (T sol/gel) of 33°C and a gelation duration of less than 15 seconds. When D-N Gel was injected into the articular cavity of rats suffering from diseases caused by collagen-induced arthritis, it turned into gels and gradually released medication into the joint fluid over a period of 72 hours. D-N Gel was found to be very effective in reducing inflammatory cytokine expression, bone degradation, and joint edema in the fluid of the knee and ankle joints. Crucially, studies for liver and kidney function as well as histology examinations revealed that D-N Gel was physiologically safe. According to the study's findings, D-N Gel has great promise for the sustained co-delivery of MTX and IND, offering RA patients a synergistic treatment. This strategy targets the fundamental causes of RA, as well as its symptoms, and it emphasizes the potential of temperature-sensitive hydrogels for regulated medication release in the management of inflammatory joint disorders [126].

The next study's authors examined the mechanism of RA pathogenesis and the role that STING (Stimulator of Interferon Genes) activation plays in that mechanism. Given that abnormal STING activity in RA has been linked to a number of different autoimmune illnesses, the study sought to investigate the implications of this finding in RA. Since double-stranded DNA (dsDNA) build-up is a specific trigger for STING in RA, the researchers created spherical polyethyleneimine-coated mesoporous polydopamine NPs (PEI-PDA@C-176 NPs) loaded with the STING antagonist C-176 for use in therapy. The biocompatible nanoparticles (NPs) exhibited a potent DNA binding capability and efficiently inhibited the STING pathway and inflammation in macrophages. PEI-PDA@C-176 NPs were injected intra-articularly into animals that had models of arthritis produced by collagen and dsDNA. This greatly reduced joint damage. According to the study, the use of drugs like PEI-PDA@C-176 NPs, which have synergistic STING inhibitory effects, may improve RA therapy regimens. These NPs have shown exceptional efficacy in reducing the inflammation and pathology linked to RA, suggesting that they could be applied as a therapeutic intervention for this inflammatory condition [127]. Researchers examined dsDNA accumulation and STING activation in human and mouse RA synovial tissues in an interesting work, and then used genetic ablation of STING in mice (Sting1+/− mice) to examine the functional impact of STING on RA (Figure 10.2). Furthermore, to provide a high affinity for DNA binding, mesoporous polydopamine (PDA) nanoparticles (NPs) were coated with specialized cationic PEI. C-176 was intended to be used for the local delivery of these NPs. The potency of these PEI-PDA@C-176 NPs lay in their combined ability to suppress STING signaling and dsDNA-induced immunological responses in macrophages, hence enabling the use of a lower therapeutic dose with notable effects. The efficacy of PEI-PDA@C-176 NPs in the treatment of RA was investigated in two models of arthritis and Sting1+/− mice [128].

Another study by Duan et al. that was discussed looked at the effectiveness of systemic transport of therapeutic siRNA to experimental arthritic joints using a nanocarrier PEI-superparamagnetic iron oxide NP (SPIO), which is composed of an iron oxide core and a PEI shell. Using in vitro synthesis, PEI-SPIO/siRNA NPs were investigated. In order to investigate tissue distribution, cellular absorption, and the therapeutic efficiency of siRNA targeting the IL-2/-15 receptor β chain (IL-2/IL-15Rβ), NPs were administered intravenously to rats suffering from arthritis. Targeted gene silencing was successfully induced in vitro using PEI-SPIOs loaded with siRNA, exhibiting enhanced siRNA stability, decreased cytotoxicity, and effective macrophage uptake. Joint macrophages and T cells readily absorbed the siRNA given by PEI-SPIO, which caused it to quickly accumulate in aching joints. For the treatment of experimental arthritis, PEI-SPIOs loaded with IL-2/

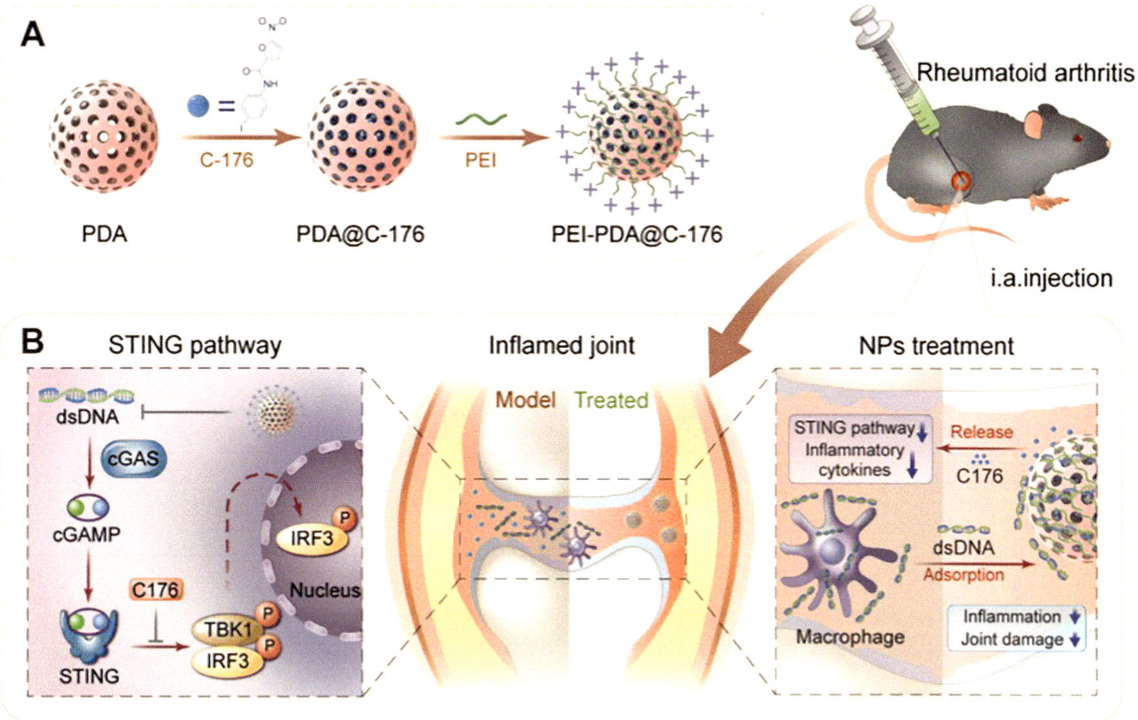

FIGURE 10.2 A schematic illustration of PEI-PDA@C-176 NPs as a RA therapy via suppressing the STING pathway. (A) The PDA NPs were loaded with the STING antagonist C-176, and the surface of the PDA@C176 NPs was further decorated with PEI to increase DNA binding affinity. (B) The mechanisms by which PEI-PDA@C-176 NPs treat RA. The chemical can adsorb STING stimulation dsDNA while also inhibiting STING palmitoylation. As a consequence, PEI-PDA@C-176 NPs have shown a significant effect in reducing RA inflammation and pathogenesis. I. a., intra-articular; dsDNA, double-stranded DNA [128]. [© 2022 The Authors. Publishing services by Elsevier B.V. on behalf of KeAi Communications Co. Ltd. This is an open-access article distributed under the terms of the Creative Commons Attribution License (CC BY).]

IL-15RβsiRNA alone were effective. On the other hand, a magnetic field combined with PEI-SPIO/IL-2/IL-15RβsiRNA had an additional anti-inflammatory effect. Consequently, systemic siRNA administration in RA can be made possible using PEI-functionalized SPIOs, and therapeutic efficacy can be further increased by applying an external magnetic field [129].

10.3.1.2 PLGA

Trujillo-Nolasco et al.'s work was to create a novel medication delivery method for radio synovectomy, a surgery that includes injecting a β-emitting radionuclide – 177Lu in this case-intraarticularly to diminish inflammation of the synovial tissue. PLGA polymer-based drug-encapsulating NPs were used in the drug delivery system. These NPs are designed to deliver anti-inflammatory drugs such as MTX and HA, with the express goal of treating arthritic disorders. Using DOTA (1,4,7,10-tetraazacyclododecane-1,4,7,10-tetraacetic acid), the scientists were able to create lutetium-177 (177Lu) with radiotherapeutic characteristics (Figure 10.3). The goal was to create a clever drug delivery system that could specifically target cells in synovial tissue that overexpress CD44 and hyaluronan. Chemical characterization of the proposed polymeric NP system was carried out, along with MTX encapsulation and radio labeling. The produced NPs mean diameter, as determined by the data, was 167.6 nm ± 57.4, suggesting a limited and monomodal dispersion. Effective conjugation of HA and DOTA to the polymeric nano system was established by the spectroscopic investigation. MTX has a 95.2% encapsulation rate and a 6% loading efficiency. Using immediate thin-layer chromatography (ITLC), the radiochemical purity of 177Lu-DOTA-HA-PLGA(MTX) NPs was found to be 96 ± 2%. In summary, 177Lu-DOTA-HA-PLGA(MTX) NPs have been successfully produced by researchers as a biocompatible polymeric drug delivery method. With 177Lu acting as a radiotherapeutic component and MTX functioning as a disease-modifying antirheumatic medication (DMARD), the NPs showed characteristics appropriate for radio synovectomy and subsequent targeted antirheumatic treatment [130].

Park et al. set out to develop a co-delivery method for glucocorticoids, specifically dexamethasone, and small interfering RNA (siRNA) that would target human

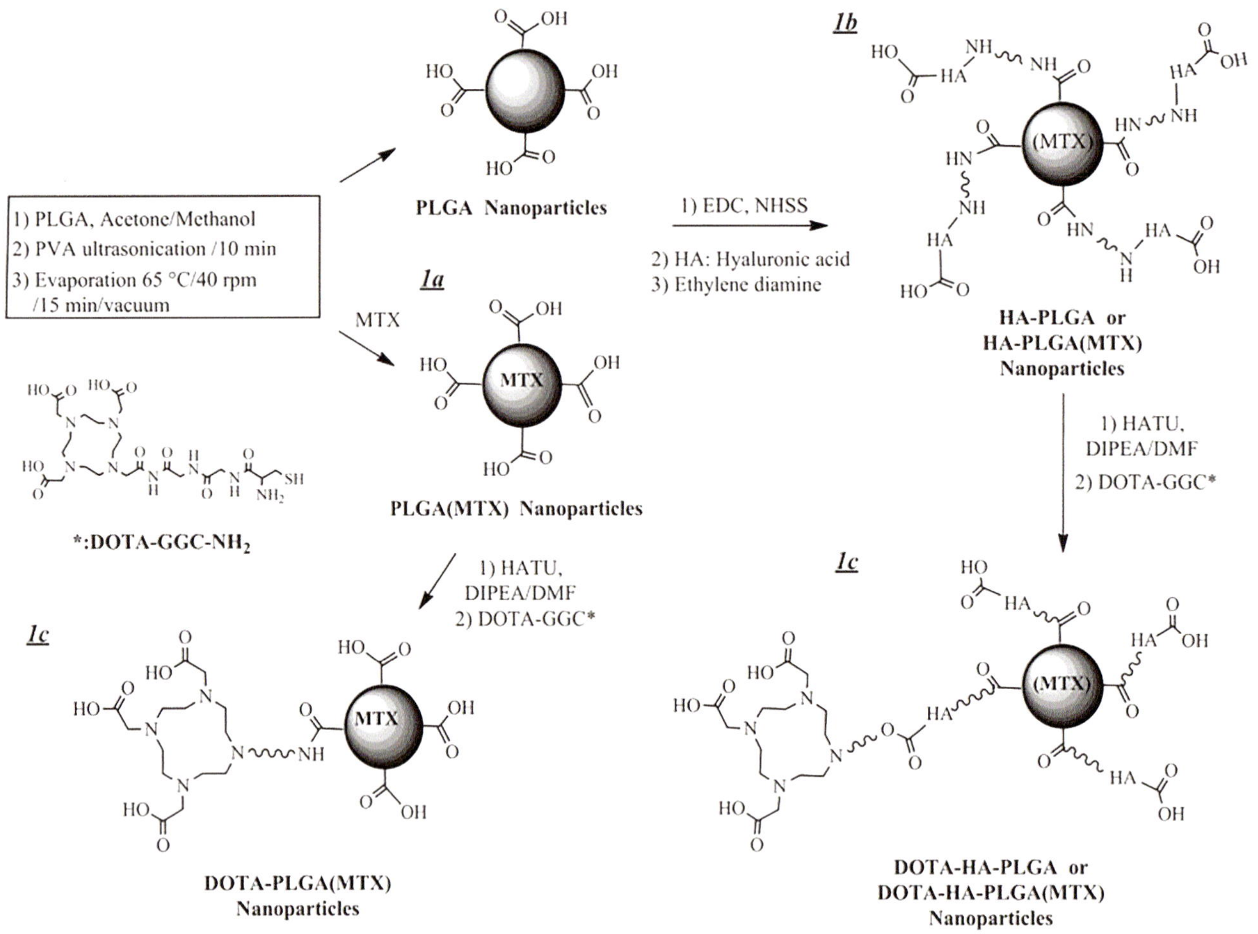

FIGURE 10.3 Schematic synthesis of (a) PLGA(MTX) nanoparticles, (b) HA-PLGA or HA-PLGA(MTX) nanoparticles, (c) DOTA-PLGA(MTX), and (d) DOTA-HA-PLGA(MTX) nanoparticles [130] (Copyright, 2019, Elsevier).

chondrocyte cells (C28/I2) in order to limit the synthesis of genes and proteins connected to arthritis. By encasing dexamethasone into PLGA NPs and then further complexing them with PEI and siRNA, the strategy allowed for the delivery of glucocorticoids and siRNA into the target cells at the same time. The researchers employed GFP siRNA to assess the co-delivery system's effectiveness and confirm successful transfection using green fluorescent protein (GFP) expression. After pretreating the cells with TNF-α to mimic conditions associated with arthritis, the expression of molecules relevant to arthritis, including COX-2 and iNOS, was upregulated. The study found that the expression of genes and proteins linked to arthritis was significantly reduced by dexamethasone-loaded PLGA NPs complexed with COX-2 siRNA transfection. Numerous methods, including real-time qPCR, RT-polymerase chain reaction (PCR), immunoblotting, immunohistochemistry, and immunofluorescence imaging, were used to assess this decline. Overall, the findings suggest that the co-delivery approach may be able to successfully lower the expression of variables associated with arthritis in chondrocytes [131]. A study by Présumey et al. said that small interfering RNAs (siRNAs) intended to target tumor necrosis factor α (TNF-α) were to be delivered using PLGA microspheres. This technique aimed to reduce TNF-α in an efficient and long-lasting manner in order to cure RA. The researchers assessed the PLGA microspheres for shape, mean diameter, siRNA dispersion, encapsulation efficiency, and in vitro release kinetics after producing them using a modified multiple emulsion-solvent evaporation technique. Using the murine monocytic cell line J774, the effectiveness of this delivery technique was assessed in vitro and in vivo using a preclinical RA model. The siRNAs were encapsulated in PLGA microspheres that demonstrated a high encapsulation efficiency and a slow and extended release of anti-TNF-α siRNAs. Interestingly, when injected intra-articularly, these PLGA microspheres effectively reduced TNF-α expression in arthritic joints. The findings of this study suggest that siRNAs targeting TNF-α may be delivered to RA patients using PLGA microspheres [132]. Developing a dependable UPLC-MS/MS method to measure triptolide (TPL) concentrations in a range of biological samples, including the kidney, lung, liver, spleen, plasma, and joint, was the aim of the work conducted by L. Wang et al. High levels of precision, accuracy, and stability were required in the process. The longer retention time in joint tissue of triptolide-encapsulated microspheres (TPL-PLGA-MS) has led to a reduction in the frequency of joint cavity administrations. The impact of TPL-PLGA-MS was shown to be significantly more sustained than that of the solution dosage form. Pharmacokinetic features, such as MTR(0-t), MTR (0-1), T1/2, Tmax, AUC(0-t), and AUC (0-1) were among the pharmacokinetic attributes in which the TPL-PLGA-MS group performed better than the solution group. The microsphere composition effectively prevented the medication's release from the joint cavity. Furthermore, TPL-PLGA-MS showed a strong potential to lower the expression of inflammatory factors such hs-CRP, TNF-α, IL-1, and IL-6. This implies that TPL-PLGA-MS has potential as a RA treatment when injected into the articular cavity [133].

10.3.1.3 PEG

Click PEG-Dex, a new linear multifunctional PEG-dexamethasone (Dex) molecule, was synthesized by Liu et al. using a Cu(I)-catalyzed Huisgen 1,3-dipolar cycloaddition, which is also referred to as A click reaction. An acid-labile hydrazone bond allowed Dex to connect to click PEG, which created a pathophysiological framework for regulated drug release. Adjuvant-induced arthritis (AA) in rats was modeled utilizing click PEG-Dex conjugates to test the drug delivery platform's flexibility. Rat arthritic joints were the conjugates' intended target for arthroscopy joints, as demonstrated by in vivo optical imaging. A single click injection of PEG-Dex decreased the inflammation in the ankle joints of AA rats in an extended therapy experiment, and the effect persisted for almost 15 days. However, arthritis symptoms were only momentarily alleviated by therapy with an equivalent dose of free dexamethasone phosphate sodium (free Dex); the symptoms returned as soon as the medication was stopped. The novel polymer-drug conjugates were shown to have superior anti-inflammatory and disease-modifying capabilities in ankle joints of the click PEG-Dex and free Dex therapy groups based on histological and bone mineral density assessments. This suggests that click PEG-Dex might be a helpful way for arthritis patients to take their medications in order to get improved treatment outcomes and long-lasting comfort [134]. In a fascinating work, Sujitha et al. investigated the potential therapeutic benefits of PEGylated liposomal berberine (PEG-BBR) against fibroblast-like synoviocytes (FLS) using an adjuvant-induced rat model of arthritic (AA) bone disintegration caused by Wnt1/β-catenin. In this investigation, the role of microRNA-23a (miR-23a) in these therapeutic benefits was also investigated. PEG-BBR and PEG-miR-23a internalization was effectively demonstrated both in vitro and in vivo using confocal microscopy and near-infrared imaging (NIR), with retention lasting up to 48 hours. The specific absorption of PEG-BBR into the inflammatory joint area significantly reduced the amount of bone deterioration in rats and the expression of important Wnt1 signaling mediators in both genes and proteins. PEG-BBR treatment decreased the β-catenin, Dvl-1, FZD4, and LRP5 expression levels brought on by CYLD in FLS cells. By blocking these components, the RANKL/OPG axis was changed, improving calcium retention and lowering bone loss. Similar to DKK-1, an inhibitor of LRP5/Wnt signaling, PEG-BBR administration similarly stopped the transcriptional activation of β-catenin and greatly decreased the synthesis of LRP5 protein. Additionally, the study demonstrated that miR-23a altered the levels of LRP5 expression by RNA interference. The results imply that berberine and miR-23a are therapeutically effective in the pathophysiology of RA, which may open up a therapeutic window for Wnt1/β-catenin signaling [135]. He and associates created PEGylated lipid NPs, which react to matrix metalloproteinase (MMP) to specifically target regions of arthritis with medicine delivery. Pain, inflammation, and swelling in the joints are the main symptoms of RA. The overexpression of MMPs in RA inflammatory areas was addressed in order to increase the therapeutic efficacy of anti-inflammatory medications. Tri glycerol monostearate (TGMS) and 1,2-diste aroyl-sn-glycero-3-phospho-ethanolamine-poly (ethylene glycol) (DSPE-PEG2000) were coassembled to create the PEGylated lipid NPs. TGMS's ester bond was engineered to be cleavable by MMPs, with the PEG chain functioning as a surrogate layer. In inflammatory joints linked to RA, lipid nanoparticles (NPs) demonstrated improved blood circulation, improved biocompatibility, and favorable dispersion. Dex is an anti-inflammatory drug that was one of these PEGylated lipid NPs that reacted to MMP. When the NPs were given intravenously into arthritic animals, the MMPs induced a rapid release of Dex. As a result, there was a discernible reduction in joint swelling and a suppression of the tissues' production of TNF-α and IL-1β. According to the study, a more efficient method of delivering medication for the treatment of RA is by MMP-responsive PEGylated lipid NPs, which preferentially target inflammatory areas and release the medication when MMPs are overexpressed [136].

Apart from earlier research, Q. Wang et al. sought to tackle the difficulties related to conventional liposomes in the management of RA. Although liposomes show great promise as drug delivery vehicles, their short half-life and physicochemical instability in blood arteries are drawbacks. The researchers used a thin-film hydration technique to create polymerized stealth liposomes in order to get around these obstacles. 1,2-distearoyl-sn-glycero-3-phosph oethanolamine-poly (ethylene glycol) (DSPE-PEG2000) and 1,2-bis(10,12-tricosadiynoyl)-sn-glycero-3-phosphocho line (DC8,9PC) were the components of the polymerized stealth liposomes. The PEG chains on the liposome surface functioned as a stealth layer, and UV radiation caused the DC8,9PC molecules in the liposome bilayer to cross-link. This design aimed to improve stability, extend circulation time, and enhance targeted liposome delivery to inflamed joints. The polymerized stealth liposomes containing Dex demonstrated long blood circulation periods and preferential accumulation in inflamed joints. As a result, there was a drop in proinflammatory cytokines (TNF-α and IL-1β) in the tissues around the joints, which slowed the progression of RA and reduced swelling. According to the research, polymerized stealth liposomes may prove to be a useful drug delivery system for a range of medical uses, especially when it comes to the treatment of inflammatory diseases like RA [137].

10.3.1.4 PVA

Nornberg et al. created films for the transdermal delivery of MTX in order to lessen the side effects associated with administering the medication orally or subcutaneously for the treatment of RA. PVA, cationic starch (CSt), and chondroitin sulfate (ChS) were the materials used to make the films. To evaluate the formulation's efficacy in vivo, a mouse model of radiation arthritis was employed. MTX was added to the mixture and made available. The films were made simply under moderate conditions, and it was determined that the drug loading distribution and efficiency of the ones loaded with 5% MTX were adequate. The majority of the MTX was found in the medium, according to studies on the penetration of pig skin layers, suggesting the potential for systemic absorption. In vivo studies using a mouse model of arthritis revealed that MTX-loaded films effectively treated and reduced RA symptoms, including changes in inflammatory processes, oxidative stress, and nociceptive behaviors. MTX was applied transdermally with pharmacological activity equivalent to oral therapy by using CSt/PVA and CSt/PVA/ChS films. These findings suggest that the films generated could work well as MTX transdermal delivery materials, providing a practical and noninvasive alternative for the treatment of RA [138]. Another study looked at the possible interactions and effects of coatings of amino-functionalized amino-polyvinyl alcohol with superparamagnetic iron oxide NPs (a-PVA-SPION) on human immune cells. The study's objective was to investigate potential negative effects on immune cell types, including monocytes and macrophages, as well as on cell survival and activity in human whole blood. The researchers evaluated the impact of a-PVA-SPION on human immune cells using a multiplex suspension array, flow cytometry, and transmission electron microscopy (TEM), among other analytical methods. Finding out if a-PVA-SPION is safe and suitable for use in imaging applications, specifically in the early diagnosis of RA, was the main objective of the study. The investigation's findings showed that a-PVA-SPION had no negative effects on the general survival of human immune cells. On the other hand, its effect on cytokine release has been shown. Interestingly, exposure to a-PVA-SPION increased the percentage of viable macrophages, with a greater effect early in the differentiation process. Furthermore, TEM showed that a-PVA-SPION was endocytosed by monocytes and macrophages. The results of the study elucidated the connection between human immune cells and a-PVA-SPION, emphasizing how important it is to consider this connection when considering the use of these NPs in therapeutic contexts, particularly when diagnosing RA [139]. An et al. undertook an intriguing investigation with the goal of addressing issues related to the therapeutic therapy of RA by co-delivering the aptamer Apt1-67, which specifically inhibits TNF receptor 1, and the interleukin-6 receptor (IL-6R) inhibitor tocilizumab (TCZ). The unique technique used detachable microneedles (MN) for drug administration, with potential benefits such as continuous release of TCZ and rapid release of Apt1-67. The use of methacrylate groups that grafted HA for the needle tip filling and polyvinyl alcohol/polyvinyl pyrrolidone for the needle body filling allowed TCZ and Apt1-67 to display the right releasing qualities. The results of the investigation showed that TCZ and Apt1-67 could be released into the surrounding environment through the skin and were efficiently dispersed within the microneedles. When comparing the efficacy of single medication microneedles (TCZ/Apt1-67@MN) with combination therapy given by microneedles, in vivo experiments using a mouse model of collagen-induced arthritis (CIA) demonstrated superior efficacy. Compared to subcutaneous injection, TCZ/Apt1-67@MN dramatically reduced bone degeneration and improved symptoms in CIA mice. The results of the study not only suggest a novel approach for co-delivering drugs with distinct release patterns, but they also provide a means of enhancing pharmacological efficacy in the management of RA [140].

10.3.2 Hepatitis

Hepatitis B is an acute liver illness caused by the hepatitis B virus (HBV), a major global health threat. Both acute and chronic infections brought on by HBV have the potential to induce cirrhosis and liver cancer-related mortality [141]. An estimated 325 million people worldwide suffer from a chronic HBV infection, according to recent data from the World Health Organization [142]. Hepatitis is now seen as a serious public health issue that needs immediate attention as a result. Short half-lives, low bioavailability, and frequent dosing were among the disadvantages of the medication used to treat hepatitis B in its conventional dosage form [143]. To attain the best possible drug concentration at the site of infection, targeted administration to the liver can be considered, which will result in high efficacy and low toxicity.

10.3.2.1 PEI

In this work, Saffari et al. developed a Quartz Crystal Microbalance (QCM) biosensor that enables label-free detection of the hepatitis B surface antigen, or HBsAg. The anti-HBsAg antibodies used in the biosensor were surface-modified, thiolated-PEI, and then covalently bonded to the gold electrode. After the QCM biosensor modified by PEI was optimized using Response Surface Methodology (RSM) to demonstrate good immobilization yields, it was selected for further study. Contact angle (CA) measurements, AFM, ATR-FT-IR (Attenuated Total Reflectance Fourier Transform Infrared Spectroscopy), Field Emission Scanning Electron Microscopy (FE-SEM), and other techniques were used to describe the PEI-modified surface. Surface roughness and hydrophilicity improved after PEI coating, and FT-IR confirmed that PEI had been properly coated onto the gold surface. Antibody immobilization yield increased significantly by up to 80% as a result of RSM tuning. The QCM biosensor produced results comparable to the Enzyme-Linked Immunosorbent Assay (ELISA) due to its large dynamic range of 1-1 x 10^3 ng/mL, Limit of Detection (LOD) of 3.14 ng/mL, Limit of Quantification (LOQ) of 9.52 ng/mL, and good detection capabilities in human blood. In the spike recovery test using human sera, the biosensor had a mean accuracy of 91%, and it was totally regenerable using 50 mM NaOH and 1% SDS. The QCM biosensor has an excellent dynamic range, sensitivity, selectivity, stability, renewability, and cost-effectiveness, among other advantageous features that indicate its potential for timely and noninvasive hepatitis B biomarker monitoring [144]. Somiya et al. examined the synthesis of stable chemicals with polyplex, DNA, PEI, and bio-nanocapsules (BNCs). The researchers have shown that, when combined with BNCs, lipoplex – a cationic liposome – and DNA work well to transfer the HBV surface antigen L protein to human hepatic cells. With a diameter of roughly 240 nm and a -ve potential of +3.54 mV, the recently synthesized BNC-polyplex complexes demonstrated good capabilities for in vivo applications, making them more suited than polyplex alone. The phosphate group of DNA and the amine group of PEI have a molar ratio of 40. These complexes demonstrated remarkable transfection effectiveness in human liver cells at this N/P ratio. The results of the investigation showed that BNC-polyplex complexes outperformed polyplex alone in terms of transfection efficiency when bafilomycin A1 was utilized to reduce endosomal acidification. This implies that a combination of polyplex proton sponge activity and BNC fusogenic activity allowed the complexes to escape from endosomes. Furthermore, it was discovered that the cytotoxicity of BNC-polyplex complexes was equivalent to that of polyplex with a same N/P ratio. According to the research, BNC-polyplex complexes are a useful tool for delivering genes for human liver-specific gene therapy, offering hepatic cell lines a targeted and efficient form of treatment [145].

Uchida et al. discovered in their study that using PEI-conjugated magnetic beads to modify virus concentration parameters resulted in a more effective concentration of a variety of viruses. In comparison to other cationic polymers, the PEI beads showed better viral adsorption, especially in weakly acidic environments. By using mass spectrometric analysis, it was possible to identify the co-adsorption of many serum proteins, including immunoglobulin M (IgM), complement type 3, and kind 4. This implies the possibility of both direct and indirect virus adsorption by the beads through the formation of immunological complexes. The research looked at the viruses hepatitis A (HAV), hepatitis B (HBV), and hepatitis C (HCV), and it demonstrated how the sensitivity of virus detection may be increased using the PEI bead-based viral concentration technique. The researchers also found that HBV was effectively concentrated after anti-HBV IgM was added, indicating the versatility and effectiveness of the PEI bead-based technique in concentrating a wide range of viruses [146].

10.3.2.2 PLGA

Zheng et al. developed a delivery system for the hepatitis B surface antigen (HBsAg) using mannose-modified polylactic acid nanoparticles (PLA NNPs) and water-soluble microneedle arrays. The researchers used flow cytometry to look at how the different-sized MNPs affected the development of dendritic cells. Next, arrays of water-soluble microneedles were coated with HBsAg/MNPs, a combination of the appropriate size MNPs and HBsAg. The study's findings demonstrated that PLGA NPs with an average size of about 800 nm were the most effective at encouraging the maturation of dendritic cells. After being loaded onto water-soluble microneedle arrays, targeted PLGA NPs carrying HBsAg showed a distinct distribution with a maximum height of roughly 280 μm and a drug load of 0.98 ± 0.05 μg/mg. These drug-containing microneedle arrays possessed outstanding mechanical and biosafety characteristics. According to in vivo immune response evaluations, subcutaneous distribution of the microneedle arrays raised specific IL-12 and IFN-γ release, splenocyte proliferation, and elevated IgG production in mice. The study provides novel approaches for noninjectable vaccine delivery methods and validates the viability of using soluble composited microneedles for hepatitis B vaccination [147]. Zhu et al. developed a delivery technique utilizing mannose-modified polylactic acid nanoparticles (PLA NPs) loaded with hepatitis B surface antigen (HBsAg) protein to boost the immunological response to the hepatitis B vaccine. The traditional vaccine's use of an aluminum adjuvant mostly induces a Th2-type response, which results in robust antibody production but poor cellular immune responses. MNPs were engineered in this study to specifically target antigen-presenting cells located just below the skin's surface, facilitating the progressive and efficient presentation of HBsAg to lymphocytes. Persistent immunity was made possible by the special qualities of PLGA. In vitro absorption tests revealed that MNPs were more effectively internalized by RAW 264.7 cells and bone marrow-derived dendritic cells (BMDCs). Mice's cellular immunity rose but their humoral immunity was maintained when they received subcutaneous MNP injections. Because MNPs differ in several aspects from their parent PLGA NPs, preliminary studies suggest that they could be a promising vehicle for delivering the hepatitis B immunization and eliciting a strong humoral and cellular immune response [148]. Enteraceavir (Baraclude®) is the usual treatment for chronic infections with the HBV. In this work, the form, saturation solubility, in vitro release, and in vivo efficacy of entecavir sustained injectable biodegradable in-situ implant (ISI) fabricated from PLGA polymer were assessed. Entecavir was injected intramuscularly into one-way rats, and an ANOVA was used to evaluate the pharmacokinetic properties of the injection. TEM pictures were taken 24 hours after the materials were

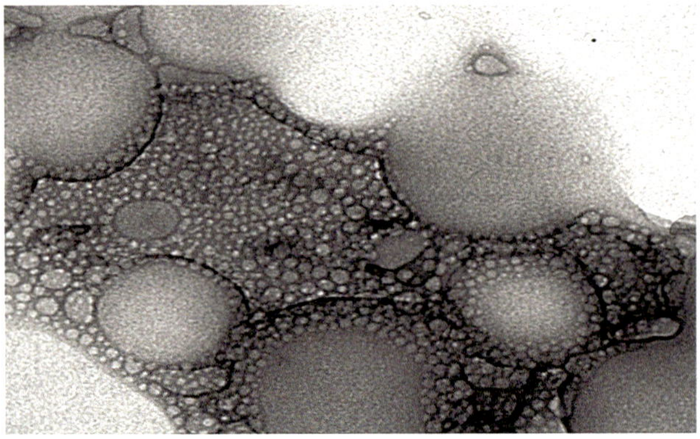

(A)

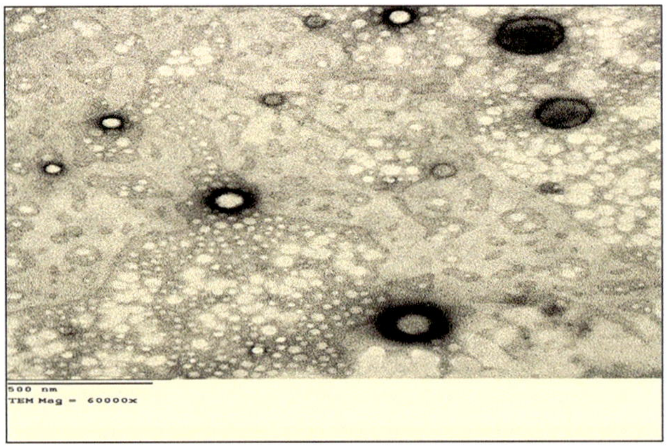

(B)

FIGURE 10.4 TEM images of (A): 40% PLGA in triacetin–ISI formulation and (B): 300 mg PLGA-5% PVA NPs formulation [149] (Copyright, 2020, Elsevier).

treated with Sörensen phosphate buffer (pH 7.4). As can be seen in (Figure 10.4), the surface of the ISI was porous, but the surface of the NPs loaded with entecavir was smooth and spherical. The entrapment efficacy of entecavir NPs ranged from 47.03% to 77.84%. Based on the findings of the pharmacokinetic studies, the C max of pure entecavir decreased significantly ($p < 0.05$) to 16.70 and 172.85 ng/ml for the ISI and NPs formulations, respectively. For both the ISI and NP formulations, the pure entecavir AUC0-∞ value increased significantly, rising from 211.32 ng/ml/h to 1520.64 and 2363.52 ng/ml/h, respectively. The effectiveness of intramuscular ISI and NP formulations loaded with entecavir for the treatment of chronic hepatitis B was demonstrated in this study [149]. These formulations exhibit sustained release characteristics.

10.3.2.3 PEG

It is imperative that the pursuit of enhancing treatment accessibility and refining approaches for optimal outcomes persists, within the confines of the present guidelines for antiviral medication. Despite research showing that antiviral medication works in inactive HBsAg carriers (IHCs), these individuals are currently not included by the guidelines. Song et al. used a systematic review and meta-analysis to study the topic of HBsAg clearance and conversion in IHCs. An examination of data from 11 distinct trials, totaling 1029 IHCs, revealed that after 48 weeks of PEGylated interferon (Peg-IFN) treatment, a significant 47% HBsAg clearance rate and a 26% conversion rate were discovered in that study. Conversely, the control group's HBsAg clearance rate of 1.54% was much lower. Further research indicated that lower baseline HBsAg levels and longer treatment durations were significant contributors to higher clearance rates. Consequently, Aixin Song's research emphasizes the value of antiviral medications in the clinical treatment of IHCs and emphasizes the necessity of considering this hitherto ignored population when developing treatment regimens [150]. Hui et al. investigated alterations in the cytokine profiles of chronic hepatitis B (CHB) patients undergoing antiviral medication with entecavir (ETV) and PEGylated interferon (PEG-IFN) for patients who tested positive for hepatitis B e antigen (HBeAg). During the course of a 48-week therapy period, a cohort of 200 patients – 100 in the PEG-IFN group and 100 in the ETV group – were examined at baseline and every three months. Over the course of three to six months, the PEG-IFN group experienced significant reductions in TGF-β, IL-17A, IL-6, IL-10, and IFN-γ ($P < 0.001$) along with increases in IFN-α2 ($P < 0.001$). The ETV group had considerably reduced levels of TGF-β1 and IL-10 ($P < 0.001$). Following a three-month period, the PEG-IFN group had reduced levels of TGF-β3 and IL-6 ($P < 0.01$) along with elevated levels of TNF-α, IL-17A, and IFN-α2 relative to the ETV group ($P < 0.01$). After six months, the PEG-IFN group had higher levels of IFN-α2, IFN-γ, and TNF-α and lower levels of IL-6 and TGF-β3 ($P < 0.01$) compared to the ETV group. The greater rates of HBeAg and HBsAg elimination after antiviral therapy indicate that PEG-IFN greatly surpassed ETV in immunological control and functional cure, as reported by Hui et al. [151]. The complex impacts on infected populations at different phases of the infection are poorly understood when it comes to assessing the therapeutic benefits of PEGylated interferon alpha (peg-IFNα) therapy for HBV, particularly when compared to nucleotide analog (NAs) therapy. To ascertain the clinical efficacy of peg-IFNα, two distinct cohorts of HBV-infected individuals were examined: patients with chronic hepatitis B (CHB) and IHC. In this real-world study, 328 HBV-infected patients were treated with peg-IFNα monotherapy or combination therapy with NAs for IHCs, and peg-IFNα added-on NAs therapy for CHB patients. HBsAg loss was the main effective outcome as of Week 24. According to the findings, the IHC group had 39.50% (n = 47/119) Kaplan–Meier cumulative rates of HBsAg reduction at Week 24, while the CHB group had 28.71% (n = 60/209) ($p < .05$). Following Propensity Score Matching (PSM), the percentages of HBsAg loss were 32.63% (n = 31/95) and 36.84% (n = 35/95), respectively ($p > .05$. In both the IHC and CHB groups, there were greater rates of HBsAg clearance among patients whose baseline HBsAg level was less than 100 IU/ml.

In both groups, the HBsAg loss at Week 24 was found to be predicted by the initial level of HBsAg and its drop from baseline to Week 12. Therefore, Zhang et al.'s research suggests that in CHB patients getting NA treatment in addition to IHCs, the effectiveness of HBsAg clearance is rather strong in the early stages of peg-IFNα therapy. During peg-IFNα treatment, there was a noticeable downward trend in both groups' HBsAg levels [152].

In a study by Yaici et al., the critical importance of achieving HBsAg reduction for definitive clearance and long-lasting response in chronic hepatitis delta (CHD) was investigated. The study suggests that the optimal targets for CHD treatment should be the production of anti-HBs and a complete response (CR), which is defined by the absence of HBsAg and HDV RNA. Notably, the ideal duration of treatment for congestive heart failure remains unknown. The article includes two noteworthy cases of CHD cirrhosis patients who were treated with Peg-IFNα-2a + tenofovir disoproxil fumarate for a long time until they attained the critical threshold of HBsAg reduction. Surprisingly, both patients achieved a CR after 46 and 55 months of therapy, respectively. These findings highlight the potential advantages of a tailored approach and extended treatment duration motivated by HBsAg decrease in increasing the likelihood of a CR in CHD patients. Yaici et al.'s research provides significant new insights into the evolution of CHD treatment [153].

10.3.2.4 PVA

El-Shafai et al. carried out a thorough investigation into the cooperative fusion of nano-chitosan (Cs) and nano-PVA to create a sandwich nanohybrid (SNH), which has uses in enhanced water treatment and sofosbuvir (SOF) adsorption. This study looked at SNH's capacity to oxidize organic dyes like methyl orange (MO), methylene blue (MB), and rhodamine B (RhB) and produce reactive oxygen species (ROS). Using UV–visible spectroscopy (UV-Vis), the study also looked at how SNH affected the release of SOF in blood and within cells at pH 7.4 and pH 6.8, respectively. Important findings included the determination of the loading constant (Kb) and binding constant (Kb), which were found to be 0.0024 min^{-1} and 0.0035 min^{-1}, respectively. At pH 7.4 and 6.8, 42.6% and 74.7%, respectively, of SOF's release efficiency were measured. The results showed that the photocatalytic efficacy against the organic dyes MO, MB, and RhB was 2.4%, 1%, and 42%, respectively. For cytotoxicity tests, HepG2 and MDA-MB-231 cell lines were employed. The outcomes demonstrated concentration-dependent effects, with greater concentrations resulting in a reversal of this tendency and clearly apparent, yellow-colored cell lysis. The viability of HepG2 was not affected by lower concentrations. Additionally, the study evaluated SNH's antioxidant potential using the FRAP methodology, providing important insights into the diverse properties and potential applications of the synthesized SNH [154]. Shawky et al. investigated the self-assembly of the anti-hepatitis C virus (HCV) medication sofosbuvir (SOF) in a study using bio-polymeric NPs, specifically PVA NPs and chitosan NPs (Cs NPs). Several analytical techniques, including energy-dispersive X-ray spectroscopy (EDX), SEM, UV-Vis spectrophotometer, and Fourier Transmittance Infrared (FT-IR) were successfully used to characterize the new composites. Fluorescence and UV-Vis spectroscopy methods demonstrated that putting sofosbuvir onto the surface of the biopolymer NPs enhanced the drug's efficacy. Monitoring the drug release from the biopolymer NPs surface at pH 7.4 and following the releasing process with UV-Vis spectra gave additional proof of SOF's increased efficiency. The binding constants (Kb) for PVA NPs and Cs were reported to be 0.3613 min^{-1} and 0.000055, respectively. Furthermore, for the hybrid nanocomposites Cs@SOF and PVA@SOF, the KSV values were recorded at 0.0014 and 7.16 min^{-1}, respectively. The estimated SOF incorporation rates (k) for PVA NPs and Cs on the surface of biopolymer nanomolecules were 0.0165 and 0.00812, respectively. Interestingly, the observed values of (n), which were 0.74 and 0.86 for PVA NPs and Cs, respectively, were relatively close to unity. Following a 24-hour period, it was determined that the surface of Cs NPs had released 0.09 mg of SOF, whereas PVA NPs had released 0.7 mg. PVA@SOF and Cs@SOF had release efficiencies of 73% and 56.5%, respectively. Although more investigation and clinical trials are required to validate this, these results suggest that PVA/SOF and Cs/SOF hybrid nanocomposites offer promising spectroscopic features that could lead to their development as pharmacological candidates [155].

Another study developed and used dual-release dissolving microneedles (DMNs) to deliver a hepatitis B vaccine, which needs to be administered in multiple doses for optimal immunological response. Polyvinylpyrrolidone (PVP) and PVA for whole skin dissolve, PLA tips for gradual release, and a carboxy-methyl cellulose (CMC) coating formulation for bolus release comprised the dissolving basis for the DMNs, also known as HBsAg-PLA/CMC-DMNs. Hepatitis B surface antigen (HBsAg) release from PLA tips and the CMC coating formulation was investigated in vitro. The study employed in vivo testing to evaluate the immunological efficacy and delivery efficiency of HBsAg-PLA/CMC-DMNs compared to traditional intramuscular (IM) administration and the two injections required for HBsAg-coated microneedle (CMNs) administration. Since the PVA/PVP base dissolved completely in 10 minutes, all of the microneedle tips could be introduced into the skin, and the HBsAg-PLA/CMC-DMNs effectively penetrated the skin. The HBsAg in the PLA tips showed continuous release for a period of more than 55 days; however, all of the HBsAg in the CMC coating formulation was released in 20 minutes during the in vitro release assay. It's interesting to note that the antibody titer generated by a single dosage of HBsAg-PLA/CMC-DMNs was either higher or equivalent to the antibody titer generated by two doses given using conventional IM and CMN methods. This dual-release microneedle technology demonstrated that two formulations could be administered simultaneously with a single injection, potentially increasing the immunological efficacy of HBsAg, which typically requires multiple doses. Furthermore, the delivery of different treatments that require multiple doses may benefit from the use of this dual-release microneedle technology. The research that yielded the results was done by Kim et al. [156]. Shrestha et al. investigated PVA-coated chitosan microparticles as a potential oral vaccine delivery system for a recombinant hepatitis B vaccine. Following the application of PVA coating on chitosan microparticles generated via the ionotropic gelation process, the recombinant hepatitis B vaccination was administered. The optimal loading capacities and efficiencies of the microparticles, with typical diameters ranging from 100 to 410 nm, were found to be approximately 3.4% and 74%, respectively. Studies on in vitro release demonstrate that antigens were continuously released by the produced microparticles. The microparticles showed resilience to gastric models and efficiently released the antigen in the intended intestinal environment. Furthermore, rats that were orally administered the PVA-coated chitosan hepatitis B microparticle vaccine experienced a seroprotective immune response that was similar to what was produced by a traditional intramuscular vaccination. According to the study, chitosan microparticles coated with PVA could be used as an oral vaccine delivery technology for hepatitis B, providing a choice for immunization without the need for needles [157].

10.3.3 Nephropathy

Nephropathy is the term for the deterioration of renal function [158]. The latter stage of nephropathy is known as end-stage renal disease (ESRD). The Centers for Disease Control and Prevention (CDC) states that diabetes is the most common cause of ESRD. In 2011, it was projected that 26 million Americans had diabetes, and of those with ESRD due to diabetes, over 200,000 were either on chronic renal dialysis or had undergone kidney transplantation. Both type 1 and type 2 diabetes can induce diabetic nephropathy, while type 1 is more likely to culminate in ESRD. There are five phases to the diabetic nephropathy. The fifth stage is ESRD. Going from one level to the next could take several years [159].

10.3.3.1 PEI

Li et al. examined the possible impacts of RNA interference (RNAi) caused by jet PEI on renal interstitial fibrosis in rats suffering from obstructive nephropathy, as well as the RNAi's capacity to stop the re-expression of the PAX2 (Paired Box 2) gene. Based on the whole mRNA sequence of the PAX2 gene, a negative control and four pairs of small interfering RNA (siRNA) interference sequences were designed and synthesized. Next, these siRNAs were delivered into the renal capsule of a mouse model of unilateral ureteral obstruction (UUO) by use of an in vivo jet PEI delivery vector. Western blot and real-time quantitative PCR were used to measure the expression of PAX2 mRNA and protein in renal tissue, respectively. A negative control labeled with carboxyfluorescein (FAM) was made in order to confirm in vivo transfection. Renal interstitial fibrosis was seen under a light microscope following PAX2 gene silencing. Green fluorescence surrounding renal tubule epithelial cells showed that the transfection was successful three days after the therapy. It was demonstrated that all siRNA interference sequences inhibited the expression of the PAX2 gene, with siRNA3 exhibiting the most suppression. The inhibition led to a 55% downregulation of PAX2 mRNA and an 81% downregulation of PAX2 protein. Following PAX2 gene silencing, there was a notable improvement in interstitial fibrosis and renal tubular damage. By efficiently inhibiting the expression of PAX2 in tubular epithelial cells in the UUO model, siRNA may be used as a therapy to slow the pathological advancement of obstructive nephropathy [160]. The challenges of employing an NP delivery system to treat renal illness are discussed in the Wang et al. study. The authors focus on increasing renal cellular absorption, decreasing fast urine output, and optimizing NP size for glomerular filtration membrane penetration. They suggest a two-step nanoparticular cascade that combines size control and enhanced renal cellular absorption to overcome these difficulties. Renal-targeted rhein (RH)-loaded LipoNPs (KLPPR) are lipid layers that have had their yolk-shell structure altered by the kidney-targeting peptide (KTP), and their cores are based on polycaprolactone-polyethyleneimine (PCL-PEI). Because of its size, which varies from 30 to 80 nm, KLPPR was able to more easily pass through the glomerular filtration barrier and enter the kidney. The KTP modification improved renal cellular absorption and endocytosis via a nonlysosomal route. KLPPR, with an average size of 59.5±6.2 nm, demonstrated significant RH loading, sustained release, stability, and biocompatibility. The particles exhibited rapid cellular uptake in HK-2 cells. When streptozocin-induced diabetic nephropathy (DN) in mice was injected intravenously, KLPPR resulted in decreased urine excretion and increased kidney-targeted distribution. Reductions in blood creatinine, urea nitrogen, and kidney index values indicated that this tactic improved renal physiological function. Furthermore, KLPPR aided in the recovery of renal function by preventing TGF-β1 and fibronectin from being released and accumulating. Ultimately, the most effective therapeutic results for DN and the efficient targeting of diseased kidneys were shown by KLPPR's two-step nanoparticular cascade of size regulation and quicker cellular absorption. This work offers a feasible lipo NP delivery approach for the therapy of renal disorders [161].

10.3.3.2 PEG

In Tong et al.'s work, a poly (ethylene glycol)-b-brush poly(l-lysine) polymer, or PEG-b-(PELG50-g-PLL3), was synthesized and evaluated as a nanocarrier for the continuous subcutaneous injection of exenatide into the abdominal cavity. It was found that exenatide made electrostatic interactions with the PEG-b-(PELG50-g-PLL3) polymer at pH 7.4, suggesting that it is a feasible choice for maintaining

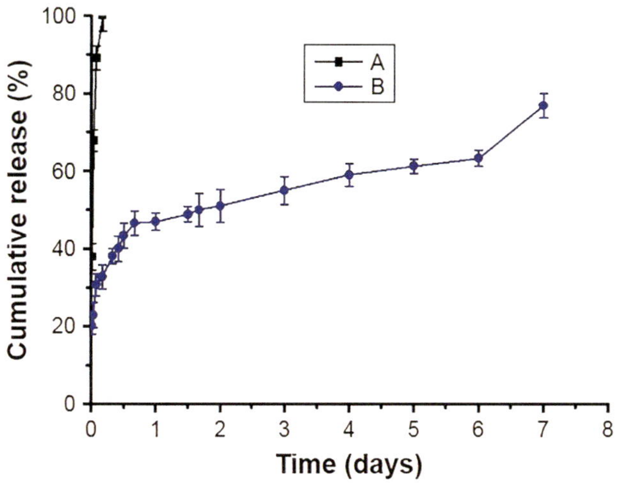

FIGURE 10.5 *In vitro* release of exenatide. (A) Cumulative release profile of free exenatide and (B) exenatide from exenatide–Peg-b-(Pelg50-g-Pll3) complexes. Abbreviation: Peg-b-(Pelg50-g-Pll3), poly(ethylene glycol)-b-brush poly(l-lysine) polymer [162]. [This is an open-access article distributed under the terms of the Creative Commons Attribution-NonCommercial License (http://creativecommons.org/licenses/by-nc/4.0/), which permits unrestricted noncommercial use, distribution, and reproduction in any medium, provided the original work is properly cited.]

drug administration throughout time. The isoelectric point of extenatide is 4.86. The polymer was physicochemically characterized using gel permeation chromatography, TEM, SEM, NP size measurement, and nuclear magnetic resonance spectroscopy. In just 7 days, 20.16%–76.88% of the total exenatide was released from the PEG-b-(PELG50-g-PLL3) polymer, based on in vitro release experiments (Figure 10.5). The researchers determined the best formulation for in vivo testing to assess blood glucose control in models of diabetic rats using free exenatide as a reference. The persistent hypoglycemia effect lasted for seven days after the formulation was subcutaneously administered into diabetic rats, indicating that it had a longer-acting hypoglycemic effect than free exenatide. In addition to their hypoglycemic effects, the exenatide-block-bush polymers significantly decreased DN. This was supported by decreased albumin/creatinine ratio, decreased oxidative stress injury, decreased blood cholesterol, decreased renal index, decreased albumin excretion in urine, downregulated connective tissue growth factor expression, and decreased apoptosis. According to this study's findings, PEG-b-(PELG50-g-PLL3) polymers may be useful as exenatide nanocarriers, offering effective encapsulation and prolonged release, as well as potential therapeutic advantages for the treatment of DN and hypoglycemia [162].

10.3.4 Pancreatitis

The clinical signs of acute pancreatitis, a severe case of acute inflammation, include sudden, constant stomach pain that is accompanied by nausea, vomiting, fever, and distension of the abdomen. Acute pancreatitis was characterized by inflammatory neutrophil and macrophage infiltration, vacuolar accumulation, acinar cell necrosis, and stromal edema. The pancreas showed restricted inflammation and typically little organ damage in cases of moderate acute pancreatitis (MAP). But when pancreatic inflammation spreads to other organs, it's referred to as severe acute pancreatitis (SAP), mostly seen as severe multiple organ failure (MODS), systemic inflammatory response syndrome, and pancreatic necrosis. Numerous consequences, including shock, organ failure, pancreatic abscess, pancreatic pseudocyst, and secondary infections of the urinary tract, respiratory system, and abdominal cavity, are frequently observed in cases with SAP [163].

10.3.4.1 PEG

The possible therapeutic effects of polyethylene glycol 35 (PEG35) in lessening the severity of acute necrotizing pancreatitis (ANP) and accompanying lung damage were examined in this study by Ferrero-Andrés et al. The inflammatory processes in the pancreas are the hallmark of acute pancreatitis, which can vary in severity from mild instances with low survival rates to severe cases with significant mortality. In the research, 5% sodium taurocholate injections into the biliopancreatic duct caused ANP. Due to PEG35's nonimmunogenic, nontoxic, and water-soluble characteristics, it was injected intravenously for both preventative and therapeutic purposes. Three hours following ANP induction, the degree of ANP and related lung damage was evaluated. The findings demonstrated that preemptive PEG35 treatment decreased the inflammatory processes linked to ANP, both locally and systemically. The assessment of adhesion molecules, myeloperoxidase levels, proinflammatory cytokines, and chemokines in the pancreas and lungs demonstrated this. Additionally, as shown by decreased lactate dehydrogenase (LDH) activity and apoptotic cleaved caspase-3 localization, PEG35 pretreatment protected against cell death linked to acute pancreatitis. It's interesting to note that PEG35 dramatically reduced lung damage even when given therapeutically after ANP started, indicating that it may have been used in treating the disease's consequences. In conclusion, the research validates PEG35's protective function against ANP-related inflammatory reactions and cell death. PEG35 shows promise as a therapeutic agent for the management of ANP' severe sequelae [164]. In vivo and in vitro pancreatic damage resulting from cerulein-induced acute pancreatitis was evaluated by the same research team about the preventative effects of a 35-kDa molecular weight PEG (PEG35). An abrupt inflammation of the pancreas that may also affect distant organs and/or surrounding tissues is known as acute pancreatitis (AP). Typical pathological characteristics of this illness and indicators of disease severity include inflammation and parenchymal cell death. In biological, chemical, clinical, and pharmacological contexts, PEG is a widely utilized class of water-soluble, nonimmunogenic, nontoxic polymers. Three groups consisting of control, cerulein-induced AP, and PEG35 treatment were randomly assigned to Wistar rats. Cerulein (50 μg/kg/bw) was administered intraperitoneally five times per hour to induce AP, while saline solution was given to control rats. Ten minutes before each cerulein injection, 10 mg/kg of PEG35 was administered intraperitoneally. Pancreatic tissue and blood were removed for analysis after AP induction. PEG35 was added to AR42J pancreatic acinar cells in increasing amounts before their exposure to cerulein, staurosporine, or TNFα. Pancreatic edema, histological anomalies, plasma lipase levels, and lactate dehydrogenase activity were used to gauge the severity of AP. The pancreatic and AR42J-treated cells' gene expression of markers linked to inflammation was evaluated to gauge the severity of the inflammatory response. *In vivo* and *in vitro* pancreatic damage models were used to assess inflammation-induced cell death. PEG35 administration effectively reduced pancreatic damage by lowering lipase levels and tissue edema in cerulein-induced AP mice. Cerulein injection lowered the pancreatic inducible nitric oxide synthase enzyme and cytokine-related gene expression, which in turn lowered the inflammatory response that accompanied it. Conversely, there was a notable increase in interleukin 10 mRNA expression in pancreatic tissue. PEG35 therapy prevented inflammation-induced cell death in cerulein-stimulated animals by reducing lactate dehydrogenase activity and altering pancreatic levels of the apoptosis regulator protein BCL-2. When TNFα, cerulein, or staurosporine was administered to pancreatic acinar cells, PEG35 treatment significantly and dose-dependently reduced proinflammatory markers and cell death. PEG35 was found to lessen the inflammatory response and associated cell death in cells treated with AR42J and cerulein-induced AP, hence minimizing pancreatic damage. Regarding AP management, PEG35 may be a helpful option. PEG35 might be a useful choice for AP management [165].

10.4 FUTURE PERSPECTIVE AND CHALLENGES IN SYNTHETIC POLYMER-DERIVED DRUG DELIVERY

The future of drug delivery systems derived from synthetic polymers is anticipated to be at the crossroads of innovation and difficulty, as scientists and drug developers strive to achieve the highest standards in precision medicine while managing the intricacies of biocompatibility, regulatory frameworks, and the inherent variability of human illnesses. As we learn more about the potential that synthetic polymers have to transform medication delivery, it becomes clear that overcoming challenges and utilizing novel materials and technologies will be just as important as advancing the field. Synthetic polymer development, from traditional platforms such as PLGA and PEG to more advanced stimuli-responsive and targeted delivery systems, is indicative of a larger trend in health care toward customization and accuracy [166]. With the ability to release their therapeutic payloads at precisely the right position and time in response to specific physiological triggers like pH variations, temperature fluctuations, or enzyme activity, these innovative materials maximize efficacy while minimizing negative effects. However, there are many obstacles in the way of these cutting-edge drug delivery systems getting from the bench to the bedside. Ensuring that synthetic polymers are biocompatible and biodegradable is essential because any residue must be appropriately metabolized or removed by the body to prevent harmful effects over an extended period [167]. This means that designing polymers for functionality and durability while making sure they can decompose into nontoxic byproducts requires a careful balance. In addition, demanding quality control and reproducibility requirements must be met during the scale-up and manufacturing of polymer-based drug delivery systems [168]. This can be difficult because polymer synthesis involves complex chemistry, as can the subsequent drug encapsulation or conjugation procedures.

Accurate polymer-based delivery system targeting is a major additional challenge. Targeting ligands and molecular recognition elements has advanced, but getting accurate distribution to disease areas without off-target consequences is still a significant challenge [169]. This is especially relevant to cancer therapy, as minimizing collateral harm requires the capacity to distinguish between healthy and malignant cells. Furthermore, the dynamic nature of diseases like cancer – which are marked by changing phenotypes and the rise of treatment resistance – highlights the necessity for flexible delivery systems with multiple functions that can adjust to shifting disease profiles. An additional level of complication arises from the regulatory approval process, wherein innovative drug delivery systems have to pass stringent testing to prove their safety and effectiveness over a range of periods [170]. This frequently calls for intensive preclinical and clinical testing, which can be expensive and time-consuming. Furthermore, the regulatory environment is always changing, placing more and more focus on real-world evidence and patient-centered outcomes. This calls for proactive and flexible approaches to data analysis and clinical trial design. Given these difficulties, interdisciplinary cooperation in the development of smarter, more flexible, and more efficient drug delivery systems is expected to be a hallmark of the future of synthetic polymer-derived drug delivery. This collaboration will draw on advancements in nanotechnology, materials science, molecular biology, and computational modeling. Genomics and proteomics-driven personalized medicine will be essential in steering the creation of medication delivery systems customized to each patient's unique profile. In addition to technological innovation, this will call for a paradigm change in the way medications are created, authorized, and administered, with a stronger focus on patient-centered care and holistic illness management.

Furthermore, when designing and developing synthetic polymer-based drug delivery systems, sustainability, and environmental effects are becoming increasingly important factors to take into account. There is an increasing need for biodegradable materials, lifetime assessments, and greener synthesis techniques to reduce the environmental impact of pharmaceutical goods as the focus on environmental health becomes more widespread [171]. This is indicative of a larger movement in health care toward sustainability, where the environmental costs of cutting-edge medicine delivery methods must be balanced against their advantages. Although synthetic polymer-derived drug delivery systems have great potential to improve patient outcomes and treatments in the future, achieving this promise will need coordinated efforts to overcome the many obstacles that stand in the way. Our capacity to innovate not only at the level of specific materials and technologies but also in the way we handle integrating these systems into a larger health-care framework, balancing efficacy, safety, regulatory compliance, and sustainability, will be critical to our endeavor's success. The ultimate objective, which remains unwavering as we traverse this challenging terrain, is to use synthetic polymers to provide patients all around the world with safer, more efficient, and more customized treatments.

10.5 CONCLUSION

A ray of hope for medical science is the potential for synthetic polymer-derived drug delivery systems to transform the treatment of complicated illnesses like pancreatitis, hepatitis, RA, and nephropathy. The introduction of these novel drug delivery methods is surpassing conventional treatment approaches, which are hampered by issues like resistance, poor absorption, high dosage requirements, lack of specificity, and negative effects. At the forefront of this revolution are polymers such as PEG, PLA, and PLGA, which by improving medication delivery efficacy, targeting accuracy, and side effect minimization, provide a new paradigm in illness management. These polymers' special structural qualities allow for the customized release of medications, resulting in a regulated and long-lasting therapeutic effect that should greatly enhance patient compliance and treatment results overall. For example, these devices can directly deliver anti-inflammatory drugs to inflamed joints in the context of RA, therefore significantly decreasing systemic adverse effects. In the struggle against hepatitis, directing antiviral medications to the liver could boost efficacy while sparing healthy tissues. Localized delivery systems have the potential to considerably attenuate pancreatic damage and alleviate symptoms in pancreatitis, while focused delivery of renoprotective medicines in nephropathy may stabilize the illness and delay its progression. The interplay between formulation techniques, biological settings, and polymer chemistry is important in unlocking these achievements, heralding a trend toward personalized medicine where treatments are tuned to particular patient needs. The transition from laboratory to clinic is fraught with difficulties despite the bright future; these include the complexity of disease causes, patient variability, and the arduous process of clinical validation. Furthermore, the legal frameworks and economic factors will be crucial in deciding the availability and extensive integration of these cutting-edge drug delivery systems. Nevertheless, the coordinated effort by researchers, medical professionals, and business partners to get over these obstacles is evidence of their shared dedication to improving patient care. The investigation of drug delivery systems developed from synthetic polymers represents, all things considered, a quantum leap forward in medical therapy, providing a comprehensive strategy to battle diseases that have long plagued the medical community. The future of drug delivery is set to change significantly with the potential to deliver therapies that are safer, more effective, and more in line with patient-centered care by using the power of polymers such as PEG, PLA, and PLGA. The idea of using these state-of-the-art systems in actual clinical settings is becoming more and more real as research progresses, offering a new era of illness management marked by previously unheard-of levels of accuracy, effectiveness, and patient happiness.

ABBREVIATIONS

RA	Rheumatoid arthritis
PEG	Polyethylene glycol
PLA	Polylactic acid
PLGA	Polylactic-co-glycolic acid
PVA	Poly (vinyl acetate)
PAE	Poly(amino ester)
PEI	Poly(ethylenimine)
BPEI	Branched PEI
PGA	Poly(glycolic acid)
PLLA	Poly(L-lactic acid)
PDLA	Poly(D lactic acid)
PDLLA	Poly-DL lactic acid
PE	Polyethylene
PP	Polypropylene
PTFE	Poly(tetrafluoroethylene)
PVC	Poly(vinyl chloride)
PPDO	Poly(p-dioxanone)
MTX	Methotrexate
HA	Hyaluronic acid
NPs	Nanoparticles
IND	Indomethacin
STING	Stimulator of Interferon Genes

REFERENCES

1. Juillerat-Jeanneret L, Aubert JD, Mikulic J, Golshayan D et al (2018) Fibrogenic disorders in human diseases: from inflammation to organ dysfunction. *Journal of Medicinal Chemistry* 61(22):9811–40.
2. Isoherranen K, O'Brien JJ, Barker J, Dissemond J, Hafner J, Jemec GB, Kamarachev J, Läuchli S, Montero EC, Nobbe S, Sunderkötter C et al (2019) Atypical wounds. Best clinical practice and challenges. *Journal of Wound Care* 28(Sup6):S1–92.
3. Sahu T, Ratre YK, Chauhan S, Bhaskar LV, Nair MP, Verma HK et al (2021) Nanotechnology based drug delivery system: Current strategies and emerging therapeutic potential

for medical science. *Journal of Drug Delivery Science and Technology* 63:102487.
4. Patra JK, Das G, Fraceto LF, Campos EV, Rodriguez-Torres MD, Acosta-Torres LS, Diaz-Torres LA, Grillo R, Swamy MK, Sharma S, Habtemariam S et al (2018) Nano based drug delivery systems: recent developments and future prospects. *Journal of Nanobiotechnology* 16(1):1–33.
5. Harun-Or-Rashid M, Aktar MN, Hossain MS, Sarkar N, Islam MR, Arafat ME, Bhowmik S, Yusa SI et (2023) Recent Advances in Micro-and Nano-Drug Delivery Systems Based on Natural and Synthetic Biomaterials. *Polymers* 15(23):4563.
6. Nguyen MK, Alsberg E et al (2014) Bioactive factor delivery strategies from engineered polymer hydrogels for therapeutic medicine. *Progress in Polymer Science* 39(7):1235–65.
7. Andretto V, Rosso A, Briançon S, Lollo G et al (2021) Nanocomposite systems for precise oral delivery of drugs and biologics. *Drug Delivery and Translational Research* 11:445–70.
8. Zhang XQ, Xu X, Bertrand N, Pridgen E, Swami A, Farokhzad OC et al (2012) Interactions of nanomaterials and biological systems: Implications to personalized nanomedicine. *Advanced Drug Delivery Reviews* 64(13):1363–84.
9. Pethe AM, Yadav KS. et al (2019) Polymers responsiveness and cancer therapy. *Artificial Cells, Nanomedicine, and Biotechnology* 47(1):395–405.
10. Van Vlerken LE, Amiji MM et al (2006) Multi-functional polymeric nanoparticles for tumour-targeted drug delivery. *Expert Opinion on Drug Delivery* 3(2):205–16.
11. Chen YL, Bao CJ, Duan JL, Xie Y, Lu WL et al (2023) Overcoming biological barriers by virus-like drug particles for drug delivery. *Advanced Drug Delivery Reviews* 203:115134.
12. Longmire M, Choyke PL, Kobayashi H et al (2008) Clearance properties of nano-sized particles and molecules as imaging agents: considerations and caveats. *Nanomedicine* 3(5):703–17.
13. Nyström AM et al (2012) *Update on Polymer Based Nanomedicine*. Smithers Rapra.
14. Wang Y, Wang C, Li K, Song X, Yan X, Yu L, He Z et al (2021) Recent advances of nanomedicine-based strategies in diabetes and complications management: Diagnostics, monitoring, and therapeutics. *Journal of Controlled Release* 330:618–40.
15. Xu R, Fang Y, Zhang Z, Cao Y, Yan Y, Gan L, Xu J, Zhou G et al (2023) Recent advances in biodegradable and biocompatible synthetic polymers used in skin wound healing. *Materials* 16 (15):5459.
16. Rahimi M, Charmi G, Matyjaszewski K, Banquy X, Pietrasik J et al (2021) Recent developments in natural and synthetic polymeric drug delivery systems used for the treatment of osteoarthritis. *Acta Biomaterialia* 123:31–50.
17. Paolini MS, Fenton OS, Bhattacharya C, Andresen JL, Langer R et al (2019) Polymers for extended-release administration. *Biomedical Microdevices* 21:1–24.
18. Prajapati VD, Jani GK, Kapadia JR et al (2015) Current knowledge on biodegradable microspheres in drug delivery. *Expert Opinion on Drug Delivery*;12(8):1283–99.
19. Abou-ElNour M, Ishak RA, Tiboni M, Bonacucina G, Cespi M, Casettari L, Soliman ME, Geneidi AS et al (2019) Triamcinolone acetonide-loaded PLA/PEG-PDL microparticles for effective intra-articular delivery: Synthesis, optimization, in vitro and in vivo evaluation. *Journal of Controlled Release* 309:125–44.
20. Katchalski E et al (1951) Poly-α-amino acids. *Advances in Protein Chemistry* 6:123–85.
21. Luckachan GE, Pillai CK et al (2011) Biodegradable polymers-a review on recent trends and emerging perspectives. *Journal of Polymers and the Environment* 19:637–76.
22. Hyun J, Eom J, Song J, Seo I, Um SH, Park KM, Bhang SH et al (2021) Poly (amino ester)-Based Polymers for Gene and Drug Delivery Systems and Further Application toward Cell Culture System. *Macromolecular Bioscience* 21(8):2100106.
23. Nishiyama N, Kanayama N, Jang WD, Yamasaki Y, Kataoka K et al (2006) PEGylated gene nanocarriers based on block catiomers bearing ethylenediamine repeating units directed to remarkable enhancement of photochemical transfection. *Journal of Controlled Release* 115(2):208–15.
24. Mbarushimana PC et al (2018) *Anionic polymerization of activated aziridines*. The University of Alabama
25. Zu H, Gao D et al (2021) Non-viral vectors in gene therapy: Recent development, challenges, and prospects. *The AAPS Journal* 23(4):78.
26. Yemul O, Imae T et al (2008) Synthesis and characterization of poly (ethyleneimine) dendrimers. *Colloid and Polymer Science* 286:747–52.
27. Cao M, Wang J, Wang Y et al (2007) Surface patterns induced by Cu2+ ions on BPEI/PAA layer-by-layer assembly. *Langmuir* 23(6):3142–9.
28. Neu M, Fischer D, Kissel T et al (2005) Recent advances in rational gene transfer vector design based on poly (ethylene imine) and its derivatives. *The Journal of Gene Medicine: A Cross-Disciplinary Journal for Research on the Science of Gene Transfer and Its Clinical Applications* 7(8):992–1009.
29. De Oliveira FA, Albuquerque LJ, Delecourt G, Bennevault V, Guégan P, Giacomelli FC et al (2021) Current Designs of Polymeric Platforms Towards the Delivery of Nucleic Acids Inside the Cells with Focus on Polyethylenimine. *Current Gene Therapy* 21(5):431–51.
30. Rueda-Gensini L, Cifuentes J, Castellanos MC, Puentes PR, Serna JA, Muñoz-Camargo C, Cruz JC et al (2020) Tailoring iron oxide nanoparticles for efficient cellular internalization and endosomal escape. *Nanomaterials* 10(9):1816.
31. Maurisse R, De Semir D, Emamekhoo H, Bedayat B, Abdolmohammadi A, Parsi H, Gruenert DC et al (2010) Comparative transfection of DNA into primary and transformed mammalian cells from different lineages. *BMC Biotechnology* 10(1):1–9.
32. Saha NC, Ghosh AK, Garg M, Sadhu SD et al (2022) Flexible Packaging Material—Manufacturing Processes and Its Application. In *Food Packaging: Materials, Techniques and Environmental Issues*. Singapore: Springer Nature Singapore pp. 47–87.
33. Niu L, Xia Z et al (2012) Comparison of the different emulsion polymerization routes between acrylic and vinyl acetate monomers using poly (vinyl alcohol) as the sole stabilizer. *Colloid and Polymer Science* 290:699–707.
34. Rivera-Hernandez G, Antunes-Ricardo M, Martínez-Morales P, Sanchez ML et al (2021) Polyvinyl alcohol based-drug delivery systems for cancer treatment. *International Journal of Pharmaceutics* 600:120478.
35. Tripathi S, Mehrotra GK, Dutta PK et al (2009) Physicochemical and bioactivity of cross-linked chitosan–PVA film for food packaging applications. *International Journal of Biological Macromolecules* 45(4):372–6.
36. Pavaloiu RD, Stoica-Guzun A, Stroescu M, Jinga SI, Dobre T et al (2014) Composite films of poly (vinyl alcohol)–chitosan–bacterial cellulose for drug controlled

release. *International Journal of Biological Macromolecule* 68:117–24.
37. Djonlagic J, Nikolic MS et al (2011) Biodegradable polyesters: Synthesis and physical properties. *A Handbook of Applied Biopolymer Technology Synthesis, Degradation & Application* 20:164–6.
38. Premkumar J, SonicaSree K, Sudhakar T et al (2021) *Polymers in biomedical use. InHandbook of polymer and ceramic nanotechnology*. Cham: Springer International Publishing (pp. 1329–1355).
39. Banerjee A, Chatterjee K, Madras G et al (2014) Lipase mediated enzymatic degradation of polydioxanone in solution. *Polymer Degradation and Stability* 110:284–9.
40. Ho MP, Wang H, Lee JH, Ho CK, Lau KT, Leng J, Hui D et al (2012) Critical factors on manufacturing processes of natural fibre composites. *Composites Part B: Engineering* 43(8):3549–62.
41. Darie-Niță RN, Râpă M, Frąckowiak S et al (2022) Special features of polyester-based materials for medical applications. *Polymers* 14(5):951.
42. Qin H, Wang K et al (2019) Study on preparation and performance of PEG-based polyurethane foams modified by the chitosan with different molecular weight. *International Journal of Biological Macromolecules* 140:877–85.
43. Fu Z, Cui J, Zhao B, Shen SG, Lin K et al (2021) An overview of polyester/hydroxyapatite composites for bone tissue repairing. *Journal of Orthopaedic Translation* 28:118–30.
44. Park JH, Lee S, Kim JH, Park K, Kim K, Kwon IC et al (2008) Polymeric nanomedicine for cancer therapy. *Progress in Polymer Science* 33(1):113–37.
45. D'souza AA, Shegokar R et al (2016) Polyethylene glycol (PEG): a versatile polymer for pharmaceutical applications. *Expert Opinion on Drug Delivery* 13(9):1257–75.
46. Pelosi C, Tinè MR, Wurm FR et al (2020) Main-chain water-soluble polyphosphoesters: Multi-functional polymers as degradable PEG-alternatives for biomedical applications. *European Polymer Journal* 141:110079.
47. Veronese FM, Mero A et al (2008) The impact of PEGylation on biological therapies. *BioDrugs* 22:315–29.
48. Ryan SM, Mantovani G, Wang X, Haddleton DM, Brayden DJ et al (2008) Advances in PEGylation of important biotech molecules: delivery aspects. *Expert Opinion on Drug Delivery* 5(4):371–83.
49. Shi D, Beasock D, Fessler A, Szebeni J, Ljubimova JY, Afonin KA, Dobrovolskaia MA et al (2022) To PEGylate or not to PEGylate: Immunological properties of nanomedicine's most popular component, polyethylene glycol and its alternatives. *Advanced Drug Delivery Reviews* 180:114079.
50. Jem KJ, Tan B et al (2020) The development and challenges of poly (lactic acid) and poly (glycolic acid). *Advanced Industrial and Engineering Polymer Research* 3(2):60–70.
51. Ghanbarzadeh B, Almasi H et al (2013) Biodegradable polymers. *Biodegradation-life of Science* 2013:141–85.
52. Low YJ, Andriyana A, Ang BC, Zainal Abidin NI et al (2020) Bioresorbable and degradable behaviors of PGA: Current state and future prospects. *Polymer Engineering & Science* 60(11):2657–75.
53. Manocha B, Margaritis A et al (2008) Production and characterization of γ-polyglutamic acid nanoparticles for controlled anticancer drug release. *Critical Reviews in Biotechnology* 28(2):83–99.
54. Davoodi P, Lee LY, Xu Q, Sunil V, Sun Y, Soh S, Wang CH et al (2018) Drug delivery systems for programmed and on-demand release. *Advanced Drug Delivery Reviews* 132:104–38.
55. Benatti AC, Pattaro AF, Rodrigues AA, Xavier MV, Kaasi A, Barbosa MI, Jardini AL, Maciel Filho R, Kharmandayan P et al (2019) Bioreabsorbable polymers for tissue engineering: PLA, PGA, and their copolymers. In*Materials for biomedical engineering* (pp. 83–116).
56. Sharma M et al (2021) *Biodegradable Polymers: Materials and Their Structures*. CRC Press.
57. Nampoothiri KM, Nair NR, John RP et al (2010) An overview of the recent developments in polylactide (PLA) research. *Bioresource Technology* 101(22):8493–501.
58. Prajapati SK, Jain A, Jain A, Jain S et al (2019) Biodegradable polymers and constructs: A novel approach in drug delivery. *European Polymer Journal* 120:109191.
59. Tyler B, Gullotti D, Mangraviti A, Utsuki T, Brem H et al (2016) Polylactic acid (PLA) controlled delivery carriers for biomedical applications. *Advanced Drug Delivery Reviews* 107:163–75.
60. Jain RA et al (2000) The manufacturing techniques of various drug loaded biodegradable poly (lactide-co-glycolide) (PLGA) devices. *Biomaterials* 21(23):2475–90.
61. Patel CM, Patel MA, Patel NP, Prajapati PH, Patel CN et al (2010) Poly lactic glycolic acid (PLGA) as biodegradable polymer. *Research Journal of Pharmacy and Technology* 3(2):353–60
62. Chavan YR, Tambe SM, Jain DD, Khairnar SV, Amin PD et al (2022) Redefining the importance of polylactide-co-glycolide acid (PLGA) in drug delivery. In *Annales Pharmaceutiques Françaises*. Elsevier Masson. (Vol. 80, No. 5, pp. 603–616).
63. Alizadeh-Osgouei M, Li Y, Wen C et al (2019) A comprehensive review of biodegradable synthetic polymer-ceramic composites and their manufacture for biomedical applications. *Bioactive Materials* 4:22–36.
64. Zeb A, Gul M, Nguyen TT, Maeng HJ et al (2022) Controlled release and targeted drug delivery with poly (lactic-co-glycolic acid) nanoparticles: Reviewing two decades of research. *Journal of Pharmaceutical Investigation* 52(6):683–724.
65. Kapoor DN, Bhatia A, Kaur R, Sharma R, Kaur G, Dhawan S et al (2015) PLGA: a unique polymer for drug delivery. *Therapeutic Delivery* 6(1):41–58.
66. Stewart SA, Domínguez-Robles J, Donnelly RF, Larrañeta E et al (2018) Implantable polymeric drug delivery devices: classification, manufacture, materials, and clinical applications. *Polymers* 10(12):1379.
67. Gumargalieva KZ, Zaikov GE, Polishchuk AY, Adamyan AA, Vinokurova TI et al (2002) Biocompatibility and Biodegradation of Polyolefins. *International Polymer Science and Technology* 29(9):60–72.
68. Barnetson A et al (1997) *Advances in polymers*. iSmithers Rapra Publishing.
69. Gasmi S et al (2021) Relationship between structure, properties and UV/Heat protection behavior of two different types of polyethylene manufactured in Qatar, Qspace Institutional Repository, Project Document, http://hdl.handle.net/10576/20826
70. Huseyin AV, Hassanin A, Hamouda T, Kiliç A et al (2019) High performance fibers: a review on current state of art and future challenges. *Eskişehir Osmangazi Üniversitesi Mühendislik ve Mimarlık Fakültesi Dergisi* 27(2):130–55.
71. Maitz MF et al (2015) Applications of synthetic polymers in clinical medicine. *Biosurface and Biotribology* 1(3):161–76.
72. Baican M, Stoleru E, Vasile C et al (2020) Cellular response to synthetic polymers. In *Handbook of Biomaterials Biocompatibility*, Woodhead Publishing (pp. 269–319).
73. Fabbri P, Messori M et al (2017) Surface modification of polymers: chemical, physical, and biological routes.

In *Modification of polymer properties*, William Andrew Publishing (pp. 109–130).
74. Wang Q, Huang J, Lai Y et al (2016) Smart drug delivery strategies based on porous nanostructure materials. Smart Drug Delivery System; Sezer, AD, Ed.; *InTech: Rijeka*, Croatia 63–90.
75. Wang XY, Gao Y, Tang Y et al (2023) Sustainable Developments in Polyolefin Chemistry: Progress, Challenges, and Outlook. *Progress in Polymer Science* 25:101713.
76. Dargaville TR, George GA, Hill DJ, Whittaker AK et al (2003) High energy radiation grafting of fluoropolymers. *Progress in Polymer Science* 28(9):1355–76.
77. Wentrup-Byrne E, George K et al (2010) Selected polymeric materials for orbital reconstruction. In *Biomaterials and Regenerative Medicine in Ophthalmology*. Woodhead Publishing (pp. 473–495).
78. Brauker JH, Carr-Brendel VE, Martinson LA, Crudele J, Johnston WD, Johnson RC et al (1995) Neovascularization of synthetic membranes directed by membrane microarchitecture. *Journal of Biomedical Materials Research* 29(12):1517–24.
79. Khan W, Muntimadugu E, Jaffe M, Domb AJ et al (2014) Implantable medical devices. *Focal Controlled Drug Delivery*, 33–59.
80. Babensee JE, Anderson JM, McIntire LV, Mikos AG et al (1998) Host response to tissue engineered devices. *Advanced drug delivery review* 33(1–2):111–39.
81. Dubey KA, Chaudhari CV, Bhardwaj YK, Varshney L et al (2017) Polymers, blends and nanocomposites for implants, scaffolds and controlled drug release applications. *Advances in Biomaterials for Biomedical Applications* 1–44.
82. Wypych G et al (2020) *PVC degradation and stabilization*. Elsevier.
83. Rahman M, Brazel CS et al (2004) The plasticizer market: an assessment of traditional plasticizers and research trends to meet new challenges. *Progress in Polymer Science* 29(12):1223–48
84. Latini G, Ferri M, Chiellini F et al (2010) Materials degradation in PVC medical devices, DEHP leaching and neonatal outcomes. *Current Medicinal Chemistry* 17(26):2979–89.
85. Horne DC et al (2021) The investigation of di-(2-ethylhexyl) phthalate (DEHP) plasticiser migration and extraction from medical grade poly vinyl chloride by in vitro and ex vivo methods.
86. Greiner TO, Volkmann AS, Hildenbrand S, Wodarz R, Perle N, Ziemer G, Rieger M, Wendel HP, Walker T et al (2012) DEHP and its active metabolites: leaching from different tubing types, impact on proinflammatory cytokines and adhesion molecule expression. Is there a subsumable context?. *Perfusion* 27(1):21–9.
87. Ma Y, Liao S, Li Q, Guan Q, Jia P, Zhou Y et al (2020) Physical and chemical modifications of poly (vinyl chloride) materials to prevent plasticizer migration-Still on the run. *Reactive and Functional Polymers* 147:104458.
88. Deb PK, Kokaz SF, Abed SN, Paradkar A, Tekade RK et al (2019) Pharmaceutical and biomedical applications of polymers. In *Basic fundamentals of drug delivery*. Academic press (pp. 203–267).
89. Skov AL, Yu L et al (2018) Optimization techniques for improving the performance of silicone-based dielectric elastomers. *Advanced Engineering Materials* 20(5):1700762.
90. Byczyński Ł et al (2013) Effect of different polyethers on surface and thermal properties of poly (urethane-siloxane) copolymers modified with side-chain siloxane. *Journal of Thermal Analysis and Calorimetry* 114:397–408.
91. Maitz MF et al (2015) Applications of synthetic polymers in clinical medicine. *Biosurface and Biotribology* 1(3):161–76.
92. Rahimi A, Mashak A et al (2013) Review on rubbers in medicine: Natural, silicone and polyurethane rubbers. *Plastics, Rubber and Composites*. 42(6):223–30.
93. Shields IV CW, White JP, Osta EG, Patel J, Rajkumar S, Kirby N, Therrien JP, Zauscher S et al (2018) Encapsulation and controlled release of retinol from silicone particles for topical delivery. *Journal of Controlled Release* 278:37–48.
94. Kumar A, Pillai J et al (2018) Implantable drug delivery systems: An overview. *Nanostructures for the Engineering of Cells, Tissues and Organs* 1:473–511.
95. Chen L, Zhang S, Duan Y, Song X, Chang M, Feng W, Chen Y et al (2024) Silicon-containing nanomedicine and biomaterials: materials chemistry, multi-dimensional design, and biomedical application. *Chemical Society Reviews* 53:1167–1315.
96. Kazanskiy NL, Khonina SN, Butt MA et al (2024) A review on flexible wearables-Recent developments in non-invasive continuous health monitoring. *Sensors and Actuators A: Physical* 2:114993.
97. Krol P et al (2007) Synthesis methods, chemical structures and phase structures of linear polyurethanes. Properties and applications of linear polyurethanes in polyurethane elastomers, copolymers and ionomers. *Progress in Materials Science* 52(6):915–1015.
98. Gunatillake PA, Adhikari R, Felton G et al (2011) Biodegradable polyurethanes: design, synthesis, properties and potential applications. *Biodegradable Polymers: Processing, Degradation and Applications* 431–70.
99. Maitz MF et al (2015) Applications of synthetic polymers in clinical medicine. *Biosurface and Biotribology* 1(3):161–76.
100. Xie F, Zhang T, Bryant P, Kurusingal V, Colwell JM, Laycock B et al (2019) Degradation and stabilization of polyurethane elastomers. *Progress in Polymer Science* 90:211–68.
101. Farokhi M, Mottaghitalab F, Fatahi Y, Khademhosseini A, Kaplan DL et al (2018) Overview of silk fibroin use in wound dressings. *Trends in Biotechnology* 36(9):907–22.
102. Nair DM. *Biodegradable hydrogels from silk sericin & Gelatin: development and characterization for medical applications* (Doctoral dissertation, M. Phil Thesis).
103. Shakiba M, Rezvani Ghomi E, Khosravi F, Jouybar S, Bigham A, Zare M, Abdouss M, Moaref R, Ramakrishna S et al (2021) Nylon—A material introduction and overview for biomedical applications. *Polymers for advanced technologies* 32(9):3368–83.
104. Maitz MF et al (2015) Applications of synthetic polymers in clinical medicine. *Biosurface and Biotribology* 1(3):161–76.
105. Vilar G, Tulla-Puche J, Albericio F et al (2010) Polymers and drug delivery systems. *Current Drug Delivery* 9(4):367–94.
106. García JM, García FC, Serna F, de la Peña JL et al (2010) High-performance aromatic polyamides. *Progress in Polymer Science* 35(5):623–86.
107. Tan MJ, Owh C, Chee PL, Kyaw AK, Kai D, Loh XJ et al (2016) Biodegradable electronics: cornerstone for sustainable electronics and transient applications. *Journal of Materials Chemistry* 4(24):5531–58.
108. Jaeblon T et al (2010) Polymethylmethacrylate: properties and contemporary uses in orthopaedics. *JAAOS-Journal of the American Academy of Orthopaedic Surgeons* 18(5):297–305.
109. Tunc EP et al (2007) Finite element analysis of heat generation from different light-polymerization sources during cementation of all-ceramic crowns. *The Journal of Prosthetic Dentistry* 97(6):366–74.

110. Holly FJ, Refojo MF et al (1975) Wettability of hydrogels I. Poly (2-hydroxyethyl methacrylate). *Journal of Biomedical Materials Research* 9(3):315–26.
111. Corsaro C, Neri G, Santoro A, Fazio E et al (2021) Acrylate and methacrylate polymers' applications: second life with inexpensive and sustainable recycling approaches. *Materials* 15(1):282.
112. Lima CS, Balogh TS, Varca JP, Varca GH, Lugão AB. Camacho-Cruz L, Bucio E, Kadlubowski SS et al (2020) An updated review of macro, micro, and nanostructured hydrogels for biomedical and pharmaceutical applications. *Pharmaceutics* 12(10):970.
113. Kass LE, Nguyen J et al (2022) Nanocarrier-hydrogel composite delivery systems for precision drug release. *Wiley Interdisciplinary Reviews: Nanomedicine and Nanobiotechnology* 14(2):e1756.
114. Virmani T, Kumar G, Sharma A, Pathak K, Akhtar MS, Afzal O, Altamimi AS et al (2023) Amelioration of cancer employing chitosan, its derivatives, and chitosan-based nanoparticles: Recent updates. *Polymers* 15(13):2928.
115. Bettencourt A, Almeida AJ et al (2012) Poly (methyl methacrylate) particulate carriers in drug delivery. *Journal of Microencapsulation* 29(4):353–67.
116. Martins JA, Lach AA, Morris HL, Carr AJ, Mouthuy PA et al (2020) Polydioxanone implants: a systematic review on safety and performance in patients. *Journal of Biomaterials Applications* 34(7):902–16.
117. Ten Brink T, Damanik F, Rotmans JI, Moroni L et al (2024) Unravelling & Harnessing The Immune Response at The Cell-Biomaterial Interface for Tissue Engineering Purposes. *Advanced Healthcare Materials* 13:2301939.
118. Feng Q, Jiang W, Sun K, Sun K, Chen S, Zhao L, Dai K, Ma N et al (2011) Mechanical properties and in vivo performance of a novel sliding-lock bioabsorbable poly-p-dioxanone stent. *Journal of Materials Science: Materials in Medicine* 22:2319–27.
119. Larrañaga A, Lizundia E et al (2019) A review on the thermomechanical properties and biodegradation behaviour of polyesters. *European Polymer Journal* 121:109296.
120. Vilar G, Tulla-Puche J, Albericio F et al (2012) Polymers and drug delivery systems. *Current Drug Delivery* 9(4):367–94.
121. Jarak I, Silva I, Domingues C, Santos AI, Veiga F, Figueiras A et al (2022) Nanofiber carriers of therapeutic load: current trends. *International Journal of Molecular Sciences* 23(15):8581.
122. Ding Y, Li W, Zhang F, Liu Z, Zanjanizadeh Ezazi N, Liu D, Santos HA et al (2019) Electrospun fibrous architectures for drug delivery, tissue engineering and cancer therapy. *Advanced Functional Materials* 29(2):1802852.
123. Arif ZU, Khalid MY, Sheikh MF, Zolfagharian A, Bodaghi M et al (2022) Biopolymeric sustainable materials and their emerging applications. *Journal of Environmental Chemical Engineering* 10(4):108159.
124. Syed A, Devi VK et al (2019) Potential of targeted drug delivery systems in treatment of rheumatoid arthritis. *Journal of Drug Delivery Science and Technology* 53:101217.
125. Zhong S, Liu P, Ding J, Zhou W et al (2021) Hyaluronic acid-coated MTX-PEI nanoparticles for targeted rheumatoid arthritis therapy. *Crystals* 11(4):321.
126. Yin N, Guo X, Sun R, Liu H, Tang L, Gou J, Yin T, He H, Zhang Y, Tang X et al (2020) Intra-articular injection of indomethacin–methotrexate in situ hydrogel for the synergistic treatment of rheumatoid arthritis. *Journal of Materials Chemistry B* 8(5):993–1007.
127. Jin X, Wang W, Zhao X, Jiang W, Shao Q, Chen Z, Huang C et al (2023) The battle between the innate immune cGAS-STING signaling pathway and human herpesvirus infection. *Frontiers in Immunology* 14.
128. Shen H, Jin L, Zheng Q, Ye Z, Cheng L, Wu Y, Wu H, Jon TG, Liu W, Pan Z, Mao Z et al (2023) Synergistically targeting synovium STING pathway for rheumatoid arthritis treatment. *Bioactive Materials* 24:37–53.
129. Duan J, Dong J, Zhang T, Su Z, Ding J, Zhang Y, Mao X et al (2014) Polyethyleneimine-functionalized iron oxide nanoparticles for systemic siRNA delivery in experimental arthritis. *Nanomedicine* 9(6):789–801.
130. Trujillo-Nolasco RM, Morales-Avila E, Ocampo-Garcia BE, Ferro-Flores G, Gibbens-Bandala BV, Escudero-Castellanos A, Isaac-Olive K et al (2019) Preparation and in vitro evaluation of radiolabeled HA-PLGA nanoparticles as novel MTX delivery system for local treatment of rheumatoid arthritis. *Materials Science and Engineering* 103:109766.
131. Park JS, Yang HN, Jeon SY, Woo DG, Kim MS, Park KH et al (2012) The use of anti-COX2 siRNA coated onto PLGA nanoparticles loading dexamethasone in the treatment of rheumatoid arthritis. *Biomaterials* 33(33):8600–12.
132. Présumey J, Salzano G, Courties G, Shires M, Ponchel F, Jorgensen C, Apparailly F, De Rosa G et al (2012) PLGA microspheres encapsulating siRNA anti-TNFalpha: efficient RNAi-mediated treatment of arthritic joints. *European Journal of Pharmaceutics and Biopharmaceutics* 82(3):457–64.
133. Wang L, Che K, Liu Y et al (2021) Pharmacokinetics, distribution and efficacy of triptolide PLGA microspheres after intra-articular injection in a rat rheumatoid arthritis model. *Xenobiotica* 51(6):703–15.
134. Liu XM, Quan LD, Tian J, Laquer FC, Ciborowski P, Wang D et al (2010) Syntheses of click PEG– dexamethasone conjugates for the treatment of rheumatoid arthritis. *Biomacromolecules* 11(10):2621–8.
135. Sujitha S, Dinesh P, Rasool M et al (2020) Berberine encapsulated PEG-coated liposomes attenuate Wnt1/β-catenin signaling in rheumatoid arthritis via miR-23a activation. *European Journal of Pharmaceutics and Biopharmaceutics* 149:170–91.
136. He L, Fan D, Liang W, Wang Q, Fang J et al (2020) Matrix metalloproteinase-responsive PEGylated lipid nanoparticles for controlled drug delivery in the treatment of rheumatoid arthritis. *ACS Applied Bio Materials* 3(5):3276–84.
137. Wang Q, Qin X, Fang J, Sun X et al (2021) Nanomedicines for the treatment of rheumatoid arthritis: State of art and potential therapeutic strategies. *Acta Pharmaceutica Sinica B* 11(5):1158–74.
138. Nornberg AB, Martins CC, Cervi VF, Sari MH, Cruz L, Luchese C, Wilhelm EA, Fajardo AR et al (2022) Transdermal release of methotrexate by cationic starch/poly (vinyl alcohol)-based films as an approach for rheumatoid arthritis treatment. *International Journal of Pharmaceutics* 611:121285.
139. Strehl C, Gaber T, Maurizi L, Hahne M, Rauch R, Hoff P, Häupl T, Hofmann-Amtenbrink M, Poole AR, Hofmann H, Buttgereit F et al (2015) Effects of PVA coated nanoparticles on human immune cells. *International Journal of Nanomedicine* 8:3429–45.
140. An M, Shi M, Su J, Wei Y, Luo R, Sun P, Zhao Y et al (2022) Dual-drug loaded separable microneedles for efficient rheumatoid arthritis therapy. *Pharmaceutics* 14(7):1518.
141. Revill PA, Chisari FV, Block JM, Dandri M, Gehring AJ, Guo H, Hu J, Kramvis A, Lampertico P, Janssen HL, Levrero M et al (2019) A global scientific strategy to cure hepatitis B. *The lancet Gastroenterology & Hepatology* 4(7):545–58.
142. Smith S, Harmanci H, Hutin Y, Hess S, Bulterys M, Peck R, Rewari B, Mozalevskis A, Shibeshi M, Mumba M, Le LV

et al (2019) Global progress on the elimination of viral hepatitis as a major public health threat: An analysis of WHO Member State responses 2017. *JHEP Reports* 1(2):81–9.
143. Rivkina A, Rybalov S et al (2002) Chronic hepatitis B: current and future treatment options. *Pharmacotherapy: The Journal of Human Pharmacology and Drug Therapy* 22(6):721–37.
144. Saffari Z, Sepahi M, Ahangari-Cohan R, Khoobi M, Hamidi-Fard M, Ghavidel A, Aghasadeghi MR, Norouzian D et al (2023) A quartz crystal microbalance biosensor based on polyethylenimine-modified gold electrode to detect hepatitis B biomarker. *Analytical Biochemistry* 661:114981.
145. Somiya M, Yoshimoto N, Iijima M, Niimi T, Dewa T, Jung J, Kuroda SI et al (2012) Targeting of polyplex to human hepatic cells by bio-nanocapsules, hepatitis B virus surface antigen L protein particles. *Bioorganic & Medicinal Chemistry* 20(12):3873–9.
146. Uchida E, Kogi M, Oshizawa T, Furuta B, Satoh K, Iwata A, Murata M, Hikata M, Yamaguchi T et al (2007) Optimization of the virus concentration method using polyethyleneimine-conjugated magnetic beads and its application to the detection of human hepatitis A, B and C viruses. *Journal of Virological Methods* 143(1):95–103.
147. Zheng X, Zhu J, Zheng C, Tan Z, Ji Z, Tao J, Zhao Y, Ji Z, Hu Y et al (2023) Dissolving Microneedle Arrays as a Hepatitis B Vaccine Delivery System Adjuvanted by APC-Targeted Poly (Lactic-co-Glycolic Acid)(PLGA) Nanoparticles. *AAPS PharmSciTech* 24(1):42.
148. Zhu J, Qin F, Ji Z, Fei W, Tan Z, Hu Y, Zheng C et al (2020) Mannose-modified PLGA nanoparticles for sustained and targeted delivery in hepatitis B virus immunoprophylaxis. *AAPS PharmSciTech* 21:1–9.
149. Ayoub MM, Jasti B, Elantouny NG, Elnahas H, Ghazy FE et al (2020) Comparative study of PLGA in-situ implant and nanoparticle formulations of entecavir; in-vitro and in-vivo evaluation. *Journal of Drug Delivery Science and Technology* 56:101585.
150. Song A, Lin X, Lu J, Ren S, Cao Z, Zheng S, Hu Z, Li H, Shen C, Chen X et al (2021) Pegylated interferon treatment for the effective clearance of hepatitis b surface antigen in inactive HBsAg carriers: A meta-analysis. *Frontiers in Immunology* 12:779347.
151. Li MH, Lu HH, Chen QQ, Lin YJ, Zhan ZE, Yao LU, Zhang L, Dong JP, Wei YI, Yao XI et al (2021) Changes in the cytokine profiles of patients with chronic hepatitis B during antiviral therapy. *Biomedical and Environmental Sciences* 34(6):443–53.
152. Zhang W, Xing M, Sun W, Chen J, Xie N, Cai Y, Wang Y, Li N, Jiang Y, Zhang F, Wang Y et al (2023) Early clinical efficacy of pegylated interferon treatment in patients with different phases of chronic HBV infection: A real-world analysis. *Journal of Viral Hepatitis*, 30(5).
153. Yaici L, Gatouillat G, Andreoletti L, N'Guyen Y, Hentzien M, Gordien E, Bani-Sadr F et al (2023) Chronic hepatitis delta cirrhosis cured by adapting PEG-IFNα-2a+ tenofovir disoproxil fumarate treatment duration until HBsAg loss. *Clinics and Research in Hepatology and Gastroenterology* 47(7):102148.
154. El-Shafai NM, Shawky S, El-Mehasseb IM, El-Kemary MA et al (2021) Sandwich nanohybrid of chitosan-polyvinyl alcohol for water treatment and Sofosbuvir drug delivery for anti-hepatitis C virus (HCV). *International Journal of Biological Macromolecules* 190:927–39.
155. Shawky S, El-Shafai NM, El-Mehasseb IM, Shoueir KR, El-Kemary MA et al (2021) Spectroscopic study of self-assembly of anti-hepatitis C virus sofosbuvir drug with bio-polymeric nanoparticles for improving the drug release effect. *Spectrochimica Acta Part A: Molecular and Biomolecular Spectroscopy* 261:120008.
156. Kim JS, Choi JA, Kim JC, Park H, Yang E, Park JS, Song M, Park JH et al (2020) Microneedles with dual release pattern for improved immunological efficacy of Hepatitis B vaccine. *International Journal of Pharmaceutics* 591:119928.
157. Shrestha B, Rath JP et al (2014) Poly (vinyl alcohol)-coated chitosan microparticles act as an effective oral vaccine delivery system for hepatitis B vaccine in rat model. *IET Nanobiotechnology* 8(4):201–7.
158. Lopez-Giacoman S, Madero M et al (2015) Biomarkers in chronic kidney disease, from kidney function to kidney damage. *World Journal of Nephrology* 4(1):57.
159. Anumudu SJ, Fadem SZ et al (2021) Nephrology in the United States of America. *Nephrology Worldwide* 173–86.
160. Li L, Wu Y, Wang C, Zhang W et al (2012) Inhibition of PAX2 gene expression by siRNA (polyethylenimine) in experimental model of obstructive nephropathy. *Renal Failure* 34(10):1288–96.
161. Wang G, Li Q, Chen D, Wu B, Wu Y, Tong W, Huang P et al (2019) Kidney-targeted rhein-loaded liponanoparticles for diabetic nephropathy therapy via size control and enhancement of renal cellular uptake. *Theranostics* 9(21):6191.
162. Tong F et al (2017) Preparation of exenatide-loaded linear poly (ethylene glycol)-brush poly (l-lysine) block copolymer: potential implications on diabetic nephropathy. *International Journal of Nanomedicine* 12: 4663–78.
163. Muniraj T, Gajendran M, Thiruvengadam S, Raghuram K, Rao S, Devaraj P et al (2012) Acute pancreatitis. *Disease-a-Month* 58(3):98–144.
164. Ferrero-Andrés A, Panisello-Roselló A, Serafín A, Roselló-Catafau J, Folch-Puy E et al (2020) Polyethylene glycol 35 (PEG35) protects against inflammation in experimental acute necrotizing pancreatitis and associated lung injury. *International Journal of Molecular Sciences*. 2020 Jan 30;21(3):917.
165. Ferrero-Andrés A, Panisello-Roselló A, Roselló-Catafau J, Folch-Puy E. Polyethylene glycol 35 ameliorates pancreatic inflammatory response in cerulein-induced acute pancreatitis in rats. *World J Gastroenterol*. 2020 Oct 21;26(39):5970–5982.
166. Zhang Y, Wu BM et al (2023) Current Advances in Stimuli-Responsive Hydrogels as Smart Drug Delivery Carriers. *Gels* 9(10):838.
167. Tyagi P, Agate S, Velev OD, Lucia L, Pal L et al (2022)A critical review of the performance and soil biodegradability profiles of biobased natural and chemically synthesized polymers in industrial applications. *Environmental Science & Technology* 56(4):2071–95.
168. Operti MC, Bernhardt A, Grimm S, Engel A, Figdor CG, Tagit O et al (2021) PLGA-based nanomedicines manufacturing: Technologies overview and challenges in industrial scale-up. *International Journal of Pharmaceutics* 605:120807.
169. Khalili L, Dehghan G, Sheibani N, Khataee A et al (2022) Smart active-targeting of lipid-polymer hybrid nanoparticles for therapeutic applications: Recent advances and challenges. *International Journal of Biological Macromolecules* 213:166–94.
170. Adepu S, Ramakrishna S et al (2021) Controlled drug delivery systems: current status and future directions. *Molecules* 26(19):5905.
171. Kar S, Sanderson H, Roy K, Benfenati E, Leszczynski J et al (2021) Green chemistry in the synthesis of pharmaceuticals. *Chemical Reviews* 122(3):3637–710.

11 Synthetic Polymers for Immunomodulatory Drug Delivery

Atul Kabra, Ritika Puri, Ruchika Kabra, and Deepika Bhatia
University Institute of Pharma Sciences, Chandigarh University, Mohali, India

11.1 INTRODUCTION

The body's defense mechanism against illnesses and infections is called immunity. This complex mechanism is crucial for maintaining health and involves a coordinated effort among various cells, tissues, and organs to combat harmful pathogens like bacteria, viruses, fungi, and parasites [1, 2]. By identifying and neutralizing these invaders, the immune system ensures the organism's survival and health. Immunity is generally categorized in two types: nonspecific (innate) and specific (adaptive) immunity [3].

Immunity can be categorized into two primary types: cellular immunity and humoral immunity. Each type is essential for defending the body against infections and diseases, utilizing distinct mechanisms and components. Cellular immunity, or cell-mediated immunity, involves immune cells that directly target and destroy infected or abnormal cells. This process is mainly carried out by T lymphocytes (T cells), which develop in the thymus and move through the blood and lymphatic system. Humoral immunity involves B lymphocytes (B cells) producing antibodies, as shown in Figure 11.1. These antibodies travel through the blood and lymphatic system, binding to specific antigens to neutralize them or tag them for destruction [4, 5].

11.1.1 Immune Systems

The immune system stands as a remarkable biological defense mechanism, tirelessly shielding our bodies from harmful intruders. It operates through an intricate web of cells, organs and tissues, forming the frontline defense against a spectrum of pathogens including virus, fundi, bacteria, and parasites. Central to its functioning are white blood cells, each specializing in distinct roles [6, 7]. Macrophages, for instance, engulf and break down foreign substances, while T cells coordinate targeted attacks on infected cells and regulate overall immune responses. B cells contribute by producing antibodies that tag invaders for elimination by other immune cells. The immune system's ability to distinguish between self and nonself is a crucial defense mechanism that protects the body's tissues from inadvertent damage. This defense is especially important in situations like autoimmune illness. Equally impressive is its adaptability [8]. Exposure to pathogens or vaccines allows the immune system to develop immunological memory. This memory enables faster and more effective responses upon subsequent encounters with familiar threats, providing long-lasting defense. Various factors influence the immune system's health, including diet, physical activity, sleep patterns, and stress levels. Maintaining a balanced lifestyle is essential to support its optimal function, ensuring robust protection against infections and diseases [9].

11.2 IMMUNOMODULATORS

Immunomodulators are substances or agents capable of adjusting the body's immune response, either by enhancing it to strengthen defenses against infections and diseases or by suppressing it to prevent harmful overreactions, such as in autoimmune disorders or organ transplants. Their role is pivotal in maintaining a balanced and efficient immune system. These substances regulate various aspects of immune activity, ensuring that responses are appropriately targeted and controlled [10].

11.2.1 Organizing Immunomodulators by Class

In terms of clinical application, immunomodulators are divided into three groups.

11.2.1.1 Immunosuppressants

The substances, known as immunosuppressants, are frequently employed in medical contexts to prevent the rejection of transplanted organs, such as after an organ transplant, and to treat autoimmune disorders like rheumatoid arthritis and lupus. They function by decreasing the proliferation of B cells and T cells, which helps to moderate immune responses that might otherwise cause harm to tissues or lead to organ rejection, as summarized in Table 11.1 [11].

11.2.1.2 Immunostimulants

These substances improve the immune system's function, strengthening its ability to combat infections or diseases. They encompass a variety of substances, including natural compounds like specific vitamins (such as vitamin C, vitamin D), herbal supplements like echinacea, or synthetic drugs crafted to stimulate the activity of immune

DOI: 10.1201/9781003477389-11

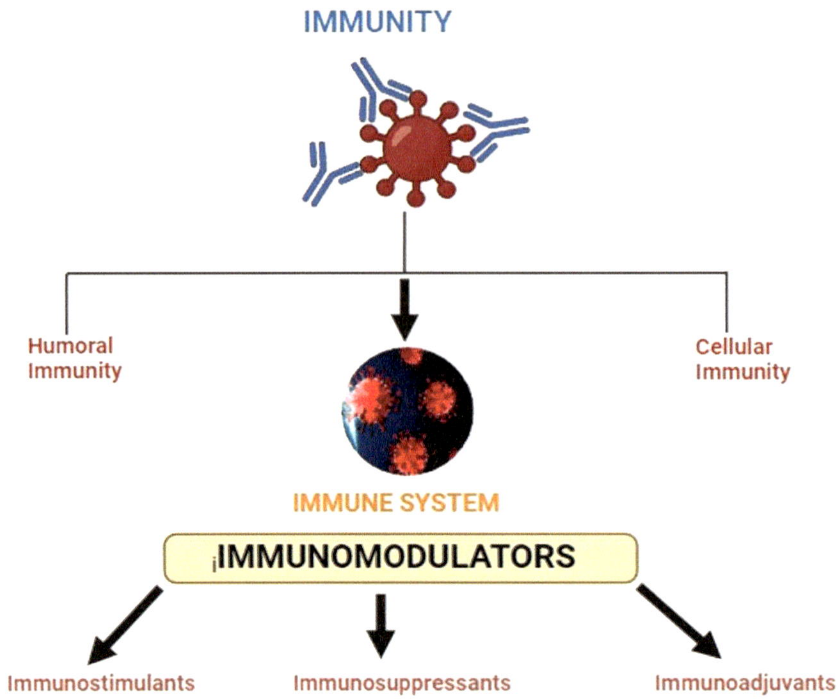

FIGURE 11.1 Immunity and immunomodulators.

TABLE 11.1
Classification of Immunosuppressants with Examples and Their Limitation/Side Effects

S. no.	Class	Examples	Limitations/Side Effects	References
1		Cyclosporine, Tacrolimus	Liver toxicity, kidney damage, hypertension, neuronal damage, hyperglycemia	[15]
2	Antiproliferative Agents	Azathioprine, Mycophenolate mofetil	Bone marrow suppression, gastrointestinal disturbances, hepatotoxicity, increased risk of infections, lymphoproliferative disorders	[16]
3		Sirolimus, Everolimus	Increase lipid profile, and increased risk of infections	[17]
4	Corticosteroids	Prednisone, Methylprednisolone	Hyperglycemia, hypertension, osteoporosis, weight gain, increased risk of infections, cataracts, mood swings	[18]
5	Alkylating Agents	Cyclophosphamide, Chlorambucil	Bone marrow suppression, increased risk of malignancies, hemorrhagic cystitis, gastrointestinal disturbances, alopecia	[19]
6	Antimetabolites	Methotrexate	Hepatotoxicity, bone marrow suppression, gastrointestinal disturbances, pulmonary toxicity, teratogenicity	[20]
7	Janus Kinase (JAK) Inhibitors	Tofacitinib, Baricitinib	Increased risk of infections, thrombosis, gastrointestinal perforations, hyperlipidemia, hepatotoxicity	[21]
8	Biologics	Basiliximab, Rituximab	Infusion reactions, increased risk of infections, cytokine release syndrome, potential for developing antibodies against the drug	[22]

cells. Immunostimulants are employed at times when the immune system needs bolstering, particularly during periods of increased vulnerability to infections, as summarized in Table 11.2 [12].

11.2.1.3 Immunoadjuvants

They can be categorized as specific immune stimulators and are used to increase the effectiveness of vaccines. It is possible for immunoadjuvants to function as effective immunomodulators [13]. A major difficulty for vaccine developers is that it has been suggested that they might be utilized to distinguish between humoral and cellular helper T1 (Th1) and helper T2 (Th2) cells, as well as immunoprotective, immunodestructive, and reaginic IgE vs. IgG type immune responses, as summarized in Table 11.3 [14].

TABLE 11.2
Classification of Immunostimulants with Examples and Their Limitation/Side Effects

S. no.	Class	Examples	Limitations/Side Effects	References
1	Cytokines	Interferons (IFN-α, IFN-β, IFN-γ)	Nephrotoxicity, hypertension, neurotoxicity, hyperglycemia, increased risk of infections, hepatotoxicity	[23]
2	Factors Stimulate Colonies	Filgrastim, Sargramostim	Bone pain, leukocytosis, splenomegaly, injection site reactions, fever	[24]
3	Immunomodulatory Drugs	Thalidomide, Lenalidomide	Teratogenicity, thromboembolism, neuropathy, hematologic toxicity, fatigue	[25]
4	Toll-Like Receptor Agonists	Imiquimod, CpG Oligodeoxynucleotides	Local skin reactions, flu-like symptoms, fatigue, potential for autoimmune reactions	[26]
5	Vaccines	BCG Vaccine, HPV Vaccine	Injection site reactions, mild fever, muscle aches, rare allergic reactions, potential for exacerbation of autoimmune disorders	[27]
6	Interleukins	IL-2 (Aldesleukin)	Capillary leak syndrome, hypotension, renal dysfunction, flu-like symptoms, mental status changes	[28]

TABLE 11.3
Classification of Immunoadjuvants with Examples and Their Limitation/Side Effects

S. No.	Class	Examples	Limitations/Side Effects	References
1	Alum (Aluminum Salts)	Aluminum hydroxide, Aluminum phosphate	Local injection site reactions, granulomas, potential for long-term accumulation, rarely associated with macrophagic myofasciitis	[29]
2	Saponins	QS-21 (component of AS01 adjuvant system)	Local injection site reactions, mild fever, flu-like symptoms, potential for hemolysis at high doses	[30, 31]
3	Bacterial Derivatives	BCG (used in bladder cancer treatment), Flagellin	Local injection site reactions, fever, flu-like symptoms, potential for severe inflammatory responses	[32]

11.3 CHALLENGES ASSOCIATED WITH IMMUNOMODULATORY DRUG DELIVERY

Delivering immunomodulatory drugs, whether immunosuppressants or immunostimulants, presents unique challenges due to the intricacy of the immune system and the critical balance required to achieve therapeutic effects without causing significant side effects [33, 34]. This section explores the specific challenges faced in the delivery of these drugs, considering both immunosuppressive and immunostimulatory agents.

11.3.1 Achieving Target Specificity

11.3.1.1 Off-Target Effects

A significant challenge with immunosuppressants is their need to selectively target specific immune components responsible for unwanted immune responses while sparing the broader immune system [35–37]. Nonspecific targeting can result in generalized immunosuppression, heightening susceptibility to infections and cancers.

11.3.1.2 Tissue Penetration

Another critical challenge is ensuring that immunosuppressants effectively reach targeted tissues or organs at therapeutic levels while minimizing exposure to nontarget tissues. Inadequate penetration into tissues can lead to less-than-optimal therapeutic results [38–40].

11.3.1.3 Minimizing Side Effects

11.3.1.3.1 Nephrotoxicity and Hepatotoxicity

Several immunosuppressive agents, like calcineurin inhibitors, are known for their potential nephrotoxicity and hepatotoxicity [41, 42]. It is imperative to develop delivery systems that can mitigate these side effects to ensure long-term patient safety.

11.3.1.3.2 Cardiovascular Risk

Extended use of specific immunosuppressants may heighten the likelihood of cardiovascular diseases [43, 44]. Striking a balance between immunosuppression and cardiovascular health remains a delicate and continuous challenge.

11.3.1.4 Managing Long-Term Use

11.3.1.4.1 Resistance Development

As treatment progresses, patients may develop resistance to immunosuppressive agents, requiring adjustments in therapy or dosage [45, 46]. Vigilant monitoring and effective management of this resistance are crucial to maintain therapeutic effectiveness over time.

11.3.1.4.2 Cumulative Toxicity

Prolonged use of immunosuppressants can result in cumulative toxicity, impacting multiple organs and systems [47]. It is essential to conduct ongoing safety monitoring and explore the creation of alternatives with reduced toxicity.

11.4 IMMUNOSTIMULANTS

These are agents that enhance the immune response, beneficial in treating conditions like infections, cancer, and immunodeficiencies [48–50]. Delivering them effectively involves ensuring optimal stimulation, preventing excessive activation, and addressing patient-specific variability.

11.4.1 Ensuring Effective Stimulation

11.4.1.1 Dose Optimization

Determining the ideal dosage to effectively enhance the immune response without inducing excessive activation is complex [51]. Insufficient dosing may result in inadequate therapeutic benefits, whereas excessive dosing can lead to adverse reactions.

11.4.1.2 Delivery to Target Cells

Ensuring the effective delivery of immunostimulants to the precise immune cells requiring activation is essential. Targeting these cells specifically while preventing widespread systemic activation presents a major challenge [52].

11.4.2 Avoiding Overactivation

11.4.2.1 Autoimmunity

Immunostimulants have the potential to induce autoimmune reactions when they excessively activate the immune system. To mitigate this risk, precise dosing and targeted delivery strategies must be carefully implemented [53, 54].

11.4.2.2 Cytokine Storms

Excessive immune stimulation can result in cytokine storms, characterized by an excessive production of cytokines that induces severe inflammation and tissue damage [55, 56]. Maintaining a balance in immune stimulation to prevent such overreactions is crucial.

11.4.3 Patient Variability

11.4.3.1 Genetic Differences

The genetic backgrounds of patients can greatly influence how they respond to immunostimulants. Tailoring dosing and delivery approaches may be necessary to accommodate these individual differences [57, 58].

11.4.3.2 Health Status

Patients' health conditions, encompassing underlying illnesses and current immune function, significantly impact their response to immunostimulants [59–61]. Customizing treatments based on individual health profiles is essential to achieve optimal outcomes.

11.4.4 Delivery Mechanisms

11.4.4.1 Nanotechnology and Biomaterials

Advancements in delivery systems, such as nanoparticles and biodegradable polymers, present a promising avenue for augmenting the precision and effectiveness of immunosuppressants and immunostimulants [62–64]. Yet, ensuring the biocompatibility, safety, and regulatory approval of these systems poses ongoing challenges.

11.4.4.2 Controlled Release

It is crucial to achieve controlled-release profiles that sustain therapeutic drug levels over time without causing spikes that may result in side effects [65]. This necessitates the use of advanced delivery systems capable of finely regulating drug release.

11.4.5 Patient Adherence

11.4.5.1 Complexity of Regimens

Complex dosing regimens, especially those requiring frequent administration or multiple drugs, can negatively impact patient adherence [66]. Simplifying these regimens through innovative delivery systems is crucial for improving compliance.

11.4.5.2 Side Effects Management

Managing side effects effectively to enhance patient comfort and adherence is vital [67]. Reducing the incidence and severity of side effects through improved drug formulations and delivery methods can significantly impact treatment success [68].

11.5 POLYMERS IN DRUG DELIVERY

Polymers are large molecules with chemical bonding of repeating structural units in a function called monomers. It shows high molecular weight there in daily life materials and plays a huge role in the advancement era of today such as packaging, electronics, and health-care industries [69]. Various polymers, as depicted in Figure 11.2, play a crucial part in the design of drug delivery systems in the pharmaceutical industry which can regulate the release of a drug per unit of time, improve the stability, and possess target specificity. They act as carriers, holding medications within their structure and releasing them gradually.

The following essential requirements must be fulfilled, even though the precise physiochemical properties of the polymers used in a medicinal formulation can vary significantly [70].

1. The solubility in water
2. The ability to degrade or be biocompatible

3. The ability to conjugate a medication or for multifaceted development utilizing a biological macromolecule

Furthermore, it is quite preferable that the polymer utilized is of inexpensive price and well-established pharmacological profile, even though it is observed that conjugates of polymers and drugs are taken as new chemical entities (NCEs) in accordance with regulations and such that admission to must first undergo the standard approval procedures marketplace.

Particularly synthetic polymers are extremely adjustable, enabling the creation of systems that satisfy certain pharmacokinetic and pharmacodynamic specifications. These polymers can be designed to have specific delivery, biodegradable qualities, and a response to pH, temperature, or enzymes, among other external stimuli [71]. They are extremely versatile and can be shaped into a variety of drug delivery systems, such as dendrimers, hydrogels, micelles, and nanoparticles, to meet a range of therapeutic requirements.

11.5.1 Advantages of Synthetic Polymers [72]

Synthetic polymers used in the formulation of various delivery system possess certain advantages:

1. Persistent Therapeutic Effect: Synthetic polymers may allow drugs to be sputtered in such a manner that pharmacologically active drugs will be within therapeutic levels in the body for a longer time.
2. Localized Drug Administration: It is also possible with the use of polymers to deliver certain drugs to certain tissues or cells, which implies less occurrence of unwanted effects and possibly better results. This principle should be followed in drug development, especially in oncology where drugs have to penetrate tumor cells without damaging ichor.
3. Protection from Environmental Degradation: It is known that the encapsulation of drugs in polymers will protect them from factors such as enzymes, pH and other more adverse conditions. This concerns especially the more fragile molecules like proteins and peptides, which may be sensitive to degradation.
4. Enhanced Bioavailability: The low water solubility of the majority of hydrophobic drugs, and consequently their bioavailability is the main challenge in to show their efficacy. Increasing dispersion of such polymers within the body or improving polymer intrinsic microbicidal activity, would have a positive impact on the solubility and distribution of such drugs.
5. Safety of Materials Used for Drug Delivery: Several synthetic polymers that are used for drug delivery have also been noted to be biocompatible and biodegradable, making it ideal when conducting drug delivery.

11.5.2 Synthetic Polymers in Drug Delivery

Polymers used in a drug delivery system can be categorized on the basis of their chemical structure, degradability, and interaction with biological environments. Some commonly used synthetic polymers in the pharmaceuticals are discussed in the subsection.

11.5.2.1 Polyesters

Polyesters are among the most widely used synthetic polymers for drug delivery, thanks to their biocompatibility and biodegradability. They break down into nontoxic byproducts, such as lactic acid and glycolic acid, which are naturally metabolized by the body. Notable synthetic polyesters employed in drug delivery include:

- **Poly(lactic-co-glycolic acid) (PLGA)**: The Food and Drug Administration has approved PLGA, a copolymer of lactic acid and glycolic acid, for use in drug delivery applications. PLGA is commonly used for drug encapsulation. PLGA nanoparticles or microspheres have been applied to tissue engineering, cancer treatment, and vaccinations. They can offer prolonged or regulated drug release. Lamivudine (LAM)-loaded PLGA NPs (PLGA-NPs-LAM) were created by Ekinci et al. in 2022 and radiolabeled with radioactive technetium (99mTc) to aid in the diagnosis of lung cancer. The findings showed that 99mTc radiolabeling of PLGA-NPs-LAM significantly enhanced absorption by A-549 cells, indicating that it is a viable

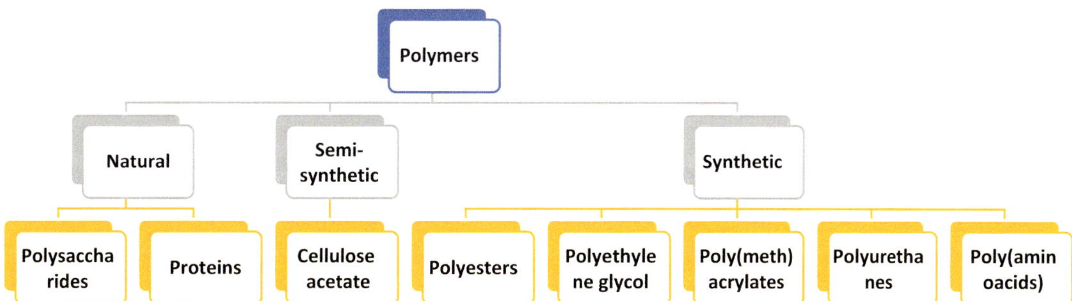

FIGURE 11.2 Classification of polymers.

substitute agent for lung cancer imaging [73]. PLGA is not a suitable polymer due to a few disadvantages, such as low entrapment efficiency, early burst release, instability of the entrapped protein, and incomplete release profile [74–76]. PLGA can be copolymerized with different polymers to solve these drawbacks. Numerous PLGA-PEG block copolymers, such as PLGA-PEG, PLGA-PEG-PLGA, and PEG-PLGA PEG, have been created [77].

- **Poloxamers**: Nonionic triblock copolymers are called poloxamers. The polypropylene oxide (PPO) that makes up the poloxamer's center portion is hydrophobic in nature, whereas the polyethylene oxide (PEO) that surrounds it is hydrophilic. Another commercial name for ploximers is pluronics. These thermosensitive polymers, which come in a variety of commercial grades like poloxamer F-68, F127, and others, have been thoroughly researched for their potential to deliver therapeutic proteins over an extended period [78].
- **Polylactic acid (PLA)**: Made from lactic acid, PLA is a biodegradable polyester. It is utilized in drug-eluting stents and controlled-release formulations, where the medicine is progressively released as the body absorbs it [79].
- **Polycaprolactone (PCL)**: utilized in long-term drug administration applications, PCL is a slow-degrading polyester. Its slow degradation rate makes it suitable for the sustained release of drugs over months or even years [80].

11.5.2.2 Polyethylene Glycol (PEG)

PEG is a hydrophilic polymer that is commonly used to enhance the solubility, stability, and circulation time of drugs. PEGylation, the process of attaching PEG chains to drugs or nanoparticles, increases the half-life of drugs in the bloodstream by reducing their renal clearance and protecting them from enzymatic degradation. PEG is often used in the formulation of nanocarriers, micelles, and liposomes for cancer therapy and targeted delivery [81].

11.5.2.3 Poly(amino acids)

Poly(amino acids) are synthetic polymers that mimic natural peptides and proteins. These polymers are biodegradable and biocompatible, making them suitable for drug delivery systems, especially in the delivery of proteins and peptides. Polymers such as poly(L-glutamic acid) (PGA) and poly(L-lysine) (PLL) have been used in the development of drug conjugates, nanoparticles, and hydrogels for targeted drug delivery [82].

11.5.2.4 Polyurethanes

Polyurethanes as synthetic elastomers have unique numerous characteristics such as good mechanical strength, elasticity, and biocompatibility [83]. Such constructs include drug-eluting stents, wound dressings and implantable drug delivery systems. Such properties make them suitable for modification with a view to varying the delivery systems of various drugs.

11.5.2.5 Polyvinyl Alcohol (PVA)

PVA is another type of water-soluble polymer that is utilized in the area of drug delivery such as in hydrogels films and microspheres. It is biocompatible and can be designed to achieve sustained drug delivery. The use of PVA is common in ocular drug delivery systems, wound healing and transdermal drug delivery systems. In a study by Prausnitz and Langer [84], polyacrylamide hydrogels were tested as transdermal drug device systems. The study concluded that the hydrogel patches could be used to deliver drugs through the skin over a long period, providing a non-invasive option to other systems of drug delivery such as oral and injectable methods.

11.5.2.6 Poly(meth)acrylates

Poly(meth)acrylates also known as PMMA polymers like polymethyl methacrylate and or poly(hydroxyethyl methacrylate) PHEMA belong to a group of synthetic polymers and are used in sustained and controlled drug delivery systems [85]. These polymers possess advantageous properties such as stability and formation of solid matrices and are therefore useful when incorporated into drug-eluting implants and topical formulations. PMMA bone cement is also used in orthopedics for the purpose of drug delivery systems for local applications [86].

11.6 MODULATION OF IMMUNE RESPONSE THROUGH SYNTHETIC POLYMERS

Synthetic polymers play a crucial role in modulating the immune response by enhancing the delivery and efficacy of immunomodulatory agents, such as vaccines and therapeutic drugs [87, 88]. This modulation can occur through several mechanisms, which can be broadly categorized into targeted delivery, adjuvant properties, and controlled release. There are several mechanisms of immune modulation.

11.6.1 Targeted Delivery

Cell Targeting: Synthetic polymers can be designed to target specific immune cells, such as dendritic cells (DCs), macrophages, or T cells. By attaching ligands (e.g., antibodies or peptides) to the polymer, the delivery system can enhance uptake by these immune cells, leading to a more robust immune response [89, 90].

Antigen Presentation: Polymers can encapsulate antigens and facilitate their uptake by antigen-presenting cells (APCs). Once internalized, these cells process the antigens and present them on major histocompatibility complex (MHC) molecules, activating T cells [91].

11.6.2 Adjuvant Properties

Enhanced Immunogenicity: Some synthetic polymers can act as adjuvants, which are substances that enhance the body's immune response to an antigen. They can do this by stimulating innate immune receptors (like toll-like receptors), promoting the production of cytokines and chemokines [92–94].

Depot Effect: Polymers can create a localized depot of the immunogenic material, providing sustained exposure to the immune system and leading to a stronger and longer-lasting immune response [95].

Controlled Release

Sustained Release of Antigens: By tuning the degradation rate of biodegradable polymers, drugs or antigens can be released in a controlled manner over time, ensuring a prolonged immune stimulation [96].

Temporal Regulation: Controlled release can also help in managing the timing of immune response, which is particularly important in vaccination strategies [97].

11.6.3 Stimuli-Responsive Release

Smart Polymers: Some polymers respond to specific stimuli (pH, temperature, or enzymes) to release their payload in a controlled manner. This allows for the targeted release of immunomodulatory agents at the site of infection or inflammation [98].

11.7 SYNTHETIC POLYMERS USED IN IMMUNOMODULATORY DRUG DELIVERY

Polymers intended for biomedical applications should ideally be biodegradable and should not produce harmful (toxic) effects upon degradation within the host [99]. In the field of immunology, polymers serve various roles, including as adjuvants, attenuators, stimulators, or suppressors of immune responses. As adjuvants, polymers are promising candidates for boosting immune responses to purified recombinant or weak antigens, facilitating the development of animal disease models with minimal bias in the resulting immune reactions [100]. They have the potential to induce a broad spectrum of cytokines and are effective as carriers in vaccination studies, enhancing specific immune responses when administered alongside an antigen. Conversely, as attenuators, polymers can diminish the adhesion or binding of host immune cells to implanted surfaces, potentially reducing unwanted immune reactions. Additionally, they can function as activators or suppressors of response mediated by immune system, depending on the desired therapeutic outcome, as described in Table 11.4 [101].

Polymeric nanoparticles (NPs) have become valuable assets in immunology because of their distinctive characteristics and capacity to influence immune responses. These nanoparticles can be specifically designed to deliver therapeutic agents efficiently, thereby either boosting or dampening immune reactions as needed. Table 11.5 represents the immunomodulatory potential of polymeric nanoparticles.

TABLE 11.4
Synthetic Polymers Used in Immunomodulatory Drug Delivery

Polymer Name	Abbreviation	Type	Properties	Applications	References
Polyethyleneimine	PEI	Cationic Polymer	High charge density biocompatible	Gene delivery, drug formulation	[102]
Polyethylene glycol	PEG	Water-soluble polymer	Water-soluble polymer	Drug delivery systems, cosmetics and personal care, surface modification, medical devices	[103]
Polyvinyl pyrrolidone	PVP	Water-soluble polymer	Biocompatible excellent solubility	Coatings, medical devices, food industry	[104]
Poly(2-(dimethylamino)ethyl methacrylate)	PDMAEMA	Cationic polymer	Biocompatible	Nanoparticle formation, tissue engineering	[105]
Hydroxypropyl methyl cellulose	HPMC	Cellulose derivative	Gel-formingsoluble in cold water	Drug formulations, food industry,	[106]
Poly(lactide-co-glycolide acid)	PLGA	Biodegradable copolymer	Biodegradable adjustable degradation rate	drug delivery, tissue engineering	[107]
Monomethoxy (polyethylene glycol)-poly(d,l-lactide-co-glycolide)-poly(l-lysine)	mPEG-PLGA-PLL	Amphiphilic copolymer	Enhances solubility improves biocompatibility	Drug delivery systems	[108]
Polylactide	PLA	Biodegradable polymer	Biodegradable good mechanical properties	Sutures tissue engineering	[109]

TABLE 11.5
Polymeric Nanoparticle as Immunomodulatory Potential

S. No.	Polymer Used	Therapeutic Agent	Study (*In Vitro*/*In Vivo*)	Outcome	References
1	PEI	Mycophenolate mofetil (MMF)	*In vitro* and *ex vivo*	MSN-NPs ↑ MMF bioavailability in kidney ↓ Lymphatic cell numbers, lymph node size, and ↓ inflammation in kidney glomerular and tubulointerstitial regions	[110]
2	PEG-PLGA	Rapamycin	*In vitro*	↑bioavailability and sustained effects	[111]
3	Laden polymer	Cyclosporine	*In vivo*	↑bioavailability	[112]
4	Polyvinyl pyrrolidone (PVP)	Deflazacort (DEF)	*In vivo*	↑therapeutic effect of DEF	[113]
5	PDMAEMA and PLGA	Dexamethasone	*In vitro* and *in vivo*	↑effects of dexamethasone and ↑Treg expansion both *in vitro* and *in vivo*	[114]
6	HPMC	Doxycycline	*In vitro* and *in vivo*	↑antitumor activity	[115]
7	PLGA	Sparfloxacin and tacrolimus	*In vitro* and *in vivo*	↑antibacterial toward Gram-positive and Gram-negative bacteria, ↓ inflammation	[116]
8	mPEG-PLGA-PLL)	miR-125a	*In vivo*	↑biocompatibility and ↑circulatory time of miRNA in vivo	[117]
9	PEG-PLA	Tacrolimus	*In vivo*	↓toxic side effects of soluble vehicle, but also ↑survival time of rat with liver transplantation.	[118]
10	PEG-PLGA)	Cyclosporine A (CsA)	In vitro	↓T-cell proliferation and production of inflammatory cytokines, ↓reduced toxicity	[119]
11	PVP	Cyclosporin A (Cs-A)	In vitro CsA activity in vitro inhibitory effect on interleukin-2 (IL-2) production of Jurkat cells	↑ dispersion of poorly soluble drugs Cs-A	[120]
12	Polylactide	Cyclosporin A	In vitro	↑targeted delivery and ↓organ toxicity	[121]
13	PLGA	Monophosphoryl lipid A	In vivo BALB/c mice	↑systemic and mucosal immune responses	

11.8 CONCLUSION AND FUTURE PROSPECTS

The integration of synthetic polymers in immunomodulatory drug delivery represents a significant advancement in biomedical engineering and therapeutic science. In this chapter, we have explored the varied variety of synthetic polymers and their applications in improving the delivery and effectiveness of immunomodulatory drugs. The capability of customizing the physiochemical properties of these polymers allows precise control over how drugs are released, targeting specific cells or tissues, and minimizing unwanted immune responses. Innovations in polymer chemistry have led to the creation of sophisticated delivery systems such as nanoparticles, hydrogels, and micelles, each playing a role in finely tuning the modulation of the immune system. Despite notable progress, several challenges persist. These include thoroughly understanding the long-term biocompatibility and potential toxicity of these materials. Furthermore, there is an ongoing need to develop more efficient and economical manufacturing processes to facilitate the transition from research to clinical use. Future research directions should emphasize integrating synthetic polymers with emerging technologies like personalized medicine and advanced diagnostics. This integration promises to further enhance the therapeutic capabilities of these innovative drug delivery systems.

REFERENCES

1. Peng G, Fadeel B (2022) Understanding the bidirectional interactions between two-dimensional materials, microorganisms, and the immune system. *Advanced Drug Delivery Reviews* 188:114422.
2. Quiros-Roldan E, Sottini A, Natali PG, et al (2024) The impact of immune system aging on infectious diseases. *Microorganisms* 12(4):775.
3. Schirrmacher V (2023) Bone marrow: The central immune system. *Immuno* 3(3):289–329.
4. Mukherjee J, Das PK, Banerjee D, et al (2023) Immune system. In: *Textbook of Veterinary Physiology* (pp. 89–110). Singapore: Springer Nature Singapore.
5. Ferrari C, Barili V, Varchetta S, et al (2020) Immune mechanisms of viral clearance and disease pathogenesis during viral hepatitis. In: *The Liver: Biology and Pathobiology* (pp. 821–850).
6. Christophers E, Schröder JM (2022) Evolution of innate defense in human skin. *Experimental Dermatology* 31(3):304–311.
7. Huus KE, Ley RE (2021) Blowing hot and cold: Body temperature and the microbiome. *Msystems* 6(5):1128.

8. Keselowsky BG, Acharya A, Lewis JS (2020) Innate and adaptive immunity: The immune response to foreign materials. In: *Biomaterials Science* (pp. 747–775). Academic Press.
9. Soni SS, Rodell CB (2021) Polymeric materials for immune engineering: Molecular interaction to biomaterial design. *Acta Biomaterialia* 133:139–152.
10. Shah AA, Kumar V, Gupta A (2021) Immunomodulation via phyto active compounds: A promising therapy for future medical systems. *Journal of Medical Pharmaceutical and Allied Sciences* 10(2):120–125.
11. Parlakpinar H, Gunata M (2021) Transplantation and immunosuppression: A review of novel transplant-related immunosuppressant drugs. *Immunopharmacology and Immunotoxicology* 43(6):651–665.
12. Kumar P, Rai S, Verma S K, Prakash, et al. (2024) Classification, mode of action and uses of various immunomodulators. *In Immunomodulators and human health* (pp. 3–38). Singapore: Springer Nature Singapore.
13. Zhu X, Zhu J (2020) CD4 T helper cell subsets and related human immunological disorders. *International Journal of Molecular Sciences* 21(21):8011.
14. Petkar KC, Patil SM, Chavhan SS, et al (2021) An overview of nanocarrier-based adjuvants for vaccine delivery. *Pharmaceutics* 13(4):455.
15. Farouk SS, Rein JL (2020) The many faces of calcineurin inhibitor toxicity—What the FK? *Advances in Chronic Kidney Disease* 27(1):56–66.
16. Shakya A, Sarma R, Ghimire N, et al (2022) Therapeutic drug monitoring and toxicology of immunosuppressant. In: *Recent Advances in Therapeutic Drug Monitoring and Clinical Toxicology* (pp. 181–196). Cham: Springer International Publishing.
17. Wang D, Eisen HJ (2022) Mechanistic target of rapamycin (mTOR) inhibitors. In: *Pharmacology of Immunosuppression* (pp. 53–72). Cham: Springer International Publishing.
18. Galati A, Brown ES, Bove R, et al (2021) Glucocorticoids for therapeutic immunosuppression: Clinical pearls for the practicing neurologist. *Journal of the Neurological Sciences* 430:120004.
19. Scorer M, Golberg O, Harman K (2023) Cyclophosphamide. In: *Handbook of Systemic Drug Treatment in Dermatology* (pp. 112–119). CRC Press.
20. Akman AU (2021) Methotrexate induced hepatotoxicity and antioxidants. *Sabuncuoglu Serefeddin Health Sciences* 3(1):22–35.
21. Nash P, Lim I, Marabani M (2021) A comparison of Janus kinase inhibitor safety in rheumatoid arthritis. *International Journal of Rheumatic Diseases* 24:3–14.
22. Baldo BA (2022) Immune-and non-immune-mediated adverse effects of monoclonal antibody therapy: A survey of 110 approved antibodies. *Antibodies* 11(1):17.
23. Bagheri A, Moezzi SM, Mosaddeghi P, et al (2021) Interferon-inducer antivirals: Potential candidates to combat COVID-19. *International Immunopharmacology* 91:107245.
24. Benna M, Guy JB, Bosacki C, et al (2020) Chemoradiation and granulocyte-colony or granulocyte macrophage-colony stimulating factors (G-CSF or GM-CSF): Time to think out of the box? *The British Journal of Radiology* 93(1109):20190147.
25. Charliński G, Vesole DH, Jurczyszyn A (2022) Principles of prevention and management of adverse events of immunomodulatory drugs in the treatment of multiple myeloma. *Biuletyn Polskiego Towarzystwa Onkologicznego Nowotwory* 7(4):270–280.
26. Jeon D, Hill E, McNeel DG (2024) Toll-like receptor agonists as cancer vaccine adjuvants. Human Vaccines & Immunotherapeutics 20(1):2297453.
27. Akarsu S, Polat F (2024) Possible cutaneous adverse effects of anti-infective vaccinations. In: *Frontiers in Anti-Infective Drug Discovery*: Volume 10 (pp. 116).
28. Roelfsema F, Yang R, Veldhuis JD (2021) Interleukin-2 transiently inhibits pulsatile growth hormone secretion in young but not older healthy men. *The Journal of Clinical Endocrinology & Metabolism* 106(10):2855–2864.
29. Gherardi RK, Aouizerate J, Cadusseau J, et al (2016) Aluminum adjuvants of vaccines injected into the muscle: Normal fate, pathology and associated disease. *Morphologie* 100(329):85–94.
30. Milicic A, Reinke S, Fergusson J, et al (2022) Adjuvants, immunomodulators, and adaptogens. In: *Vaccinology and Methods in Vaccine Research* (pp. 223–280). Academic Press.
31. Cui Y, Ho M, Hu Y, et al (2024) Vaccine adjuvants: Current status, research and development, licensing, and future opportunities. *Journal of Materials Chemistry B* 12(17):4118–4137.
32. Luchner M, Reinke S, Milicic A (2021) TLR agonists as vaccine adjuvants targeting cancer and infectious diseases. *Pharmaceutics* 13(2):142.
33. Feng X, Xu W, Li Z, et al (2019) Immunomodulatory nanosystems. *Advanced Science* 6(17):1900101.
34. Adams DH, Sanchez-Fueyo A, Samuel D (2015) From immunosuppression to tolerance. *Journal of Hepatology* 62(1):S185.
35. Thell K, Hellinger R, Schabbauer G, et al (2014) Immunosuppressive peptides and their therapeutic applications. *Drug Discovery Today* 19(5):645–653.
36. Andersen MH (2014) The targeting of immunosuppressive mechanisms in hematological malignancies. *Leukemia* 28(9):1784–1792.
37. Carson WF, Cavassani KA, Dou Y, et al (2011) Epigenetic regulation of immune cell functions during post-septic immunosuppression. *Epigenetics* 6(3):273–283.
38. Liu D, Gao S, Zhai Y, et al (2022) Research progress of tumor targeted drug delivery based on PD-1/PD-L1. *International Journal of Pharmaceutics* 616:121527.
39. Abed S, Turner R, Serniuck N, et al (2021) Cell-specific drug targeting in the lung. *Biochemical Pharmacology* 190:114577.
40. Isazadeh H, Oruji F, Shabani S, et al (2023) Advances in siRNA delivery approaches in cancer therapy: challenges and opportunities. *Molecular Biology Reports* 50(11):9529–9543.
41. Issa N, Kukla A, Ibrahim HN (2013) Calcineurin inhibitor nephrotoxicity: A review and perspective of the evidence. *American Journal of Nephrology* 37(6):602–612.
42. Naesens M, Kuypers DR, Sarwal M (2009) Calcineurin inhibitor nephrotoxicity. *Clinical Journal of the American Society of Nephrology* 4(2):481–508.
43. Marcén R (2009) Immunosuppressive drugs in kidney transplantation: Impact on patient survival, and incidence of cardiovascular disease, malignancy and infection. *Drugs* 69:2227–2243.
44. Miller LW (2002) Cardiovascular toxicities of immunosuppressive agents. *American Journal of Transplantation* 2(9):807–818.
45. Cronin II DC, Faust TW, Brady L, et al (2000) Modern immunosuppression. *Clinics in Liver Disease* 4(3):619–655.

46. Fishman JA (2013) Opportunistic infections—coming to the limits of immunosuppression? *Cold Spring Harbor Perspectives in Medicine* 3(10).
47. Bamoulid J, Staeck O, Halleck F, et al (2015) The need for minimization strategies: Current problems of immunosuppression. *Transplant International* 28(8):891–900.
48. Paul S, El Bethel Hmar HK (2020) Strengthening immunity with immunostimulants: A. *Current Trends in Pharmaceutical Research* 7(1):35–64.
49. Bruce TJ, Brown ML (2017) A review of immune system components, cytokines, and immunostimulants in cultured finfish species. *Open Journal of Animal Sciences* 7(03):267.
50. Krensky AM, Vincenti F, Bennett WM (2006) Immunosuppressants, tolerogens, and immunostimulants. In: *Goodman & Gilman's The Pharmacological Basis of Therapeutics*. 11th ed. New York: McGraw-Hill Companies, Inc, pp. 1405–1431.
51. Kosinsky Y, Dovedi SJ, Peskov K, et al (2018) Radiation and PD-(L)1 treatment combinations: Immune response and dose optimization via a predictive systems model. *Journal for Immunotherapy of Cancer* 6:1–15.
52. Zhuang J, Holay M, Park JH, et al (2019) Nanoparticle delivery of immunostimulatory agents for cancer immunotherapy. *Theranostics* 9(25):7826.
53. Carballido JM, Regairaz C, Rauld C, et al (2020) The emerging jamboree of transformative therapies for autoimmune diseases. *Frontiers in Immunology* 11:472.
54. Pisetsky DS (2023) Pathogenesis of autoimmune disease. *Nature Reviews Nephrology* 19(8):509–524.
55. Tang XD, Ji TT, Dong JR, et al (2021) Pathogenesis and treatment of cytokine storm induced by infectious diseases. *International Journal of Molecular Sciences* 22(23):13009.
56. Song P, Li W, Xie J, et al (2020) Cytokine storm induced by SARS-CoV-2. *Clinica Chimica Acta* 509:280–287.
57. Stakheyeva M, Riabov V, Mitrofanova I, et al (2017) Role of the immune component of tumor microenvironment in the efficiency of cancer treatment: Perspectives for the personalized therapy. *Current Pharmaceutical Design* 23(32):4807–4826.
58. Stergiopoulos GM, Iankov I, Galanis E (2024) Personalizing oncolytic immunovirotherapy approaches. *Molecular Diagnosis & Therapy* 28(2):153–168.
59. Wu D, Lewis ED, Pae M, et al (2019) Nutritional modulation of immune function: Analysis of evidence, mechanisms, and clinical relevance. *Frontiers in Immunology* 9:3160.
60. Fridman WH, Zitvogel L, Sautès–Fridman C et al (2017) The immune contexture in cancer prognosis and treatment. *Nature Reviews Clinical Oncology* 14(12):717–734.
61. Nash A, Aghlara-Fotovat S, Hernandez A, et al (2021) Clinical translation of immunomodulatory therapeutics. *Advanced Drug Delivery Reviews* 176:113896.
62. Feng X, Xu W, Li Z, et al (2019) Immunomodulatory nanosystems. Advanced *Science* 6(17):1900101.
63. Zhuang J, Holay M, Park JH, et al (2019) Nanoparticle delivery of immunostimulatory agents for cancer immunotherapy. *Theranostics* 9(25):7826.
64. Xie YQ, Wei L, Tang L (2018) Immunoengineering with biomaterials for enhanced cancer immunotherapy. *Wiley Interdisciplinary Reviews: Nanomedicine and Nanobiotechnology* 10(4):E1506.
65. Uhrich KE, Cannizzaro SM, Langer RS, et al (1999) Polymeric systems for controlled drug release. *Chemical Reviews* 99(11):3181–3198.
66. Pantuzza LL, Ceccato MDGB, Silveira MR, et al (2017) Association between medication regimen complexity and pharmacotherapy adherence: A systematic review. *European Journal of Clinical Pharmacology* 73:1475–1489.
67. DiMatteo MR, Haskard-Zolnierek KB, Martin LR (2012) Improving patient adherence: A three-factor model to guide practice. *Health Psychology Review* 6(1):74–91.
68. Davis ME, Chen Z, Shin DM (2008) Nanoparticle therapeutics: An emerging treatment modality for cancer. *Nature Reviews Drug Discovery* 7(9):771–782.
69. Rowe RC, Sheskey PJ, Owen SC (2005) *Handbook of Pharmaceutical Excipients*. 5th ed. Pharmaceutical Press; American Pharmacists Association; Grayslake, IL: Washington, D.C., p. 850.
70. Twaites B, Alarcon DLH, Alexander C (2005) Synthetic polymers as drugs and therapeutics. *Journal of Materials Chemistry* 15(4):441
71. Sung YK, Kim SW (2020) Recent advances in polymeric drug delivery systems. *Biomater Res* 24:12.
72. Liechty WB, Kryscio DR, Slaughter BV, et al (2010) Polymers for drug delivery systems. *Annual Review of Chemical and Biomolecular Engineering* 1:149–173.
73. Ekinci M, Öztürk AA, Santos-Oliveira R, et al (2022) The use of Lamivudine-loaded PLGA nanoparticles in the diagnosis of lung cancer: Preparation, characterization, radiolabeling with 99mTc and cell binding. *Journal of Drug Delivery Science and Technology* 69:103139.
74. Neuse EW (2008) Synthetic polymers as drug-delivery vehicles in medicine. *Metal-Based Drugs* 2008: 469531.
75. Diwan M, Park TG (2003) Stabilization of recombinant interferon-alpha by pegylation for encapsulation in PLGA microspheres. *International Journal of Pharmaceutics* 252:111–122.
76. Zentner GM, Rathi R, Shih C, et al (2001) Biodegradable block copolymers for delivery of proteins and water-insoluble drugs. *Journal of Controlled Release* 72:203–215.
77. Jeong B, Choi YK, Bae YH, et al (1999) New biodegradable polymers for injectable drug delivery systems. *Journal of Controlled Release* 62:109–114.
78. Kabanov AV, Alakhov VY (2002) Pluronic block copolymers in drug delivery: From micellar nano containers to biological response modifiers. *Critical Reviews™ in Therapeutic Drug Carrier Systems* 19:1–72.
79. Zheng CH, Liang WQ, Yu HY, et al(2004) Evaluation of different methods to determine the loading of proteins in PLGA microspheres. *Pharmazie* 59:232–233.
80. Johnson MT, Komane LL, N'Da DD, et al (2005) Polymeric drug carriers functionalized with pairwise arranged hydroxyl and/or carboxyl groups for platinum chelation. *Journal of Applied Polymer Science* 96(1):10–19.
81. Lee S, Youn YS, Lee SH, et al (2006) PEGylated glucagon-like peptide-1 displays preserved effects on insulin release in isolated pancreatic islets and improved biological activity in db/db mice. *Diabetologia* 49:1608–1611.
82. Meirim MG, Neuse EW, Caldwell G (1998) Water-soluble polymer-ferrocene conjugates based on polyamide carriers containing intrachain-type secondary amine functions as binding sites. *Journal of Inorganic and Organometallic Polymers* 8(4):225–236.
83. Yu L, Zhang Z, Zhang H, et al (2010) Biodegradability and biocompatibility of thermoreversible hydrogels formed from mixing a sol and a precipitate of block copolymers in water. *Biomacromolecules* 11:2169–2178.
84. Prausnitz MR, Langer R (2008) Transdermal drug delivery. *Nature Biotechnology* 26(11):1261–1268.
85. Xue X, Li D, Yu J, et al (2013) Phenyl linker-induced dense PEG conformation improves the efficacy of C-terminally monoPEGylated staphylokinase. *Biomacromolecules* 14:331–341.

86. Sosnik A, Sefton MV (2005) Semi-synthetic collagen/poloxamine matrices for tissue engineering. *Biomaterials* 26:7425–7435.
87. Říhová B (2002) Immunomodulating activities of soluble synthetic polymer-bound drugs. *Advanced Drug Delivery Reviews* 54(5):653–674.
88. Ben-Akiva E, Witte SE, Meyer RA, Rhodes KR, Green JJ (2019) Polymeric micro-and nanoparticles for immune modulation. *Biomaterials Science* 7(1):14–30.
89. Lee NK, Kim SN, Park CG (2021) Immune cell targeting nanoparticles: A review. *Biomaterials Research* 25(1):44.
90. Soni SS, Rodell CB (2021) Polymeric materials for immune engineering: Molecular interaction to biomaterial design. *Acta Biomaterialia* 133:139–152.
91. Baljon JJ, Wilson JT (2022) Bioinspired vaccines to enhance MHC class-I antigen cross-presentation. *Current Opinion in Immunology* 77:102215.
92. Soares E, Borges O (2022) Polymer-based adjuvant requirements for vaccines. In: *System Vaccinology* (pp. 177–204). Academic Press.
93. Nooraei S, Sarkar Lotfabadi A, Akbarzadehmoallemkolaei M, Rezaei N (2023) Immunogenicity of different types of adjuvants and nano-adjuvants in veterinary vaccines: A comprehensive review. *Vaccines* 11(2):453.
94. Díaz-Dinamarca DA, Salazar ML, Castillo BN, et al (2022) Protein-based adjuvants for vaccines as immunomodulators of the innate and adaptive immune response: Current knowledge, challenges, and future opportunities. *Pharmaceutics* 14(8):1671.
95. Ibeanu N, Egbu R, Onyekuru L, et al (2020) Injectables and depots to prolong drug action of proteins and peptides. *Pharmaceutics* 12(10):999.
96. Luzuriaga MA, Shahrivarkevishahi A, Herbert FC, et al (2021) Biomaterials and nanomaterials for sustained release vaccine delivery. *Wiley Interdisciplinary Reviews: Nanomedicine and Nanobiotechnology* 13(6):E1735.
97. Roth GA, Picece VC, Ou BS, et al (2022) Designing spatial and temporal control of vaccine responses. *Nature Reviews Materials* 7(3):174–195.
98. Pacifici N, Bolandparvaz A, Lewis JS (2020) Stimuli-responsive biomaterials for vaccines and immunotherapeutic applications. *Advanced Therapeutics* 3(11):2000129.
99. Ulery BD, Nair LS, Laurencin CT (2011) Biomedical applications of biodegradable polymers. *Journal of Polymer Science Part B: Polymer Physics* 49(12):832–864.
100. Bergmann-Leitner ES, Leitner WW (2014) Adjuvants in the driver's seat: how magnitude, type, fine specificity, and longevity of immune responses are driven by distinct classes of immune potentiators. *Vaccines* 2(2):252–296.
101. Ogra PL, Faden H, Welliver RC (2001) Vaccination strategies for mucosal immune responses. *Clinical Microbiology Reviews* 14(2):430–445.
102. Wang X, Niu D, Hu C, et al (2015) Polyethyleneimine-based nanocarriers for gene delivery. *Current Pharmaceutical Design* 21(42):6140–6156.
103. D'souza AA, Shegokar R (2016) Polyethylene glycol (PEG): A versatile polymer for pharmaceutical applications. *Expert Opinion on Drug Delivery* 13(9):1257–1275.
104. Teodorescu M, Bercea M (2015) Poly (vinylpyrrolidone)–a versatile polymer for biomedical and beyond medical applications. *Polymer-Plastics Technology and Engineering* 54(9):923–943.
105. Farshbaf M, Davaran S, Zarebkohan A, et al (2018) Significant role of cationic polymers in drug delivery systems. *Artificial Cells, Nanomedicine, and Biotechnology* 46(8):1872–1891.
106. Vlaia L, Coneac G, Olariu I, et al (2016) Cellulose-derivatives-based hydrogels as vehicles for dermal and transdermal drug delivery. *Emerging Concepts in Analysis and Applications of Hydrogels* 2:64.
107. Avgoustakis K (2005) Polylactic-co-glycolic acid (PLGA). In: *Encyclopedia of Biomaterials and Biomedical Engineering*, 1(1):1–11.
108. He Z, Sun Y, Cao J, et al (2016) Degradation behavior and biosafety studies of the mPEG–PLGA–PLL copolymer. *Physical Chemistry Chemical Physics* 18(17):11986–11999.
109. Benatti ACB, Pattaro AF, Rodrigues AA, et al (2019) Bioreabsorbable polymers for tissue engineering: PLA, PGA, and their copolymers. In: *Materials for Biomedical Engineering* (pp. 83–116). Elsevier.
110. Wu W, Liu R, Guo J, et al (2024) Modulation of immunosuppressive effect of rapamycin via microfluidic encapsulation within PEG-PLGA nanoparticles. *Journal of Biomaterials Applications* 38(7):821–833.
111. Ganugula R, Babalola KT, Heyns IM, Arora M, Agarwal SK, Mohan C, Kumar MR (2024) Lymph node targeting of cyclosporine ameliorates ocular manifestations in a mouse model of systemic lupus erythematosus (SLE) via PD-L1. *Nano Today* 57:102359.
112. Abdelkader DH, Belal AM, Elkordy EA, Sarhan NI, Essa EA (2024) Fabrication and in-vivo evaluation of polyvinyl pyrrolidone/poloxamer 188 hybrid nanofibers of deflazacort. *International Journal of Pharmaceutics* 655:123997.
113. Khiewkamrop P, Kaewraemruaen C, Manipuntee C, Saengruengrit C, Insin N, Leelahavanichkul A, Ritprajak P (2023) Immunosuppressive polymeric nanoparticles targeting dendritic cells alleviate lupus disease in Fcgr2b-/- mice by mediating antigen-specific immune tolerance. *International Journal of Molecular Sciences* 24(9):8313.
114. Alshaman R, Alattar A, El-Sayed RM, Gardouh AR, Elshaer RE, Elkazaz AY, Zaitone SA (2022) Formulation and characterization of doxycycline-loaded polymeric nanoparticles for testing antitumor/antiangiogenic action in experimental colon cancer in mice. *Nanomaterials* 12(5):857.
115. Yang Y, Ding Y, Fan B, Wang Y, Mao Z, Wang W, Wu J (2020) Inflammation-targeting polymeric nanoparticles deliver sparfloxacin and tacrolimus for combating acute lung sepsis. *Journal of Controlled Release* 321:463–474.
116. Zhang J, Chen C, Fu H, Yu J, Sun Y, Huang H, Duan Y (2020) MicroRNA-125a-loaded polymeric nanoparticles alleviate systemic lupus erythematosus by restoring effector/regulatory T cells balance. *ACS Nano* 14(4):4414–4429.
117. Xu W, Ling P, Zhang T (2014) Toward immunosuppressive effects on liver transplantation in rat model: tacrolimus loaded poly(ethylene glycol)-poly(D,L-lactide) nanoparticle with longer survival time. *International Journal of Pharmaceutics* 460(1–2):173–180.
118. Tang L, Azzi J, Kwon M, Mounayar M, Tong R, Yin Q, Cheng J (2012) Immunosuppressive activity of size-controlled PEG-PLGA nanoparticles containing encapsulated cyclosporine A. *Journal of Transplantation* 2012:896141.
119. Takebe G, Takagi T, Suzuki M, Hiramatsu M (2011) Preparation of polymeric nanoparticles of cyclosporin A using infrared pulsed laser. *International Journal of Pharmaceutics* 414(1–2):244–250.
120. Azzi J, Tang L, Moore R, Tong R, El Haddad N, Akiyoshi T, Abdi R (2010) Polylactide-cyclosporin A nanoparticles for targeted immunosuppression. *The FASEB Journal* 24(10):3927.
121. Sarti F, Perera G, Hintzen F, Kotti K, Karageorgiou V, Kammona O, Bernkop-Schnürch A (2011) In vivo evidence of oral vaccination with PLGA nanoparticles containing the immunostimulant monophosphoryl lipid A. *Biomaterials* 32(16):4052–4057.

12 Synthetic Polymers for Designing 3D-Printed Drug Delivery Systems

Vimal Arora and Neelam Poonia
University Institute of Pharma Sciences, Chandigarh University, Mohali, India

Sanjeev Mittal
Swami Vivekanand College of Pharmacy, Banur, Patiala (Punjab), India

12.1 INTRODUCTION

Inflammatory diseases encompass a variety of diseases – *viz.*, rheumatoid arthritis, inflammatory bowel disease, and various autoimmune conditions, present significant challenges in terms of treatment efficacy and patient compliance [1]. The management of inflammatory diseases poses inherent challenges, including the need for precise drug delivery to inflamed tissues, variability in patient responses, and the requirement for complex multidrug regimens [2]. Traditional drug delivery methods often fall short in addressing these issues, leading to suboptimal treatment outcomes and increased risk of side effects [3]. However, the three-dimensional (3D) printing technology provides a transformative solution to above mentioned challenges via enabling the fabrication of highly precise and customizable drug delivery systems, 3D printing permits creation of patient-specific formulations that release therapeutic agents with controlled kinetics [4]. This approach minimizes systemic exposure, enhances treatment efficacy, and provides the capability to incorporate multiple drugs within a single dosage form [5, 6]. Figure 12.1 illustrates the disease benefits associated with the utilization of 3D printing in development of delivery vehicles.

The flexibility of this novel technique, coupled with its rapid prototyping capabilities, not only addresses the inherent problems in inflammatory disease management but also opens new avenues for the design of tailored and effective therapeutic delivery systems. Moreover, robust, precise manipulation and layer-by-layer deposition make it possible to fabricate complex drug delivery devices with customized topologies [7]. With the use of 3D printing, researchers can create customized implants, scaffolds, patches, microparticles, etc., that surpass the constraints of conventional manufacturing techniques [8]. The number of publications in PubMed has continuously increased in the last two decades (Figure 12.2).

The fusion of synthetic polymers with 3D technique in design of delivery vehicles signifies a paradigm shift in the pharmaceutical industry [9]. Synthetic polymers and 3D printing work together synergistically to overcome obstacles of conventional manufacturing approaches and offer a flexible platform that can be customized to meet the needs of particular patients and drugs [10]. It helps in improving accuracy and creates new opportunities for medication distribution customization, which is a big step in the direction of more individualized and efficient pharmaceutical solutions [11]. Formulation scientists can employ different synthetic polymers such as Gelucire®, poly(glycerol sebacate), polyurethane acrylate, and polylactic acid (PLA) for developing biodegradable, biocompatible and stimuli-responsive dosage forms [12]. They are the excellent choice for developing complicated carrier systems owing to this potential to entrap a variety of medicinal compounds. Drug release kinetics, stability, and bioavailability could be easily tailored by judiciously choosing and combining polymers to maximize therapeutic effects [13].

This chapter deals with 3D printed drug delivery systems, emphasizing the critical role that synthetic polymers play in determining the formulation and functionality of these systems. Moreover, the opportunities and difficulties that come with bringing these cutting-edge technologies from the lab to the clinic while navigating the constantly changing regulatory environment will also be covered [14].

12.2 PATHOGENESIS OF INFLAMMATORY DISORDERS

Various factors are involved in the pathogenesis of inflammation, such as infections, foreign irritants, and various inflammatory mediators. The inherent immune system, involving immune cells (dendritic cells, mast cells, neutrophils, and lymphocytes) plays a pivotal role in inflammatory responses [15]. Additionally, stromal cells (fibroblasts, epithelial, endothelial) are implicated in inflammation [16]. Various stimuli activate inflammatory pathways, such as MAPK, NFκ-B, and JAK-signal transducer and activator of transcription, resulting in the secretion of inflammatory

DOI: 10.1201/9781003477389-12

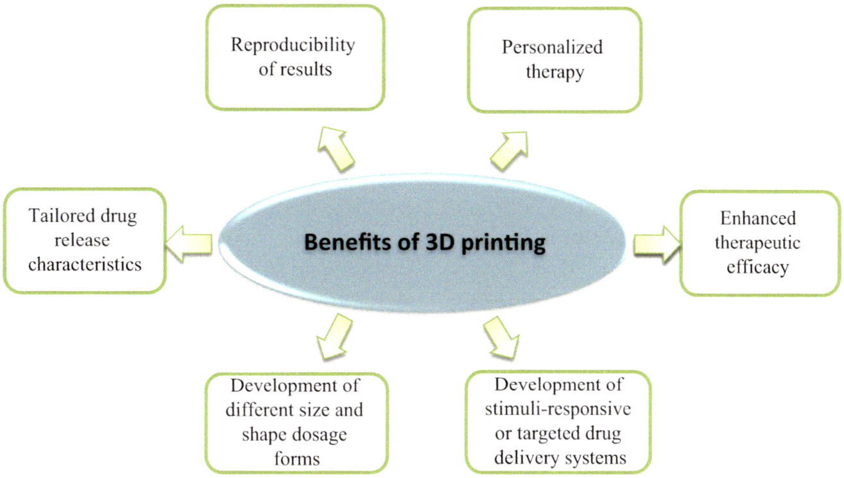

FIGURE 12.1 Various advantages offered by 3D printing.

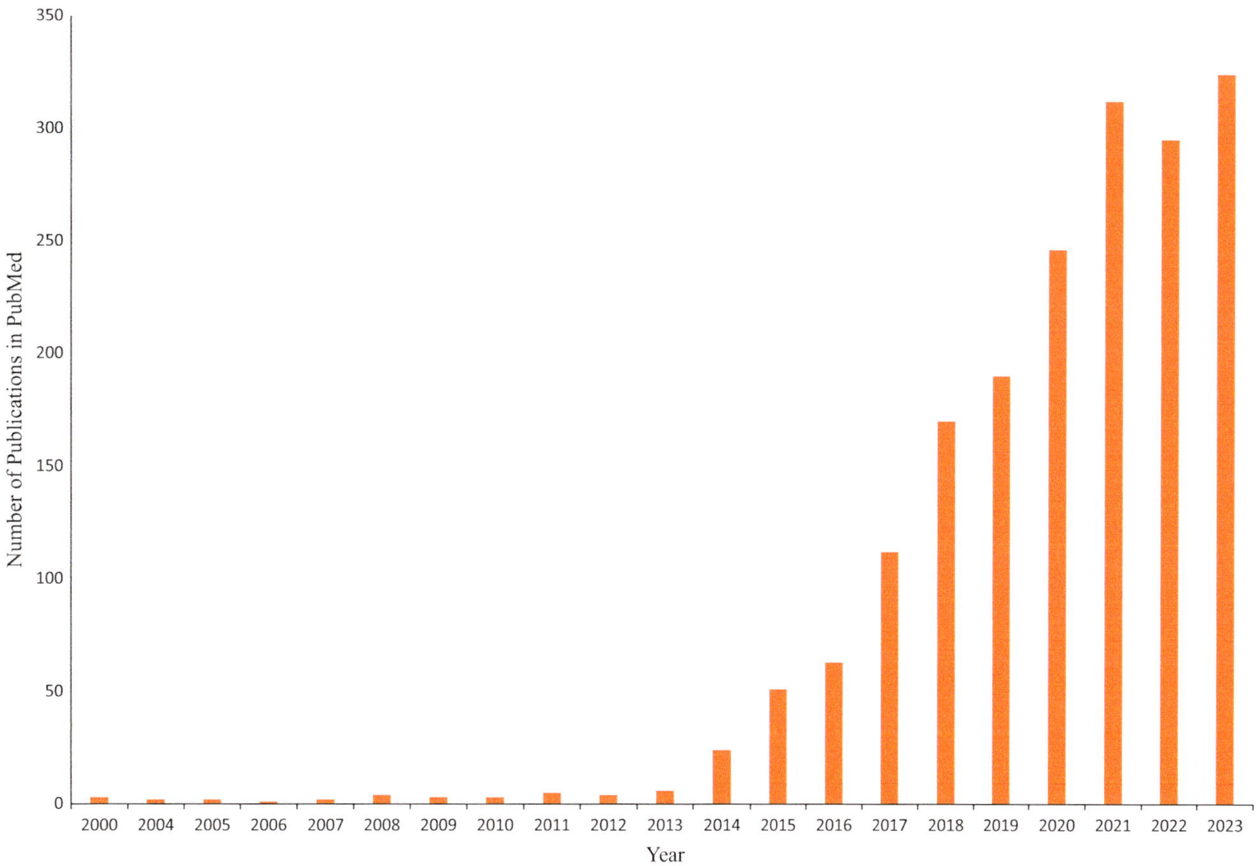

FIGURE 12.2 Trend in publications indexed in PubMed related to the application of "3D technique in drug delivery."

cytokines, interleukins, and enzymes. Reactive oxygen species (ROS) – namely, hydroxyl/oxygen radicals, H_2O_2, and hypochlorous acid – have detrimental effects on cellular metabolism and tissue injury. Inflammatory processes contribute significantly to cardiovascular disorders, rheumatoid arthritis, inflammatory bowel diseases, cancer, etc. [17, 18].

12.2.1 Rheumatoid Arthritis

It is a multifactorial autoimmune ailment marked by chronic inflammation of the synovial membrane within joints and subsequent physical impairment [19]. Genetic susceptibility, particularly involving the HLA-DRB1 gene, in conjunction with environmental triggers such as smoking, contributes to the breakdown of immune tolerance

[20]. Scientific evidence suggests that the activation of cluster of differentiation (CD4+) T cells by antigens leads to an augmented production of cytokines [21]. This cascade effect intensifies hyperimmune responses, inflammatory reactions, and the activity of adhesion molecules. B cells produce autoantibodies, including anti-citrullinated protein antibodies and rheumatoid factor. The ensuing immune complex formation exacerbates synovial inflammation [22]. Key inflammatory mediators – *viz*., TNF-α, IL-1, IL-6 – play major roles in perpetuating inflammation, recruiting immune cells, and activating osteoclasts, thereby initiating joint damage. The synovium becomes infiltrated with macrophages, T cells, and B cells, leading to pannus formation and subsequent erosion of cartilage and bone [23]. Enzymes like matrix metalloproteinases (MMPs) further contribute to joint destruction. Clinical investigations have substantiated the reduction of NAD+ precursors, such as nicotinamide and tryptophan, in individuals suffering with this disease [24].

12.2.2 Inflammatory Bowel Diseases

This disease represents a persistent inflammatory condition affecting the gastrointestinal tract termed as ulcerative colitis or Crohn's disorder characterized by inflammation in the large intestine and rectum, or digestive tract lining, respectively [25]. The infiltration of pathogens into the lamina propria *via* ulcerated mucosa results in the perturbation of the intestinal membrane, triggering the activation of immune cells, notably T cells [26]. This activation leads to an upregulation of cytokines and interleukin-12, originating a chronic inflammatory state. Moreover, adhesion molecules – *viz*., ICAM, VCAM, and MMP – are also associated with disease pathogenesis. Elevated levels of these molecules in the mucosa are linked to cell migration and a subsequent inflammatory state [27]. MMPs responsible for extracellular matrix cleavage and disruption of intestinal barrier, contribute significantly to disease progression. MMP-9 hinders the differentiation of goblet cells and impedes fibrosis, whereas MMP-13 induces the generation of TNF-α and endostatin, further exacerbating the inflammatory response [28, 29].

12.2.3 Cancer

The pathogenesis of cancer manifests as a complicated process involving unregulated proliferation of cells. This aberrant cellular behavior is often initiated by genetic mutations, which can arise due to various factors, including exposure to carcinogens, genetic predisposition, and altered DNA repair mechanisms [30, 31]. Oncogenes (enhance cell proliferation) and tumor suppressor genes (inhibit cell proliferation) play critical roles in regulating cell proliferation [32]. Additionally, chronic inflammation has been recognized as a key contributor to the initiation and growth of cancer [33]. Inflammatory mediators released in response to infections, tissue damage, or other stimuli can create a microenvironment that enhance the survival and growth of cancer cells [34]. Angiogenesis, the formation of new blood vessels, is stimulated, ensuring a blood supply that supports the growing tumor. The tumor microenvironment, comprising immune cells, fibroblasts, and extracellular matrix components, further influences cancer cell behavior [35]. Evading the immune system, resisting cell death (apoptosis), and the capability to invade surrounding tissues and metastasize to other sites are hallmarks of cancer progression [36].

12.2.4 Respiratory Diseases

The pathogenesis of respiratory disorders, encompassing conditions – namely, chronic obstructive pulmonary disease (COPD) and asthma – are complex process dictated by genetic, environmental, and immunological factors [37]. Genetic susceptibility, marked by specific gene variants, interacts with environmental exposures, including allergens and pollutants, triggering immune system activation [38]. T cells, notably Th2 cells, initiate an inflammatory response, releasing various cytokines. Inflammatory mediators – namely, TNF-α, IL-1, IL-6, IL-8, and eicosanoids – play crucial roles in bronchoconstriction and airway inflammation [39]. Mast cell activation contributes to histamine release, aggravating bronchoconstriction. Prolonged inflammation leads to changes in the airways, characterized by structural alterations including increased smooth muscle mass, mucus hypersecretion, and fibrosis [40]. Neutrophil and macrophage infiltration, observed prominently in COPD, further aggravates tissue damage. A comprehensive understanding of these immunological and molecular processes is necessary for designing targeted therapeutic interventions aimed at mitigating inflammation, preserving airway function, and improving clinical outcomes in respiratory diseases [41].

12.2.5 Cardiovascular Diseases

Numerous clinical investigations have revealed an association between inflammatory biomarkers – *viz*., high-sensitivity C-reactive protein – correlating with prognosis of heart disorders [42]. Furthermore, a proinflammatory condition exhibits features of persistent low-grade inflammation, contributing to heart related issues such as myocardial infarction and stroke [43]. The redox-responsive cofactor tetrahydrobiopterin (BH4) also exerts deleterious effects on cardiovascular disease. BH4 leads to production of endothelial nitric oxide synthase (eNOS). A decline in BH4 levels negatively impacts eNOS, leading to vasoconstriction [44]. Consequently, BH4 therapy proves effective in partially restoring endothelium-dependent coronary flow [45]. Moreover, adhesion molecules play significant part in cardiovascular disease by inducing leukocyte infiltration, fostering endothelial cell proliferation, among other processes, thereby increasing the likelihood of adverse cardiac events. Additionally, conditions such as diabetes, hypertension, and hypercholesterolemia elevate endothelial-associated Xanthine oxidoreductase activity, leading to the generation of free radicals derived from uric acid and eliciting inflammatory responses, consequently promoting atherosclerosis [46].

12.3 PRESENT CLINICAL REGIMENS FOR ANTI-INFLAMMATORY DISORDERS

The application of salicylate-containing plant extracts for alleviating pain, inflammation, and fever has been documented. Approximately 150 years ago, Felix Hoffman synthesized aspirin by acetylating salicylic acid. Aspirin acts as an inhibitor for cyclooxygenase (COX) enzymes responsible for the synthesis of different inflammatory mediators. Various nonsteroidal anti-inflammatory drugs, such as ibuprofen, serve as frontline agents to alleviate pain and control inflammation by targeting prostaglandin production [47].

Additionally, synthetic derivatives of natural cortisol find widespread use in treating various inflammatory ailments. Despite associated side effects, glucocorticoids play a pivotal role in management of inflammation. Nevertheless, the ongoing challenge for pharmaceutical chemists is to develop more efficacious therapeutic agents with minimal side effects for addressing acute inflammatory signs and symptoms, along with long-term consequences of chronic inflammation [48, 49].

A new class of drugs, disease-modifying antirheumatic drugs (DMARDs), exemplified by methotrexate, aim to modify the underlying disease processes and impede progression, particularly in autoimmune disorders like rheumatoid arthritis. Biologics, including TNF inhibitors and other targeted agents, selectively intervene in immune pathways implicated in inflammation [50, 51]. Immunosuppressants like azathioprine find utility in mitigating overactive immune responses. Emerging anti-interleukin therapies, such as IL-6 or IL-17 inhibitors, present targeted options for conditions like rheumatoid arthritis and psoriasis. Monoclonal antibodies, exemplified by rituximab, target specific immune cells in disorders like rheumatoid arthritis and vasculitis [52].

Clinical regimens for several other inflammatory disorders encompass a diverse array of therapeutic strategies tailored to the distinctive nature of each diseased condition. For instance, in cardiovascular diseases, statins, antiplatelet agents, and drugs targeting the renin-angiotensin system play pivotal roles in managing atherosclerosis and preventing cardiovascular events [53]. Inflammatory bowel diseases necessitate the administration of 5-aminosalicylates, corticosteroids, immunomodulators, and biological agents, aiming to modulate the immune response and maintain remission. Respiratory disorders like asthma and COPD are managed with inhaled corticosteroids, bronchodilators, and leukotriene modifiers, targeting airway inflammation and bronchoconstriction [54, 55]. Cancer treatment modalities range from chemotherapy, which targets rapidly dividing cells, to targeted therapies and immunotherapy that selectively address specific cancer-related molecules and enhance the immune response against malignant cells [56]. These evolving regimens highlight the importance of individualized care, considering factors such as diagnosis, disease stage, and patient characteristics with regular monitoring to optimize therapeutic efficacy.

12.4 FUNDAMENTALS OF SYNTHETIC POLYMERS

Synthetic polymers find wide applications in biomedical applications and are formed through polymerization reactions. They serve a major role in 3D printing, contributing to the creation of diverse objects and structures [57]. These polymers are artificially synthesized in laboratories and offer a range of desirable properties for additive manufacturing processes. This approach permits the fabrication of diverse delivery vehicles with specific physicochemical properties to address the challenges associated with inflammatory disorders [58]. PLA and polyethylene glycol (PEG) are frequently employed owing to their biocompatibility and biodegradability characteristics. Polycaprolactone (PCL) provides a slow degradation rate, making it suitable for sustained drug release. Polyvinyl alcohol (PVA), is hydrophilic in nature and acts as supportive material in the development of complex drug delivery vehicles. Additionally, the biodegradable and versatile nature of poly(lactic-co-glycolic acid) (PLGA) facilitates encapsulation of a number of drugs. These synthetic polymers enable the development of precisely controlled drug delivery vehicles, offering potential advancements in personalized medicine and targeted therapeutic applications [59, 60].

12.5 3D PRINTING TECHNIQUES

Additive manufacturing has been investigated in industrial research for the last few decades. Basic procedure for 3D printing is depicted in Figure 12.3. Substantial progress has been made in refining fundamental printing techniques, exploring new materials, and enhancing postprocessing methodologies. Notably, the range of printable materials has expanded significantly, enabling researchers to 3D print diverse substances, including viable cells, polymers with adjustable modulus, and even materials meeting the stringent standards for spacecraft applications, etc. [61].

Various 3D printing technologies utilize distinct materials, encompassing polymers, ceramics, metals, and composites [62]. Fused filament fabrication method highly employed for 3D printing predominantly employs polymers and polymer-based composites. Notably, polymers are highly used in most of 3D printing modalities because of their widespread applicability in generating intricate structures and functional components, depicting their crucial role in advancing additive manufacturing capabilities [63].

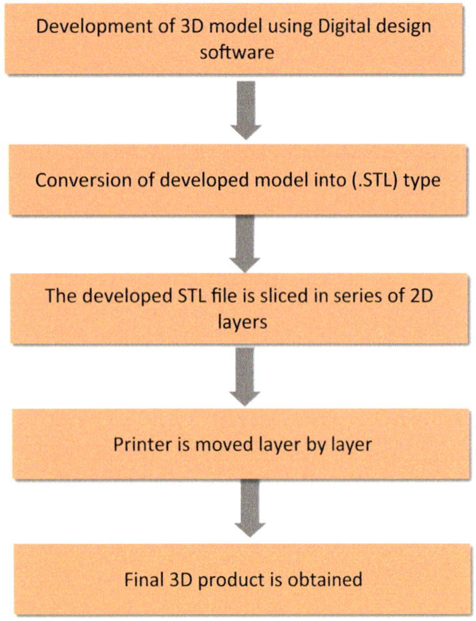

FIGURE 12.3 Basic procedure for 3D printing.

3D printing technique employs development of a 3D model *via* CAD software followed by conversion of that model into file encoded in STL (Surface Tessellation Language) format. The STL file undergoes a segmentation process wherein it is sliced into two-dimensional (2D) layers. These layers serve as a blueprint for arranging the printing path of the 3D objects [64]. Various 3D printing techniques – *viz.*, material extrusion, stereolithography, material jetting, and powder bed fusion – are well-developed for creating 3D objects using polymers and polymer composites (Figure 12.4) as discussed below:

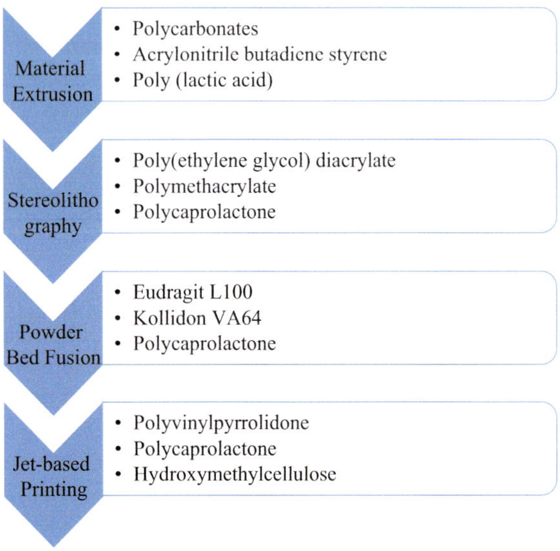

FIGURE 12.4 Different polymers exploited for drug delivery employing the 3D printing technique.

12.5.1 Material Extrusion

It is a 3D printing process wherein material is dispensed via nozzle for building a 3D object layer by layer (Figure 12.5A). "Fused filament fabrication" is a highly used material extrusion method due to its versatility & less cost. Herein, a solid filament undergoes melting at the hot end, transitioning into a semi-liquid state. This molten material is extruded via nozzle, depositing layers on a platform. Thermoplastic materials like polycarbonates, acrylonitrile butadiene styrene, PLA, etc., are employed in this method [65, 66].

Direct ink writing is another material extrusion technique wherein extrusion of a viscous material, often in paste or liquid forming ink is done through a nozzle or a syringe to create 3D structures layer by layer. This technique permits for precise control over the deposition of the ink, allowing the development of complex geometries. This methodology is widely recognized for its high resolution reaching 200 micrometers [67, 68].

12.5.2 Stereolithography

Stereolithography utilizes photopolymers, specifically epoxy- or acrylic-based resins that undergo solidification through the action of ultraviolet light (Figure 12.5B). Two primary machine setups, namely top-down and bottom-up VAT photopolymerization (VP) have been utilized. In the top-down configuration, UV light traverses a predetermined path above the liquid resin tank, inducing solidification of the polymer in the uppermost layer by forming crosslinked networks. Photopolymers typically comprise monomers/oligomers and photo-initiators [69]. To enhance the photo-initiation process and ensure the production of better quality products, diluent and chain transfer agents can be incorporated wherein former serve to control the resin's viscosity, ensuring proper wetting and impregnation, while latter alter the crosslink network. The top-down VP method exhibits high resolution, ranging from 6 to 140 micrometers, facilitating the fabrication of precise objects within the hundred-micrometer scale. However, this technique is time-consuming for large sample size due to its low printing speed [70]. Furthermore, top-down printing is relatively costly. In the bottom-up technique, the UV light source is located below the resin container. The object is built by gradually lifting the platform from the bottom of the resin container after each layer is cured. The cured layers adhere to the platform, and the curing is carried out till complete a 3D object is formed. This method may offer faster printing speeds, as the cured layers are not subject to the potential drag forces in the liquid resin [71].

12.5.3 Powder Bed Fusion

This technique stands out as a highly flexible technology, gaining prominence in mass production applications. This

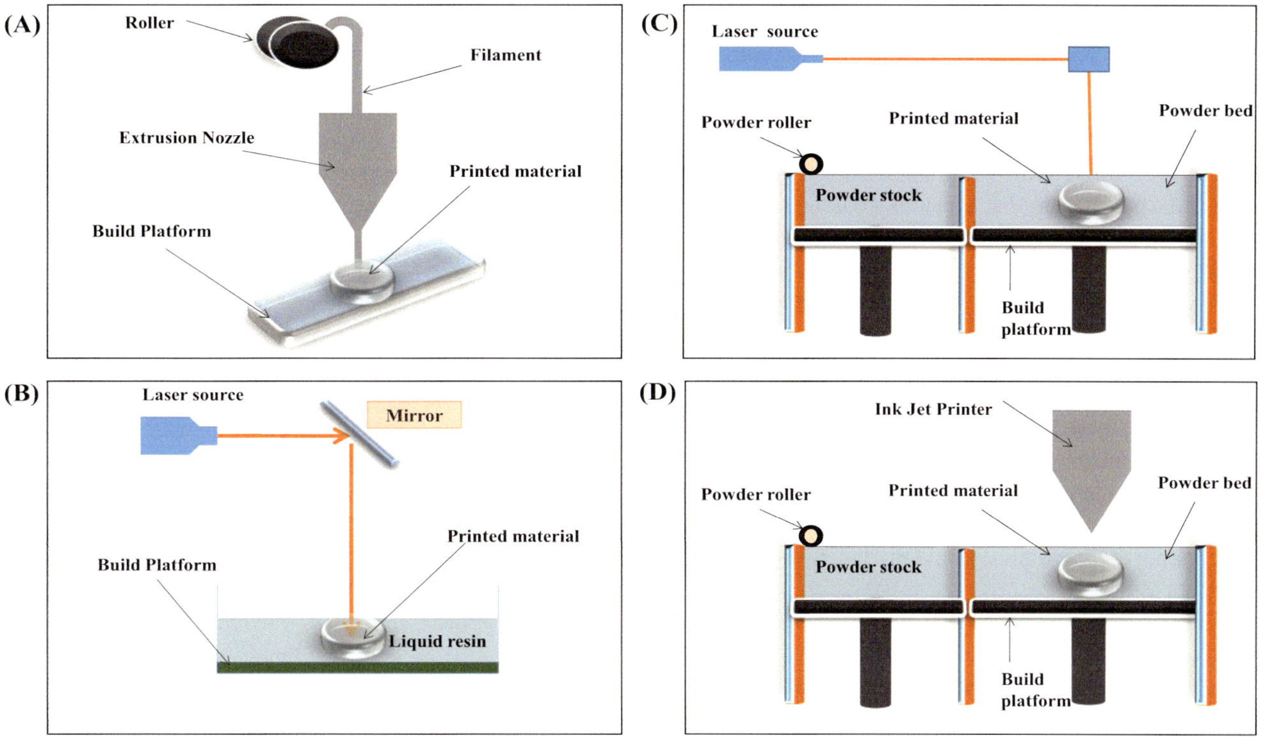

FIGURE 12.5 Schematic illustration of (A) material extrusion, (B) stereo lithography, (C) powder bed fusion, and (D) material jet and binding jet process.

method employs powdered materials and utilizes a selective application of thermo-energy, typically in the form of lasers, to construct 3D shapes (Figure 12.5C) [72]. The laser traverses the powder surface, selectively sintering the particles and fusing them together, creating the desired configuration for each 2D layer. Throughout the printing process, the temperature of the powder bed is maintained just below the melting point. This approach reduces the heat needed for fusion while being above the crystallization threshold to impede crystallization [73].

12.5.4 Material Jet and Binding Jet Processes

In jet-based additive manufacturing processes, material jetting and binder jetting represent prominent methodologies introduced in the 1990s. This technique fabricates 3D objects by selectively placing droplets of build materials [74]. Noteworthy for its precision, material jet is one of most accurate 3D printing methods, achieving high resolutions. Furthermore, it possesses the capability to print multimaterial products utilizing multiple nozzles. The polymer utilized in this technique should exhibit shear-thinning characteristics to enable fluidization and seamless flow through the narrow nozzle. Subsequently, the polymer should rapidly regain its modulus after escaping the nozzle to maintain the structural integrity of the developed object [75, 76].

Conversely, the binder jet technique involves dispensing a liquid binder over a powder bed, wherein powder serves as the primary constituent in the final product (Figure 12.5D). The procedural steps closely resemble those of powder bed fusion, as previously discussed, with the exception of the curing process. In this methodology, chemical reactions occur between the powder and the binder, eliminating the need for thermal processing in contrast to powder bed fusion [77, 78].

12.6 POLYMER SELECTION CRITERIA

The selection of polymers for the management of inflammatory diseases is a critical factor for design and development of effective therapeutic delivery systems demanding a comprehensive evaluation of diverse criteria to optimize efficacy, safety, and precision in drug delivery [79]. Biocompatibility is one of the foremost considerations, and PEG emerges as a preeminent choice due to its capacity to alleviate immunogenic responses and prolong the circulation of nanoparticles. Furthermore, the degradability of polymers plays a pivotal role, exemplified by PLGA, which, through hydrolysis, yields metabolically benign by-products – lactic and glycolic acid – ensuring the biocompatibility of the delivery system [80].

Drug loading and release kinetics are also key factors, and polymers such as poly(N-isopropylacrylamide) exhibit responsiveness to temperature fluctuations, enabling controlled drug release in response to the dynamic inflammatory milieu. For sustained drug release, PLA offers a solution with their ability to provide prolonged and controlled drug delivery profiles [81, 82].

Various other smart polymers such as "poly (N-isopropylacrylamide-co-acrylic acid)," introduces a new dimension to drug delivery systems. These polymers respond to changes in pH or temperature, allowing for on-demand release of therapeutic agents in response to the intrinsic conditions of inflammation. Biodegradable hydrogels, such as those derived from hyaluronic acid, serve a dual purpose by creating a 3D matrix that not only supports localized drug delivery but also aids in tissue repair [83].

Cellular uptake is a crucial determinant of drug efficacy, and polymers such as chitosan facilitate efficient internalization, ensuring targeted drug delivery to specific cell types involved in inflammatory processes. Moreover, mechanical properties of polymers, including poly(ethylene glycol) diacrylate (PEGDA), enable their application in 3D-printed platforms for tissue engineering. The use of various polymers in advanced 3D techniques exemplifies a sophisticated and comprehensive strategy in managing the complex challenges associated with inflammatory diseases, highlighting the imperative need for a judicious selection of polymers [84].

A consensus among various extrusion-based printing techniques underscores the necessity for materials to exhibit shear-thinning behavior. Shear-thinning fluids depicting a decrease in apparent viscosity at high shear rates are highly preferred, as opposed to shear-thickening fluids that are prone to clogging printer heads at higher shear rates. The presence of particles influences the flow behavior of polymers, hindering polymer chain movement. Parameters such as particle shape, size, and aspect ratio significantly impact rheological performance. For instance, hydrophilic water-soluble polymers like PVP can enhance the viscosity of hydrophilic polymers. Lubricants, such as silicon carbide whiskers or calcium stearate, play a role in reducing inter-particle friction and improving gel flowability [85].

A prominent challenge while using 3D printing materials lies in the limited array of permissible polymers. Unlike conventional techniques such as extrusion, injection molding, casting, or spinning, which can handle variety of polymers, 3D printing is restricted to processing only a fraction of small and macromolecule types. For example, the prevalent resins employed in 3D printing platforms are predominantly "acrylate-, urea, urethane-, or epoxy-based monomers." Nevertheless, the exploration of functional resins capable of fulfilling mechanical, thermal, dielectric, optical, and biological requirements demands further investigation [86].

12.7 DIFFERENT POLYMERS EMPLOYED IN 3D PRINTING

3D printing technology exhibits versatility in processing various materials, including liquid polymer solutions, monomers, gels, particle colloids, etc. The mechanical characteristics of composite materials depend on the intrinsic characteristics of their constituent phases, synergistically contributing to enhanced mechanical attributes [86, 87]. To enhance the mechanical, electrical, and functional characteristics of 3D printed polymer products, various filling materials are incorporated, ranging from nanoscale materials (e.g., CNT, graphene, nanoparticles) to macro-sized fibers (e.g., carbon/glass fibers). High-strength fibers contribute to structural reinforcement, while materials like ceramics, metals, metal oxides, and living cells introduce functionalities such as electrical and thermal conductivity, magnetic and optical properties [88].

Various polymers employed in 3D printing are categorized into three categories as follows:

12.7.1 Thermoplastic Polymers

Thermoplastics, frequently employed in 3D printing, undergo softening upon heating, allowing for reshaping, and solidify upon cooling [89]. Amorphous polymers possess a randomly ordered molecular structure, facilitating thermoforming, while semi-crystalline polymers exhibit a highly ordered arrangement with a sharp melting point. All thermoplastics exhibit reversible processing ability, rendering them highly suitable for extrusion-based 3D printing. Standard polymers like PLA, polyamide-nylon, etc., are common feedstocks for fused deposition modeling, owing to their low melting temperatures and favorable mechanical properties [90, 91].

12.7.2 Thermosets

These polymers experience irreversible hardening as a result of the curing of a liquid prepolymer/resin *via* heat or radiation. Higher crosslinking densities in thermosets correspond to increased mechanical strength, hardness, and resistance to thermal/chemical attack [92]. Photopolymers, a subset of thermosets, are initiated by UV radiation, resulting in rapid polymerization [93]. Various acrylate polymers, such as acrylate polyesters/silicone/urethanes, are commonly used in VP and droplet-based 3D printing [94].

12.8 HYDROGELS

These polymers characterized by a 3D crosslinked structure are favored for extrusion-based 3D printing [95]. Classification of hydrogels includes homopolymers, copolymers, and semi-interpenetrating networks. Hydrogels find applicability in bioengineering, therapeutic delivery and agriculture owing to their high elasticity and softness [96]. Successful 3D printing of hydrogels necessitates appropriate rheological properties and mechanical integrity. Viscosity is influenced by monomer concentration and macromolecular chain length, while shear-thinning behavior is achieved through a self-assembly process [97].

12.9 APPLICATION OF SYNTHETIC POLYMERS IN DRUG DELIVERY

The use of synthetic polymers in therapeutic delivery represents a transformative frontier in pharmaceutical research, offering innovative solutions to address various medical challenges from arthritis and inflammatory bowel diseases to cancer and respiratory disorders as discussed the following sections.

12.9.1 Arthritis

Traditional therapeutic approaches employed for treatment of rheumatoid arthritis primarily involve orally administered or injectable formulations, emphasizing the critical necessity for an efficacious topical intervention [98, 99]. An investigation discloses a pioneering triple-layered transdermal system designed using 3D printing for the effective treatment of arthritis. The transdermal patches feature an innovative composition, comprising the core of nanofibers/nanoparticles of diclofenac conjugated with polyvinyl alcohol. The core was sandwiched between polycaprolactone nanofibers and a 3D printed sodium alginate-based hydrogel layer containing sodium hyaluronate and rosuvastatin-encapsulated nanocapsules. The developed delivery system facilitates the controlled release of chemically conjugated drugs at inflamed sites upon exposure to skin-secreted esterases. Rosuvastatin loaded nanocapsules exhibited a size of less than 50 nm and demonstrated the release of approximately 97% over a five-day period. *In vitro* assessments confirmed the effectiveness of the fabricated platform, as demonstrated by a significant decrease in interleukin-6 and tumor necrosis factor-alpha levels in ELISA. Additionally, histological examination of the rats' ankle joints provided visual confirmation of disease symptom alleviation [100]. A report mentioned development of a biomimetic microstructured hydrogel using an emulsion-type photosensitive resin *via* "digital light processing" 3D printing technology. The water-oil biphasic hydrogel ink facilitated entrapment of both hydrophilic and lipophilic bioactive agents, resulting in scaffolds with sustained release capability. Developed scaffolds revealed superior mechanical characteristics and minimal swelling attributed to the rapid curing facilitated by the hydrophobic photo-initiator. These scaffolds were utilized as substitutes for meniscal and labral tissue for porcine model exhibiting promising potential for preventing secondary osteoarthritis [101]. In another investigation, a regenerative approach for cartilage was fabricated using a "chitosan hydrogel/3D-printed poly(ε-caprolactone) hybrid" containing synovium-derived stem cells that can recruit "tetrahedral framework nucleic acid" introduced into the articular region due to the electrostatic nature of the polymer. Nucleic acid promotes stem cells proliferation and chondrogenic differentiation, leading to improved cartilage regeneration and effectively alleviating osteoarthritis [102].

12.9.2 Inflammatory Bowel Diseases

3D printing emerges as a groundbreaking technology, offering a versatile design that mitigates dosing errors and presents a cost-effective, mold-free process for the preparation of suppositories crucial in the management of bowel diseases [103]. For instance in a study, 3D printed suppositories containing budesonide and tofacitinib citrate were developed using Gelucire® 44/14 polymer for use in acute ulcerative colitis. The poorly water-soluble nature of the drugs is addressed by taking advantage of the suppositories' self-emulsifying ability. The study explored different drug doses, and the formulations demonstrated consistent dissolution and disintegration behaviors irrespective of the doses. The results revealed the flexibility of 3D printing technology and proved its high potential for tailored drug delivery, allowing dose adjustments based on disease progression [104]. In another study, the same technology was exploited to develop antibody loaded suppositories for bowel diseases employing Gelucire® (48/16 or 44/14) polymer. A mixture of polymer with coconut oil and/or water were evaluated for loading and binding efficiency of infliximab. It was observed that aqueous infliximab solution mixed with polymer solution (48/16) was capable of undergo extrusion technique. Also, it was noticed that both oil and water individually facilitated high loading efficiencies, but the higher binding efficiency of infliximab to antigen was observed in the usage of coconut oil in ELISA test. These findings highlight 3D printing as a potential platform for biopharmaceutical dosage forms, addressing compliance concerns associated with injectables [105]. Tacrolimus loaded suppositories fabricated using the mixture of Gelucire® 44/14 polymer and coconut oil *via* semisolid extrusion 3D printing also demonstrated superior therapeutic effectiveness in a murine model of inflammatory bowel disorder. Histological analysis confirmed disease remission after treatment with developed suppositories [106].

Exploring cutting-edge advancements in pharmaceutical formulation, Pistone et al., investigated direct powder extrusion technology to develop budesonide-loaded solid oral formulations for treating pediatric eosinophilic colitis. Budesonide's limitations, including low aqueous solubility and poor bioavailability, prompted development of 3D printed mini-tablets tailored to pediatric needs. Incorporating hydroxypropyl-β-cyclodextrin polymer within powder mix facilitated drug solubility and Eudragit FS 30D coating over the surface of tablets facilitated pH responsive release for site-specific delivery to the colon region [107]. Another group of researchers developed colon-targeted tablet incorporating oxaliplatin through the synergistic application of nanotechnology and fused deposition technique. Utilizing the hot-melt extrusion technique, Eudragit L100-55 filament laden with drug-encapsulating alginate nanoparticles was formulated and subsequently 3D printed into tablets ensuring uniform drug content. The effectiveness of 3D printed tablets against tumors was compared with direct compressed tablets containing drug-loaded nanoparticles in CT-26 tumor-bearing mice. Notably, 3D printed tablets demonstrated superior antitumor effects compared to tablets formulated using a conventional technique that could be due to their nonselective drug release. The results of the study emphasize the prospective of integrating nanoscience and 3D printing for the development of site-specific systems [108]. The capability of the fused deposition technique for successful development of pulsatile-release Chronotopic™ system has been reported by Melocchi and his coworkers. The formulation was composed of drug core enveloped by a coat of swellable/soluble hydrophilic polymers coating, which impart a programmable lag phase before release of therapeutics. This type of system facilitates colon-targeting owing to a gastro-resistant film to hinder premature interaction of the inner core with gastric fluids, allowing for the establishment of a lag phase synchronized with the transit time through the small intestine [109].

12.9.3 Cancer

Polymers are finding wide applicability in palliation therapy for dysphagia in inoperable esophageal malignancies that commonly involves the use of stents. Existing metallic stents can lead to difficulties, mainly tumor ingrowth and relocation of stent into the stomach. To address these issues, a flexible polymer stent was developed utilizing 3D printing optimizing spiral parameters through finite element analysis. The spiral stents were compared with nonspiral controls. In an *ex vivo* porcine esophagus, the stent with spirals exhibited higher self-expansion and anti-migration forces. Biodegradability tests in neutral and acidic buffers indicated gradual degradation without compromising compressive strength over six weeks. Moreover, cytotoxicity assessments on human esophagus epithelial cells revealed their non-toxic profile. Results revealed the potential of 3D-printed stent composed of PLA and thermoplastic polyurethane with spirals in addressing complications associated with esophageal malignancies [110]. In a study, 3D-printed scaffolds were developed to overcome issues with bone metastasis such as severe adverse effects of conventional systemic chemotherapy and requirement of bone grafting to manage defects after surgical interventions. Addressing these challenges, a localized chemotherapeutic delivery system was designed that can also function as potential bone substitutes. The developed doxorubicin loaded porous 3D-printed scaffolds demonstrated 60%–75% sustained release over seven days. This release exceptionally decreased metabolic activity and proliferation in both LAPC4 and patient-derived spine cells, whereas scaffolds without drug load showed no effect. These nanoporous scaffolds represent a promising, cost-effective approach for localized chemotherapeutic delivery tailored to individual patients at tumor resection sites [111]. Another investigation focused on development of a bioadhesive vaginal film with the aim to decrease dosing intervals and increase patient compliance of those suffering from cervical cancer. A hot-melt extrusion technique was employed to fabricate disulfiram-loaded, sustained release films. 3D printing speed and processing temperature were judicially optimized to deal with the thermosensitive nature of a bioactive compound. Films composed of poly(ethylene oxide) depicted high drug content and notable mechanical strength determined using sheep cervical tissue. Also, the *in vitro* drug release study revealed sustained release of bioactive up to 24 h suggesting its wide applicability in cervical cancer [112].

12.10 RESPIRATORY DISORDERS

3D printing is increasingly applied in the management of respiratory disorders through the fabrication of 3D printed medical devices and models. This innovative approach involves the creation of prototypes of lungs, enabling medical specialists to gain a comprehensive understanding of diseased conditions. These prototypes hold the potential to revolutionize the diagnosis and treatment of respiratory organ disorders, offering a unique avenue for personalized inhaled medicines through advanced 3D printing techniques [113]. In this context, an investigation was carried out to create a 3D printed scaffold with both biological and mechanical compatibility using chitosan/polycaprolactone bioink for the purpose of lung tissue engineering. The "Design-Expert software" was exploited to systematically explore various compositions for 3D printing. Outcomes revealed that the printability of the scaffolds was predominantly governed by the amount of chitosan, while alterations in polycaprolactone content did not exhibit a significant impact on printability. The scaffolds demonstrated exceptional features in terms of swelling, degradation, and mechanical behavior. Furthermore, the scaffolds exhibited favorable cell adhesion, non-toxicity, and *in vitro* cell biocompatibility. Scaffold prepared using 4:1 ratio of chitosan/polycaprolactone displayed superior activity in supporting MRC-5 cell culture and thus could act as an ideal candidate for lung tissue engineering [114].

In an investigation, a novel 3D printed "tracheobronchial bioresorbable airway splint" was developed that demonstrated improved survival efficiency in a porcine animal model under severe conditions of tracheobronchomalacia [115]. In another study, airway splints for management of tracheobronchomalacia in pediatric patients were developed using 3D printing. They were found to be a viable substitute for alleviating airway collapse in patients [116]. Furthermore, this novel technique has been demonstrated in the treatment of lung cancer. A prototype (lung tumor movement simulator) was developed using 3D printing to understand tumor dynamics during respiration. This simulator serves as a valuable tool in radiotherapy for lung cancer, providing insights into tumor movements during respiration for enhanced precision [117]. The multifaceted applications of 3D printing in respiratory disease management describe its transformative capability in advancing personalized medicine and enhancing therapeutic outcomes.

12.11 CARDIOVASCULAR DISEASES

The demand for small diameter synthetic-vascular grafts arises in bypass surgeries when autologous vessels are not available due to prior surgeries. Thrombosis poses a primary challenge leading to the failure of these synthetic-vascular grafts in revascularization procedures. Consequently, the advancement of biodegradable vascular grafts with the capability of localized and sustained antithrombotic drug release represents a significant step in combating heart disorders [118]. An investigation carried out by Dominguez-Robles et al. elucidates the utilization of extrusion-based 3D printing technology for crafting acetylsalicylic acid loaded tubular grafts. For this, poly(caprolactone) and drug were incorporated to fabricate tubular grafts. Additionally, rifampicin was introduced to the formulation for preventing vascular prosthesis infections. Comprehensive characterization of the produced grafts was conducted using various techniques, followed by assessments of drug release, antiplatelet and antimicrobial potential. Results indicated sustained drug release for two weeks. Tubular grafts with 10% (w/w) drug demonstrated lower platelet adhesion compared to blank counter parts and those with 5% (w/w) drug. Furthermore, tubular grafts with 1% (w/w) rifampicin effectively inhibited *Staphylococcus aureus* growth [119]. In another study, vascular grafts encapsulating clopidogrel were designed *via* digital light technique. A composite blend consisting of PLA-polyurethane acrylate, polycaprolactone along with a drug was formulated to attain specific mechanical and therapeutic attributes suitable for vascular grafts. Utilizing 3D printing technology, the composite material enabled precise control over the vascular grafts' shape and dimensions. The grafts were characterized by employing various analytical techniques. Subsequently, an assessment of drug release kinetics indicated that grafts having the highest concentration of the drug (20% w/w) exhibited sustained drug release for a period extending up to 27 days. Also drug-loaded grafts exhibited a considerable reduction in platelet deposition on their surfaces compared to blank materials devoid of the drug. The hemolysis percentage for all samples remained below 5%. Additionally, the 3D-printed materials facilitated a conducive environment for cellular attachment, viability, and growth [120]. A novel system comprising nanocellulose, poly(glycerol sebacate), and polypyrrole using direct ink printing has been formulated as a therapeutic intervention for myocardial infarction. The developed patches possessed sufficient elasticity, electrical conductivity, and the ability to support cell growth and were observed to be biocompatible with H9c2 cardiomyoblasts. Two therapeutic agents (3i-1000 and curcumin) were loaded within patches and drug release studies revealed a gradual *in vitro* degradation of the cardiac patches, preventing burst release of drugs and indicating their suitability for prolonged therapeutic applications. These systems could open avenues for myocardial regeneration after infarction [121].

12.12 REGULATORY AND SAFETY CONSIDERATIONS

Despite the promising prospects of 3D printing technology in healthcare, substantial technical and regulatory challenges hinder its clinical applications. A major drawback lies in the absence of any regulatory guideline governing the fabrication of pharmaceuticals using the 3D printing technique. Existing regulations from the Food and Drug Administration (FDA) primarily focus on application of this technology for medical devices and prosthetics, neglecting drug production. However, now the Technology Team within the FDA has initiated efforts to address this gap, aiming to inspire technological innovation, particularly in 3D printing, within pharmaceutical research. Various regulatory agencies including "Health Canada, Therapeutic Goods Administration (Australia), European Medicines Agency," etc., discussed the opportunities presented by 3D printing technique under the "International Coalition of Medicines Regulatory Authorities" and most of them have begun evolving new regulatory frameworks or implementing legislation to facilitate innovative product development [122, 123].

The technologies employed in 3D printing exploit various factors like heat, solvents, and light to fabricate intricate 3D shapes that can influence the stability and quality of printed drugs, necessitating quality control measures. Therefore, despite noteworthy advancement in proving the efficacy of this novel technology in personalized medicine, there remains reluctance in its implementation due to regulatory and stakeholder concerns. Collaboration between regulators, health practitioners, and technology developers is necessary to accelerate the implementation of 3D printing technology, transforming it from theoretical and applied research into practical applications [124, 125].

In conclusion, 3D printing in drug delivery offers an innovative approach, bringing manufacturing closer to patients and enabling personalized therapy and promising development of safer and more effective treatments. Although still in its early stages, 3D printing is poised to revolutionize drug manufacturing, offering flexibility and transforming drug delivery systems.

12.13 CONCLUSION AND FUTURE PERSPECTIVES

The integration of 3D printing and pharmaceutical research has shown remarkable potential in advancing drug delivery systems across various therapeutic areas. Various studies discussed in this chapter highlight innovative approaches in the fabrication of solid dosage forms, emphasizing the use of diverse polymers and complex formulations. Three-dimensional printing has facilitated the production of drug-loaded nanocapsules within 3D printed tablets, depicting the adaptability and precision of this technology. Furthermore, the application of this novel approach in cardiovascular interventions, arthritis treatments, and cancer therapies demonstrates the versatility of this technology in addressing specific challenges associated with each medical condition. From developing biodegradable vascular grafts with localized drug release capabilities to creating transdermal systems for arthritis treatment, these

TABLE 12.1
Application of 3D Printing in Management of Inflammatory Disorders

Polymer employed	Disease	Therapeutic agent	Technique	Formulation	Outcome	Reference
Chitosan/polycaprolactone	Lung tissue engineering	NA	Fused Filament deposition	3D printed scaffold	✓ Scaffold with a chitosan/polycaprolactone ratio of 4:1, exhibited superior MRC-5 cell line growth, proliferation, and migration.	[114]
Cyclodextrin and Eudragit FS 30D	Pediatric eosinophilic colitis	Budesonide	Material extrusion	Mini-tablets	✓ Incorporating hydroxypropyl-β-cyclodextrin polymer within powder mix improved drug solubility.	[107]
Eudragit L100-55	Colon Cancer	Oxaliplatin	Material extrusion	Tablets	✓ Anti-cancer activity of 3D-printed tablets was superior to the drug solution.	[108]
Gelucire® (48/16)	Ulcerative colitis	Budesonide and tofacitinib citrate	Material extrusion	Suppositories	✓ Different suppositories with varied dose demonstrated similar dissolution and disintegration profile. ✓ Feasibility of additive manufacturing technique to titrate the drug doses depending upon disease progression.	[104]
Gelucire® (48/16 or 44/14)	Inflammatory bowel disease	Infliximab	Material extrusion	Suppositories	✓ The drug was successfully loaded into the printing ink of polymer and was able to withstand the extrusion process.	[105]
Gelucire® (44/14)	Inflammatory bowel disease	Tacrolimus	Material extrusion	Suppositories	✓ The developed formulation depicted superior efficacy in a murine model of inflammatory bowel disease.	[106]
Hydroxypropylmethyl cellulose	NA	Theophlline	Material extrusion	Hydrogels	✓ The optimal formulation, HPMC K4M 12% w/w hydrogel, exhibited excellent extrudability and shape retention, enabling flexible dosage combinations and extended drug release over 12 hours.	[126]
Mannitol, trehalose and xylitol	Hyperglycemia	Insulin	Stereolithography	Microneedle arrays	✓ 3D-printed microneedle arrays demonstrated superior skin penetration with minimal forces (2–5 N) compared to metal arrays. ✓ Animal studies revealed effective hypoglycemia control.	[127]
Nanofibrillated cellulose, poly(glycerol sebacate), and polypyrrole	Myocardial infarction	3i-1000 and curcumin	Direct ink eriting	Cardiac patches	✓ Developed patches were found to be biocompatible with suitable mechanical strength.	[121]
Poly(caprolactone)	Thrombosis	Acetylsalicylic acid and Rifampicin	Material extrusion	Tubular grafts	✓ The developed tubular grafts demonstrated superior antiplatelet and antimicrobial activity.	[119]
PLA-polyurethane acrylate, polycaprolactone	Cardiovascular disorders	Clopidogrel	Digital light processing	Vascular grafts	✓ Developed drug-loaded vascular grafts (20% w/w) provided sustained drug release for 27 days. ✓ Fabricated grafts significantly reduced platelet deposition and depicted enhanced cellular attachment, viability, and growth compared to blank materials.	[120]
Polyvinyl alcohol, sodium alginate, sodium hyaluronate	Rheumatoid arthritis	Diclofenac, rosuvastatin	Material extrusion	Triple-layered transdermal patch	✓ Substantial reduction in inflammatory markers (IL-6 and TNF-α) in rats was observed in rheumatoid model. ✓ Rheumatoid arthritis symptoms were observed to be alleviated in rat ankle joints.	[100]

Poly(ε-caprolactone)	Articular cartilage injury	Synovial mesenchymal stem cell, tetrahedral framework nucleic acid	Dynamic light scattering	Hydrogel	✓ Significant improvement in cartilage regeneration and repair of defects.	[102]
Polycaprolactone, poly(ethylene glycol) diacrylate and poly(ethylene glycol)	Acne	Salicylic acid	Fused Deposition modeling and stereolithography	Drug-loaded devices	✓ Stereolithography printing was found to be more suitable for manufacturing anti-acne devices with salicylic acid due to higher resolution, better drug loading, and no drug degradation compared to fused deposition modeling.	[128]
PLA and thermoplastic polyurethane	Esophageal malignancies	NA	Material extrusion	Polymer stent with spirals	✓ Mechanical properties of the stent were modulated by adjusting the ratios of both polymers, allowing for customization of the stent's flexibility and self-expansion characteristics.	[110]
PORO LAY FOMM 60, 40	Prostate cancer	Doxorubicin	Material extrusion	3D-printed scaffolds	✓ Scaffolds exhibited a release capacity of 60%–75% for the loaded drug for a seven-day period. ✓ Significant reduction in metabolic activity and proliferation in both LAPC4 and patient-derived cells.	[111]
Poly(ethylene oxide)	Cervical cancer	Disulfiram	Material extrusion	Bioadhesive vaginal film	✓ The bioadhesion study, utilizing sheep cervical tissue, revealed promising adhesive peak forces. ✓ Developed films sustained drug release for up to 24 hours.	[112]
Polycaprolactone	Tracheobronchomalac a	NA	Stereolithography	Airway splints	✓ Infants with severe tracheobronchomalacia, treated with patient-specific 3D-printed external airway splints, demonstrated resolution of life-threatening complications.	[116]
Polydimethylsiloxane	Tissue engineering	Cytokine cocktail	NA	Tracheal patch	✓ The hydrogel-coated implants demonstrated superior survival rates (100% vs. 33%) and mitigated inflammatory responses. ✓ Enhanced tissue integration and reduced adverse immune reactions in a rat tracheal defect model.	[129]
Polycaprolactone	Tracheal defects	Extracellular matrix	NA	Tracheal grafts	✓ In a porcine model, 3D-printed bioengineered tracheal grafts successfully reconstructed anterior tracheal defects, demonstrating structural integrity, respiratory mucosal coverage, and vascularity.	[130]
Poly-lactic-co-glycolic acid	Breast cancer	5-fluorouracil and NVP-BEZ235	Jet printing	Scaffold	✓ Facilitated sustained release of 5-fluorouracil and NVP-BEZ235. ✓ Effective suppression of breast tumor growth in orthotropic breast cancer model.	[131]
Polyvinyl alcohol	NA	Prednisolone	Fused deposition modeling	Tablets	✓ 3D printing allowed precise control of tablet mass, achieving a strong correlation between target and achieved doses with an accuracy range of 88.7%–107%.	[132]

(Continued)

TABLE 12.1 (Continued)

Polymer employed	Disease	Therapeutic agent	Technique	Formulation	Outcome	Reference
Polyvinylpyrrolidine	NA	Paracetamol	Material extrusion	Tablets	✓ Tablets with an exceptionally high drug loading (80% w/w) were successfully prepared using 3D printing. ✓ The tablets exhibited an immediate release profile with disintegration in less than 60 seconds.	[133]
Poly D,L-lactic acid	Osteomyelitis	Levofloxacin and tobramycin	Material extrusion	Multidrug implant	✓ The implant provided sustained delivery of the drugs in tandem, resembling a tumor chemotherapy program in an animal model of chronic osteomyelitis.	[134]
Polyvinylpyrrolidone	Tuberculosis	Quercetin	Material extrusion	3D-printed patch	✓ Optimized patch formulation exhibited improved pharmacokinetic profiles and sustained drug delivery for 18 days in rats after a single skin application.	[135]
Poly(lactic-co-glycolic acid)	Cancer	Paclitaxel	Piezoelectric inkjet printing	Microparticles	✓ Microparticles with different well-defined shapes including circles, grids, honeycombs, and rings were successfully prepared. ✓ Release of drug was found to be dependent on geometry of the particles.	[136]
Polyvinylpyrrolidine	Cardiovascular diseases	Aspirin and hydrochlorothiazide, pravastatin, atenolol, and ramipril	Material extrusion	Polypill	✓ This developed polypill demonstrated potential in delivering complex medication regimens into a single personalized tablet. ✓ XRPD and ATR-FTIR confirmed no drug-excipient interactions.	[137]
Soluplus® (polyvinyl caprolactame-polyvinyl acetate-polyethylene)	Cancer	5-fluororacil, curcumin and cisplatin	Inkjet printing	Transdermal microneedle	✓ 3D printing enabled effective coating of drugs on microneedles. ✓ Solubilizing property of polymer facilitated increased release rate of drugs across the porcine skin.	[138]

studies demonstrate the potential of this technique in tailoring drug delivery solutions to individual patient needs. The exploration of 3D printing in palliation therapy for esophageal malignancies and the development of bioadhesive vaginal films for cervical cancer treatment reveal the technology's potential in creating patient-specific solutions for challenging conditions. In the context of respiratory disorders, there is a promising avenue for the development of personalized drug delivery systems that can enhance treatment efficacy and patient compliance. The intersection of 3D printing and pharmaceuticals is continuously expanding, offering novel possibilities for tailoring treatments to individual patient needs and revolutionizing the landscape of drug delivery systems (Table 12.1). The outlined studies provide a glimpse into the current state of the field, with future perspectives poised to unlock even more innovative applications and solutions.

ABBREVIATIONS

ABS	Acrylonitrile Butadiene Styrene
3D	Three-dimensional
COPD	Chronic obstructive pulmonary disease
CD4+	Cluster of differentiation
eNOS	Endothelial nitric oxide synthase
HLA-DRB1	Human leukocyte antigen class II histocompatibility, D related beta chain
ICAM	Intercellular adhesion molecule
JAK	Janus kinase
MAPK	Mitogen-activated protein kinase
MMPs	Matrix metalloproteinases
NF*κ*-B	Nuclear factor kappa-light-chain-enhancer of activated B cells
PLGA	Poly(lactic-co-glycolic acid)
ROS	Reactive oxygen species
TNF-*α*	Tumor necrosis factor-alpha
VCAM	Vascular cell adhesion molecule

REFERENCES

1. Li P, Zheng Y, and Chen X. Drugs for autoimmune inflammatory diseases: From small molecule compounds to anti-TNF biologics. *Frontiers in Pharmacology* 2017;8:460.
2. Bennett JM, Reeves G, Billman GE, and Sturmberg JP. Inflammation–nature's way to efficiently respond to all types of challenges: Implications for understanding and managing 'the epidemic' of chronic diseases. *Frontiers in Medicine* 2018;5:316.
3. Deng Z, and Liu S. Inflammation-responsive delivery systems for the treatment of chronic inflammatory diseases. *Drug Delivery and Translational Research* 2021;11:1475–97.
4. Trenfield SJ, Awad A, Madla CM, Hatton GB, Firth J, Goyanes A, Gaisford S, and Basit AW. Shaping the future: Recent advances of 3D printing in drug delivery and healthcare. *Expert Opinion on Drug Delivery* 2019;16(10):1081–94.
5. Ahmad J, Garg A, Mustafa G, Mohammed AA, and Ahmad MZ. 3D Printing Technology as a Promising Tool to Design Nanomedicine-Based Solid Dosage Forms: Contemporary Research and Future Scope. *Pharmaceutics* 2023;15(5):1448.
6. Jain K, Shukla R, Yadav A, Ujjwal RR, and Flora SJ. 3D printing in development of nanomedicines. *Nanomaterials* 2021);11(2):420.
7. Vanderburgh J, Sterling JA, and Guelcher SA. 3D printing of tissue engineered constructs for in vitro modeling of disease progression and drug screening. *Annals of Biomedical Engineering* 2017 45(1):164–79.
8. Jamróz W, Szafraniec J, Kurek M, and Jachowicz R. 3D printing in pharmaceutical and medical applications–recent achievements and challenges. *Pharmaceutical Research* 2018 35:1–22.
9. Vithani K, Goyanes A, Jannin V, Basit AW, Gaisford S, and Boyd BJ. An overview of 3D printing technologies for soft materials and potential opportunities for lipid-based drug delivery systems. *Pharmaceutical Research* 2019;36:1–20.
10. Park S, Shou W, Makatura L, Matusik W, and Fu KK. 3D printing of polymer composites: Materials, processes, and applications. *Matter* 2022;5(1):43–76.
11. Vora LK, Gholap AD, Jetha K, Thakur RR, Solanki HK, and Chavda VP. Artificial intelligence in pharmaceutical technology and drug delivery design. *Pharmaceutics* 2023;15(7):1916.
12. Wang Y, Wu H, Wang Z, Zhang J, Zhu J, Ma Y, Yang Z, and Yuan Y. Optimized synthesis of biodegradable elastomer pegylated poly (glycerol sebacate) and their biomedical application. *Polymers* 2019;11(6):965.
13. Wen H, Jung H, and Li X. Drug delivery approaches in addressing clinical pharmacology-related issues: Opportunities and challenges. *The AAPS Journal* 2015;17:1327–40.
14. Singh S, Kumar M, Choudhary D, Chopra S, and Bhatia A. 3D printing technology in drug delivery: Polymer properties and applications. *Journal of Dispersion Science and Technology* 2023 46:1–35.
15. Chen L, Deng H, Cui H, Fang J, Zuo Z, Deng J, Li Y, Wang X, and Zhao L. Inflammatory responses and inflammation-associated diseases in organs. *Oncotarget* 2018;9(6):7204.
16. Dainichi T, Kabashima K, Ivanov II, Goto Y. Regulation of Immunity by Non-Immune Cells. *Frontiers in Immunology* 2021;12:770847. September 21.
17. Abdulkhaleq LA, Assi MA, Abdullah R, Zamri-Saad M, Taufiq-Yap YH, Hezmee MN The crucial roles of inflammatory mediators in inflammation: A review. *Veterinary World* 2018;11(5):627. May.
18. Bansal P, Verma D, Singh Tamber S, Sharma R An update on pathogenesis of inflammatory disorders with its management. *RPS Pharmacy and Pharmacology Reports* 2023;2(2):rqad011. April 1, 2023.
19. Radu AF, Bungau SG Management of rheumatoid arthritis: an overview. *Cells* 2021;10(11):2857. October 23, 2021.
20. Guo Q, Wang Y, Xu D, Nossent J, Pavlos NJ, Xu J Rheumatoid arthritis: pathological mechanisms and modern pharmacologic therapies. *Bone Research* 2018;6(1):15. April 27, 2018.
21. Kondo N, Kuroda T, Kobayashi D Cytokine networks in the pathogenesis of rheumatoid arthritis. *International Journal of Molecular Sciences* 2021;22(20):10922. October 10, 2021.
22. Elazazy O, Midan HM, Shahin RK, Elesawy AE, Elballal MS, Sallam AA, Elbadry AM, Elrebehy MA, Bhnsawy A, Doghish AS. Long non-coding RNAs and rheumatoid arthritis: *Pathogenesis and Clinical Implications. Pathology-Research and Practice* 2023 246:154512. May 8, 2023.

23. Yap HY, Tee SZ, Wong MM, Chow SK, Peh SC, Teow SY Pathogenic role of immune cells in rheumatoid arthritis: implications in clinical treatment and biomarker development. *Cells* 2018;7(10):161. October 9, 2018.
24. Cabral-Pacheco GA, Garza-Veloz I, Castruita-De la Rosa C, Ramirez-Acuna JM, Perez-Romero BA, Guerrero-Rodriguez JF, Martinez-Avila N, Martinez-Fierro ML The roles of matrix metalloproteinases and their inhibitors in human diseases. *International Journal of Molecular Sciences* 2020;21(24):9739. December 20.
25. Padoan A, Musso G, Contran N, Basso D Inflammation, autoinflammation and autoimmunity in inflammatory bowel diseases. *Current Issues in Molecular Biology* 2023;45(7):5534–57. June 30, 2023.
26. Kellermann L, Riis LB A close view on histopathological changes in inflammatory bowel disease, a narrative review. *Digestive Medicine Research* 2021;4, no. 3: 1–15.
27. Borowitz SM. The epidemiology of inflammatory bowel disease: Clues to pathogenesis? *Frontiers in Pediatrics* 2023;10:1103713. January 17, 2023.
28. Saez A, Herrero-Fernandez B, Gomez-Bris R, Sánchez-Martinez H, Gonzalez-Granado JM. Pathophysiology of inflammatory bowel disease: innate immune system. *International Journal of Molecular Sciences* 2023;24(2):1526. January 12, 2023.
29. Guan Q. A comprehensive review and update on the pathogenesis of inflammatory bowel disease. *Journal of Immunology Research* 2019 30:2019. December 1, 2019.
30. Zhao H, Wu L, Yan G, Chen Y, Zhou M, Wu Y, Li Y. Inflammation and tumor progression: signaling pathways and targeted intervention. *Signal Transduction and Targeted Therapy* 2021;6(1):263. July 12, 2021.
31. Uzuner E, Ulu GT, Gürler SB, Baran Y. The role of MiRNA in cancer: pathogenesis, diagnosis, and treatment. In *MiRNomics: MicroRNA Biology and Computational Analysis* 2022:375–422.
32. Kontomanolis EN, Koutras A, Syllaios A, Schizas D, Mastoraki A, Garmpis N, Diakosavvas M, Angelou K, Tsatsaris G, Pagkalos A, Ntounis T. Role of oncogenes and tumor-suppressorgenes in carcinogenesis: a review. *Anticancer Research* 2020;40(11):6009–15. November 1, 2020.
33. Qian S, Golubnitschaja O, Zhan X. Chronic inflammation: key player and biomarker-set to predict and prevent cancer development and progression based on individualized patient profiles. *EPMA Journal* 2019;10:365–81. December 2019.
34. Singh N, Baby D, Rajguru JP, Patil PB, Thakkannavar SS, Pujari VB. *Inflammation and cancer. Annals of African Medicine* 2019;18(3):121. July.
35. Franco PI, Rodrigues AP, de Menezes LB, Miguel MP. Tumor microenvironment components: Allies of cancer progression. *Pathology-Research and Practice* 2020;216(1):152729. January 1, 2020.
36. Saman H, Raza SS, Uddin S, Rasul K. Inducing angiogenesis, a key step in cancer vascularization, and treatment approaches. *Cancers* 202012(5):1172. May 6, 2020.
37. Mehta M, Dhanjal DS, Paudel KR, Singh B, Gupta G, Rajeshkumar S, Thangavelu L, Tambuwala MM, Bakshi HA, Chellappan DK, Pandey P. Cellular signalling pathways mediating the pathogenesis of chronic inflammatory respiratory diseases: an update. *Inflammopharmacology* 2020;28:795–817. August.
38. Czarnecka-Chrebelska KH, Mukherjee D, Maryanchik SV, Rudzinska-Radecka M. *Biological and Genetic Mechanisms of COPD, Its Diagnosis, Treatment, and Relationship with Lung Cancer. Biomedicines* 2023;11(2):448. February 3, 2023.
39. Kany S, Vollrath JT, Relja B Cytokines in inflammatory disease. *International Journal of Molecular Sciences* 2019;20(23):6008. November 28, 2019.
40. Banafea GH, Bakhashab S, Alshaibi HF, NatesanPushparaj P, Rasool M. The role of human mast cells in allergy and asthma. *Bioengineered* 2022;13(3):7049–64. March 1, 2022.
41. Costanzo G, Costanzo GA, Del Moro L, Nappi E, Pelaia C, Puggioni F, Canonica GW, Heffler E, Paoletti G Mast Cells in Upper and Lower Airway Diseases: Sentinels in the Front Line. *International Journal of Molecular Sciences* 2023;24(11):9771. June 5, 2023.
42. Boncler M, Wu Y, Watala C The multiple faces of C-reactive protein—physiological and pathophysiological implications in cardiovascular disease. *Molecules* 2019;24(11):2062. May 30, 2019.
43. Castro AR, Silva SO, Soares SC. The use of high sensitivity C-reactive protein in cardiovascular disease detection. *Journal of Pharmacy & Pharmaceutical Sciences* 2018;21:496–503. November 29, 2018.
44. Daiber A, Xia N, Steven S, Oelze M, Hanf A, Kröller-Schön S, Münzel T, Li H New therapeutic implications of endothelial nitric oxide synthase (eNOS) function/dysfunction in cardiovascular disease. *International Journal of Molecular Sciences* 2019;20(1):187. January 7, 2019.
45. Feng Y, Feng Y, Gu L, Liu P, Cao J, Zhang S. The critical role of tetrahydrobiopterin (BH4) metabolism in modulating radiosensitivity: BH4/NOS axis as an angel or a devil. *Frontiers in Oncology* 2021;11:720632. August 27, 2021.
46. Lu Y, Cui X, Zhang L, Wang X, Xu Y, Qin Z, Liu G, Wang Q, Tian K, Lim KS, Charles CJ. The functional role of lipoproteins in atherosclerosis: *Novel directions for diagnosis and targeting therapy. Aging and Disease* 2022;13(2):491. April 2022.
47. Sinniah A, Yazid S, Flower RJ From NSAIDs to glucocorticoids and beyond. *Cells* 2021;10(12):3524. December 14, 2021.
48. Timmermans S, Souffriau J, Libert C. A general introduction to glucocorticoid biology. *Frontiers in Immunology* 2019;10:1545. July 4, 2019.
49. Bruscoli S, Febo M, Riccardi C, Migliorati G. Glucocorticoid therapy in inflammatory bowel disease: mechanisms and clinical practice. *Frontiers in Immunology* 2021;12:691480. June 3, 2021.
50. Kerrigan SA, McInnes IB. Reflections on 'older' drugs: learning new lessons in rheumatology. *Nature Reviews Rheumatology*. 2020 Mar;16(3):179–83.
51. Benjamin O, Goyal A, Lappin SL. *Disease modifying anti-rheumatic drugs* (DMARD).
52. Jung SM, Kim WU. Targeted immunotherapy for autoimmune disease. *Immune Network*. 2022 Feb;22(1).
53. Mehta JK, Kaur G, Buttar HS, Bagabir HA, Bagabir RA, Bagabir SA, Haque S, Tuli HS, Telessy IG. Role of the renin–angiotensin system in the pathophysiology of coronary heart disease and heart failure: *Diagnostic biomarkers and therapy with drugs and natural products. Frontiers in Physiology*. 2023 Feb 23;14:1034170.
54. Cai Z, Wang S, Li J. Treatment of inflammatory bowel disease: a comprehensive review. *Frontiers in Medicine*. 2021 Dec 20;8:765474.
55. Park J, Cheon JH. Updates on conventional therapies for inflammatory bowel diseases: 5-aminosalicylates, corticosteroids, immunomodulators, and anti-TNF-α. *The Korean Journal of Internal Medicine*. 2022 Sep;37(5):895.

56. Sharma A, Campbell M, Yee C, Goswami S, Sharma P. Immunotherapy of cancer. *In Clinical Immunology* 2019 Jan 1 (pp. 1033–1048). Elsevier.
57. Bozkurt Y, Karayel E. 3D printing technology; methods, biomedical applications, future opportunities and trends. *Journal of Materials Research and Technology*. 2021 Sep 1;14:1430–50.
58. Borandeh S, van Bochove B, Teotia A, Seppälä J. Polymeric drug delivery systems by additive manufacturing. *Advanced Drug Delivery Reviews*. 2021 Jun 1;173:349–73.
59. Ilyas RA, Zuhri MY, Norrrahim MN, Misenan MS, Jenol MA, Samsudin SA, Nurazzi NM, Asyraf MR, Supian AB, Bangar SP, Nadlene R. Natural fiber-reinforced polycaprolactone green and hybrid biocomposites for various advanced applications. *Polymers*. 2022 Jan 3;14(1):182.
60. Dimitrova M, Vlahova A, Kalachev Y, Zlatev S, Kazakova R, Capodiferro S. Recent Advances in 3D Printing of Polymers for Application in Prosthodontics. *Polymers*. 2023 Nov 24;15(23):4525.
61. Jamróz W, Szafraniec J, Kurek M, Jachowicz R. 3D printing in pharmaceutical and medical applications–recent achievements and challenges. *Pharmaceutical Research*. 2018 Sep;35:1–22.
62. Jandyal A, Chaturvedi I, Wazir I, Raina A, Haq MI. 3D printing–A review of processes, materials and applications in industry 4.0. *Sustainable Operations and Computers*. 2022 Jan 1;3:33–42.
63. Bekas DG, Hou Y, Liu Y, Panesar A. 3D printing to enable multifunctionality in polymer-based composites: A review. *Composites Part B: Engineering*. 2019 Dec 15;179:107540.
64. Pirozzi MA, Jacob D, Pálsson T, Gargiulo P, Helgason T, Jónsson Jr H. State of the art in 3D printing. In *Handbook of Surgical Planning and 3D Printing*. 2023 Jan 1:3–6.
65. Oleff A, Küster B, Stonis M, Overmeyer L. Process monitoring for material extrusion additive manufacturing: a state-of-the-art review. *Progress in Additive Manufacturing*. 2021 Dec;6(4):705–30.
66. Penumakala PK, Santo J, Thomas A. A critical review on the fused deposition modeling of thermoplastic polymer composites. *Composites Part B: Engineering*. 2020 Nov 15;201:108336.
67. Saadi MA, Maguire A, Pottackal NT, Thakur MS, Ikram MM, Hart AJ, Ajayan PM, Rahman MM. Direct ink writing: a 3D printing technology for diverse materials. *Advanced Materials*. 2022 Jul;34(28):2108855.
68. Shahzad A, Lazoglu I. Direct ink writing (DIW) of structural and functional ceramics: Recent achievements and future challenges. *Composites Part B: Engineering*. 2021 Nov 15;225:109249.
69. Huang J, Qin Q, Wang J. A review of stereolithography: Processes and systems. Processes. 2020 Sep 11;8(9):1138.
70. Zakeri S, Vippola M, Levänen E. A comprehensive review of the photopolymerization of ceramic resins used in stereolithography. *Additive Manufacturing*. 2020 Oct 1;35:101177.
71. Deshmane S, Kendre P, Mahajan H, Jain S Stereolithography 3D printing technology in pharmaceuticals: a review. *Drug Development and Industrial Pharmacy*. 2021 Sep 2;47(9):1362–72.
72. Singh DD, Mahender T, Reddy AR. Powder bed fusion process: A brief review. *Materials Today: Proceedings*. 2021 Jan 1;46:350–5.
73. Bartlett JL, Li X. An overview of residual stresses in metal powder bed fusion. *Additive Manufacturing*. 2019 May 1;27:131–49.
74. Sen K, Mehta T, Sansare S, Sharifi L, Ma AW, Chaudhuri B. Pharmaceutical applications of powder-based binder jet 3D printing process-a review. *Advanced Drug Delivery Reviews*. 2021 Oct 1;177:113943.
75. Gülcan O, Günayd NK, Tamer A. The state of the art of material jetting-a critical review. *Polymers*. 2021 Aug 23;13(16):2829.
76. Elkaseer A, Chen KJ, Janhsen JC, Refle O, Hagenmeyer V, Scholz SG. Material jetting for advanced applications: A state-of-the-art review, gaps and future directions. *Additive Manufacturing*. 2022 Nov 7 33:103270.
77. Ziaee M, Crane NB. Binder jetting: A review of process, materials, and methods. *Additive Manufacturing*. 2019 Aug 1;28:781–801.
78. Gibson I, Rosen D, Stucker B, Khorasani M. Binder jetting. *Additive Manufacturing Technologies*. 2021 44:237–52.
79. Calori IR, Braga G, de Jesus PD, Bi H, Tedesco AC. Polymer scaffolds as drug delivery systems. *European Polymer Journal*. 2020 Apr 15;129:109621.
80. Pardeshi SR, Nikam A, Chandak P, Mandale V, Naik JB, Giram PS. Recent advances in PLGA based nanocarriers for drug delivery system: a state of the art review. *International Journal of Polymeric Materials and Polymeric Biomaterials*. 2023 Jan 2;72(1):49–78.
81. Arefin AM, Khatri NR, Kulkarni N, Egan PF. Polymer 3D printing review: Materials, process, and design strategies for medical applications. *Polymers*. 2021 May 6;13(9):1499.
82. Ahmad A, Abbas A, Hussain G, Al-Abbasi O, Alkahtani M, Altaf K. Performance evaluation of 3D printed polymer heat exchangers: influence of printing temperature, printing speed and wall thickness with consideration of surface roughness. *The International Journal of Advanced Manufacturing Technology*. 2023 Oct;128(7-8):3627–47.
83. Grieco M, Ursini O, Palamà IE, Gigli G, Moroni L, Cortese B. HYDRHA: Hydrogels of hyaluronic acid. New biomedical approaches in cancer, neurodegenerative diseases, and tissue engineering. *Materials Today Bio*. 2022 Oct 8:100453.
84. Zhou LY, Fu J, He Y. A review of 3D printing technologies for soft polymer materials. *Advanced Functional Materials*. 2020 Jul;30(28):2000187.
85. Huang J, Chen Q, Jiang H, Zou B, Li L, Liu J, Yu H. A survey of design methods for material extrusion polymer 3D printing. *Virtual and Physical Prototyping*. 2020 Apr 2;15(2):148–62.
86. Singh S, Ramakrishna S, Berto F. 3D Printing of polymer composites: *A short review. Material Design & Processing Communications*. 2020 Apr;2(2):e97.
87. Shahbazi M, Jäger H, Ettelaie R, Mohammadi A, Kashi PA. Multimaterial 3D printing of self-assembling smart thermo-responsive polymers into 4D printed objects: *A review. Additive Manufacturing*. 2023 Jun 5;71:103598.
88. Han B, Sharma S, Nguyen TA, Li L, Bhat KS (eds). Fiber-reinforced nanocomposites: Fundamentals and applications. Elsevier; 2020 Mar 13.
89. Valino AD, Dizon JR, EsperaJr AH, Chen Q, Messman J, Advincula RC. Advances in 3D printing of thermoplastic polymer composites and nanocomposites. *Progress in Polymer Science*. 2019 Nov 1;98:101162.
90. Schönhoff LM, Mayinger F, Eichberger M, Reznikova E, Stawarczyk B. 3D printing of dental restorations: Mechanical properties of thermoplastic polymer materials. *Journal of the Mechanical Behavior of Biomedical Materials*. 2021 Jul 1;119:104544.
91. Penumakala PK, Santo J, Thomas A. A critical review on the fused deposition modeling of thermoplastic polymer

composites. *Composites Part B: Engineering*. 2020 Nov 15;201:108336.
92. Zhou Y, Zhu L, Zhai J, Yang R, Guo X. Analysis of the mechanical behavior of polyurethane thermoset elastomers based on hydrogen bonding between different crosslinking point structures. *Polymer*. 2023 Oct 20;285:126356.
93. Joseph B, Sam RM, Tharayil A, Sagarika VK, Kalarikkal N, Thomas S. Photopolymers for 3D printing. In *Polymers for 3D Printing*. 2022 Jan 1 (pp. 145–154). William Andrew Publishing.
94. Gerdroodbar AE, Alihemmati H, Zeighami M, Bodaghi M, Kouzani AZ, Pourabbas B, Zolfagharian A. Vat polymerization 3D printing of composite acrylate photopolymer-based coated glass beads. *Materials Research Express*. 2023 Aug 29;10(8):085306.
95. Xu C, Dai G, Hong Y. Recent advances in high-strength and elastic hydrogels for 3D printing in biomedical applications. *Actabiomaterialia*. 2019 Sep 1;95:50–9.
96. Chen Z, Zhao D, Liu B, Nian G, Li X, Yin J, Qu S, Yang W. 3D printing of multifunctional hydrogels. *Advanced Functional Materials*. 2019 May;29(20):1900971.
97. Kaliaraj GS, Shanmugam DK, Dasan A, Mosas KK. Hydrogels—A Promising Materials for 3D Printing Technology. *Gels*. 2023 Mar 22;9(3):260.
98. Al-Shalawi FD, Hanim MA, Ariffin MK, Kim CL, Brabazon D, Calin R, Al-Osaimi MO. Biodegradable synthetic polymer in orthopaedic application: A review. *Materials Today: Proceedings*. 2023 Jan 9.
99. Huang X, Wang Z, Wang H, Chen D, Tong L. Novel strategies for the treatment of osteoarthritis based on biomaterials and critical molecular signaling. *Journal of Materials Science & Technology*. 2023 Jan 11 149: 42–55.
100. Elshabrawy HA, Dena AS, El-Sherbiny IM. Triple-layered platform utilizing electrospunnanofibers and 3D-printed sodium alginate-based hydrogel for effective topical treatment of rheumatoid arthritis. *International Journal of Biological Macromolecules*. 2024 Jan 4:129195.
101. Ding G, He Y, Shi Y, Maimaitimin M, Zhang X, Huang H, Huang W, Yu R, Wang J. Sustained-Drug-Release, Strong, and Anti-Swelling Water-Lipid Biphasic Hydrogels Prepared via Digital Light Processing 3D Printing for Protection against Osteoarthritis: Demonstration in a Porcine Model. *Advanced Healthcare Materials*. 2023 Mar 21:2203236.
102. Li P, Fu L, Liao Z, Peng Y, Ning C, Gao C, Zhang D, Sui X, Lin Y, Liu S, Hao C. Chitosan hydrogel/3D-printed poly (ε-caprolactone) hybrid scaffold containing synovial mesenchymal stem cells for cartilage regeneration based on tetrahedral framework nucleic acid recruitment. *Biomaterials*. 2021 Nov 1;278:121131.
103. Liu Y, Huang J, Li S, Li Z, Chen C, Qu G, Chen K, Teng Y, Ma R, Wu X, Ren J. Advancements in hydrogel-based drug delivery systems for the treatment of inflammatory bowel disease: a review. *Biomaterials Science*. 2024 12(4), 837–62.
104. Awad A, Hollis E, Goyanes A, Orlu M, Gaisford S, Basit AW. 3D printed multi-drug-loaded suppositories for acute severe ulcerative colitis. *International Journal of Pharmaceutics: X*. 2023 Dec 1;5:100165.
105. Awad A, Goyanes A, Orlu M, Gaisford S, Basit AW. 3D printed infliximab suppositories for rectal biologic delivery. *International Journal of Pharmaceutics: X*. 2023 Dec 1;5:100176.
106. Seoane-Viaño I, Gómez-Lado N, Lázare-Iglesias H, García-Otero X, Antúnez-López JR, Ruibal Á, Varela-Correa JJ, Aguiar P, Basit AW, Otero-Espinar FJ, González-Barcia M. 3D printed tacrolimus rectal formulations ameliorate colitis in an experimental animal model of inflammatory bowel disease. *Biomedicines*. 2020 Dec 2;8(12):563.
107. Pistone M, Racaniello GF, Rizzi R, Iacobazzi RM, Arduino I, Lopalco A, Lopedota AA, Denora N. Direct cyclodextrin based powder extrusion 3D printing of budesonide loaded mini-tablets for the treatment of eosinophilic colitis in pediatric patients. *International Journal of Pharmaceutics*. 2023 Feb 5;632:122592.
108. Mirdamadian SZ, Varshosaz J, Minaiyan M, Taheri A. 3D printed tablets containing oxaliplatin loaded alginate nanoparticles for colon cancer targeted delivery. An in vitro/in vivo study. *International Journal of Biological Macromolecules*. 2022 Apr 30;205:90–109.
109. Melocchi A, Uboldi M, Briatico-Vangosa F, Moutaharrik S, Cerea M, Foppoli A, Maroni A, Palugan L, Zema L, Gazzaniga A. The Chronotopic™ system for pulsatile and colonic delivery of active molecules in the era of precision medicine: Feasibility by 3D printing via fused deposition modeling (FDM). *Pharmaceutics*. 2021 May 20;13(5):759.
110. Lin M, Firoozi N, Tsai CT, Wallace MB, Kang Y. 3D-printed flexible polymer stents for potential applications in inoperable esophageal malignancies. *Actabiomaterialia*. 2019 Jan 1;83:119–29.
111. Ahangar P, Akoury E, Ramirez Garcia Luna AS, Nour A, Weber MH, Rosenzweig DH. Nanoporous 3D-printed scaffolds for local doxorubicin delivery in bone metastases secondary to prostate cancer. *Materials*. 2018 Aug 21;11(9):1485.
112. Almotairy A, Alyahya M, Althobaiti A, Almutairi M, Bandari S, Ashour EA, Repka MA. Disulfiram 3D printed film produced via hot-melt extrusion techniques as a potential anticervical cancer candidate. *International Journal of Pharmaceutics*. 2023 Mar 25;635:122709.
113. Almstrand AC, Bredberg A, Runström Eden G, Karlsson H, Assenhöj M, Koca H, Olin AC, Tinnerberg H. An explorative study on respiratory health among operators working in polymer additive manufacturing. *Frontiers in Public Health*. 2023 Apr 17;11:1148974.
114. Rezaei FS, Khorshidian A, Beram FM, Derakhshani A, Esmaeili J, Barati A. 3D printed chitosan/polycaprolactone scaffold for lung tissue engineering: hope to be useful for COVID-19 studies. *RSC advances*. 2021;11(32):19508–20.
115. Zopf DA, Flanagan CL, Wheeler M, Hollister SJ, Green GE. Treatment of severe porcine tracheomalacia with a 3-dimensionally printed, bioresorbable, external airway splint. *JAMA Otolaryngology–Head & Neck Surgery*. 2014 Jan 1;140(1):66–71.
116. Morrison RJ, Hollister SJ, Niedner MF, Mahani MG, Park AH, Mehta DK, Ohye RG, Green GE. Mitigation of tracheobronchomalacia with 3D-printed personalized medical devices in pediatric patients. *Science translational medicine*. 2015 Apr 29;7(285):285ra64-.
117. Quiñones DR, Soler-Egea D, González-Pérez V, Reibke J, Simarro-Mondejar E, Pérez-Feito R, García-Manrique JA, Crispín V, Moratal D. Open source 3D printed lung tumor movement simulator for radiotherapy quality assurance. *Materials*. 2018 Jul 30;11(8):1317.
118. Shen Y, Cui J, Yu X, Song J, Cai P, Guo W, Zhao Y, Wu J, Gu H, Sun B, Mo X. Recent advances in three-dimensional printing in cardiovascular devices: Bench and bedside applications. *Smart Materials in Medicine*. 2023 Jul 17. *5*(1): 36–51.
119. Domínguez-Robles J, Diaz-Gomez L, Utomo E, Shen T, Picco CJ, Alvarez-Lorenzo C, Concheiro A, Donnelly RF, Larrañeta E. Use of 3D printing for the development

of biodegradable antiplatelet materials for cardiovascular applications. *Pharmaceuticals*. 2021 Sep 11;14(9):921.
120. Adhami M, Picco CJ, Detamornrat U, Anjani QK, Cornelius VA, Robles-Martinez P, Margariti A, Donnelly RF, Domínguez-Robles J, Larrañeta E. Clopidogrel-loaded vascular grafts prepared using digital light processing 3D printing. *Drug Delivery and Translational Research*. 2023 Dec 5:1–5.
121. Ajdary R, Ezazi NZ, Correia A, Kemell M, Huan S, Ruskoaho HJ, Hirvonen J, Santos HA, Rojas OJ. Multifunctional 3D-printed patches for long-term drug release therapies after myocardial infarction. *Advanced Functional Materials*. 2020 Aug;30(34):2003440.
122. McDonald J Regulatory considerations for devices manufactured using additive manufacturing technologies. *Polymer-Based Additive Manufacturing: Biomedical Applications*. 2019 2019:243–54.
123. Witten CM, McFarland RD, Simek SL. Concise review: the US Food and Drug Administration and regenerative medicine. *Stem Cells Translational Medicine*. 2015 Dec 1;4(12):1495–9.
124. Trenfield SJ, Goyanes A, Telford R, Wilsdon D, Rowland M, Gaisford S, Basit AW. 3D printed drug products: Non-destructive dose verification using a rapid point-and-shoot approach. *International Journal of Pharmaceutics*. 2018 Oct 5;549(1-2):283–92.
125. Mohammed AA, Algahtani MS, Ahmad MZ, Ahmad J, Kotta S. 3D printing in medicine: Technology overview and drug delivery applications. *Annals of 3D Printed Medicine*. 2021 Dec 1;4:100037.
126. Cheng Y, Qin H, Acevedo NC, Jiang X, Shi X. 3D printing of extended-release tablets of theophylline using hydroxypropyl methylcellulose (HPMC) hydrogels. *International Journal of Pharmaceutics*. 2020 Dec 15;591:119983.
127. Economidou SN, Pere CP, Reid A, Uddin MJ, Windmill JF, Lamprou DA, Douroumis D. 3D printed microneedle patches using stereolithography (SLA) for intradermal insulin delivery. *Materials Science and Engineering: C*. 2019 Sep 1;102:743–55.
128. Goyanes A, Det-Amornrat U, Wang J, Basit AW, Gaisford S. 3D scanning and 3D printing as innovative technologies for fabricating personalized topical drug delivery systems. *Journal of Controlled Release*. 2016 Jul 28;234:41–8.
129. Barthès J, Lagarrigue P, Riabov V, Lutzweiler G, Kirsch J, Muller C, Courtial EJ, Marquette C, Projetti F, Kzhyskowska J, Lavalle P. Biofunctionalization of 3D-printed silicone implants with immunomodulatory hydrogels for controlling the innate immune response: *An in vivo model of tracheal defect repair*. *Biomaterials*. 2021 Jan 1;268:120549.
130. Rehmani SS, Al-Ayoubi AM, Ayub A, Barsky M, Lewis E, Flores R, Lebovics R, Bhora FY. Three-dimensional-printed bioengineered tracheal grafts: preclinical results and potential for human use. *The Annals of Thoracic Surgery*. 2017 Sep 1;104(3):998–1004.
131. Yang Y, Qiao X, Huang R, Chen H, Shi X, Wang J, Tan W, Tan Z. E-jet 3D printed drug delivery implants to inhibit growth and metastasis of orthotopic breast cancer. *Biomaterials*. 2020 Feb 1;230:119618.
132. Skowyra J, Pietrzak K, Alhnan MA. Fabrication of extended-release patient-tailored prednisolone tablets via fused deposition modeling (FDM) 3D printing. *European Journal of Pharmaceutical Sciences*. 2015 Feb 20;68:11–7.
133. Khaled SA, Alexander MR, Wildman RD, Wallace MJ, Sharpe S, Yoo J, Roberts CJ. 3D extrusion printing of high drug loading immediate release paracetamol tablets. *International Journal of Pharmaceutics*. 2018 Mar 1;538(1-2):223–30.
134. Wu W, Ye C, Zheng Q, Wu G, Cheng Z. A therapeutic delivery system for chronic osteomyelitis via a multi-drug implant based on three-dimensional printing technology. *Journal of Biomaterials Applications*. 2016 Aug;31(2):250–60.
135. Chaudhari VS, Malakar TK, Murty US, Banerjee S. Extruded filaments derived 3D printed medicated skin patch to mitigate destructive pulmonary tuberculosis: Design to delivery. *Expert Opinion on Drug Delivery*. 2021 Feb 1;18(2):301–13.
136. Lee BK, Yun YH, Choi JS, Choi YC, Kim JD, Cho YW. Fabrication of drug-loaded polymer microparticles with arbitrary geometries using a piezoelectric inkjet printing system. *International Journal of Pharmaceutics*. 2012 May 10;427(2):305–10.
137. Khaled SA, Burley JC, Alexander MR, Yang J, Roberts CJ. 3D printing of five-in-one dose combination polypill with defined immediate and sustained release profiles. *Journal of Controlled Release*. 2015 Nov 10;217:308–14.
138. Uddin MJ, Scoutaris N, Klepetsanis P, Chowdhry B, Prausnitz MR, Douroumis D. Inkjet printing of transdermal microneedles for the delivery of anticancer agents. *International Journal of Pharmaceutics*. 2015 Oct 30;494(2):593–602.

13 Clinical Trials and Regulatory Issues of Synthetic Polymers Employed in Inflammatory Diseases

Nitin Verma, Komal Thapa, Neha Kanojia, Prarit Chandel, and Pulkit Rohilla
Chitkara University, Himachal Pradesh, India

13.1 INTRODUCTION

The immune system must be protected from harmful stimuli and injuries because inflammation can detect and contain invading pathogens and toxins [1]. The innate immune system controls acute inflammation. This kind of reaction is meant to hasten the healing of inflammation and often lasts for a few hours to several days. The process changes to a chronic (pathological) phase, which can last for weeks to years and is characterized by protracted inflammatory episodes, when the short-term inflammatory reaction is too strong or unable to manage the inflammation-related stimuli [2, 3]. The innate immune system also contributes to chronic inflammatory illnesses because repeated acute inflammatory reactions induce tissue damage despite the fact that adaptive immune system cells are frequently associated with chronic inflammation [4]. A growing number of people are interested in developing medicines that prevent this initial response because of the long-lasting significance of acute inflammation in long-term conditions. Although the current course of treatment can minimize symptoms, it cannot ensure a full recovery, and the quality of life of patients is significantly impaired by surgery, persistent pain, and incapacity to work [5]. It is interesting to note that while many of the methods for delivering drugs to tumors may also be applied to inflammatory locations, little attention has been paid to the development of synthetic polymer-based inflammation-targeted therapy. In this chapter, we highlight key elements of the acute inflammatory response and examine therapies based on synthetic polymers designed to control or prevent this process [6]. As of right now, the only clinical treatments available are fluids, antibiotics, and systemic steroidal medications, which try to completely eradicate inflammation as opposed to only managing it. This chapter provides a quick summary of the current situation regarding clinical studies and regulatory matters concerning synthetic polymers in the context of inflammatory disorders.

13.1.1 Acute Inflammation

Pathogenic infections or exogenous factors such as mechanical trauma, ischemia-reperfusion, or chemicals can cause acute inflammation. Proteins called pattern recognition receptors (PRRs) are either expressed on innate immune cells or circulate in the blood [7, 8].

Local neutrophils, dendritic cells (DCs), and macrophages are the first cells in the affected tissue area to engage with invasive infections. Simultaneously, inflammatory cytokines are released by activated endothelial cells, such as interleukins (IL-1, IL-6, IL-8, IL-12, and IL-17), tumor necrosis factor (TNF), and interferon-γ (IFNγ). These cytokines build up in the bloodstream and activate leukocytes, or white blood cells [9]. Neutrophils play a crucial function in pathogen removal and inflammation resolution because they are the first circulating white blood cells drawn to the contaminated tissue region. Roughly 60%–70% of all systemic white blood cells in humans are neutrophils, and they locate inflammation by first acting to cytokine and chemokine releases. They then progressively move down the endothelium through the formation of mild adhesive bonds with proteins produced on the surface, such as molecules of active selectin [10].

The elevated integrin expression of endothelium is the reason for this. At this point, the membrane surface of active neutrophils is cleared of L-selectin, also known as CD62L, and P-selectin glycoprotein ligand (PSGL1). After entering the body, neutrophils use a number of defense mechanisms to stop infectious agents [11, 12]. The mechanism of synthetic polymers in acute inflammation is shown in Figure 13.1.

13.2 VARIOUS TARGETING APPROACHES FOR SYNTHETIC-BASED POLYMERS USED IN INFLAMMATION

Synthetic polymers are increasingly used in the treatment of inflammatory diseases due to their versatility and ability to be engineered for specific therapeutic applications. A separate treatment approach can be taken advantage of by particle propensity of medicines to interact with immune cells. In a melanoma in vivo model, for instance, mice given poly(lactic-co-glycolic acid) (PLGA) particles loaded with the chemotherapeutic medication paclitaxel

DOI: 10.1201/9781003477389-13

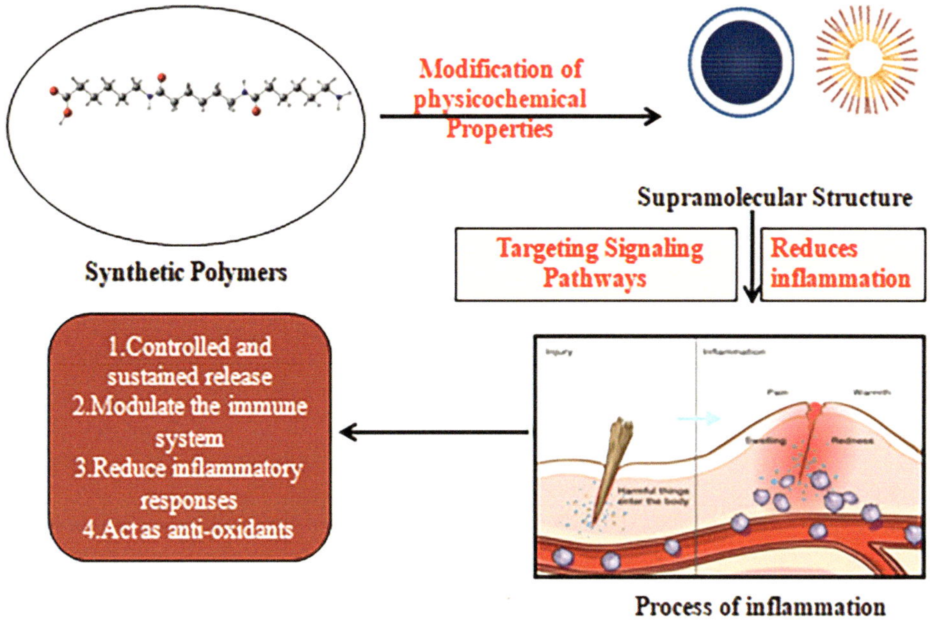

FIGURE 13.1 Depiction of mechanism of synthetic polymers in inflammation.

along with the chemokine CXCL1 showed a substantial decrease in tumor size in comparison to all other control groups. The decrease in tumor burden was attributed to neutrophil uptake of the particles and chemotaxis to the CXCL1-treated tumor. This suggests that particle-based therapies may have potential applications in immune-cell-related disorders [13, 14].

13.2.1 Tissue-Specific Targeting

When a steroid hormone is placed into a PLGA nanoparticle, immunosuppressive cytokines can be supplied alongside it to treat rheumatoid arthritis. This is an example of how polymeric particles affect tissue through tissue draining. This immunosuppressive strategy altered DCs in the draining lymph node, significantly alleviating rheumatoid arthritis symptoms in mice [15–17]. More precisely, in a multiple sclerosis mouse model, co-delivery of autoantigens and cytokines via a PLGA-based particle therapy produced the antigen-specific immune tolerance required to lessen disease manifestation and progression.

13.2.2 Vasculature Targeting

Key players in and regulators of inflammatory responses are vascular endothelial cells. The endothelium quickly modifies its phenotypic to support different phases of the inflammatory response during acute inflammation. The primary mechanism by which activated endothelial cells aid in leukocyte capture and extravasation into infected or injured tissue is described in [18–20].

13.3 CONVENTIONAL APPROACHES TO TARGET INFLAMMATION

In an effort to reduce inflammation in various disorders using different medications, current research is mapping the many pathways. Among them, some medications, such as captopril and Raf kinase inhibitory protein [21], produce anti-inflammatory effect by blocking NF-κB activation; Raf kinase inhibitory protein blocks the complete system. Several substances, including thalidomide, α-HTNFRFc, and acetaminophen (paracetamol) demonstrate TNF-αantagonistic action [22, 23].

13.3.1 Nonsteroidal Anti-Inflammatory Drugs (NSAIDs)

From the humble aspirin to the more recent ascent and decline of selective COX-2 inhibitors, NSAIDs constitute a noteworthy class of medications with substantial therapeutic applications to treat inflammatory diseases like osteoarthritis and rheumatoid arthritis [24]. However, research on a vital panacea for the treatment of inflammation is still ongoing. COX-1 and COX-2 are the two main kinds of cyclooxygenase (COX) that are now in existence. It is more likely to be inhibited by analgesics such as acetaminophen (paracetamol), as anticipated by partial characterization and cloning [25, 26].

NSAIDs are categorized as a class of medication because of their capacity to inhibit prostaglandin G/H synthase's COX activity and prevent prostaglandins from being in the metabolism of arachidonic acid. H_2 then forms a complex cascade of signaling lipids, including prostaglandins and thromboxane, which is the main prostanoid metabolite in platelets [27, 28] Recently, a number of selective COX-2 inhibitors have been discovered that have the fewest gastro-intestinal adverse effects while being just as effective as traditional NSAIDs. Both the vasoconstrictive leukotrienes and the chemo-attractant LTB4 are frequently produced by the 5–LOX metabolic pathway [29, 30].

13.3.2 Glucocorticoids

The function of glucocorticoids, a class of steroid hormones produced by the adrenal cortex in response to inflammatory stress-related stimuli that activate the hypothalamus pituitary-adrenal axis (HPA), is typically mediated by cytosolic receptors, specifically glucocorticoid receptors (GRs) [31]. GRs may also regulate the expression of certain genes through interactions with other transcription factors [32]. The anti-inflammatory effects of glucocorticoids are mediated by a negative interaction with the well-known proinflammatory cytokine expression inducers, NF-B and AP-1; however, GRs may also directly interact with NF-αB and/or AP-1. As a result, glucocorticoids stimulate GRs, which may then attach to activated NF-κB and prevent it from binding to both the nucleus and the cytoplasm [33]. Proinflammatory cytokine expression is primarily regulated by transcription factors NF-κB and AP-1, which are regulated by GRs expressed by microglial cells. It's interesting to note that dexamethasone, a synthetic glucocorticoid, has been demonstrated in rodent models to exhibit neuroprotection against toxicity induced by 1-methyl-4-phenyl-1,2,3,6-tetrahydropyridine (MPTP) or lipopolysaccharide (LPS). In these two models, it was discovered that the administration of dexamethasone inhibited the activity of microglia, which are linked to neurodegeneration [34].

13.3.3 ACE (Angiotensinogen Converting Enzyme) Inhibitors

In both human patients with essential hypertension and animal models of arterial hypertension, ACE inhibitors are beneficial in lowering blood pressure and reversing left ventricular hypertrophy [35]. Furthermore, it has been demonstrated that ACE inhibitor therapy can reverse the elevated generation of superoxide anions, NFαβ activation, proinflammatory cytokine expression [36]. An ACE inhibitor called captopril has recently been found to have anti-inflammatory properties in hypertensive animals. Its anti-inflammatory properties have also been demonstrated in a number of other diseases that involve inflammation, including colitis and arthritis [37].

13.3.4 Agonists PPAR-γ (Peroxisome Proliferator Activated Receptors) Agonists

In glial and neuronal cells of the peripheral and central nervous systems, PPAR-γ exhibits anti-inflammatory effects. It accomplishes this by interacting with the transcription factor NF-κB, which is subsequently activated by its natural ligands, PGJ2 and eicosanoids, to influence the expression of proinflammatory molecules, including iNOs, COX-2, and, ultimately, a vast class of cytokines [38]. The licensed treatments for type-2 diabetes, pioglitazone and rosiglitazone, are synthetic PPAR-γ agonists that have been shown to have a neuroprotective impact on models of various disorders [39]. Particularly, it has been demonstrated that pioglitazone reverses the loss of dopaminergic neuronal cells in mouse SNpc and the activation of microglia brought on by MPTP (1-methyl-4-phenyl-1,2,3,6-tetrahydropyridine) [40]. Additionally, pioglitazone has been demonstrated in recent research to possess anti-inflammatory characteristics and to mitigate diabetic nephropathy by preventing NF-B activation. Experimental and clinical studies have also supported the conceptual link between PPAR-γ activity and anti-atherosclerotic and anti-diabetic effects, and a number of drugs, including telmisartan, have been shown to lower inflammation and have an agonistic effect on PPAR [41, 42].

13.4 FACTORS INFLUENCING SYNTHETIC POLYMERS IN INFLAMMATORY DISEASE TREATMENT

For the purpose of treating inflammatory disorders, material, size, shape, surface chemistry, and deformability of polymeric particles can all be changed to guarantee that they interact with immune cells in an efficient manner.

13.4.1 Material

Particle buildup and by-product release are influenced by material selection, and these factors can influence immune system regulation and inflammation responses. Anti-inflammatory polymers can also be polymerized to create polymeric particles. It is possible to functionalize biodegradable polymers with various anti-inflammatory drugs, such as ibuprofen, naproxen, and aspirin [43–49].

13.4.2 Surface Modifications

Particulate systems' surface chemistry and coatings have a significant impact on how well carriers interact with immune cells, as well as how well they clear particles and have a therapeutic effect. PEGylated particles demonstrate a longer residence time when administered via other routes, such as the pulmonary route, in addition to intravenous delivery [50]. As a result, for at least a month, PEGylated particles demonstrated enhanced retention in the lungs and little inflammatory response [64]. While polyethylene glycol (PEG) is generally acknowledged to be immunologically innocuous and to rarely cause allergies, certain PEGylated drug formulations have the potential to initiate. The activation of the complement could, in a rare instance, result in life-threatening allergies. Furthermore, because PEGylation is widely used in pharmaceutical research, PEG-specific antibodies that compromise potential of PEGylation use have been discovered [51]. For instance, it has been observed that repeated doses of PEGylated particles, especially liposome carriers, enhance their removal from the blood circulation in mice and rats. As a result, biodegradable polymers like poly(glutamic acid) (PGA) and ionic liquid coatings are being investigated as PEG substitutes. Despite their critical roles in immune response modulation and pathogen clearance, neutrophils – as contrasted to macrophages and monocytes – have received less emphasis regarding the immune-evasive effects of particle PEGylation on their absorption [52]. It was found that elements found in human plasma are responsible for PEGylated particles' diminished ability to evade the immune system [66]. Determining the ideal ligand surface density is crucial in order to avoid nonspecific targeting effects or inadequate endothelium targeting. Due to increased multivalent interactions, high ligand density on the particle surface generally decreases particle detaching forces.

13.4.3 Deformation

When designing therapeutics based on particles, one adjustable factor is particle deformability [53]. Therefore, the goal of recent research has been to comprehend how particle deformability affects the creation of particle-based medication carriers. For instance, in vitro, J774 macrophages absorb stiffer (3,000 kPa) PEG-based nanoparticles more quickly than softer (10 kPa) ones. For up to four hours, soft particles also exhibit a higher persistence in the blood; beyond that, the difference is significantly diminished [54]. Similarly, compared to softer beads, micrometer-sized, rigid polyacrylamide beads with a threefold higher modulus are more likely to be phagocytosed in vitro by bone marrow-derived macrophages. Monocytes derived from bone marrow display comparable characteristics, demonstrating a threefold decrease in the uptake of soft disk-shaped particles (1.3 kPa) in contrast to their rigid counterparts (15 kPa). According to a study, deformability is a crucial factor to take into account, particularly when avoiding leukocyte clearance for immunomodulation strategies that target specific blood vessels [55].

13.5 SYNTHETIC POLYMERIC DRUG CARRIERS IN CLINICAL TRIALS

Many different kinds of studies on the administration of pharmaceutical agents have been published in recent decades. Additionally, there has been a significant increase in the development of various drug delivery methods, such as drug-carrier conjugations, transdermal, inhalers, implants, and nanoparticulate. While there are benefits and drawbacks to each kind of drug delivery system, some are already on the market, and some are starting clinical trials [56]. In the development and commercialization of synthetic polymers for the management of inflammation, regulatory oversight is essential. Thorough preclinical safety assessments are necessary from the start to guarantee adherence to legal requirements prior to moving forward with human trials. Concerns about possible immunogenicity and negative effects are addressed by these assessments, which include a thorough analysis of the synthetic polymers' pharmacokinetics, toxicity profiles, and biocompatibility. As a result, protocols carefully crafted to assess safety and efficacy endpoints must be used in clinical trials, and design and implementation must strictly conform to Good Clinical Practice (GCP) guidelines. To assess the safety of a novel scaffold biomaterial, a clinical trial has been started [57]. Patients slated for diagnostic sural nerve biopsy – a common procedure used in clinical practice – are the target population for this trial. Nerve conduction studies are performed on patients as part of the screening process in order to evaluate sural nerve function. Individuals who meet predetermined criteria and have sensory nerve action potential (SNAP) amplitudes equal to or greater than 2 μV are eligible for enrollment. This methodical approach to randomization and patient selection guarantees the validity and consistency of the trial outcomes, allowing for a thorough assessment of the novel scaffold biomaterial's safety profile [58].

13.6 CONCLUSION AND FUTURE PROSPECTIVES

Because of their great versatility, polymeric particle-based therapeutics are a great tool for developing treatment plans for inflammatory diseases. The development of particle-based therapeutics has significantly enhanced the clinical efficacy of numerous treatments, such as those for growth failure, endometriosis, cancer, gum disease, and mood disorders. Over the past 30 years, the Food and Drug Administration (FDA) has approved only 12 PLGA particle-based formulations, despite the abundance of literature on the subject and a variety of clinically available polymeric-particle-based therapeutics. Translational discrepancies between animal models and humans are the primary cause of this sharp disparity between research and clinical approval. Despite this established involvement, the challenges of simulating inflammatory signaling pathways in a test tube restrict the use of in vitro assays. Furthermore, there are species-specific differences in complement reactions, which complicates the development of accurate in vivo assays for clinical application. Optimized surface properties in conjunction with slower infusion rates can aid in the prevention of complement activation-related pseudoallergy (CARPA) reactions. Still, screening assays must be created in order to look into ways to stop CARPA reactions. The regulatory approval process presents additional challenges in the design of novel particle-based therapeutics. Particle-based medications, in contrast to systemic, carrier-free therapeutics, are generally made up of polymeric carriers, therapeutic agents, and surface modifications. Regulatory evaluation is difficult because even a small alteration to any one of these elements can have a significant impact on particle function, biodistribution, and toxicity issues. This is why the Nanotechnology Characterization Laboratory (NCL) was established by the National Cancer Institute in order to create standardized assays for characterizing particle-based therapeutics and associated toxicities191. The FDA approval process and clinical trial procedures were streamlined by the establishment of the NCL. Crucially, particle-based therapies intended to target circulating phagocytes are intrinsically prohibited by NCL guidelines, such as extended evasion of the mononuclear phagocyte system [193]. Notwithstanding these obstacles to translation, the scientific community's steady advancement and the expedited approval process have the potential to completely transform particle-based therapeutics. Particle technologies offer viable treatments for a number of severe inflammatory illnesses. Particles' physical and chemical characteristics, as well as their customizable nature, meet the need for novel clinical applications, such as the development of vaccines and the management of systemic inflammation. It is now possible to fine-tune the synthesis of polymers, but the lack of scaling-up technologies to fabricate large quantities of nonspherical particles limits the clinical translation of polymeric-based particles. The development of large-scale production workflows is necessary because the physicochemical parameters controlling the fabrication of laboratory-batch particles are frequently extremely complex. Fortunately, promising large-scale processes, such as those based on microfluidics and lithography, are presently being investigated for the fabrication of complex-shaped particles. While additional efforts are required to surmount these challenges, the swift growth of particle-based medicine presents cutting-edge remedies for worldwide issues. In summary, although synthetic polymer-related clinical trials for inflammation pose regulatory obstacles, they also provide encouraging paths for the development of new treatments. To overcome these obstacles, rigorous preclinical assessments, well-thought-out clinical trial designs, and thorough regulatory filings are required. Mechanisms for post-market surveillance are essential for keeping an eye on long-term safety and effectiveness and guaranteeing continued adherence to legal requirements. In the future, sustained cooperation between regulatory bodies and interested parties will enable global harmonization, simplifying the regulatory environment and hastening the creation of novel synthetic polymer-based anti-inflammatory treatments. Adopting cutting-edge technologies and innovative methods for drug delivery and biomaterial design has enormous potential to improve the efficacy and safety of these treatments, which will ultimately benefit patients suffering from inflammatory diseases.

REFERENCES

[1] Mogensen, T.H., 2009. Pathogen recognition and inflammatory signaling in innate immune defenses. *Clinical microbiology reviews*, *22*(2), pp. 240–273.

[2] Santacroce, L., Topi, S., Charitos, I.A., Lovero, R., Luperto, P., Palmirotta, R. and Jirillo, E., 2024. Current views about the inflammatory damage triggered by bacterial superantigens and experimental attempts to neutralize superantigen-mediated toxic effects with natural and biological products. *Pathophysiology*, *31*(1), pp. 18–31.

[3] Sparks, R. and Nichol, A., 2024. Inflammation, immunity and allergy. In *Anaesthesia & Intensive Care Medicine*.

[4] Bhol, N.K., Bhanjadeo, M.M., Singh, A.K., Dash, U.C., Ojha, R.R., Majhi, S., Duttaroy, A.K. and Jena, A.B., 2024. The interplay between cytokines, inflammation, and antioxidants: mechanistic insights and therapeutic potentials of various antioxidants and anti-cytokine compounds. *Biomedicine & Pharmacotherapy*, *178*, p. 117177.

[5] Sarfi, S., Azaryan, E. and Naseri, M., 2024. Immune system of dental pulp in inflamed and normal tissue. *DNA and Cell Biology*.

[6] Santacroce, L., Topi, S., Charitos, I.A., Lovero, R., Luperto, P., Palmirotta, R. and Jirillo, E., 2024. Current views about the inflammatory damage triggered by bacterial superantigens and experimental attempts to neutralize superantigen-mediated toxic effects with natural and biological products. *Pathophysiology*, *31*(1), pp. 18–31.

[7] Antar, S.A., Mahmoud, A.M., Abdo, W., Gad, C. and Al-Karmalawy, A.A., 2023. A comprehensive overview of organ inflammatory responses: Genesis, possible mechanisms, and mediators of inflammation. *Pharmaceutical Sciences*, *29*(4), pp. 397–416.

[8] Land, W.G., 2023. Perspectives of the danger/injury model of immunology as applied to antigen-related human disorders. In *Damage-Associated Molecular Patterns in Human Diseases: Volume 3: Antigen-Related Disorders* (pp. 3–44). Cham: Springer International Publishing.

[9] Shafqat, A., Khan, J.A., Alkachem, A.Y., Sabur, H., Alkattan, K., Yaqinuddin, A. and Sing, G.K., 2023. How neutrophils shape the immune response: reassessing their multifaceted role in health and disease. *International Journal of Molecular Sciences*, *24*(24), p. 17583.

[10] Shankar, J., Thakur, R., Clemons, K.V. and Stevens, D.A., 2024. Interplay of cytokines and chemokines in Aspergillosis. *Journal of Fungi*, *10*(4), p. 251.

[11] Ales, E. and Sackstein, R., 2023. The biology of E-selectin ligands in leukemogenesis. *Advances in Cancer Research*, *157*, pp. 229–250.

[12] He, J., Ni, Z. and Li, Z., 2023. Intercellular adhesion molecule 1 and selectin l play crucial roles in ulcerative colitis. *Medicine*, *102*(49), p. e36552.

[13] Farasati Far, B., Naimi-Jamal, M.R., Sedaghat, M., Hoseini, A., Mohammadi, N. and Bodaghi, M., 2023. Combinational system of lipid-based nanocarriers and biodegradable polymers for wound healing: an updated review. *Journal of Functional Biomaterials*, *14*(2), p. 115.

[14] Suresh, P.K., Banerjee, A. and Singh, P., 2023. Promises and challenges for targeting the immunological players in the tumour micro-environment–Critical determinants for NP-based therapy. *Open Nano*, *10*, p. 100134.

[15] Kumari, S., Goyal, A., Sönmez Gürer, E., Algın Yapar, E., Garg, M., Sood, M. and Sindhu, R.K., 2022. Bioactive loaded novel nano-formulations for targeted drug delivery and their therapeutic potential. *Pharmaceutics*, *14*(5), p. 1091.

[16] Phan, N.M., Nguyen, T.L., Shin, H., Trinh, T.A. and Kim, J., 2023. ROS-Scavenging Lignin-Based Tolerogenic Nanoparticle Vaccine for Treatment of Multiple Sclerosis. *ACS nano*, *17*(24), pp. 24696–24709.

[17] Huang, Y., Guo, X., Wu, Y., Chen, X., Feng, L., Xie, N. and Shen, G., 2024. Nanotechnology's frontier in combatting infectious and inflammatory diseases: prevention and treatment. *Signal Transduction and Targeted Therapy*, *9*(1), p. 34.

[18] Li, X., Li, C., Zhang, W., Wang, Y., Qian, P. and Huang, H., 2023. Inflammation and aging: signaling pathways and intervention therapies. *Signal Transduction and Targeted Therapy*, *8*(1), p.239.

[19] Al-Soudi, A., Kaaij, M.H. and Tas, S.W., 2017. Endothelial cells: From innocent bystanders to active participants in immune responses. *Autoimmunity reviews*, *16*(9), pp. 951–962.

[20] Pazhani, P., Arumugam, S., Dharmian, J.P. and Medapatti, V.P., 2024. Anti–intercellular adhesion molecule-1 (ICAM-1) nanoparticles for damaged endothelium to treat venous thromboembolism. *Journal of Dispersion Science and Technology*, 2024: 1–13.

[21] Sokic-Milutinovic, A. and Milosavljevic, T., 2024. Inflammatory Bowel Disease: From Conventional Immunosuppression to Biologic Therapy. *Digestive Diseases*, 42(4), pp. 325–335.

[22] Lee, J., Olivieri, C., Ong, C., Masterson, L.R., Gomes, S., Lee, B.S., Schaefer, F., Lorenz, K., Veglia, G. and Rosner, M.R., 2022. Raf Kinase Inhibitory Protein regulates the cAMP-dependent protein kinase signaling pathway through a positive feedback loop. *Proceedings of the National Academy of Sciences*, *119*(25), p. e2121867119.

[23] Filippone, A., Mannino, D., Casili, G., Lanza, M., Paterniti, I., Cuzzocrea, S., Capra, A.P., Colarossi, L., Giuffrida, D., Lombardo, S.P. and Esposito, E., 2022. Protein kinase inhibitors as a new target for immune system modulation and brain cancer management. *International Journal of Molecular Sciences*, *23*(24), p. 15693.

[24] Lee, J.K., Abbas, A.A., Cheah, T.E., Simanjuntak, R.N., Sockalingam, S. and Roohi, S., 2023. Topical nonsteroidal anti-inflammatory drugs for management of osteoarthritis pain: A consensus recommendation. *Journal of Orthopaedic Research®*, *41*(9), pp. 1916–1924.

[25] Nakata, K., Hanai, T., Take, Y., Osada, T., Tsuchiya, T., Shima, D. and Fujimoto, Y., 2018. Disease-modifying effects of COX-2 selective inhibitors and non-selective NSAIDs in osteoarthritis: a systematic review. *Osteoarthritis and cartilage*, *26*(10), pp. 1263–1273.

[26] Govindaswamy, S., Ponnusamy, K. S. I., Anil, V., Venugopal, J. and Sundaram, D.P., 2024. A systematic review and meta-analysis of selective cyclooxygenase-2 inhibitors and non-selective non-steroidal anti-inflammatory drugs for acute gout. *Journal of Pharmacology and Pharmacotherapeutics*, *15*(2), 122–132.

[27] Colella, F., Garcia, J.P., Sorbona, M., Lolli, A., Antunes, B., D'Atri, D., Barré, F.P., Oieni, J., Vainieri, M.L., Zerrillo, L. and Capar, S., 2020. Drug delivery in intervertebral disc degeneration and osteoarthritis: Selecting the optimal platform for the delivery of disease-modifying agents. *Journal of controlled release*, *328*, pp. 985–999.

[28] Colella, F., Garcia, J.P., Sorbona, M., Lolli, A., Antunes, B., D'Atri, D., Barré, F.P., Oieni, J., Vainieri, M.L., Zerrillo, L. and Capar, S., 2020. Drug delivery in intervertebral disc degeneration and osteoarthritis: Selecting the optimal platform for the delivery of disease-modifying agents. *Journal of controlled release*, *328*, pp. 985–999.

[29] Timur, U.T., Caron, M.M., Jeuken, R.M., Bastiaansen-Jenniskens, Y.M., Welting, T.J., van Rhijn, L.W., van Osch, G.J. and Emans, P.J., 2020. Chondroprotective actions of selective COX-2 inhibitors in vivo: a systematic review. *International journal of molecular sciences*, *21*(18), p. 6962.

[30] Wang, B., Wu, L., Chen, J., Dong, L., Chen, C., Wen, Z., Hu, J., Fleming, I. and Wang, D.W., 2021. Metabolism pathways of arachidonic acids: mechanisms and potential therapeutic targets. *Signal Transduction and Targeted Therapy*, *6*(1), p. 94.

[31] Al-Mosawi, R.H. and Alebadi, N.A.M., 2023. Usage of corticosteroids as therapeutic agents in diseases. *Medical Journal of Babylon*, *20*(3), pp. 447–450.

[32] Meduri, G.U. and Chrousos, G.P., 2020. General adaptation in critical illness: glucocorticoid receptor-alpha master regulator of homeostatic corrections. *Frontiers in Endocrinology*, *11*, p. 161.

[33] Roy, R., 2021. *Integrative Physiologic Pharmacodynamics: Using Biomarkers to Integrate Inflammatory Pathways in the Acute Glucocorticoid Drug Response and Diurnal Cortisol Response* (Doctoral dissertation, State University of New York at Binghamton).

[34] Chilakamarthi, U., Mahadik, N.S., Bhattacharyya, T., Gangadhar, P.S., Giribabu, L. and Banerjee, R., 2024. Glucocorticoid receptor mediated sensitization of colon cancer to photodynamic therapy induced cell death. *Journal of Photochemistry and Photobiology B: Biology*, *251*, p. 112846.

[35] Zhu, Y., Chen, C., Dai, Z., Wang, H., Zhang, Y., Zhao, Q., Xue, Y. and Shen, Q., 2024. Identification, screening and molecular mechanisms of natural stable angiotensin-converting enzyme (ACE) inhibitory peptides from foxtail millet protein hydrolysates: a combined in silico and in vitro study. *Food & Function*, *15*(15), pp. 7782–7793.

[36] Gregory, K.S., Cozier, G.E., Schwager, S.L., Sturrock, E.D. and Acharya, K.R., 2024. Structural insights into the inhibitory mechanism of angiotensin-I-converting enzyme by the lactotripeptides IPP and VPP. *FEBS letters*, *598*(2), pp. 242–251.

[37] Chen, B., Wang, X., Zhang, J. and Wang, L., 2024. Peptidomics-based study of antihypertensive activity: discovery of novel ACE inhibiting peptides from peanut yogurt. *Food & Function*, *15*(12), pp. 6705–6716.

[38] Pearson, A., Koprivica, M., Eisenbaum, M., Ortiz, C., Browning, M., Vincennie, T., Tinsley, C., Mullan, M., Crawford, F. and Ojo, J., 2024. PPARγ activation ameliorates cognitive impairment and chronic microglial activation in the aftermath of r-mTBI. *Journal of Neuroinfl ammation*, *21*(1), p. 194.

[39] Liu, C.H., Lee, T.H., Lin, Y.S., Sung, P.S., Wei, Y.C. and Li, Y.R., 2020. Pioglitazone and PPAR-γ modulating treatment in hypertensive and type 2 diabetic patients after ischemic stroke: a national cohort study. *Cardiovascular Diabetology*, *19*, pp. 1–13.

[40] Ha, J., Choi, D.W., Kim, K.J., Kim, K.Y., Nam, C.M. and Kim, E., 2023. Pioglitazone use and reduced risk of dementia in patients with diabetes mellitus with a history of ischemic stroke. *Neurology*, *100*(17), pp. e1799–e1811.

[41] Tseng, C.H., 2023. Rosiglitazone does not affect the risk of inflammatory bowel disease: a retrospective cohort study in Taiwanese type 2 diabetes patients. *Pharmaceuticals*, *16*(5), p. 679.

[42] Sangnim, T., Puri, V., Dheer, D., Venkatesh, D.N., Huanbutta, K. and Sharma, A., 2024. Nanomaterials in the

wound healing process: New insights and advancements. *Pharmaceutics*, *16*(3), p. 300.

[43] Escobar, K., Garrido-Miranda, K.A., Pulido, R., Naveas, N., Manso-Silván, M. and Hernandez-Montelongo, J., 2023. Coatings of cyclodextrin/citric-acid biopolymer as drug delivery systems: A review. *Pharmaceutics*, *15*(1), p. 296.

[44] Talha, M., 2024. Dopamine functionalized coatings for corrosion protection of metallic implants and advanced drug delivery: a review. *Corrosion Reviews*, *42*(2), pp. 163–183.

[45] Opriş, O., Mormile, C., Lung, I., Stegarescu, A., Soran, M.L. and Soran, A., 2024. An overview of biopolymers for drug delivery applications. *Applied Sciences*, *14*(4), p. 1383.

[46] Pal, R., Pandey, P., Anand, A., Saxena, A., Thakur, S.K., Malakar, R.K. and Kumar, V., 2023. The pharmaceutical polymer's; a current status in drug delivery: A comprehensive review. *Journal of Survey in Fisheries Sciences*, 10: 3682–3692.

[47] Schappo, F.B., Assunção, L.S., Ribeiro, C.D.F. and Nunes, I.L., 2023. PLGA nanoparticles: Current uses and emerging trends. *Poly (lactic-co-glycolic acid)(PLGA) Nanoparticles for Drug Delivery*, pp. 477–506.

[48] Puricelli, C., Gigliotti, C.L., Stoppa, I., Sacchetti, S., Pantham, D., Scomparin, A., Rolla, R., Pizzimenti, S., Dianzani, U., Boggio, E. and Sutti, S., 2023. Use of poly lactic-co-glycolic acid nano and micro particles in the delivery of drugs modulating different phases of inflammation. *Pharmaceutics*, *15*(6), p. 1772.

[49] Kotla, N.G., Langlois, J.B., Fisch, A., Kramer, I. and Halleux, C., 2023. Surface modified cationic PLGA microparticles as long-acting injectable carriers for intra-articular small molecule drug delivery. *European Journal of Pharmaceutics and Biopharmaceutics*, *193*, pp. 96–104.

[50] Graván, P., Aguilera-Garrido, A., Marchal, J.A., Navarro-Marchal, S.A. and Galisteo-González, F., 2023. Lipid-core nanoparticles: Classification, preparation methods, routes of administration and recent advances in cancer treatment. *Advances in Colloid and Interface Science*, *314*, p. 102871.

[51] Guadarrama-Escobar, O.R., Serrano-Castañeda, P., Anguiano-Almazán, E., Vázquez-Durán, A., Peña-Juárez, M.C., Vera-Graziano, R., Morales-Florido, M.I., Rodriguez-Perez, B., Rodriguez-Cruz, I.M., Miranda-Calderón, J.E. and Escobar-Chávez, J.J., 2023. Chitosan nanoparticles as oral drug carriers. *International Journal of Molecular Sciences*, *24*(5), p. 4289.

[52] Irvin-Choy, N.D.S., Nelson, K.M., Gleghorn, J.P. and Day, E.S., 2023. Delivery and short-term maternal and fetal safety of vaginally administered PEG-PLGA nanoparticles. *Drug Delivery and Translational Research*, *13*(12), pp. 3003–3013.

[53] Rahman, N. and Khalil, N., 2023. Synthesis and characterization of 1-phenylisatin incorporated chitosan/PEG nanoparticles: Interaction with bovine serum albumin and application in drug delivery. *Journal of Drug Delivery Science and Technology*, *86*, p. 104738.

[54] Yusuf, A., Almotairy, A.R.Z., Henidi, H., Alshehri, O.Y. and Aldughaim, M.S., 2023. Nanoparticles as drug delivery systems: A review of the implication of nanoparticles' physicochemical properties on responses in biological systems. *Polymers*, *15*(7), p. 1596.

[55] Herrera León, C., Kalacas, N.A., Mier, A., Sakhaii, P., Merlier, F., Prost, E., Maffucci, I., Montagna, V., Mora-Radó, H., Dhal, P.K. and Tse Sum Bui, B., 2023. Synthetic Peptide antibodies as TNF-α inhibitors: Molecularly imprinted polymer nanogels neutralize the inflammatory activity of TNF-α in THP-1 derived macrophages. *Angewandte Chemie*, *135*(34), p. e202306274.

[56] Coats, J.P., Cochereau, R., Dinu, I.A., Messmer, D., Sciortino, F. and Palivan, C.G., 2023. Trends in the synthesis of polymer nano-and microscale materials for bio-related applications. *Macromolecular Bioscience*, *23*(8), p. 2200474.

[57] Kyriakides, T.R., Kim, H.J., Zheng, C., Harkins, L., Tao, W. and Deschenes, E., 2022. Foreign body response to synthetic polymer biomaterials and the role of adaptive immunity. *Biomedical Materials*, *17*(2), p. 022007.

[58] Chaurasia, S. and Mishra, B., 2024. Synthetic polymers as biomaterials for the treatment of colon diseases. In *Advanced Drug Delivery Systems for Colonic Disorders* (pp. 95–130). Academic Press.

14 Synthetic Polymer-Based Drug Delivery Systems in Inflammatory Diseases

Critical Factors and Challenges

Anisha Kumari, Zahanat Muneer Mir, and Parul Thakur
Maharaja Ranjit Singh Punjab Technical University, Bathinda, India

Jitender Madan
National Institute of Pharmaceutical Education and Research, Hyderabad, India

Navjot Kanwar and Ashish Baldi
Maharaja Ranjit Singh Punjab Technical University, Bathinda, India

14.1 INTRODUCTION

Inflammatory diseases are conditions characterized by inflammation in different body sites and are immune system reactions to damage, infection, or other pathogens. Although the body needs inflammation as a normal and essential process to heal itself, chronic inflammation may cause various illnesses [1]. Inflammatory disorders may impact several organs and systems, including the digestive system (ulcerative colitis, Crohn's disease), skin (psoriasis, eczema), joints (rheumatoid arthritis), lungs (asthma), and cardiovascular system (atherosclerosis). Around the world, the prevalence of some inflammatory diseases, such as inflammatory bowel illness, asthma, and rheumatoid arthritis, is rising [2]. Rheumatoid arthritis, an inflammatory disease, affects between 0.24% and 1% of people globally, with greater incidence in industrialized nations, according to the World Health Organization. An estimated 0.5%–1% of adults are thought to have rheumatoid arthritis in India, with a greater frequency in cities [3]. An estimated several million individuals worldwide suffer from another inflammatory disease of gastrointestinal tract – i.e., ulcerative colitis and Crohn's disease – with a growing frequency in both industrialized and developing nations [4]. In metropolitan areas of India, there is also a tendency toward a rise in the incidence of these inflammatory bowel diseases. Air pollution and environmental factors are major contributors to the incidence of asthma and chronic obstructive pulmonary disease (COPD), which affect the population of India [4]. The innate immune system controls acute inflammation, which is in charge of identifying an inflammatory stimulus at the outset and concentrating on quickly containing the pathogen or damage, lasting anywhere from a few hours to many days; this kind of reaction tries to hasten the cure of inflammation [5]. A chronic (pathological) phase of inflammation, which may continue anywhere from weeks to years, is defined by protracted bouts of inflammation and occurs when the acute inflammation response cannot control the inflammatory stimuli. Repeated acute inflammatory events induce tissue damage in chronic inflammatory illnesses; thus, while cells from the adaptive immune system have been linked to chronic inflammation, the immune system's innate component is also involved [6]. The increasing interest in creating treatments targeted at squelching this acute response has partly been attributed to the realization of the enduring significance that acute inflammation plays in chronic illnesses. The usage of synthetic polymers in the management of inflammatory illnesses is growing owing to certain qualities of these polymers such as controlled medication release, biocompatibility, and biodegradability [7].

Synthetic polymers are large molecules that are composed of chemically bonded monomers or repetitive structural elements. These polymers are human-made, by joining together tiny molecules to create the lengthy chains that characterize polymers; this process is known as polymerization [8]. Because of their adaptability, durability, and adjustable qualities, synthetic polymers have many uses, they are employed in several sectors, including the building, automotive, textile, plastic, and health-care industries. Dextran, pullulan, polylactic-co-glycolic acid, polystyrene, poly(lactic acid) (PLA), and poly (e-caprolactone) are examples of common synthetic polymers. Synthetic polymers may be engineered to target and modify the inflammatory response more precisely and effectively when employed in

DOI: 10.1201/9781003477389-14

the setting of inflammatory illnesses. Anti-inflammatory medication administration is the main application of synthetic polymers in the treatment of inflammatory illnesses [9, 10]. These polymers can encapsulate medications and release them at the site of inflammation, therefore offering targeted and long-lasting treatment. This strategy enhances the treatment's overall effectiveness and reduces systemic negative effects [11]. Extensive use of synthetic polymers has, however, also sparked worries about sustainability and the effects on the environment, which has prompted continuous research in biodegradable and environmentally friendly substitutes. The adaptability, durability, mechanical strength, flexibility, elasticity, chemical resistance, electrical insulation, lightweight, transparency, moldability, thermal stability, and adjustable qualities of synthetic polymers make them appropriate for numerous applications [12]. Despite that, there are several challenges in using synthetic polymers to cure inflammatory diseases. One of the main challenges is getting targeted distribution to the diseased tissues when employing synthetic polymers to treat inflammatory diseases. It might be challenging to distribute therapeutic polymers to the affected organs or tissues of inflammatory disorders while reducing off-target effects [10]. For synthetic polymers to be safe and successful in treating inflammatory disorders, the body must be able to tolerate them and maintain their stability. The possibility of immunological responses to the synthetic polymers poses a serious risk and may reduce their therapeutic potential [13]. Creating synthetic polymers with ideal pharmacokinetic characteristics, such as suitable half-lives and clearance rates, is another difficulty in the formulation and use of these materials for the cure of inflammatory illnesses [14]. In addition, synthetic polymers face various challenges in scaling up. Reaching consistent polymerization processes on a bigger scale is one of the main challenges when scaling up, and variables that impact the attributes and quality of the polymer, such as heat and mass transfer, mixing efficiency, and reactor design, become increasingly important [15]. Furthermore, a major problem is ensuring the stability and repeatability of polymer features, such as molecular weight distribution and polymer shape. Because of the complexity of the processes involved and the possibility of fluctuations in raw materials and operating circumstances, controlling these attributes becomes very challenging when operating at an industrial scale [16]. In this chapter, we will understand the numerous features of using synthetic polymers for the cure of inflammatory diseases.

14.2 POLYMERIC DRUG DELIVERY SYSTEM IN THE TREATMENT OF INFLAMMATION

In treating inflammation, polymeric drug delivery methods have shown significant promise (Table 14.1). These systems use biocompatible and biodegradable polymers to manage the encapsulation and delivery of anti-inflammatory medications to specific locations [17]. These systems allow prolonged release of medications, which minimizes the need for frequent dosage and lowers the risk of side effects. Therapeutic concentrations are sustained for a prolonged time because of the control of drug release kinetics made possible by the development of drug delivery systems using polymers [18]. This is especially helpful for treating chronic inflammatory diseases because ongoing medication administration is necessary to control the disease's symptoms and course. Additionally, a targeted therapeutic strategy is provided by the potential of polymeric drug delivery systems to target regions of inflammation, reducing systemic exposure, and optimizing treatment effectiveness [19]. These methods can decrease the total amount of medication required while increasing the therapeutic result by augmenting drug accumulation in the inflammatory tissues. Polymeric drug delivery systems not only have the potential to target and release medications continuously, but they may also shield pharmaceuticals from degradation and clearance processes, which enhances their stability and bioavailability [17]. This may be particularly advantageous for medications often used to treat inflammation with low solubility or stability. Because of the adaptability of polymers, stimuli-responsive components may be included, allowing for the design of smart drug delivery systems that can react to certain inflammatory triggers like pH or enzyme activity changes [20]. This improves the accuracy and effectiveness of medicine administration to the inflammatory areas even more. There is a potential for using polymeric drug delivery systems to treat inflammation and enhance the effectiveness of anti-inflammatory medications [21]. Their capacity to provide prolonged, confined, and focused medication administration, in addition to improved stability and bioavailability, renders them a beneficial method for tackling the obstacles linked to the existing approaches for treating inflammation [22].

14.3 SYNTHETIC POLYMER USED IN THE TREATMENT OF INFLAMMATION

Synthetic polymers and their derivatives have significant potential for developing more effective and personalized medications for the treatment of inflammation [10]. These materials are modified to ensure that they interact with body tissues in a way that is unable to cause any undesirable effects and to function well with bodies [56]. Some examples of common synthetic polymers include PLA, poly (e-caprolactone), PLGA, polystyrene, and dextran. Additionally, we explore their various derivative forms and physiochemical properties, as well as how they could potentially be used to treat inflammatory diseases.

14.3.1 PLGA (Polylactic-co-glycolic acid)

PLGA is well-known for being biocompatible and biodegradable, and for providing an extensive variety of adjustable mechanical attributes and erosion durations. It has been extensively investigated to develop regulated drug delivery systems in both scientific and commercial

TABLE 14.1
Synthetic Polymer–Based Formulation and Its Method of Preparation

Polymer	Drug	Formulation	Method of preparation	Remarks	Ref.
PLGA (Poly-co-Glycolic Acid)	Rhoifolin (ROF)	NP	Single emulsion (O/W) solvent evaporation technique	The finding of this investigation shows a remarkably promising nanocarrier system that possesses the capacity to improve ROF's solubility and control its release. This progression will significantly expedite the delivery of ROF in the therapeutic management of inflammatory ailments.	[23]
	RA-MH (ricinoleic acid- Magnesium hydroxide)	Microspheres	O/W method	Encapsulated in RA-MH, PLGA microspheres expressed remarkable biocompatibility and demonstrated a notable anti-inflammatory efficacy.	[24]
	Epigallocatechin-3-Gallate (EG)	Microspheres	Double-emulsion evaporation method	It has been shown that PLGA microspheres are viable carriers for enhancing the loading efficiency of EG when combined with β-CD, with the added benefit of suppressing inflammation in activated BV-2 cells.	[25]
	Sparfloxacin and Tacrolimus	NPs	Emulsion-solvent evaporation method	PLGA NPs possess remarkable compatibility with cells, a minimal tendency to induce hemolysis, and negligible toxicity when administered systemically. When these NPs are applied to mice suffering from acute lung infections, their remarkable potency against both Gram-positive and Gram-negative bacteria allows them to efficiently suppress inflammation and the immune system.	[26]
	Atorvastatin and Calcium	NPs	Single emulsion and solvent evaporation technique	Calcium NPs exhibited notable efficacy in terms of their anti-inflammatory properties.	[27]
	Curcumin (CC)	NPs	Dialysis membrane method	By lowering inflammation as part of comprehensive healing, PLGA-CC NP regulated and sustained the faster wound healing process and the release of CC and lactic acid.	[28]
	Ketorolac	Microparticles	Emulsification	The PLGA/PLA-based microparticles loaded with ketorolac have potential applications in the treatment of osteoarthritis.	[29]
	Paclitaxel	Microsphere	Solvent evaporation method	PLA microspheres containing paclitaxel and ranging in size from 35 to 105 mm released paclitaxel gradually for weeks, a promising formulation for treating inflammation in arthritic conditions.	[30]
PLA	Interleukin-10	NPs	Water/oil/water Double emulsion	PLA-PEG NPs effectively deliver IL-10, extending its half-life and releasing it slowly. This maintains IL-10's anti-inflammatory properties and its ability to regulate chlamydial inflammatory mediators in vitro. Encapsulated IL-10 demonstrates its potential as an immunotherapy for bacterial, inflammatory, and autoimmune diseases.	[31]
	Curcumin (CUR)	NPs	Emulsion-diffusion Evaporation technique	Compared to free CUR, which had moderately dilated sinusoids and little residual inflammation near the central vein, PLA-PEG NPs loaded with CUR enhanced the structure of liver tissue while lowering erythrocyte congestion, inflammatory cell infiltration, and microvesicular steatosis.	[32]
	Mirna-146a	Hydrogel NPs	Double-emulsion method	The gel/NPs/miR-146a combination is effective for delivering drugs to the nasal mucosa and treating rhinitis in rats caused by ovalbumin. It has the potential to deliver nucleic acids to treat allergic rhinitis.	[33]
	Rosiglitazone	NPs	Emulsion-evaporation method	The increased anti-inflammatory effect may be due to higher drug uptake into cells when using NPs to encapsulate rosiglitazone, in contrast to the unbound molecule. Hence, these delivery methods may be used to give anti-inflammatory drugs, which might enhance the way that atherosclerosis is now treated.	[34]

(Continued)

TABLE 14.1 (Continued)

Polymer	Drug	Formulation	Method of preparation	Remarks	Ref.
	Dexamethasone-Diclofenac (Dex-Dic)	NPs	Nanoprecipitation	The enhanced anti-inflammatory properties of Dex-Dic-loaded PLA NPs over dexamethasone and diclofenac were confirmed by TNF-α suppression in vivo. Further testing and pharmacokinetic studies are necessary, as this study indicates that the twin medication Dex-Dic may be a valuable new anti-inflammatory medication.	[35]
	Etoricoxib	NPs	Single emulsification method	Etoricoxib NPs may be a suitable option and carrier for rheumatoid arthritis treatment.	[36]
	Acyclovir	Nanosphere	Solvent deposition method	PEG-coated PLA nanospheres enhanced the ocular bioavailability of acyclovir. PLA nanospheres may be used as an ophthalmic dosage delivery system to treat ocular viral infections and improve compliance and intraocular levels of antiviral agents.	[37]
	Curcumin	Nano-micelles	Thin film hydration	Due to anti-inflammatory and antioxidant properties, nano-micelles loaded with curcumin can be neuroprotective in rats suffering from cerebral ischemia/reperfusion damage in stroke.	[38]
Polystyrene	Poly (vanillin oxalate) (PVO)	NPs	The oil/water emulsion method	PVO NPs show strong antioxidant properties by removing hydrogen peroxide and stopping the production of ROS (reactive oxygen species). Moreover, in experiments conducted in laboratories and on animals, in activated macrophages, they decrease the generation of pro-inflammatory cytokines. Therefore, it seems the potential use of PVO NPs as innovative antioxidant treatments and medication delivery systems for inflammatory illnesses are rather encouraging.	[39]
	Ferulic acid	NPs	Nanoprecipitation	The utilization of intermediate diacid NPs and ferulic acid polymer as a preventive measure against the formation of macrophage foam cells signifies a significant advancement toward the progression of innovative and potent antioxidant formulations for the management of atherosclerosis.	[40]
	Aceclofenac	Nanostructured lipid carriers (NLCs)	Microemulsion	ACE-NLCs-based hydrogel formulation exhibited superior anti-inflammatory effectiveness in rat models of paw edema caused by carrageenan.	[41]
	P-Hydroxybenzyl alcohol (HBA) HBA-incorporated copolyoxalate (HPOX)	NPs	Oil-in-water emulsion method	When HPOX NPs were released into lipopolysaccharide-activated RAW 264.7 macrophage cells, the production of NO was inhibited and TNF-ά was attenuated. The remarkable features of HPOX, particularly its ability to degrade, biological compatibility, antioxidant, and anti-inflammatory properties, suggesting high potency and versatility for a variety of uses in medication delivery.	[42]
	Indomethacin	Microsphere	Solvent evaporation Process	Indomethacin release increased by adding more permeable polymer (PCL).	[43]
Pullulan	Ketoprofen	Nanoemulsion	Spontaneous emulsification solvent diffusion method	Nanoemulsions improve drug photostability and solubility. Formulation with the highest concentration of pullulan showed significant antitumor activity against glioma cells.	[44]
	Curcumin	NPs	Solvent diffusion method	Potential use of pullulan acetate nanoparticles (PAC) coated with curcumin as an effective hepatoprotective agent and address the physical and chemical limitations of curcumin, including its solubility, stability at pH values, and photostability.	[45]
	Silibinin	Nanocapsules	Deposition of preformed polymer at interfaces.	The combined utilization of silibinin-loaded nanocapsules in a gellan gum/pullulan bilayer film has the potential to mitigate the consequences induced by 2,4-dinitrochlorobenzene. This simultaneous application demonstrates a protective effect on the skin against oxidative damage, improving the model of AD atopic dermatitis's treatment effectiveness.	[46]

(*Continued*)

TABLE 14.1 (Continued)

Polymer	Drug	Formulation	Method of preparation	Remarks	Ref.
Dextran	Rapamycin	Microparticles	Single emulsion technique (water/oil)	Ac-DEX microparticles with rapamycin can suppress immune system at stages comparable to free drug in culture. Acid-sensitive Ac-DEX polymer could target macrophages and be phagocytosed in acidic phagolysosome environment.	[47]
	Dexamethasone cholesteryl butyrate-	Solid lipid NPs	Warm microemulsion method	The results imply that treating inflammatory bowel disease with dexamethasone and butyrate delivered via NPs may be advantageous.	[48]
	Protocatechuic Acid (PCA)	NPs	Coprecipitation method	By adding PCA to magnetic NPs-Dex NPs, we have developed a theranostic system that targets and improves PCA's anti-inflammatory agent efficacy through better cellular internalization by leveraging the magnetic features of the nanocarrier.	[49]
	Indomethacin	Micelles	Dialysis method	In arthritic rats, the application of indomethacin-loaded polymeric micelles may significantly reduce paw diameter, decrease arthritis index, and alleviate inflammatory symptoms.	[50]
	Interleukin 1 receptor antagonist (IL-1ra)	Microsphere	S/O/W method	The findings indicated that IL-1ra-loaded dextran/PLGA microspheres provided sustained delivery of IL-1ra, leading to enhanced bioavailability and significant protection against periodontal inflammation.	[51]
PCL	Ofloxacin	NPs	Solvent evaporation technique	A superior antibacterial activity was achieved by presenting a mucoadhesive and harmless in situ forming eye gel of ofloxacin-loaded NP formulation based on polycaprolactone.	[52]
	Clozapine (CLO)	NPs	Nanoemulsion templating method	The finding indicated that CLO-PCL NPs a potential nano-vehicles and nontoxic to model cells.	[53]
	Methotrexate (MTX)	NLCs	Solvent diffusion method	The developed MTX-NLC gel formulation showed improved anti-psoriatic activity and long-lasting release effect, making it a potential replacement for current MTX formulation in psoriasis treatment.	[54]
	Glycyrrhizic acid and budesonide	NPs	Coprecipitation method	PCL-AC-gel containing budesonide and glycyrrhizic acid (polycaprolactone-aminocellulose-gelatin) NPs with two medicines injected into them have better anti-arthritic effects and are highly biocompatible.	[55]

applications, with an emphasis on the administration of proteins, other macromolecules, and small-molecule medications [57]. PLGA is a synthetic copolymer comprising hydroxy acetic acid (glycolic acid) and alpha-hydroxy propanoic acid (lactic acid) [58]. Prominent among bioma terials, the biodegradable polymer PLGA shows a lot of potential for usage as a scaffold in tissue regeneration and drug delivery. Its ability to transfer and deliver proteins, medications, and other macromolecules including peptides, RNA, and DNA has been thoroughly studied. This Food and Drug Administration (FDA)–approved polymer family is distinguished by its strong physical strength and great biocompatibility [59]. Their intrinsic benefits over traditional devices – such as extended-release capabilities that last for days, weeks, or even months – are the reason for this attention. The incorporation of macromolecular pharmaceuticals, including human growth factors, proteins, peptides, DNA, vaccinations, and antigens has been accomplished with success using PLGA and PLGA-based NPs and microparticles. Because PLGA polymers provide unique benefits over other systems, research in this area is still moving forward at a rapid pace, even with the market's current formulations [60]. The goal of PLGA modification is to strengthen the capacity to target, degrade, and regulate drugs while also improving their stability. In an attempt to improve PLGA's hydrophilicity and other desired characteristics, attempts have been made to change its structure. It is therefore expected that this will increase protein stability after formulation, impacting the kinetics of drug release, as well as biodegradation [61, 62]. PLGA has specific physicochemical properties. It has high hydrophilicity, which facilitates better hydration. PLGA is naturally amorphous, degrades quickly, and dissolves readily in a variety of organic solvents. Surprisingly, PLGA has a noticeable softening tendency in moist environments while being hydrophobic [63]. These diverse characteristics make PLGA a material that may be utilized in numerous fields, including drug delivery and biomedicine [64]. Furthermore, PLGA has a wide range of derivatives developed for it, increasing its versatility in several applications.

Derivatives such as poly(ethylene oxide), poly(ethylene glycol), PLGA-PEG-maleimide, PLGA-gelatin, PLGA-PEG-folate, Diethylaminopropylamine-PVA-g-PLGA (DEAPA-PVAL-g-PLGA), PLGA-PLL (poly-L-lysine), PLGA-b-PEG, and PLGA-B-PEG are examples of these. The usefulness of PLGA is increased in fields like biomaterials, and tissue engineering by the unique characteristics that each derivative provides [62, 65–69].

14.3.2 Polylactic Acid

PLA is a member of the aliphatic polyester family and is often made up of α-hydroxy acids like polyglycolic acid or polymandelic acid. PLA is a very strong and elastic thermoplastic polymer that is well-known for being biodegradable and compostable. These adaptable substances may be made from yearly renewable resources and are utilized to make products for industrial packaging or biocompatible/bioabsorbable medical equipment. PLA is easily produced into molded components, films, or fibers using ordinary plastic equipment [70]. In the field of biomedicine, PLA is used because the human body readily metabolizes its intermediate lactic acid product, proving its nontoxicity. The degradation of PLA is employed for controlled drug release [71]. Maintaining an efficient medication concentration while lowering the possibility of negative effects is made possible by controlling the carrier's rate of metabolism within the body. PLA has advantageous properties that make it useful as an implant material and for medication delivery [72]. Because of its remarkable biocompatibility, there is less chance of rejection when introduced into the human body because it doesn't cause inflammatory reactions in surrounding tissues [73]. Furthermore, PLA exhibits specific physicochemical features, including a melting range of 150°C–160°C that reflects its thermal behavior [74]. Remarkably, PLA stands out as a biocompatible and biodegradable polymer, rendering it well-suited for both environmentally friendly initiatives and medical applications [75]. Its molecular weight, falling within the 32 to 290 kDa range, contributes to PLA's adaptability across a spectrum of uses, encompassing sustainable packaging and biomedical materials. PLA possesses derivatives that enhance its versatility for various applications. Examples like poly(lactic acid)-poly(ethylene glycol) (PLA-PEG), poly(lactic acid)-block-poly(2-hydroxyethyl methacrylate), poly (lactic acid)-co-malic acid, and poly (lactic acid)-co-trimethylene carbonate broaden PLA's functionalities. The incorporation of unique features by each derivative expands PLA's utility across domains such as drug delivery, materials science, and biomedical technology [76–81].

14.3.3 Polystyrene

The hydrocarbon polymer polystyrene, which is synthesized, is widely used in many different domains and is particularly important for medicines. It is applied in several ways, each with a distinct function [82, 83]. Drugs can be effectively transported via polystyrene microspheres, which are tiny spherical particles made of polystyrene that enable targeted and controlled release. They are applied in diagnostic applications as well as targeted medicine administration. Furthermore, polystyrene NPs provide more precise medication delivery due to their unique surface characteristics and minuscule size. Polystyrene solid foams offer a potential path for novel drug delivery vehicles, even if their utilization in biomedical applications is still relatively new. These polystyrene-based solid foams have the potential to improve drug delivery in the future [83]. The polystyrene has distinct physicochemical properties. Its molecular weight exceeds 4,000 g/mol, and its melting range is 240°C–260°C. Polystyrene dissolves readily in solvents but is insoluble in water [84]. Its ability to exist in both crystalline and amorphous forms contributes to its adaptability in a variety of applications [85]. The synthetic polymer polystyrene has several derivatives that increase its versatility in a variety of applications. These modifications include polystyrene-b-polyethylene glycol (PS-b-PEG), polystyrene-co-maleic anhydride (PS-MA), polystyrene-co-acrylic acid, Polystyrene-b-poly (lactic-co-glycolic acid), polystyrene-co-methacrylic acid (PS-MA), polystyrene-co-poly (ethylene oxide), and polystyrene-co-poly(4-vinylpyridine). Each derivative brings distinct functionalities, widening the scope of potential applications for polystyrene in fields like materials science, drug delivery, and nanotechnology [86–92].

14.3.4 Pullulan

The extracellular polysaccharide pullulan is highly water soluble and possesses a significant film-forming capacity but pullulan poses a problem since the created film is readily washed away when it comes into contact with water. To improve film durability, pullulan derivatization must be used to introduce hydrophobicity. Because pullulan has a lower viscosity than other polysaccharides and good adhesive qualities, it works well as a binder. It remains stable over a wide pH range (2 to 11) but breaks down at this point and may be thermostable heated to 250°C–280°C. The uses of chemical derivatization and polymer modification in pharmaceutical product development have effectively expanded despite some restrictions, such as low mechanical characteristics and high manufacturing costs [93]. Pullulan can be chemically altered to enhance its physical properties, drug delivery, anticoagulant, pH sensitivity, antimicrobial, and ability to dissolve in organic solvents. These alterations include sulfation, urethane derivatization, esterification, etherification, cationization, and periodate oxidation [94]. Physicochemical processes involving pullulan (which has nine active hydroxyl groups) produce derivatives of pullulan. Notably, pullulan research has concentrated on changing the polymer, leading to improved solubility, pH-sensitive targeting, increased potential for delivery systems, anticoagulation, and antimicrobial qualities [94]. The synthetic polymer pullulan exhibits several physicochemical properties, such as its white color, powdery consistency, melting point of 250°C, molecular mass that varies from 45 to 600 kilodaltons (kDa), and high solubility in water [95, 96]. Pullulan has been diversified through several derivatives, such as pullulan acetate, pullulan folic acid, cholesterol pullulan, pullulan-g-poly(L-lactide), pullulan deoxycholic acid, pullulan poly(dl-lactide-co-glycolide), carboxy methyl pullulan, pullulan folate conjugates, pullulan polyethyleneimine pullulan, pullulan-g-poly(L-lysine), pullulan polyethyleneimine mercaptosuccinic acid, pullulan β-cyclodextrin polyethyleneimine, pullulan stearic acid, pullulan L-lactic acid oligomer, pullulan retinoic acid pullulan, biotin modified cholesterol pullulan, O-urocanyl pullulan, and polyethyleneimine acetylated pullulan. These compounds increase pullulan's versatility for a variety of applications in materials science and medicinal products [97, 98].

14.3.5 Dextran

Dextran-based techniques provide several advantages over other polysaccharides, including extended-release, nonimmunogenicity, biocompatibility, biodegradability, and pH change responsiveness [99]. Because dextran is a renewable biological macromolecule that occurs naturally, it has strong hydrophilicity, biodegradability, and biocompatibility, making it a good option for developing targeted drug delivery systems, and biomedical and pharmaceutical industries have conducted substantial research on dextran due to its well-documented biocompatibility. Dextran and its derivatives are frequently used to reduce ischemia, reduce vascular thrombosis, and reduce inflammatory reactions, especially in situations such as organ transplantation and reperfusion damage. Dextran lowers excessive platelet activation and has ROS scavenging activity. It's beneficial for enhancing cardiac output and blood flow as well because of its capacity to expand volume [100]. Dextran demonstrates characteristics of being both biocompatible and biodegradable, and because of these properties, it can be utilized in several pharmaceutical and biological applications [101]. Numerous variants of the dextran are currently established, such as ester of dextran, oxidized dextran, diethyl aminoethyl dextran, dialdehyde dextran, carboxymethyl dextran, dextran sulfates, iron-dextran complexes, and acetylated dextran (Ac-Dex) [102]. These alterations imparted different capabilities i.e. making applications in drug administration, biomaterials, and medical therapies [103–105]. There are some derivatives of dextran like acetalated dextran (Ac-Dex), dialdehyde dextran, oxidized dextran (Odex), carboxymethyl dextran (CMD), and iron-dextran complexes [106–108] These derivatives are effective as an excipient for targeted medication delivery and due to this proximity to inflammatory tissues and tumor environments, their acid-responsive breakdown allows for the controlled release of encapsulated medication [109].

14.3.6 Poly(ε-caprolactone)

Poly(e-caprolactone) (PCL) is a highly adaptable and biocompatible polymer that is commonly used in numerous medical applications. Its application in the production of films, fibers, NPs, and microparticles offers several benefits for biological applications. PCL-based materials show great promise in vital areas, including wound healing, medication administration, and the development of implanted medical devices [110]. PCL, an aliphatic polyester that is hydrophobic, is extensively used in FDA-approved devices for tissue engineering and drug administration because of its favorable mechanical strength and distinctive thermal features. Because of this, it is particularly suitable for the development of controlled drug delivery systems. Although PCL has a long history in numerous medicinal applications, its recent incorporation into microneedles (MNs) manufacturing is a forward-thinking and exciting development [111]. PCL exhibits specific physicochemical properties, including a melting point range of 60°C to 65°C that clarifies its thermal characteristics [112]. Significantly, PCL exhibits biodegradability, which qualifies it for use in environmentally sensitive fields. Moreover, the polymer has hydrophobic characteristics by nature, meaning that it is readily resistant to water contact. PCL has extensive derivatives that have been developed to increase its versatility in a variety of applications. Some examples include poly(ε-caprolactone-co-lactide) (PCL-PLA), polycaprolactone-block-polyglycolide (PCL-b-PGA), poly (ε-caprolactone-co-D, l-lactide) (PCL-DLLA), poly (ε-caprolactone-co-ethylene oxide), and poly (ε-caprolactone-co-trimethylene carbonate). The adaptability of PCL is increased by the addition of distinct capabilities in each derivative, particularly in the areas of pharmaceutical research and related professions [113–117].

14.4 APPLICATION OF SYNTHETIC POLYMER IN THE TREATMENT OF INFLAMMATION

Synthetic polymers are becoming more and more desirable for use in medicine because of their many applications [118]. Various physical and chemical properties may be obtained by varying the monomer units, the polymerization process, and the creation of copolymers with various components at controllable quantities (Figure 14.1). Synthetic polymers have found use in medication delivery systems within the biomedical profession [119]. Furthermore, in the treatment of inflammation, synthetic polymers provide remarkable adaptability (Table 14.2). Because synthetic polymers are so suitable for our bodies, they are helpful in the fight against inflammation [120]. They possess the ability to degrade at variable rates, allowing us to regulate the release of medications. Various kinds of these polymers are used in the treatment of inflammation, either by building structures to help with tissue healing or by precisely delivering medications to specific parts of the

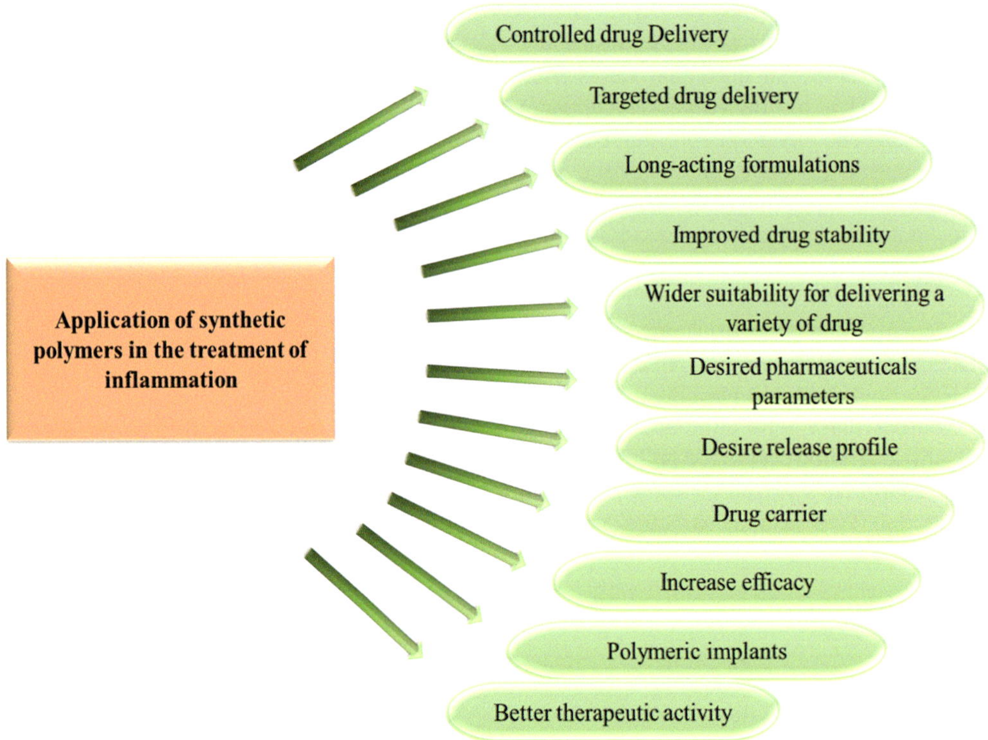

FIGURE 14.1 Application of synthetic polymers in the treatment of inflammation [120].

TABLE 14.2
Advantages and Disadvantages of Polymer-Based Drug Delivery Systems

Polymer	Advantages	Disadvantages	Ref.
PLGA	Nontoxic, nonimmunogenic, biocompatible, biodegradable	Variable degradation rates, expensive	[64]
Polycaprolactone	Elastic, bio-absorption, biocompatible, biodegradable, highly permeable	Limitations in strength and hardness, lack of functional groups, hydrophobicity, and slow rate of disintegration	[123]
Polystyrene	Stable, lightweight, cost-effective, and versatile	Nonbiodegradable, poor recycle rate	[124]
Dextran	Excellent solubility, biocompatible, biodegradable, nonimmunogenic, nontoxic, hardly affect the cell viability	Costly, limited stability	[125]
Pullulan	Sustained release, biocompatible, nonimmunogenic, chemical transformable, nontoxic, noncarcinogenic, biodegradable, nonmutagenic	Noncompatible, brittle, low heat stability	[98]
PLA	Renewable, biodegradable, biocompatible, good mechanical properties	High cost, compatibility issues, slow rate of deterioration	[126]

body [121]. These are equivalent to utilizing potent drugs that integrate with our bodies to decrease inflammation. Moreover, these materials provide the advantage of being highly customizable to fulfill particular needs because of their vast variety of chemical, physical, and biological features [122]. Current research into synthetic polymers has significant potential to develop and improve the treatment of inflammatory diseases as this field progresses. These provide the possibility of more effective and personalized treatment choices [56].

14.5 BARRIERS TO SYNTHETIC POLYMERS–BASED DRUG DELIVERY SYSTEMS

Several possible barriers might affect the efficacy of drug delivery systems containing synthetic polymers (Figure 14.2). These barriers consist of pharmaceutical delivery systems, which need synthetic polymers that are biocompatible, meaning they don't irritate the body or trigger negative immunological reactions when ingested, and immunological reactions to some synthetic polymers may

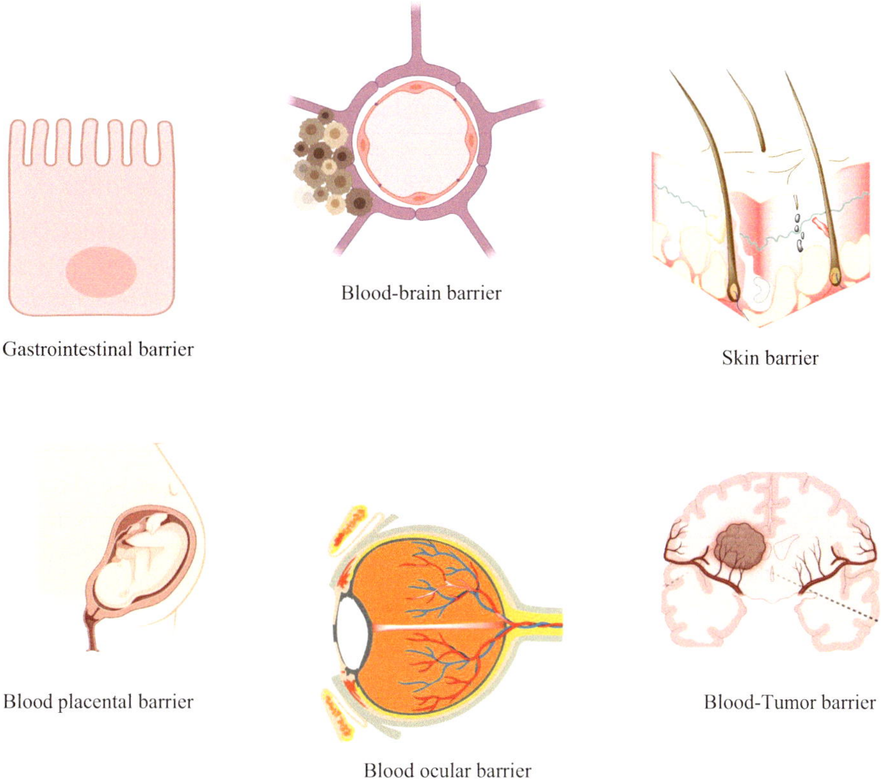

FIGURE 14.2 Biological barriers that a polymer-based drug delivery system has to cross in the human body [134].

result in inflammation or other unfavorable effects [122]. When utilized in drug delivery systems, synthetic polymers must be stable enough to retain their structure and function during storage and administration, and also have the proper degradation activity to release the drug at the appropriate pace, it may affect the drug's safety and effectiveness if the polymer breaks down too fast or too slowly [127]. The effectiveness of the delivery system depends on the efficient loading of medications into the polymer matrix and the regulated release of the medication at the intended location. Problems including ineffective drug encapsulation, drug burst release, or drug incomplete release might restrict the system's therapeutic effects [128]. The delivery method based on synthetic polymers may have an impact on cellular absorption, circulation time, and biodistribution dependent on its size and form. Ineffective delivery of drugs and less than desirable treatment results might result from poor control of these factors (size and shape). One of the main challenges is delivering the medication to the desired location with the least number of off-target effects [129]. The effective targeting of certain cells, tissues, or organs is a need for synthetic polymer–based delivery systems, and this often calls for the addition of targeting ligands or stimuli-responsive components. Substance release requires the eradication of the synthetic polymer carrier from the body by either biodegradation or elimination. Excessive retention of the polymer in the body might lead to negative consequences or disrupt regular physiological functions [130].

Several barriers may impact the efficiency of medication delivery systems based on synthetic polymers. Drug delivery methods need synthetic polymers to be biocompatible, which means they can't poison the body or trigger negative immune reactions when ingested, certain artificial polymers could cause an immunological reaction that results in inflammation or other unfavorable effects; also, synthetic polymers used for drug delivery systems are required to be stable enough to retain their structure and function throughout storage and administration and have the proper degradation kinetics to release the drug at an appropriate rate; it may affect the drug's safety and effectiveness of the polymer breaks down too fast or too slowly [122].

The effectiveness of the delivery system relies on the efficient loading of medications into the polymer matrix and the regulated release of the medication at the intended location. Problems including ineffective drug encapsulation, drug burst release, or drug incomplete release might restrict the system's therapeutic effects. In addition, the delivery method based on synthetic polymers may have an impact on cellular absorption, circulation time, and biodistribution dependent on its size and shape [128]. Ineffective medication delivery and less-than-ideal treatment results might result from poor control of these factors. One of the main challenges is delivering the medication to the desired location with the least number of off-target effects. The effective targeting of certain cells, tissues, or organs needs synthetic polymer–based delivery systems, and this often calls for the addition of targeting ligands or stimuli-responsive components [131]. One more challenge in synthetic polymer–based drug delivery is achieving the optimal loading of drugs into the polymer matrix and controlling the release of the drug at the desired rate and duration; this requires careful design and engineering of the polymer system to ensure efficient drug loading and controlled release kinetics [132]. Certain artificial polymers have the potential to cause the body to mount an immunological response, resulting in the production of antibodies or other immune responses that might impede the delivery of drugs and lower their effectiveness because minimizing the immunogenic potential of synthetic polymers is crucial [133].

14.6 METHODS OF PREPARATION AND ASSOCIATED CHALLENGES

Polymer-based drug delivery systems carry drugs throughout the body in several ways [17]. Rather than using naturally occurring polymers, these systems use synthetic polymers. By regulating the rate, time, and location of medication release, the goal is to increase therapy effectiveness and safety, the major methods are as follows.

14.6.1 Solvent Evaporation Method

The solvent evaporation method is a popular process for creating polymer micro or NPs, which have a variety of uses, including controlled drug administration [135]. The procedure starts with dispersing a polymer in a volatile solvent under monitoring (Figure 14.3). The drugs are added before or after the polymer dissolves, depending on how soluble it is in the solution [135].

14.6.2 Emulsification Method

In the pharmaceutical sector, the emulsification process is crucial because it enables targeted delivery of medications to certain tissues or organs, enhanced bioavailability, and controlled drug release [137]. Emulsification is a method of encasing drug molecules inside the dispersed phase (oil) of an emulsion, which polymers stabilize to form controlled-size droplets; in the context of polymer-based drug delivery systems, these droplets can be drug carriers, with benefits such as enhanced solubility, regulated release, and focused administration [138].

14.6.3 Coacervation Method

The coacervation technique divides a polymer solution into two different liquid phases, one is a dilute, polymer-lean equilibrium phase while the other is a dense, polymer-rich coacervate phase, changes in temperature, ionic strength, pH, or other external factors trigger this process [139]. In polymer-based drug delivery systems, coacervation is frequently utilized to encapsulate pharmaceuticals inside the coacervate phase, producing microspheres and microcapsules that are medication-filled [140]. Coacervation-based drug delivery methods often use polymers, including ethyl cellulose, gelatin, alginate, and chitosan. In addition, environmental factors play a vital role in the process of coacervation; the polymer solution may get divided into separate phases, resulting in the creation of coacervates, depending on variables including temperature, pH, and ionic strength of the solution [140].

14.6.4 Electrospinning Method

The development of a drug delivery system relies on synthetic polymers and often uses the adaptable technique of electrospinning (Figure 14.4). This method is perfect for encapsulating and administering medicinal substances because it provides exact control over the shape and properties of the resultant nanofibers [141]. Choosing a polymer is one of the important factors of the electrospinning method, PLGA, PCL, polyvinyl alcohol (PVA), and other synthetic polymers are used with electrospinning. The selection of polymers affects nanofiber mechanical properties, degradation rates, and drug release patterns. The electrospinning method encapsulates medications inside polymer nanofibers when the polymer of choice has been chosen [142]. The electrospun nanofibers can display diverse drug release kinetics, such as targeted, controlled, or sustained-release patterns (Table 14.3). This approach has great promise

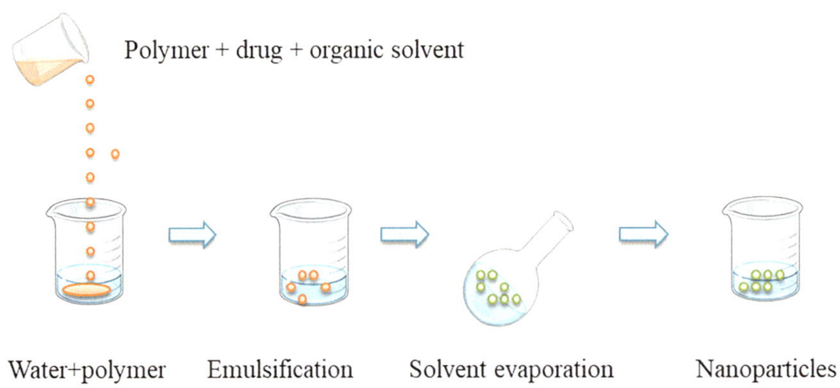

FIGURE 14.3 Solvent evaporation method [136].

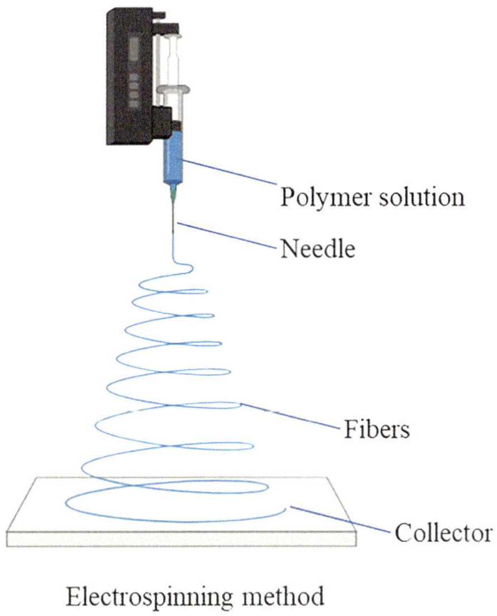

FIGURE 14.4 Electrospinning method [143].

for addressing several drug delivery-related issues, such as attaining targeted distribution, prolonged release, and improved therapeutic effects.

14.6.5 Hot-Melt Extrusion (HME) Method

HME is a flexible manufacturing method utilized in different sectors, including medicines, polymers, and food. In the area of pharmaceuticals, HME is applied to manufacture drug delivery systems and dosage forms. In the pharmaceutical sector, HME is a frequently used procedure for creating synthetic polymer–based drug delivery systems [144]. HME's fundamental process is melting a solid substance like a combination of polymers and active medicinal ingredients into a liquid form. After that, a die is used to press the molten material through to produce a continuous, consistent extrudate with the required dimensions and shape [145]. There are several benefits of using HME in synthetic polymer–based drug delivery systems; this procedure makes it practicable to produce drug delivery systems with regulated release characteristics, increase the stability of the pharmaceutical components, and enhance the bioavailability of poorly soluble medications [146].

14.7 CHALLENGES IN LAB SCALE TO LARGE SCALE AND COMMERCIALIZATION OF SYNTHETIC POLYMER–BASED DRUG DELIVERY SYSTEM

Synthetic polymers have gained significant attention as drug delivery carriers, emphasizing their importance in the field. The significant sales of pharmaceutical items that use drug delivery systems demonstrate the robust presence of polymers. Polymers have numerous applications in medicines; these are frequently utilized in latex particles, resin compositions, and liposomes or microcapsules for broad drug administration [179]. These delivery strategies have historically used polymers as a synthetic matrix to promote drug release via biodegradation or diffusion control. However, more people are realizing that polymeric methods are more flexible than passive drug release. Synthetic polymers are an essential and dynamic part of the therapeutic process and medication delivery system. Currently, the majority of drug delivery research is focused on improving therapy delivery to particular disease sites, improving biological targeting precision, and managing drug release mechanisms. These domains have great appeal in polymer technology. Medicines and polymers can form complexes or attach to one other to protect against in vivo breakdown. Additionally, by including certain functionalities in the polymer structure, the duration and specificity of release of drug can be enhanced. These areas of study look at the research methodologies being employed while acknowledging the difficulties ranging from small-scale to large-scale problems [180].

14.7.1 Lab-Scale Challenges of a Polymer-Based Drug Delivery System

The standard laboratory synthesis methods on smaller scales for producing polymeric nanoparticles (PNPs) are prone to batch-to-batch fluctuations (Figure 14.5). There are several challenges to overcome while moving from laboratory to industrial-scale manufacturing. High-yield techniques for the efficient, quick, and accurate synthesis of NPs have been developed recently, despite the difficulties in developing, optimizing, improving, manufacturing, and ensuring the sustainability connected with this method of synthesis [182]. Advantages arise from the application of biologically compatible and biodegradable qualities. PNP creation can benefit from the upscaling procedures frequently seen in metal NPs since comparable difficulties arise in both situations. Sampling plays a crucial role due to the inherent instability and pronounced tendency for aggregation in NPs. The large-scale level and nanoscale levels have different physical characteristics [183]. A statistically representative and evenly distributed particle population is necessary for accurate measurements [184]. Important to the measurement procedure is the sample preparation step, which calls for a lot of work to reduce agglomeration, which might affect the measurement outcomes of individual NPs [185].

14.7.1.1 Particle Size

The functioning of NPs in the in vivo system is greatly influenced by their size. Small-molecule-based NPs are quickly eliminated from the bloodstream by the kidneys by glomerular filtration, which has an impact on how long the NPs stay in circulation [186]. Additionally, the size of particles and different crucial variables are size variability (polydispersity), size distribution, and the particular form of the NPs [187].

TABLE 14.3
Merits and Demerits of the Method Used in the Preparation of Polymer-Based Drug Delivery System

Methods Used in Polymer-Based Drug Delivery System	Merits of Methods Used in Polymer-Based Drug Delivery System	Demerits of Methods Used in Polymer-Based Drug Delivery System	Ref.
Solvent evaporation method	• *Economical and easy to use*: This approach is simple to use in a laboratory environment and doesn't need costly equipment. • *Controlled particle size*: It is feasible to regulate the size and dispersion of particles in the finished product by changing the conditions and rate of solvent evaporation. • *Versatility*: The solvent evaporation approach works well with various substances, such as tiny molecules, polymers, and pharmaceuticals. • *Scalability*: The technique is appropriate for large-scale manufacturing since it is readily scaled up for industrial output.	• *Compatibility with solvents*: This approach may not apply to all substances because of incompatibilities. • *Residual solvent*: If the solvent is not completely evaporated, it might stay in the finished product and raise safety and legal issues. • *Environmental impact*: Using a lot of solvent may hurt the environment, especially if the solvent is flammable and pollutes the air.	[147–151]
Emulsification method	• *Controlled release*: Emulsification produces tiny emulsion droplets, possibly leading to controlled and prolonged drug release from the polymer matrix. • *Encapsulation efficiency*: By using this technique, a higher proportion of the drugs may be efficiently encapsulated inside the polymer matrix, minimizing waste and improving the therapeutic efficacy. • *Versatility*: Emulsification is an effective approach that may be used to administer a range of pharmaceutical substances since it works with a broad range of polymers and medications. • *Stability*: By preventing degradation and extending the drug's shelf life, emulsification may assist in increasing the drug's stability inside the polymer matrix.	• *Particle size distribution*: Achieving a narrow and constant particle size distribution may be problematic with emulsification, which may impact the release profile and bioavailability of the medicine. • *Emulsifier selection*: The selection of an emulsifier is a crucial step in the emulsification process. Choosing the incorrect emulsifier might result in instability, inadequate encapsulation, or disruption of the drug release properties. • *Manufacturing complexity*: The process of emulsification may be lengthy and complex, requiring exact control over variables like temperature, agitation, and formulation composition. These variations may influence the process's scalability and efficiency. • *Instability*: Emulsions may be prone to instability over time, leading to phase separation or coalescence of droplets, which can compromise the integrity of the drug delivery system.	[152–159]
Coacervation method	• *Encapsulation efficiency*: Coacervation enables high drug encapsulation efficiency inside the polymer matrix, resulting in a drug delivery system that functions well. • *Controlled release*: By designing drug delivery systems with controlled release profiles, the coacervation approach makes it possible to provide drugs continuously and for longer periods, depending on the therapeutic need. • *Compatibility with various pharmaceuticals*: It is compatible with a broad variety of pharmaceuticals, including hydrophobic and hydrophilic chemicals, making it adaptable for pharmaceutical applications. • *Drug protection*: By preventing the integrated medications from deteriorating, the coacervation process may increase their stability and prolong their shelf life. • *Ease of scale-up*: The coacervation approach can be quickly grown for industrial production, which makes it appropriate for producing polymer-based drug delivery systems on a large scale.	• *Complexity*: To obtain the appropriate encapsulation and drug release properties, the coacervation process may be technically difficult and involves an exact hold of factors consisting of temperature, pH, and polymer concentration. • *Limited compatibility with biological agents*: A possible consequence of coacervation's severe processing conditions is the loss of bioactivity in some biological medications, such as proteins and peptides. • *Possibility of aggregation*: Particle aggregation during the coacervation process might result in uneven encapsulation and reduced drug delivery system efficacy. • *Solvent use*: The use of organic solvents in the coacervation process is common, which might provide problems with solvent removal and raise toxicity issues. • *Difficulties with quality control*: It may be difficult to guarantee consistent particle size distribution and repeatability across batches, which may call for strict quality control procedures.	[140, 160–167]

(Continued)

TABLE 14.3 (Continued)

Methods Used in Polymer-Based Drug Delivery System	Merits of Methods Used in Polymer-Based Drug Delivery System	Demerits of Methods Used in Polymer-Based Drug Delivery System	Ref.
Electrospinning method	• *High surface area*: Electrospinning creates nanofibers with a high surface area: volume ratio, which may boost the drug loading capacity and promote fast drug release. • *Modified release*: By modifying the polymer solution's composition and the processing parameters, the technique enables accurate manipulation of the drug release kinetics. • *Versatility*: By working with a variety of synthetic polymers, electrospinning provides designers with flexibility when creating drug delivery systems that address certain therapeutic requirements. • *Potential for targeted delivery*: The nanofibrous structure may be engineered to provide targeted delivery of medications to particular tissues or cells, boosting therapeutic effectiveness.	• *Scale-up difficulties*: Because exact control over processing conditions is required, producing electrospun fibers on a large scale with consistent quality and attributes may be difficult. • *Restricted drug compatibility*: Some medications may lose their bioactivity or deteriorate during the electrospinning process, and not all medications are suitable for this method. • *Potential for burst release*: Inadequate design and optimization may cause electrospun drug delivery systems to experience drug burst release, which might reduce the systems' therapeutic efficacy. • *Complexity of processing*: Optimizing processing parameters may take a lot of time, and the electrospinning process requires specific equipment and expertise.	[168–174]
HME method	• *Thermally stable formulations*: Drug-polymer combinations that are thermally stable may be processed by HME, which makes it possible to create drug delivery systems with improved stability. • *Customized release profiles*: The technique allows for exact control over formulation composition and process parameters, allowing the creation of drug delivery systems with customized release profiles. • *Continuous production*: The use of HME in continuous production may result in increased productivity and reliable product quality. • *Increased bioavailability*: Spreading poorly soluble medications throughout the polymer matrix may increase the medicines' absorption and dissolution.	• *Restricted drug compatibility*: The spectrum of medications that may be used in HME is limited because some pharmaceuticals may lose their bioactivity or deteriorate when exposed to the high temperatures and shear pressures involved. • *Equipment requirements*: HME requires certain tools and knowledge, which might raise startup and running expenses. • *Problems with heat-sensitive compound processing*: Drugs that are heat-sensitive or thermolabile may provide difficulties for HME, requiring careful consideration of the processing parameters and excipients to preserve drug stability. • *Scale-up complexity*: Due to variables like heat transmission and mixing efficiency, scaling up HME methods for commercial production while preserving product uniformity may be difficult.	[146, 175–178]

14.7.1.2 Surface Properties

PNPs' surface charge is essential to their penetration. The uptake of NPs by endocytosis is facilitated by electrostatic interactions. Compared to neutrally charged NPs, macrophages were more effective in eliminating positively charged particles [188]. Since positive NPs have a stronger contact force with cells than negative or neutral ones, it was discovered that they were more harmful. The blood-brain barrier's (BBB) integrity is also affected by the charge on particles. Moreover, the NP charge has an impact on how protein coronas form, which has an impact on receptor binding [184]. According to the study, an increase in positive groups in NPs does not always translate into a corresponding rise in zeta potential, which is the characteristic that determines surface charge. Consequently, a critical indication is the surface charge density [189]. Although zeta potential and surface charge density information are essential for describing nanosystems, they are not sufficient to forecast how NPs would aggregate in a culture medium [188].

14.7.1.3 Particle Shape

The shape of NPs significantly influences their function within interactions between endothelial cells, the bloodstream, and the duration of circulation, biodistribution, and cellular uptake. In several studies, spherical NPs have been preferred. All sorts of forms, including cubes, bars, discs, and stars (Figure 14.6), can also be constructed for NPs [183]. As an illustration, in animal models, disc-shaped gold NPs outperformed rod and cube-shaped gold NPs in terms of tumor penetration. Higher tumor uptake was shown by gold nanostars with more surface contact. Using imaging methods like scanning electron

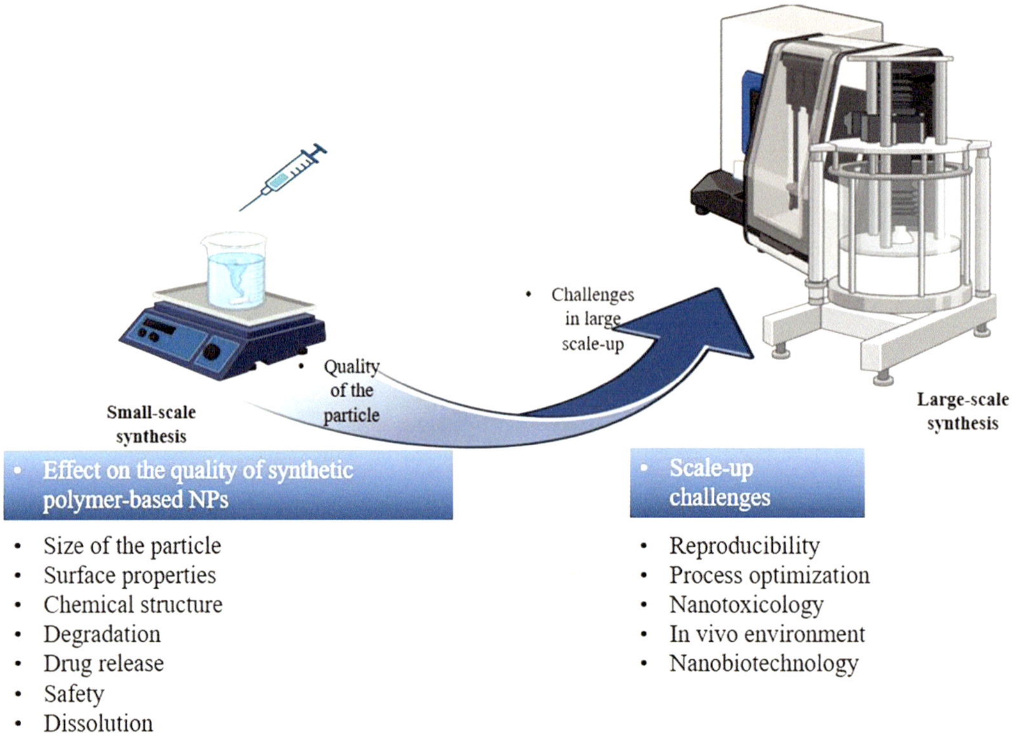

FIGURE 14.5 Scale-up challenges in synthetic polymer–based drug delivery system [181].

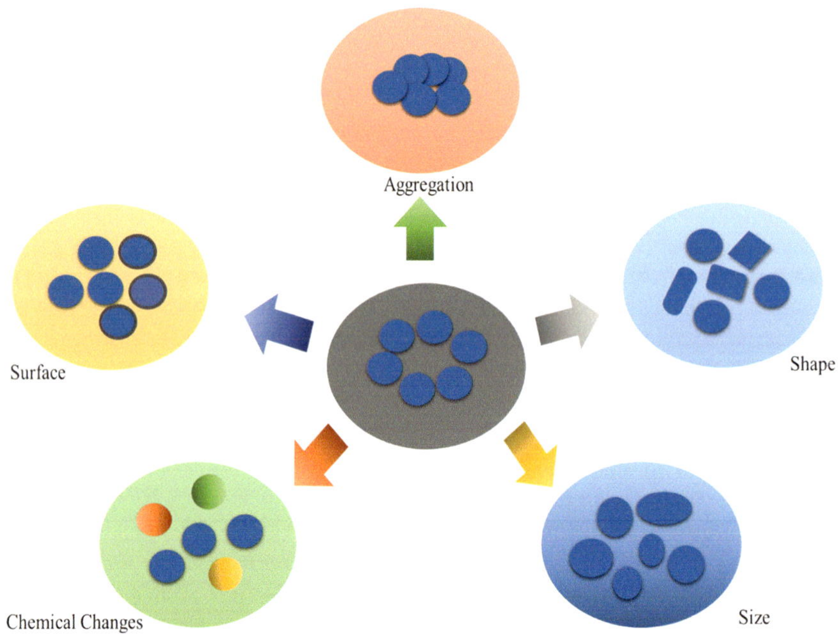

FIGURE 14.6 NPs may experience changes under different physical, chemical, or biological environmental conditions [200].

microscopy (SEM), transmission electron microscopy (TEM), and atomic force microscopy (AFM), the shape (morphology) of the particle was found. Importantly, the characteristics of the NPs may be changed by the drying and sputtering procedures used to prepare the SEM sample [187].

14.7.1.4 Stability and Degradation

It is challenging to control the stability, shape, and size distribution of biologically produced NPs. Establishing a robust and efficient nano-drug delivery system depends critically on the approach and excipient selection used in NP-based systems development [190]. A kinetic, thermal degradation approach was applied to examine the produced nanoscale entities to determine their activation energy (Ea). The product's stability and capacity to withstand fluctuations in temperature are indicated by the Ea value [191].

14.7.1.5 Drug Release

Maintaining quality performance, maintaining batch-to-batch uniformity and determining out the in vitro–in vivo correlation (IVIVC) of the drug which depend primarily on the drug release profile. For each of these methods, validation is necessary, when employing the paddle or basket method for in vitro release investigations, agitation is necessary to stop dosage forms from aggregating [192]. Test sensitivity, stability, drug solubility, and testing protocol all have an impact on the choice of release medium. To guarantee the quality of colloidal silica nanoparticles (CSNPs), in vitro release analysis is crucial, despite the lack of recognized compendial standards or current regulations. Because of the variety of testing techniques used, direct comparisons between various systems are challenging [193].

14.7.1.6 Dissolution

Higher rates of dissolution can lead to better medication absorption. When it comes to the fact that these particles change and become poisonous in aquatic environments, the breakdown of the particles is a major factor [194]. Several variables that influence the breakdown of fibers and micro-sized particles also probably influence the breakdown of nanomaterials [195]. These variables include the media's makeup, the nanomaterial's degree of dispersion, and the components' capacity to combine to form complexes with released ions. The production process of NPs has an impact on the crystalline shape, which is another important element determining dissolving [196]. It is challenging to manage the dispersion of seed crystals, which affects the formation of shaped NPs, in batch techniques used in NP manufacturing because of the lack of exact control over mixing, growth processes, and nucleation. Challenges with repeatability arise when batch processes are used without precise control over the formation of NPs [197, 198].

14.7.1.7 Safety

Most research on nanomedicine begins with an evaluation of efficacy in laboratory environments, usually with the use of animal models or cell lines. Transitioning to clinical trials presents certain difficulties, nevertheless, especially concerning determining product safety. The undesirable biodistribution and changed pharmacokinetics of NPs, which may have toxicities or adverse effects, are important factors in trial difficulties [199].

14.7.2 Large-Scale Challenges

Generally, nanomedicine products are manufactured in small quantities within a laboratory environment (Figure 14.5). However, achieving the desired product quality profile becomes more challenging when this process is scaled up for industrial production. These involve working with bigger batches and using different equipment settings for effective and quick manufacture. These include aspects such as re-dispersibility, size distribution, and size of particles. Moreover, for parenteral products, the sterilization setup varies from that employed in laboratory environments [192]. A nanomedicine product's scale-up approach from laboratory research to market involves several factors. These include things like the material's composition and Generally Regarded as Safe (GRAS) status, the toxicological effects of NP size and shape, the in vivo biodegradability of nanocarriers, and the larger-scale management of multicomponent systems [201]. Careful consideration must be provided to selecting materials, solvents, methods for developing NPs, potential financial effects, and whether or not patients and physicians will find the finished product acceptable. Sometimes, when laboratory techniques are scaled up, the desirable properties of NPs change. An increase in agitation duration and impeller speed, for instance, decreased particle size while maintaining the same entrapment effectiveness in research on the scaling up of NPs produced by the emulsion technique. To expedite the pilot batch production process from a scale-up standpoint, careful selection of a suitable NP manufacturing technology is also essential [202].

14.7.2.1 Reproducibility

Nanomaterials must be reproducible for their practical uses to be achieved. Replicating NPs presents challenges in the areas of characterization, distribution, and manufacture [203]. Some of the beneficial characteristics of NPs may be lost when production techniques from the lab are scaled up for bigger quantities. Particles' dynamic character, which allows them to transform or in reaction to various environmental factors, adds complexity [204]. One crucial factor that emphasizes the need for an accurate and consistent verification process is the amount of time that elapses between analysis and implementation [192].

14.7.2.2 Process Optimization

Nanomedicine products are in the process of being created to assess the effectiveness of enclosed therapeutics, both in vitro and in vivo. However, there is relatively less emphasis on refining the formulation development process. Considering the pharmaceutical industry's needs, it is crucial to apply optimization to the product development process early on. This is why the International Conference on Harmonization (ICH) guidelines advocate for the Quality by Design (QbD) approach in product development and optimization [205, 206].

14.7.2.3 Nanotoxicology

Even though NPs are now versatile with many different areas of use, concerns about their safety remain. Two main criteria are used to evaluate NPs: cellular absorption and toxicity. Particle size is an important factor since smaller particles often have higher levels of biological activity, chemical reactivity, and toxicity [207]. Further characteristics that influence the toxicity of NPs are their chemical constitution, porosity, surface charge, shape, surface structure, heterogeneity, rate of dissolution, solubility and aggregation, existence or nonexistence of functional groups, roughness, roughness between surface groups, and dispersion state (Figure 14.6) [208]. NPs usually accumulate in the organ phagocytic system after acute systemic exposure, which can cause toxicity of the liver, spleen, and kidneys, among other organs [209]. Furthermore, NPs have the capacity to damage DNA, which might lead to consequences, including mitochondrial malfunction, impaired membrane integrity, changes in the structure and function of proteins, as well as cell cycle arrest and death. Metal NPs and metal oxides have been demonstrated to have a deleterious effect on DNA [210]. Moreover, specific NPs, such as gold NPs, have the potential to cause kidney damage, a danger that is compounded when they combine with other substances or drugs [210].

14.7.2.4 Safety

Manufacturers must guarantee the final NP-based compositions' safety and purity per regulatory guidelines [192]. Most research on nanomedicine begins with an evaluation of efficacy in laboratory environments, usually with the use of animal models or cell lines. Transitioning to clinical trials presents certain difficulties, nevertheless, especially concerning determining product safety. The undesirable biodistribution and changed pharmacokinetics of NPs, which may have toxicities or adverse effects, are important factors in trial difficulties [199].

14.7.2.5 In vivo Environment

To understand the activity of NPs in biological systems, in vivo assessments are essential. Numerous evaluations, such as pharmacokinetics, biodistribution analysis, acute effectiveness tests, and dose-therapy response studies, offer insightful information [211]. Tumor permeability, biodistribution, and circulating blood half-life are all strongly impacted by NP size. Particle dispersion is assessed using appropriate techniques after delivery, and analytical chemistry is used to characterize nonradioactive NPs [212]. Active ingredient release may be stimulus-responsive, which can lead to problems including agglomeration that doesn't happen properly and inaccurate toxicity evaluations. NPs' ability to accumulate optimally is hindered by the reticuloendothelial and mononuclear phagocyte systems quick elimination. Rapid elimination is prevented by modifying NP size to modify pharmacokinetics. Measured in macrophages using fluorescence-labeled NPs, phagocytic uptake extrapolates in vivo behavior [213]. Spheroid multicellular models, monoculture, coculture, and "organ-on-a-chip" techniques are examples of in vitro model systems that provide insights. Before doing investigations on mammals, the chick chorioallantoic membrane (CAM) model provides an affordable bridge to evaluate the potential of nanosystems. Cytotoxicity, changed drug biodistribution, difficulties demonstrating the permeability and retention impact in tumors, and interference with complex proteins are some of the safety challenges [214].

14.7.2.6 Nanobiotechnology

In the field of industrial nanobiotechnology, strong biotechnological instruments are utilized to modify gene regulation, protein structure, and metabolic pathways to improve the fungal production of NPs [214]. This manipulation enables the production of NPs with a variety of properties, such as differences in shape, surface charge, and size. Numerous biotic and abiotic variables, including pH, temperature, and bacterial species, can affect the characteristics of NPs produced by microorganisms [215]. However, there are three main techniques for synthesizing NPs: physical, chemical, and biological. When compared to their physical and chemical equivalents, biologically based approaches provide several benefits [216]. They produce NPs with well-defined sizes and morphologies that exhibit good compatibility and prove to be economical, eco-friendly, and energy-efficient [215, 217]. On the other hand, conventional techniques have disadvantages such as excessive energy consumption, increased expenses, and the use of hazardous chemicals that pollute the environment [216].

14.8 COMMERCIALIZATION CHALLENGES AND REGULATORY APPROVAL

The growing NP production area requires improving market and commercial effectiveness. Business-wise, the nanomedicine sector is predicted to reach a valuation of USD 293.1 billion by 2022 and expand to USD 350.8 billion by 2025 [218]. Despite this positive forecasting, many challenges stand in the way of PNPs moving from the research to the clinical phases of nanomedicine. As such, it is critical to determine whether the regulatory requirements and existing techniques adequately examine nanomedicines' delivery, destiny, safety, and effectiveness [219]. Furthermore, it is challenging to get funding for nanomedicine research; therefore, before goods are commercialized, they must first demonstrate societal and economic added values, such as how much cheaper they are than existing therapies. Acquiring doctors' and patients' acceptance of nanomedicine solutions is another significant business obstacle [220]. Scaling up NP production for large-scale manufacturing is one significant challenge. The FDA has required pharmaceutical producers to switch from batch to continuous processing by 2026 in response to these advancements, which will gradually lead to the standardization of chemical flows in pharmaceutical processing [221].

Currently, regulatory authorities do not recognize simulations as an effective alternative for empirical data. Therefore, to achieve regulatory approval for drug delivery systems (DDSs), simulations must be supplemented with experimental data to showcase the safety and effectiveness of the system (Table 14.4). Consequently, rather than being used independently simulations are usually combined with

TABLE 14.4
Overcoming Strategies of Process-Related Challenges

Process-Related Challenges	Overcoming Strategies	Ref.
Biocompatibility	To improve biocompatibility, use biocompatible polymers or alter currently used polymers. During the development phase, carry out extensive biocompatibility testing.	[224, 225]
Stability and degradation	Modify the polymer's properties to increase stability and medicine compatibility. Use additives or protective coatings to stop medication deterioration.	[226, 227]
Particle size	Achieve uniform particle size by modifying milling parameters, employing advanced techniques such as microfluidics, and monitoring with dynamic light scattering.	[228, 229]
Drug release	Improve the accuracy of medication release by improving structure, composition, and addition of modifiers to polymers. Use controlled release techniques to minimize initial bursts. Analyze release dynamics in vivo and in vitro in great detail in various conditions.	[230]
Dissolution	Conduct comprehensive in vitro and in vivo analyses to examine dissolution in complex situations. Include solubility enhancers and pH-sensitive components to create polymers with increased water permeability for more efficient dissolution.	[228, 231]
Safety	To reduce harm, use safer polymers or alter those that already exist. Perform comprehensive safety testing in both laboratory and human trials, assigning materials with the least possibility to harm health priority. Employees should get safe polymer handling training.	[229, 232]
Large-scale challenges	The quality of the polymer won't be compromised by using inexpensive raw components. Maintain high levels of quality by routinely inspecting and auditing output. Adopt eco-friendly procedures, reduce waste, and replace outdated technology with more effective technologies. Overcome rules for economical, efficient manufacturing, working with specialists to ensure environmental sustainability and efficient scalability.	[233, 234]
Reproducibility	Safely obtain products from reputable vendors with a solid reputation for excellence. Establish clear guidelines, conduct routine material inspections, and provide personnel with thorough training for automated, precise procedures. With well-designed processes, frequent equipment inspections, and thorough documentation, variances may be decreased. Engage in real-time quality control monitoring, collaborate with specialists, and apply sophisticated analysis. Perform performance reviews and make necessary adjustments to procedures regularly.	[235, 236]
Process optimization	Investigate carefully to find the best conditions. Utilize sophisticated models to anticipate and control responses. Observe mixing and heat while gradually expanding. Observe closely with sophisticated equipment. To achieve continual improvement, establish consistent protocols, provide excellent training, and update processes often.	[229, 237]
In vivo environment	Make sure that the application in medicine is thoroughly tested in laboratory and living systems before implementation. Use coatings if necessary, and choose materials that won't trigger severe immunological responses. Utilize safe, slow-degrading biodegradable products. To ensure high-quality imaging, target certain tissues, test under realistic circumstances, ensure stability, and collaborate.	[238, 239]
Nanotoxicology	Observe carefully the movements and outflow of these nanomaterials throughout the body. Recognize the differences in their size, shape, and composition along with how they interact. Obtain real-time movement by using sophisticated imaging. Observe how altering their surface affects the interactions between cells. Verify whether they trigger immunological reactions or inflammation. Examine if, in some circumstances, the advantages of utilizing them exceed the hazards. Based on this assessment, establish explicit safety regulations.	[240, 241]
Regulatory approval	Examine products thoroughly while maintaining accurate records. Assist professionals to ensure that all aspects of the design and manufacturing process are safe. Maintain communication with regulatory bodies, asking for advice and being aware of what is anticipated. Get assistance from experts who are familiar with the regulations to handle complex procedures and deal with safety concerns upfront.	[13, 242]
Commercialization challenges	Investigate consumer preferences through research. Improve the performance or add extra features to your product to make it stand out. Make use of astute marketing to emphasize its worth. Reduce costs by optimizing production. Take good care of your supplies. Get tips to abide by the regulations and be aware of them. Make sure your product complies with requirements and maintain accurate records.	[243, 244]

experimental investigations [222]. The scarcity of biodegradable polymers with an extensive track record of regulatory approval and the difficulties associated with obtaining regulatory clearance for a novel polymer contributes to significant technical challenges and developmental risks when transitioning from bench-scale to industrial commercialization. Maintaining control over size and ensuring batch-to-batch reproducibility further adds to the complexities of this scale-up process [223].

14.9 CONCLUSION AND FUTURE PROSPECTS

Inflammatory illnesses, such as asthma, intestinal disease, arthritis, and many more, are important concerns for both patients and health-care providers. Traditional treatment options for these disorders are generally linked with few and undesirable side effects. However, the rapidly emerging field of polymer-based DDSs has shown the ability to transform the management of inflammatory diseases. This chapter has discussed a thorough study of synthetic polymers (PLGA, PLA, Polystyrene) with their physiochemical properties, advantages, disadvantages, and applications, utilized in conjunction with the drugs for the cure of inflammatory disorders in the form of nano-formulations (NPs, microsphere, microparticles, nanoemulsion). These drug delivery methods based on synthetic polymers have great potential to treat inflammatory illnesses, a wide variety of polymers, including PLGA, PEG, and chitosan, have special qualities that may be adjusted to meet certain drug delivery requirements. However, creating a controlled release and ensuring biocompatibility are important factors in the production of efficient polymer-based DDSs. The current study also includes lab-scale to commercial-scale challenges faced by the researchers during the production of synthetic polymer–based DDSs and their overcoming strategies. To conquer these challenges, will need additional research and creativity in polymer synthesis and formulation methods. Techniques including electrospinning, emulsion-solvent evaporation, and nanoprecipitation have shown potential in the development of innovative polymer-based DDSs that may successfully target and treat inflammatory illnesses. Notwithstanding the difficulties, the area of synthetic polymer–based DDSs has phenomenal promise for future research and development because of its ability to completely transform the treatment of inflammatory illnesses.

REFERENCES

[1] Chen, L., Deng, H., Cui, H., Fang, J., Zuo, Z., Deng, J., et al. (2018) Inflammatory responses and inflammation-associated diseases in organs. *Oncotarget*. 9 (6), 7204–7218.

[2] Placha, D. and Jampilek, J. (2021) Chronic inflammatory diseases, anti-inflammatory agents and their delivery nanosystems. *Pharmaceutics*. 13 (1), 1–27.

[3] Finckh, A., Gilbert, B., Hodkinson, B., Bae, S.-C., Thomas, R., Deane, K.D., et al. (2022) Global epidemiology of rheumatoid arthritis. *Nature Reviews Rheumatology*. 18 (10), 591–602.

[4] Kedia, S. and Ahuja, V. (2017) Epidemiology of inflammatory bowel disease in India: the great shift east. *Inflammatory Intestinal Diseases*. 2 (2), 102–115.

[5] Barton, G.M. (2008) A calculated response: Control of inflammation by the innate immune system. *The Journal of Clinical Investigation*. 118 (2), 413–420.

[6] Hasturk, H. and Kantarci, A. (2015) Activation and resolution of periodontal inflammation and its systemic impact. *Periodontology* 2000. 69 (1), 255–273.

[7] Brannon, E.R., Guevara, M.V., Pacifici, N.J., Lee, J.K., Lewis, J.S., and Eniola-Adefeso, O. (2022) Polymeric particle-based therapies for acute inflammatory diseases. *Nature Reviews Materials*. 7 (10), 796–813.

[8] Saleh, T.A. (2021) *Polymer hybrid materials and nanocomposites: Fundamentals and applications*. William Andrew.

[9] Rajak, D.K., Pagar, D.D., Menezes, P.L., and Linul, E. (2019) Fiber-reinforced polymer composites: Manufacturing, properties, and applications. *Polymers*. 11 (10), 1–37.

[10] Rahimi, M., Charmi, G., Matyjaszewski, K., Banquy, X., and Pietrasik, J. (2021) Recent developments in natural and synthetic polymeric drug delivery systems used for the treatment of osteoarthritis. *Acta Biomaterialia*. 123 31–50.

[11] Paolini, M.S., Fenton, O.S., Bhattacharya, C., Andresen, J.L., and Langer, R. (2019) Polymers for extended-release administration. *Biomedical Microdevices*. 21 1–24.

[12] Elsawy, M.A., Kim, K.-H., Park, J.-W., and Deep, A. (2017) Hydrolytic degradation of polylactic acid (PLA) and its composites. *Renewable and Sustainable Energy Reviews*. 79 1346–1352.

[13] Gaspar, R. and Duncan, R. (2009) Polymeric carriers: preclinical safety and the regulatory implications for design and development of polymer therapeutics. *Advanced Drug Delivery Reviews*. 61 (13), 1220–1231.

[14] Grund, S., Bauer, M., and Fischer, D. (2011) Polymers in drug delivery—state of the art and future trends. *Advanced Engineering Materials*. 13 (3), B61–B87.

[15] Mohanty, A.K., Wu, F., Mincheva, R., Hakkarainen, M., Raquez, J.-M., Mielewski, D.F., et al. (2022) Sustainable polymers. *Nature Reviews Methods Primers*. 2 (1), 46.

[16] Rousseau, I.A. (2008) Challenges of shape memory polymers: A review of the progress toward overcoming SMP's limitations. *Polymer Engineering & Science*. 48 (11), 2075–2089.

[17] Allen, T.M. and Cullis, P.R. (2004) Drug delivery systems: Entering the mainstream. *Science*. 303 (5665), 1818–1822.

[18] Kost, J. and Langer, R. (2012) Responsive polymeric delivery systems. *Advanced Drug Delivery Reviews*. 64 327–341.

[19] Zhang, S., Langer, R., and Traverso, G. (2017) Nanoparticulate drug delivery systems targeting inflammation for treatment of inflammatory bowel disease. *Nano Today*. 16 82–96.

[20] Nair, A., Varghese, B.A., Gopi, S., and Jacob, J. (2021) Smart drug delivery systems of natural products for inflammation: From fundamentals to the clinic. in: *Inflamm*. Nat. Prod., Elsevier, pp. 261–282.

[21] Li, C. and Wallace, S. (2008) Polymer-drug conjugates: recent development in clinical oncology. *Advanced Drug Delivery Reviews*. 60 (8), 886–898.

[22] Chiellini, F., Piras, A.M., Errico, C., and Chiellini, E. (2008) Micro/nanostructured polymeric systems for biomedical and pharmaceutical applications. *Nanomedicine*. 3 367–393.

[23] Al-Shalabi, E., Abusulieh, S., Hammad, A.M., and Sunoqrot, S. (2022) Rhoifolin loaded in PLGA nanoparticles alleviates oxidative stress and inflammation in vitro and in vivo. *Biomaterials Science*. 10 (19), 5504–5519.

[24] Kim, J.-K., Go, E.-J., Ko, K.-W., Oh, H.-J., Han, J., Han, D.K., et al. (2021) PLGA microspheres containing hydrophobically modified magnesium hydroxide particles for acid neutralization-mediated anti-inflammation. *Tissue Engineering and Regenerative Medicine*. 18 613–622.

[25] Cheng, C.-Y., Pho, Q.-H., Wu, X.-Y., Chin, T.-Y., Chen, C.-M., Fang, P.-H., et al. (2018) PLGA microspheres loaded with β-cyclodextrin complexes of epigallocatechin-3-gallate for the anti-inflammatory properties in activated microglial cells. *Polymers*. 10 (5), 1–19.

[26] Yang, Y., Ding, Y., Fan, B., Wang, Y., Mao, Z., Wang, W., et al. (2020) Inflammation-targeting polymeric nanoparticles deliver sparfloxacin and tacrolimus for combating acute lung sepsis. *Journal of Controlled Release*. 321 463–474.

[27] Arafa, M.G., Girgis, G.N.S., and El-Dahan, M.S. (2020) Chitosan-coated PLGA nanoparticles for enhanced ocular anti-inflammatory efficacy of atorvastatin calcium. *International Journal of Nanomedicine*. 15: 1335–1347.

[28] Chereddy, K.K., Coco, R., Memvanga, P.B., Ucakar, B., Des Rieux, A., Vandermeulen, G., et al. (2013) Combined effect of PLGA and curcumin on wound healing activity. *Journal of Controlled Release*. 171 (2), 208–215.

[29] Wongrakpanich, A., Khunkitchai, N., Achayawat, Y., and Suksiriworapong, J. (2023) Ketorolac-Loaded PLGA-/PLA-Based Microparticles Stabilized by Hyaluronic Acid: Effects of Formulation Composition and Emulsification Technique on Particle Characteristics and Drug Release Behaviors. *Polymers*. 15 (2), 1–16.

[30] Liggins, R.T., Cruz, T., Min, W., Liang, L., Hunter, W.L., and Burt, H.M. (2004) Intra-articular treatment of arthritis with microsphere formulations of paclitaxel: biocompatibility and efficacy determinations in rabbits. *Inflammation Research*. 53 363–372.

[31] Duncan, S.A., Dixit, S., Sahu, R., Martin, D., Baganizi, D.R., Nyairo, E., et al. (2019) Prolonged release and functionality of interleukin-10 encapsulated within PLA-PEG nanoparticles. *Nanomaterials*. 9 (8), 1–16.

[32] El-Naggar, M.E., Al-Joufi, F., Anwar, M., Attia, M.F., and El-Bana, M.A. (2019) Curcumin-loaded PLA-PEG copolymer nanoparticles for treatment of liver inflammation in streptozotocin-induced diabetic rats. *Colloids and Surfaces B: Biointerfaces*. 177 389–398.

[33] Su, Y., Sun, B., Gao, X., Liu, S., Hao, R., and Han, B. (2020) Chitosan hydrogel doped with PEG-PLA nanoparticles for the local delivery of miRNA-146a to treat allergic rhinitis. *Pharmaceutics*. 12 (10), 1–17.

[34] Giacalone, G., Tsapis, N., Mousnier, L., Chacun, H., and Fattal, E. (2018) PLA-PEG nanoparticles improve the anti-inflammatory effect of rosiglitazone on macrophages by enhancing drug uptake compared to free rosiglitazone. *Materials*. 11 (10), 1–11.

[35] Assali, M., Shawahna, R., Shareef, M., and Alhimony, I.-A. (2018) Dexamethasone-diclofenac loaded polylactide nanoparticles: Preparation, release and anti-inflammatory activity. *European Journal of Pharmaceutical Sciences*. 122 179–184.

[36] Dave, V., Srivastava, P., Sharma, S., Bajaj, J., and Tak, K. (2019) PEGylated PLA-Phospholipon 90G complex hybrid nanoparticles loaded with etoricoxib for effective treatment pain relief potential. *International Journal of Polymeric Materials and Polymeric Biomaterials*. 2: 1–13.

[37] Giannavola, C., Bucolo, C., Maltese, A., Paolino, D., Vandelli, M.A., Puglisi, G., et al. (2003) Influence of preparation conditions on acyclovir-loaded poly-d, l-lactic acid nanospheres and effect of PEG coating on ocular drug bioavailability. *Pharmaceutical Research*. 20 584–590.

[38] Li, F., Xu, Y., Li, X., Wang, X., Yang, Z., Li, W., et al. (2021) Triblock copolymer nanomicelles loaded with curcumin attenuates inflammation via inhibiting the NF-κB pathway in the rat model of cerebral ischemia. *International Journal of Nanomedicine*. 16: 3173–3183.

[39] Kwon, J., Kim, J., Park, S., Khang, G., Kang, P.M., and Lee, D. (2013) Inflammation-responsive antioxidant nanoparticles based on a polymeric prodrug of vanillin. *Biomacromolecules*. 14 (5), 1618–1626.

[40] Chmielowski, R.A., Abdelhamid, D.S., Faig, J.J., Petersen, L.K., Gardner, C.R., Uhrich, K.E., et al. (2017) Atheroinflammatory nanotherapeutics: Ferulic acid-based poly (anhydride-ester) nanoparticles attenuate foam cell formation by regulating macrophage lipogenesis and reactive oxygen species generation. *Acta Biomaterialia*. 57 85–94.

[41] Garg, N.K., Sharma, G., Singh, B., Nirbhavane, P., Tyagi, R.K., Shukla, R., et al. (2017) Quality by design (QbD)-enabled development of aceclofenac loaded-nano structured lipid carriers (NLCs): An improved dermatokinetic profile for inflammatory disorder (s). *International Journal of Pharmaceutics*. 517 (1–2), 413–431.

[42] Park, H., Kim, S., Kim, S., Song, Y., Seung, K., Hong, D., et al. (2010) Antioxidant and anti-inflammatory activities of hydroxybenzyl alcohol releasing biodegradable polyoxalate nanoparticles. *Biomacromolecules*. 11 (8), 2103–2108.

[43] Bodmeier, R. and Chen, H. (1989) Preparation and characterization of microspheres containing the anti-inflammatory agents, indomethacin, ibuprofen, and ketoprofen. *Journal of Controlled Release*. 10 (2), 167–175.

[44] Ferreira, L.M., Cervi, V.F., Gehrcke, M., da Silveira, E.F., Azambuja, J.H., Braganhol, E., et al. (2015) Ketoprofen-loaded pomegranate seed oil nanoemulsion stabilized by pullulan: Selective antiglioma formulation for intravenous administration. *Colloids and Surfaces B: Biointerfaces*. 130 272–277.

[45] Lu, L., Qi, S., Chen, Y., Luo, H., Huang, S., Yu, X., et al. (2020) Targeted immunomodulation of inflammatory monocytes across the blood-brain barrier by curcumin-loaded nanoparticles delays the progression of experimental autoimmune encephalomyelitis. *Biomaterials*. 245 1–14.

[46] Gehrcke, M., Martins, C.C., de Bastos Brum, T., da Rosa, L.S., Luchese, C., Wilhelm, E.A., et al. (2022) Novel pullulan/gellan gum bilayer film as a vehicle for silibinin-loaded nanocapsules in the topical treatment of atopic dermatitis. *Pharmaceutics*. 14 (11), 1–23.

[47] Kauffman, K.J., Kanthamneni, N., Meenach, S.A., Pierson, B.C., Bachelder, E.M., and Ainslie, K.M. (2012) Optimization of rapamycin-loaded acetalated dextran microparticles for immunosuppression. *International Journal of Pharmaceutics*. 422 (1–2), 356–363.

[48] Dianzani, C., Foglietta, F., Ferrara, B., Rosa, A.C., Muntoni, E., Gasco, P., et al. (2017) Solid lipid nanoparticles delivering anti-inflammatory drugs to treat inflammatory bowel disease: Effects in an in vivo model. *World Journal of Gastroenterology*. 23 (23), 4200–4210.

[49] Anghelache, M., Turtoi, M., Petrovici, A.R., Fifere, A., Pinteala, M., and Calin, M. (2021) Development of dextran-coated magnetic nanoparticles loaded with protocatechuic acid for vascular inflammation therapy. *Pharmaceutics*. 13 (9), 1–19.

[50] Abdollahi, A.R., Firouzian, F., Haddadi, R., and Nourian, A. (2021) Indomethacin loaded dextran stearate polymeric micelles improve adjuvant-induced arthritis in rats: Design and in vivo evaluation. *Inflammopharmacology*. 29 107–121.

[51] Lu, J., Ren, B., Wang, L., Li, M., and Liu, Y. (2020) Preparation and evaluation of IL-1ra-loaded dextran/PLGA microspheres for inhibiting periodontal inflammation in vitro. *Inflammation*. 43 168–178.

[52] Salama, A.H., AbouSamra, M.M., Awad, G.E.A., and Mansy, S.S. (2021) Promising bioadhesive ofloxacin-loaded polymeric nanoparticles for the treatment of ocular inflammation: Formulation and in vivo evaluation. *Drug Delivery and Translational Research*. 11 1943–1957.

[53] Łukasiewicz, S., Mikołajczyk, A., Szczęch, M., Szczepanowicz, K., Warszyński, P., and Dziedzicka-Wasylewska, M. (2019) Encapsulation of clozapine into

polycaprolactone nanoparticles as a promising strategy of the novel nanoformulation of the active compound. *Journal of Nanoparticle Research*. 21 1–16.

[54] Agrawal, Y.O., Mahajan, U.B., Mahajan, H.S., and Ojha, S. (2020) Methotrexate-loaded nanostructured lipid carrier gel alleviates imiquimod-induced psoriasis by moderating inflammation: formulation, optimization, characterization, in-vitro and in-vivo studies. *International Journal of Nanomedicine*. 15: 4763–4778.

[55] Ansari, M.M., Ahmad, A., Kumar, A., Alam, P., Khan, T.H., Jayamurugan, G., et al. (2021) Aminocellulose-grafted-polycaprolactone coated gelatin nanoparticles alleviate inflammation in rheumatoid arthritis: A combinational therapeutic approach. *Carbohydrate Polymers*. 258 1–15.

[56] Brusini, R., Varna, M., and Couvreur, P. (2020) Advanced nanomedicines for the treatment of inflammatory diseases. *Advanced Drug Delivery Reviews*. 157 161–178.

[57] Lanao, R.P.F., Jonker, A.M., Wolke, J.G.C., Jansen, J.A., van Hest, J.C.M., and Leeuwenburgh, S.C.G. (2013) Physicochemical properties and applications of poly (lactic-co-glycolic acid) for use in bone regeneration. *Tissue Engineering Part B: Reviews*. 19 (4), 380–390.

[58] Avgoustakis, K. (2005) Polylactic-co-glycolic acid (PLGA). *Encyclopedia of Biomaterials and Biomedical Engineering*. 1 (1), 1–11.

[59] Ruhe, P.Q., Hedberg, E.L., Padron, N.T., Spauwen, P.H.M., Jansen, J.A., and Mikos, A.G. (2003) rhBMP-2 release from injectable poly (DL-lactic-co-glycolic acid)/calcium-phosphate cement composites. *JBJS*. 85 (suppl_3), 75–81.

[60] Gaspar, M.M., Blanco, D., Cruz, M.E.M., and Alonso, M.J. (1998) Formulation of L-asparaginase-loaded poly (lactide-co-glycolide) nanoparticles: influence of polymer properties on enzyme loading, activity and in vitro release. *Journal of Controlled Release*. 52 (1–2), 53–62.

[61] Breitenbach, A., Li, Y.X., and Kissel, T. (2000) Branched biodegradable polyesters for parenteral drug delivery systems. *Journal of Controlled Release*. 64 (1–3), 167–178.

[62] Mundargi, R.C., Babu, V.R., Rangaswamy, V., Patel, P., and Aminabhavi, T.M. (2008) Nano/micro technologies for delivering macromolecular therapeutics using poly (D, L-lactide-co-glycolide) and its derivatives. *Journal of Controlled Release*. 125 (3), 193–209.

[63] Bouissou, C. (2006) *Encapsulation of an integrin-binding protein into PLGA microspheres*. University of Bath (United Kingdom).

[64] Mir, M., Ahmed, N., and ur Rehman, A. (2017) Recent applications of PLGA based nanostructures in drug delivery. *Colloids and Surfaces B: Biointerfaces*. 159 217–231.

[65] Fasehee, H., Dinarvand, R., Ghavamzadeh, A., Esfandyari-Manesh, M., Moradian, H., Faghihi, S., et al. (2016) Delivery of disulfiram into breast cancer cells using folate-receptor-targeted PLGA-PEG nanoparticles: in vitro and in vivo investigations. *Journal of Nanobiotechnology*. 14 (1), 1–18.

[66] Galindo, R., Sánchez-López, E., Gómara, M.J., Espina, M., Ettcheto, M., Cano, A., et al. (2022) Development of peptide targeted PLGA-PEGylated nanoparticles loading Licochalcone-A for ocular inflammation. *Pharmaceutics*. 14 (2), 1–17.

[67] Ajmal, G., Bonde, G.V., Mittal, P., Pandey, V.K., Yadav, N., and Mishra, B. (2023) PLGA/Gelatin-based electrospun nanofiber scaffold encapsulating antibacterial and antioxidant molecules for accelerated tissue regeneration. *Materials Today Communications*. 35 1–15.

[68] Paswan, S.K. (2023) Optimization, and evaluation of hyaluronic acid conjugated PLGA nanoparticles of etoposide for cancer treatment. *Asian Journal of Pharmaceutics (AJP)*. 17 (1), 217–235.

[69] Wang, M., Lin, Y., Gao, J., and Liu, D. (2022) DPD simulations on morphologies and structures of blank PLGA-b-PEG-b-PLGA polymeric micelles and docetaxel-loaded PLGA-b-PEG-b-PLGA polymeric micelles. *RSC Advances*. 12 (19), 12078–12088.

[70] Kaplan, D.L. (1998) Introduction to biopolymers from renewable resources. in: *Biopolym. from Renew. Resour.*, Springer, pp. 1–29.

[71] Li, G., Zhao, M., Xu, F., Yang, B., Li, X., Meng, X., et al. (2020) Synthesis and biological application of polylactic acid. *Molecules*. 25 (21), 1–18.

[72] Nottelet, B., Darcos, V., and Coudane, J. (2015) Aliphatic polyesters for medical imaging and theranostic applications. *European Journal of Pharmaceutics and Biopharmaceutics*. 97 350–370.

[73] Mitragotri, S., Burke, P.A., and Langer, R. (2014) Overcoming the challenges in administering biopharmaceuticals: formulation and delivery strategies. *Nature Reviews Drug Discovery*. 13 (9), 655–672.

[74] Sreekumar, K., Bindhu, B., and Veluraja, K. (2021) Perspectives of polylactic acid from structure to applications. *Polymers from Renewable Resources*. 12 (1–2), 60–74.

[75] Anwer, A.H., Ahtesham, A., Shoeb, M., Mashkoor, F., Ansari, M.Z., Zhu, S., et al. (2023) State-of-the-art advances in nanocomposite and bio-nanocomposite polymeric materials: A comprehensive review. *Advances in Colloid and Interface Science*. 318 1–39.

[76] Vila, A., Gill, H., McCallion, O., and Alonso, M.J. (2004) Transport of PLA-PEG particles across the nasal mucosa: effect of particle size and PEG coating density. *Journal of Controlled Release*. 98 (2), 231–244.

[77] Zhang, Y., Ni, C., Shi, G., Wang, J., Zhang, M., and Li, W. (2015) The polyion complex nano-prodrug of doxorubicin (DOX) with poly (lactic acid-co-malic acid)-block-polyethylene glycol: preparation and drug controlled release. *Medicinal Chemistry Research*. 24 1189–1195.

[78] Hu, X., He, J., Yong, X., Lu, J., Xiao, J., Liao, Y., et al. (2020) Biodegradable poly (lactic acid-co-trimethylene carbonate)/chitosan microsphere scaffold with shape-memory effect for bone tissue engineering. *Colloids and Surfaces B: Biointerfaces*. 195 1–11.

[79] Kalaimagal, G., Devarajan, S., Soundararajan, S., and Gajalakshmi, D. (2023) Curcumin loaded poly (lactic acid)–poly (propylene glycol) with boc-glycine end group copolymer in drug delivery applications. *Research Square*. 1–16.

[80] Said, N., Lau, W.J., Ho, Y.-C., Lim, S.K., Zainol Abidin, M.N., and Ismail, A.F. (2021) A review of commercial developments and recent laboratory research of dialyzers and membranes for hemodialysis application. *Membranes*. 11 (10), 1–38.

[81] Lim, W.S., Chen, K., Chong, T.W., Xiong, G.M., Birch, W.R., Pan, J., et al. (2018) A bilayer swellable drug-eluting ureteric stent: Localized drug delivery to treat urothelial diseases. *Biomaterials*. 165 25–38.

[82] Scheirs, J. and Priddy, D.B. (2003) *Modern styrenic polymers: Polystyrenes and styrenic copolymers*. Wiley Online Library.

[83] Wünsch, J.R. (2000) *Polystyrene: Synthesis, production and applications*. iSmithers Rapra Publishing.

[84] Liu, C., Liu, F., Cai, J., Xie, W., Long, T.E., Turner, S.R., et al. (2011) Polymers from fatty acids: Poly (ω-hydroxyl tetradecanoic acid) synthesis and physico-mechanical studies. *Biomacromolecules*. 12 (9), 3291–3298.

[85] Nazarzadeh, Z.E., Najafi, M.P., Azariyan, E., and Sharifian, I. (2011) Conductive and biodegradable polyaniline/starch blends and their composites with polystyrene. *Iranian Polymer Journal*. 4 (20), 319–328.
[86] Shim, A. (2015) Simulations of hard nanoparticle encapsulation during polymer micelle formation. 1–48. The Ohio State University Repository/Knowledge Bank.
[87] Datar, A., Joshi, P., and Lee, M.-Y. (2015) Biocompatible hydrogels for microarray cell printing and encapsulation. *Biosensors*. 5 (4), 647–663.
[88] da Silva Paula, M.M., de Souza, H.J., Búrigo, C., Langbehn, J.T., Baesso, A.S., da Silva, L., et al. (2017) Evaluation of the permeability of silver nanoparticles in polymer films of sulfonated polystyrene-co-acrylic acid. *Colloids and Surfaces A: Physicochemical and Engineering Aspects*. 520 78–84.
[89] Kotha, R., Kara, D.D., Roychowdhury, R., Tanvi, K., and Rathnanand, M. (2023) Polymersomes based versatile nanoplatforms for controlled drug delivery and imaging. *Advanced Pharmaceutical Bulletin*. 13 (2), 218–232.
[90] Júnior, J.A.C.N., Santos, A.M., Oliveira, A.M.S., Santos, A.B., de Souza Araujo, A.A., Frank, L.A., et al. (2023) Use of nanotechnology applied to sunscreens: Technological prospection based on patents. *Journal of Drug Delivery Science and Technology*. 91 1–20.
[91] Santore, M. (2006) Polymer vesicles in biomimetic applications. In: *APS March Meet. Abstr.*, pp. 28–102.
[92] Galogahi, F.M., Zhu, Y., An, H., and Nguyen, N.-T. (2020) Core-shell microparticles: Generation approaches and applications. *Journal of Science: Advanced Materials and Devices*. 5 (4), 417–435.
[93] Shibata, M., Asahina, M., Teramoto, N., and Yosomiya, R. (2001) Chemical modification of pullulan by isocyanate compounds. *Polymer*. 42 (1), 59–64.
[94] Agrawal, S., Budhwani, D., Gurjar, P., Telange, D., and Lambole, V. (2022) Pullulan based derivatives: Synthesis, enhanced physicochemical properties, and applications. *Drug Delivery*. 29 (1), 3328–3339.
[95] Singh, R.S., Kaur, N., Hassan, M., and Kennedy, J.F. (2021) Pullulan in biomedical research and development-A review. *International Journal of Biological Macromolecules*. 166 694–706.
[96] Ganie, S.A., Mir, T.A., Ali, A., and Li, Q. (2020) *Pullulan: properties and applications*. Pullulan: Jenny Stanford Publishing. 89–132.
[97] Ganie, S.A., Rather, L.J., and Li, Q. (2021) A review on anticancer applications of pullulan and pullulan derivative nanoparticles. *Carbohydrate Polymer Technologies and Applications*. 2 1–17.
[98] Thakur, A., Sharma, S., Naman, S., and Baldi, A. (2023) Pullulan based polymeric novel drug delivery systems: A review on current state of art and prospects. *Journal of Drug Delivery Science and Technology*. 90 1–18.
[99] Varshosaz, J. (2012) Dextran conjugates in drug delivery. *Expert Opinion on Drug Delivery*. 9 (5), 509–523.
[100] Sharma, S., Naman, S., Dwivedi, J., and Baldi, A. (2023) Dextran for Application in DDS for Lung Diseases. in: *Nat. Polym. Mater. Based Drug Deliv. Syst. Lung Dis.*, Springer, pp. 297–327.
[101] Díaz-Montes, E. (2021) Dextran: Sources, structures, and properties. *Polysaccharides*. 2 (3), 554–565.
[102] Maity, S., Saha, S., Sarkar, S., Kumar, P., Gautam, A.K., and Sonkar, A.B. (2021) Dextran-based nanomaterials in drug delivery applications. in: *Biopolym. Nanomater. Drug Deliv. Biomed. Appl.*, Elsevier, pp. 293–312.
[103] Delvart, A., Moreau, C., and Cathala, B. (2022) Dextrans and dextran derivatives as polyelectrolytes in layer-by-layer processing materials–A review. *Carbohydrate Polymers*. 293 1–74.
[104] Pandey, M.M., Kulkarni, O.P., Singh, S.B., and Madan, J. (2022) *Dextran in the delivery of Therapeutics: Chronicle of the Journey from Preclinical to Clinical Trials*. Royal Society of Chemistry.
[105] Khan, M.S., Gowda, B.H.J., Nasir, N., Wahab, S., Pichika, M.R., Sahebkar, A., et al. (2023) Advancements in dextran-based nanocarriers for treatment and imaging of breast cancer. *International Journal of Pharmaceutics*. 643 1–18.
[106] Bachelder, E.M., Beaudette, T.T., Broaders, K.E., Dashe, J., and Fréchet, J.M.J. (2008) Acetal-derivatized dextran: An acid-responsive biodegradable material for therapeutic applications. *Journal of the American Chemical Society*. 130 (32), 10494–10495.
[107] Betancor, L., Fuentes, M., Dellamora-Ortiz, G., López-Gallego, F., Hidalgo, A., Alonso-Morales, N., et al. (2005) Dextran aldehyde coating of glucose oxidase immobilized on magnetic nanoparticles prevents its inactivation by gas bubbles. *Journal of Molecular Catalysis B: Enzymatic*. 32 (3), 97–101.
[108] Weng, L., Romanov, A., Rooney, J., and Chen, W. (2008) Non-cytotoxic, in situ gelable hydrogels composed of N-carboxyethyl chitosan and oxidized dextran. *Biomaterials*. 29 (29), 3905–3913.
[109] Meenach, S.A., Kim, Y.J., Kauffman, K.J., Kanthamneni, N., Bachelder, E.M., and Ainslie, K.M. (2012) Synthesis, optimization, and characterization of camptothecin-loaded acetalated dextran porous microparticles for pulmonary delivery. *Molecular Pharmaceutics*. 9 (2), 290–298.
[110] Pawar, R., Pathan, A., Nagaraj, S., Kapare, H., Giram, P., and Wavhale, R. (2023) Polycaprolactone and its derivatives for drug delivery. *Polymers for Advanced Technologies*. 34 (10), 3296–3316.
[111] Bader, N. and Ammar, A.A. (2024) Incorporating surfactants into PCL microneedles for sustained release of a hydrophilic model drug. *International Journal of Pharmaceutics*. 652: 1–26.
[112] Wei, X., Gong, C., Gou, M., Fu, S., Guo, Q., Shi, S., et al. (2009) Biodegradable poly (ε-caprolactone)–poly (ethylene glycol) copolymers as drug delivery system. *International Journal of Pharmaceutics*. 381 (1), 1–18.
[113] Feng, Z., Wu, L., Dong, H., Liu, B., and Cheng, R. (2021) Copolyesters of ε-caprolactone and l-lactide catalyzed by a tetrabutylammonium phthalimide-N-oxyl organocatalyst. *RSC Advances*. 11 (31), 19021–19028.
[114] Hu, S., Zhao, Q., Niu, Y., Luo, C., Zhang, Z., He, B., et al. (2023) Synthesis, characterization, thermal stability, and in vitro and in vivo degradation study of polycaprolactone and polyglycolide block copolymers. *Journal of Biomaterials Science, Polymer Edition*. 34 (3), 302–314.
[115] Zhao, X., Jin, L., Shi, H., Tong, W., Gorin, D., Kotelevtsev, Y., et al. (2020) Recent advances of designing dynamic surfaces to regulate cell adhesion. *Colloid and Interface Science Communications*. 35 1–13.
[116] Martini, L.G., Collett, J.H., and Attwood, D. (2000) The release of 5-fluorouracil from microspheres of poly (ε-caprolactone-co-ethylene oxide). *Drug Development and Industrial Pharmacy*. 26 (1), 7–12.
[117] Erthal, L.C.S., Gobbo, O.L., and Ruiz-Hernandez, E. (2021) Biocompatible copolymer formulations to treat glioblastoma multiforme. *Acta Biomaterialia*. 121 89–102.

[118] Tian, H., Tang, Z., Zhuang, X., Chen, X., and Jing, X. (2012) Biodegradable synthetic polymers: Preparation, functionalization and biomedical application. *Progress in Polymer Science*. 37 (2), 237–280.
[119] Pal, K. and Banerjee, I. (2018) *Polymeric gels: Characterization, properties and biomedical applications*. Woodhead Publishing.
[120] Maitz, M.F. (2015) Applications of synthetic polymers in clinical medicine. *Biosurface and Biotribology*. 1 (3), 161–176.
[121] Sionkowska, A. (2011) Current research on the blends of natural and synthetic polymers as new biomaterials. *Progress in Polymer Science*. 36 (9), 1254–1276.
[122] Fournier, E., Passirani, C., Montero-Menei, C.N., and Benoit, J.P. (2003) Biocompatibility of implantable synthetic polymeric drug carriers: focus on brain biocompatibility. *Biomaterials*. 24 (19), 3311–3331.
[123] Dash, T.K. and Konkimalla, V.B. (2012) Poly-ɛ-caprolactone based formulations for drug delivery and tissue engineering: A review. *Journal of Controlled Release*. 158 (1), 15–33.
[124] Sharma, G.K., Joseph, S.L., and James, N.R. (2024) Recent progress in poly (3, 4-ethylene dioxythiophene): Polystyrene sulfonate based composite materials for electromagnetic interference shielding. *Advanced Materials Technologies*. 9 (1), 1–17.
[125] Li, B., Elango, J., and Wu, W. (2020) Recent advancement of molecular structure and biomaterial function of chitosan from marine organisms for pharmaceutical and nutraceutical application. *Applied Sciences*. 10 (14), 1–24.
[126] Fattahi, F.S., Khoddami, A., and Avinc, O. (2020) Sustainable, renewable, and biodegradable poly (lactic acid) fibers and their latest developments in the last decade. *Sustainability in the Textile and Apparel Industries: Sourcing Synthetic and Novel Alternative Raw Materials*. 173–194.
[127] Liechty, W.B., Kryscio, D.R., Slaughter, B. V., and Peppas, N.A. (2010) Polymers for drug delivery systems. *Annual Review of Chemical and Biomolecular Engineering*. 1 149–173.
[128] Varde, N.K. and Pack, D.W. (2004) Microspheres for controlled release drug delivery. *Expert Opinion on Biological Therapy*. 4 (1), 35–51.
[129] Ting, J.M., Porter III, W.W., Mecca, J.M., Bates, F.S., and Reineke, T.M. (2018) Advances in polymer design for enhancing oral drug solubility and delivery. *Bioconjugate Chemistry*. 29 (4), 939–952.
[130] Paramjot, Khan N.M., Kapahi, H., Kumar, S., Bhardwaj, T.R., Arora, S., et al. (2015) Role of polymer–drug conjugates in organ-specific delivery systems. *Journal of Drug Targeting*. 23 (5), 387–416.
[131] Montenegro, J.-M., Grazu, V., Sukhanova, A., Agarwal, S., de la Fuente, J.M., Nabiev, I., et al. (2013) Controlled antibody/(bio-) conjugation of inorganic nanoparticles for targeted delivery. *Advanced Drug Delivery Reviews*. 65 (5), 677–688.
[132] Yang, L. and Alexandridis, P. (2000) Physicochemical aspects of drug delivery and release from polymer-based colloids. *Current Opinion in Colloid & Interface Science*. 5 (1–2), 132–143.
[133] Říhová, B. (2002) Immunomodulating activities of soluble synthetic polymer-bound drugs. *Advanced Drug Delivery Reviews*. 54 (5), 653–674.
[134] Waheed, S., Li, Z., Zhang, F., Chiarini, A., Armato, U., and Wu, J. (2022) Engineering nano-drug biointerface to overcome biological barriers toward precision drug delivery. *Journal of Nanobiotechnology*. 20 (1), 395.
[135] O'Donnell, P.B. and McGinity, J.W. (1997) Preparation of microspheres by the solvent evaporation technique. *Advanced Drug Delivery Reviews*. 28 (1), 25–42.
[136] Umerska, A., Gaucher, C., Oyarzun-Ampuero, F., Fries-Raeth, I., Colin, F., Villamizar-Sarmiento, M.G., et al. (2018) Polymeric nanoparticles for increasing oral bioavailability of curcumin. *Antioxidants*. 7 (4), 46.
[137] Pouton, C.W. (1997) Formulation of self-emulsifying drug delivery systems. *Advanced Drug Delivery Reviews*. 25 (1), 47–58.
[138] Sajjadi, S. (2006) Nanoemulsion formation by phase inversion emulsification: on the nature of inversion. *Langmuir*. 22 (13), 5597–5603.
[139] De Kruif, C.G., Weinbreck, F., and de Vries, R. (2004) Complex coacervation of proteins and anionic polysaccharides. *Current Opinion in Colloid & Interface Science*. 9 (5), 340–349.
[140] Timilsena, Y.P., Akanbi, T.O., Khalid, N., Adhikari, B., and Barrow, C.J. (2019) Complex coacervation: Principles, mechanisms and applications in microencapsulation. *International Journal of Biological Macromolecules*. 121 1276–1286.
[141] Thakkar, S. and Misra, M. (2017) Electrospun polymeric nanofibers: New horizons in drug delivery. *European Journal of Pharmaceutical Sciences*. 107 148–167.
[142] Srikar, R., Yarin, A.L., Megaridis, C.M., Bazilevsky, A. V., and Kelley, E. (2008) Desorption-limited mechanism of release from polymer nanofibers. *Langmuir*. 24 (3), 965–974.
[143] Segala, K., Nista, S.V.G., Cordi, L., Bizarria, M.T.M., Ávila Júnior, J. De, Kleinubing, S.A., et al. (2015) Silver nanoparticles incorporated into nanostructured biopolymer membranes produced by electrospinning: a study of antimicrobial activity. *Brazilian Journal of Pharmaceutical Sciences*. 51 911–921.
[144] Hwang, I., Kang, C.-Y., and Park, J.-B. (2017) Advances in hot-melt extrusion technology toward pharmaceutical objectives. *Journal of Pharmaceutical Investigation*. 47 123–132.
[145] Stanković, M., Frijlink, H.W., and Hinrichs, W.L.J. (2015) Polymeric formulations for drug release prepared by hot melt extrusion: application and characterization. *Drug Discovery Today*. 20 (7), 812–823.
[146] Repka, M.A., Majumdar, S., Kumar Battu, S., Srirangam, R., and Upadhye, S.B. (2008) Applications of hot-melt extrusion for drug delivery. *Expert Opinion on Drug Delivery*. 5 (12), 1357–1376.
[147] Lin, J.-L., Kao, Z.-K., and Liao, Y.-C. (2013) Preserving precision of inkjet-printed features with solvents of different volatilities. *Langmuir*. 29 (36), 11330–11336.
[148] Paudel, A., Worku, Z.A., Meeus, J., Guns, S., and Van den Mooter, G. (2013) Manufacturing of solid dispersions of poorly water soluble drugs by spray drying: formulation and process considerations. *International Journal of Pharmaceutics*. 453 (1), 253–284.
[149] Li, M., Rouaud, O., and Poncelet, D. (2008) Microencapsulation by solvent evaporation: State of the art for process engineering approaches. *International Journal of Pharmaceutics*. 363 (1–2), 26–39.
[150] Mahon, A.R. (1997) The safe and efficient evaporation of a solvent from solution. *Journal of Chemical Education*. 74 (2), 245–246.
[151] Borlongan, M.C., Borlongan, M.C., and Sanberg, P.R. (2020) *The disillusioned comfort with COVID-19 and the potential of convalescent plasma and cell therapy*. SAGE Publications Inc.

[152] Zarzar, L.D., Sresht, V., Sletten, E.M., Kalow, J.A., Blankschtein, D., and Swager, T.M. (2015) Dynamically reconfigurable complex emulsions via tunable interfacial tensions. *Nature*. 518 (7540), 520–524.

[153] Ramazani, F., Chen, W., van Nostrum, C.F., Storm, G., Kiessling, F., Lammers, T., et al. (2016) Strategies for encapsulation of small hydrophilic and amphiphilic drugs in PLGA microspheres: state-of-the-art and challenges. *International Journal of Pharmaceutics*. 499 (1–2), 358–367.

[154] Kohli, K., Chopra, S., Dhar, D., Arora, S., and Khar, R.K. (2010) Self-emulsifying drug delivery systems: an approach to enhance oral bioavailability. *Drug Discovery Today*. 15 (21–22), 958–965.

[155] Li, Q., Zhan, S., Liu, Q., Su, H., Dai, X., Wang, H., et al. (2018) Preparation of a sustained-release nebulized aerosol of r-terbutaline hydrochloride liposome and evaluation of its anti-asthmatic effects via pulmonary delivery in guinea pigs. *AAPS PharmSciTech*. 19 232–241.

[156] Nakashima, T., Shimizu, M., and Kukizaki, M. (2000) Particle control of emulsion by membrane emulsification and its applications. *Advanced Drug Delivery Reviews*. 45 (1), 47–56.

[157] Tamilvanan, S. (2004) Oil-in-water lipid emulsions: implications for parenteral and ocular delivering systems. *Progress in Lipid Research*. 43 (6), 489–533.

[158] Tadros, T.F. (2016) Emulsions: Formation, stability, industrial applications. *Walter de Gruyter GmbH & Co KG*.

[159] Capek, I. (2004) Degradation of kinetically-stable o/w emulsions. *Advances in Colloid and Interface Science*. 107 (2–3), 125–155.

[160] Mora-Huertas, C.E., Fessi, H., and Elaissari, A. (2010) Polymer-based nanocapsules for drug delivery. *International Journal of Pharmaceutics*. 385 (1–2), 113–142.

[161] Johnson, N.R. and Wang, Y. (2014) Coacervate delivery systems for proteins and small molecule drugs. *Expert Opinion on Drug Delivery*. 11 (12), 1829–1832.

[162] Yan, C. and Zhang, W. (2014) Coacervation processes. in: *Microencapsul. Food Ind.*, Elsevier, pp. 125–137.

[163] Chapeau, A.-L., Hamon, P., Rousseau, F., Croguennec, T., Poncelet, D., and Bouhallab, S. (2017) Scale-up production of vitamin loaded heteroprotein coacervates and their protective property. *Journal of Food Engineering*. 206 67–76.

[164] Sing, C.E. and Perry, S.L. (2020) Recent progress in the science of complex coacervation. *Soft Matter*. 16 (12), 2885–2914.

[165] Saravanan, M. and Rao, K.P. (2010) Pectin–gelatin and alginate–gelatin complex coacervation for controlled drug delivery: Influence of anionic polysaccharides and drugs being encapsulated on physicochemical properties of microcapsules. *Carbohydrate Polymers*. 80 (3), 808–816.

[166] Melnyk, A., Wolska, L., and Namieśnik, J. (2014) Coacervative extraction as a green technique for sample preparation for the analysis of organic compounds. *Journal of Chromatography A*. 1339 1–12.

[167] Immanuel, C.D., Wang, Y., and Bianco, N. (2008) Feedback controllability assessment and control of particle size distribution in emulsion polymerisation. *Chemical Engineering Science*. 63 (5), 1205–1216.

[168] Yoo, H.S., Kim, T.G., and Park, T.G. (2009) Surface-functionalized electrospun nanofibers for tissue engineering and drug delivery. *Advanced Drug Delivery Reviews*. 61 (12), 1033–1042.

[169] Tipduangta, P., Belton, P., Fabian, L., Wang, L.Y., Tang, H., Eddleston, M., et al. (2016) Electrospun polymer blend nanofibers for tunable drug delivery: the role of transformative phase separation on controlling the release rate. *Molecular Pharmaceutics*. 13 (1), 25–39.

[170] Sill, T.J. and Von Recum, H.A. (2008) Electrospinning: applications in drug delivery and tissue engineering. *Biomaterials*. 29 (13), 1989–2006.

[171] Fong, H., Chun, I., and Reneker, D.H. (1999) Beaded nanofibers formed during electrospinning. *Polymer*. 40 (16), 4585–4592.

[172] Zare, M., Dziemidowicz, K., Williams, G.R., and Ramakrishna, S. (2021) Encapsulation of pharmaceutical and nutraceutical active ingredients using electrospinning processes. *Nanomaterials*. 11 (8), 1–15.

[173] Ding, Y., Li, W., Zhang, F., Liu, Z., Zanjanizadeh Ezazi, N., Liu, D., et al. (2019) Electrospun fibrous architectures for drug delivery, tissue engineering and cancer therapy. *Advanced Functional Materials*. 29 (2), 1–126.

[174] Li, D., McCann, J.T., Xia, Y., and Marquez, M. (2006) Electrospinning: a simple and versatile technique for producing ceramic nanofibers and nanotubes. *Journal of the American Ceramic Society*. 89 (6), 1861–1869.

[175] Crowley, M.M., Zhang, F., Koleng, J.J., and McGinity, J.W. (2002) Stability of polyethylene oxide in matrix tablets prepared by hot-melt extrusion. *Biomaterials*. 23 (21), 4241–4248.

[176] Censi, R., Gigliobianco, M.R., Casadidio, C., and Di Martino, P. (2018) Hot melt extrusion: Highlighting physicochemical factors to be investigated while designing and optimizing a hot melt extrusion process. *Pharmaceutics*. 10 (3), 1–27.

[177] Repka, M.A., Bandari, S., Kallakunta, V.R., Vo, A.Q., McFall, H., Pimparade, M.B., et al. (2018) Melt extrusion with poorly soluble drugs–An integrated review. *International Journal of Pharmaceutics*. 535 (1–2), 68–85.

[178] Li, Y., Pang, H., Guo, Z., Lin, L., Dong, Y., Li, G., et al. (2014) Interactions between drugs and polymers influencing hot melt extrusion. *Journal of Pharmacy and Pharmacology*. 66 (2), 148–166.

[179] Ravi Kumar, M.N. V. and Kumar, N. (2001) Polymeric controlled drug-delivery systems: perspective issues and opportunities. *Drug Development and Industrial Pharmacy*. 27 (1), 1–30.

[180] Alexander, C. (2001) Synthetic polymer systems in drug delivery. *Expert Opinion on Emerging Drugs*. 6 (2), 345–363.

[181] Almeida, A., Günday-Türeli, N., and Sarmento, B. (2021) A scale-up strategy for the synthesis of chitosan derivatives used in micellar nanomedicines. *International Journal of Pharmaceutics*. 609 121151.

[182] Avramović, N., Mandić, B., Savić-Radojević, A., and Simić, T. (2020) Polymeric nanocarriers of drug delivery systems in cancer therapy. *Pharmaceutics*. 12 (4), 1–17.

[183] Ghomrasni, N.B., Chivas-Joly, C., Devoille, L., Hochepied, J.-F., and Feltin, N. (2020) Challenges in sample preparation for measuring nanoparticles size by scanning electron microscopy from suspensions, powder form and complex media. *Powder Technology*. 359 226–237.

[184] Peltonen, L., Singhal, M., and Hirvonen, J. (2020) Principles of nanosized drug delivery systems. in: *Nanoeng. Biomater. Adv. Drug Deliv.*, Elsevier, pp. 3–25.

[185] Peltonen, L. (2018) Practical guidelines for the characterization and quality control of pure drug nanoparticles and nano-cocrystals in the pharmaceutical industry. *Advanced Drug Delivery Reviews*. 131 101–115.

[186] Zhang, W., Mehta, A., Tong, Z., Esser, L., and Voelcker, N.H. (2021) Development of polymeric nanoparticles for

blood–brain barrier transfer—strategies and challenges. *Advanced Science*. 8 (10), 1–32.
[187] Langevin, D., Raspaud, E., Mariot, S., Knyazev, A., Stocco, A., Salonen, A., et al. (2018) Towards reproducible measurement of nanoparticle size using dynamic light scattering: Important controls and considerations. *NanoImpact*. 10 161–167.
[188] Sukhanova, A., Bozrova, S., Sokolov, P., Berestovoy, M., Karaulov, A., and Nabiev, I. (2018) Dependence of nanoparticle toxicity on their physical and chemical properties. *Nanoscale Research Letters*. 13 1–21.
[189] Baer, D.R. (2018) The chameleon effect: characterization challenges due to the variability of nanoparticles and their surfaces. *Frontiers in Chemistry*. 6 1–7.
[190] Cavuslar, O., Nakay, E., Kazakoglu, U., Abkenar, S.K., Ow-Yang, C.W., and Acar, H.Y. (2020) Synthesis of stable gold nanoparticles using linear polyethyleneimines and catalysis of both anionic and cationic azo dye degradation. *Materials Advances*. 1 (7), 2407–2417.
[191] Yaqoob, A.A., Sekeri, S.H., Othman, M.B.H., Ibrahim, M.N.M., and Feizi, Z.H. (2021) Thermal degradation and kinetics stability studies of oil palm (Elaeis Guineensis) biomass-derived lignin nanoparticle and its application as an emulsifying agent. *Arabian Journal of Chemistry*. 14 (6), 1–18.
[192] Paliwal, R., Babu, R.J., and Palakurthi, S. (2014) Nanomedicine scale-up technologies: feasibilities and challenges. *Aaps Pharmscitech*. 15 (6), 1527–1534.
[193] Herdiana, Y., Wathoni, N., Shamsuddin, S., and Muchtaridi, M. (2022) Drug release study of the chitosan-based nanoparticles. *Heliyon*. 8 1–16.
[194] Mazumder, S., Dewangan, A.K., and Pavurala, N. (2017) Enhanced dissolution of poorly soluble antiviral drugs from nanoparticles of cellulose acetate based solid dispersion matrices. *Asian Journal of Pharmaceutical Sciences*. 12 (6), 532–541.
[195] Turan, N.B., Erkan, H.S., Engin, G.O., and Bilgili, M.S. (2019) Nanoparticles in the aquatic environment: Usage, properties, transformation and toxicity—A review. *Process Safety and Environmental Protection*. 130 238–249.
[196] Hedberg, J., Blomberg, E., and Odnevall Wallinder, I. (2019) In the search for nanospecific effects of dissolution of metallic nanoparticles at freshwater-like conditions: A critical review. *Environmental Science & Technology*. 53 (8), 4030–4044.
[197] López-Lorente, Á.I., Valcárcel, M., and Mizaikoff, B. (2014) Continuous flow synthesis and characterization of tailor-made bare gold nanoparticles for use in SERS. *Microchimica Acta*. 181 1101–1108.
[198] Utembe, W., Potgieter, K., Stefaniak, A.B., and Gulumian, M. (2015) Dissolution and biodurability: Important parameters needed for risk assessment of nanomaterials. *Particle and Fibre Toxicology*. 12 1–12.
[199] Wu, L.-P., Wang, D., and Li, Z. (2020) Grand challenges in nanomedicine. *Materials Science and Engineering: C*. 106 1–7.
[200] Casals, E., Gonzalez, E., and Puntes, V.F. (2012) Reactivity of inorganic nanoparticles in biological environments: insights into nanotoxicity mechanisms. *Journal of Physics D: Applied Physics*. 45 (44), 443001.
[201] Buzea, C., Pacheco, I.I., and Robbie, K. (2007) Nanomaterials and nanoparticles: sources and toxicity. *Biointerphases*. 2 (4), MR17–MR71.
[202] Colombo, A.P., Briançon, S., Lieto, J., and Fessi, H. (2001) Project, design, and use of a pilot plant for nanocapsule production. *Drug Development and Industrial Pharmacy*. 27 (10), 1063–1072.
[203] Panariello, L., Damilos, S., du Toit, H., Wu, G., Radhakrishnan, A.N.P., Parkin, I.P., et al. (2020) Highly reproducible, high-yield flow synthesis of gold nanoparticles based on a rational reactor design exploiting the reduction of passivated Au (III). *Reaction Chemistry & Engineering*. 5 (4), 663–676.
[204] Baer, D.R., Munusamy, P., and Thrall, B.D. (2016) Provenance information as a tool for addressing engineered nanoparticle reproducibility challenges. *Biointerphases*. 11 (4), 1–10.
[205] Svenson, S. (2015) The dendrimer paradox–high medical expectations but poor clinical translation. *Chemical Society Reviews*. 44 (12), 4131–4144.
[206] Svenson, S. (2012) Clinical translation of nanomedicines. *Current Opinion in Solid State and Materials Science*. 16 (6), 287–294.
[207] Zielińska, A., Carreiró, F., Oliveira, A.M., Neves, A., Pires, B., Venkatesh, D.N., et al. (2020) Polymeric nanoparticles: Production, characterization, toxicology and ecotoxicology. *Molecules*. 25 (16), 1–20.
[208] Najahi-Missaoui, W., Arnold, R.D., and Cummings, B.S. (2020) Safe nanoparticles: Are we there yet? *International Journal of Molecular Sciences*. 22 (1), 1–21.
[209] Weiss, M., Fan, J., Claudel, M., Sonntag, T., Didier, P., Ronzani, C., et al. (2021) Density of surface charge is a more predictive factor of the toxicity of cationic carbon nanoparticles than zeta potential. *Journal of Nanobiotechnology*. 19 (1), 1–19.
[210] Seabra, A.B. and Durán, N. (2015) Nanotoxicology of metal oxide nanoparticles. *Metals*. 5 (2), 934–975.
[211] Kornberg, T.G., Stueckle, T.A., Antonini, J.M., Rojanasakul, Y., Castranova, V., Yang, Y., et al. (2017) Potential toxicity and underlying mechanisms associated with pulmonary exposure to iron oxide nanoparticles: Conflicting literature and unclear risk. *Nanomaterials*. 7 (10), 1–26.
[212] Hoshyar, N., Gray, S., Han, H., and Bao, G. (2016) The effect of nanoparticle size on in vivo pharmacokinetics and cellular interaction. *Nanomedicine*. 11 (6), 673–692.
[213] Jain, A.K. and Thareja, S. (2019) In vitro and in vivo characterization of pharmaceutical nanocarriers used for drug delivery. *Artificial Cells, Nanomedicine, and Biotechnology*. 47 (1), 524–539.
[214] Guilger-Casagrande, M. and Lima, R. De (2019) Synthesis of silver nanoparticles mediated by fungi: A review. *Frontiers in Bioengineering and Biotechnology*. 7 1–16.
[215] Marooufpour, N., Alizadeh, M., Hatami, M., and Asgari Lajayer, B. (2019) Biological synthesis of nanoparticles by different groups of bacteria. *Microbial Nanobionics: Volume 1*, State-of-the-Art. 63–85.
[216] Bovone, G., Guzzi, E.A., and Tibbitt, M.W. (2019) *Flow-based reactor design for the continuous production of polymeric nanoparticles*. Wiley Online Library.
[217] Messaoudi, O. and Bendahou, M. (2020) Biological synthesis of nanoparticles using endophytic microorganisms: Current development. *Nanotechnology and the Environment*. 1–19. http://doi.org/10.5772/intechopen.93734
[218] Herdiana, Y., Wathoni, N., Shamsuddin, S., and Muchtaridi, M. (2022) Scale-up polymeric-based nanoparticles drug delivery systems: Development and challenges. *OpenNano*. 7 1–17.
[219] Albalawi, F., Hussein, M.Z., Fakurazi, S., and Masarudin, M.J. (2021) Engineered nanomaterials: The challenges and

opportunities for nanomedicines. *International Journal of Nanomedicine*. 2021: 161–184.

[220] Agrahari, V. and Hiremath, P. (2017) Challenges associated and approaches for successful translation of nanomedicines into commercial products. *Nanomedicine*. 12 (8), 819–823.

[221] Wojnicki, M., Luty-Błocho, M., Hessel, V., Csapó, E., Ungor, D., and Fitzner, K. (2018) Micro droplet formation towards continuous nanoparticles synthesis. *Micromachines*. 9 (5), 1–12.

[222] Salahshoori, I., Golriz, M., Nobre, M.A.L., Mahdavi, S., Malekshah, R.E., Javdani-Mallak, A., et al. (2023) Simulation-based approaches for drug delivery systems: Navigating advancements, opportunities, and challenges. *Journal of Molecular Liquids*. 395 1–57.

[223] Mao, S., Guo, C., Shi, Y., and Li, L.C. (2012) Recent advances in polymeric microspheres for parenteral drug delivery–part 1. *Expert Opinion on Drug Delivery*. 9 (9), 1161–1176.

[224] de Moraes Porto, I.C.C. (2012) Polymer biocompatibility. *Polymerization*. 2012 47–63.

[225] Ranganathan, B., Miller, C., and Sinskey, A. (2018) Biocompatible synthetic and semi-synthetic polymers-a patent analysis. *Pharmaceutical Nanotechnology*. 6 (1), 28–37.

[226] Sazali, N., Ibrahim, H., Jamaludin, A.S., Mohamed, M.A., Salleh, W.N.W., and Abidin, M.N.Z. (2020) Degradation and stability of polymer: A mini review. in: *IOP Conf. Ser. Mater. Sci. Eng.*, IOP Publishing, pp. 1–16.

[227] Visan, A.I., Popescu-Pelin, G., and Socol, G. (2021) Degradation behavior of polymers used as coating materials for drug delivery—A basic review. *Polymers*. 13 (8), 1–36.

[228] Viswanathan, P., Muralidaran, Y., and Ragavan, G. (2017) Challenges in oral drug delivery: a nano-based strategy to overcome. in: *Nanostructures Oral Med.*, Elsevier, pp. 173–201.

[229] Andonova, V. (2017) Synthetic polymer-based nanoparticles: Intelligent drug delivery systems. in: *Acrylic Polym. Healthc.*, IntechOpen, pp. 1–26.

[230] Kamaly, N., Yameen, B., Wu, J., and Farokhzad, O.C. (2016) Degradable controlled-release polymers and polymeric nanoparticles: mechanisms of controlling drug release. *Chemical Reviews*. 116 (4), 2602–2663.

[231] Sosnik, A. and Augustine, R. (2016) Challenges in oral drug delivery of antiretrovirals and the innovative strategies to overcome them. *Advanced Drug Delivery Reviews*. 103 105–120.

[232] Hamid Akash, M.S., Rehman, K., and Chen, S. (2015) Natural and synthetic polymers as drug carriers for delivery of therapeutic proteins. *Polymer Reviews*. 55 (3), 371–406.

[233] Driguez, P.-A., Potier, P., and Trouilleux, P. (2014) Synthetic oligosaccharides as active pharmaceutical ingredients: Lessons learned from the full synthesis of one heparin derivative on a large scale. *Natural Product Reports*. 31 (8), 980–989.

[234] Tundisi, L.L., Ataide, J.A., Costa, J.S.R., de Freitas Coêlho, D., Liszbinski, R.B., Lopes, A.M., et al. (2023) Nanotechnology as a tool to overcome macromolecules delivery issues. *Colloids and Surfaces B: Biointerfaces*. 222 1–40.

[235] Luxenhofer, R. (2015) Polymers and nanomedicine: considerations on variability and reproducibility when combining complex systems. *Nanomedicine*. 10 (20), 3109–3119.

[236] Tran, S., DeGiovanni, P.-J., Piel, B., and Rai, P. (2017) Cancer nanomedicine: a review of recent success in drug delivery. *Clinical and Translational Medicine*. 6 1–21.

[237] Krska, S.W., DiRocco, D.A., Dreher, S.D., and Shevlin, M. (2017) The evolution of chemical high-throughput experimentation to address challenging problems in pharmaceutical synthesis. *Accounts of Chemical Research*. 50 (12), 2976–2985.

[238] Kumar, A., Sood, A., Agrawal, G., Thakur, S., Thakur, V.K., Tanaka, M., et al. (2023) Polysaccharides, proteins, and synthetic polymers based multimodal hydrogels for various biomedical applications: A review. *International Journal of Biological Macromolecules*. 247: 1–26.

[239] Gao, P. and Shi, Y. (2012) Characterization of supersaturatable formulations for improved absorption of poorly soluble drugs. *The AAPS Journal*. 14 703–713.

[240] Ahmad, A., Imran, M., and Sharma, N. (2022) Precision nanotoxicology in drug development: Current trends and challenges in safety and toxicity implications of customized multifunctional nanocarriers for drug-delivery applications. *Pharmaceutics*. 14 (11), 1–31.

[241] Somwanshi, S.B., Bhoye, U.R., Dolas, R.T., Kotade, K.B., Dhamak, K.B., and Gaware, V.M. (2016) Understanding Nanotoxicology and Its Implications for Overcoming Challenges in the Development of Nanoparticles. *American Journal of Pharm Tech Research*. 6 (3), 103–121.

[242] Mühlebach, S. (2018) Regulatory challenges of nanomedicines and their follow-on versions: a generic or similar approach? *Advanced Drug Delivery Reviews*. 131 122–131.

[243] Farjadian, F., Ghasemi, A., Gohari, O., Roointan, A., Karimi, M., and Hamblin, M.R. (2019) Nanopharmaceuticals and nanomedicines currently on the market: challenges and opportunities. *Nanomedicine*. 14 (1), 93–126.

[244] Hua, S., De Matos, M.B.C., Metselaar, J.M., and Storm, G. (2018) Current trends and challenges in the clinical translation of nanoparticulate nanomedicines: Pathways for translational development and commercialization. *Frontiers in Pharmacology*. 9 1–48.

15 Future Prospects of Synthetic Polymers in Drug Delivery in Combating Inflammatory Diseases

Alan Raj
Manipal Academy of Higher Education, Manipal, India

K. Rahul and C. Sarath Chandran
Government Medical College Kannur, Pariyaram, India

Angel Treasa Alex
Manipal Academy of Higher Education, Manipal, India

15.1 INTRODUCTION

Since the 1980s, research on polymeric drug delivery has significantly advanced the field of biomedicine. These polymers have a profound impact on several domains of the healthcare sector, including drug delivery, immunological compatibility, organ transplantation, and artificial organ development. Specifically, using polymers in designing and developing novel drug delivery systems has transformed the treatment of various diseases [1–4].

Polymers are widely used in drug delivery, including in the development of controlled release, sustained release, and targeted drug delivery platforms. They are also used to protect the loaded drugs from unfavorable environments, ensuring that the drugs are delivered at the desired site in an undiluted form. Many approved drugs have challenging physicochemical properties, such as poor solubility and permeability, and polymers are used as carriers to ensure therapeutic efficacy despite their properties [6–11]. Certain polymers can mitigate immunogenicity, which benefits biological drugs such as proteins. The availability of biocompatible and biodegradable polymers has further expanded their application to the development of novel drug delivery systems beyond traditional boundaries.

All biological systems share the idea of polymers, which are created by joining multiple simple structural components to create a three-dimensional structure, which includes intracellular strings, cytoskeleton, and proteins in the soft extracellular matrix. Additionally, keratin found in the skin and hair is an example of a polymer at the human-environment interface, whereas insects have silk-like polymers for external construction. Natural polymers have been used since the beginning of civilization and have numerous applications in the medical field, such as suture materials or controlled drug delivery [1, 2, 4–6, 11].

Recently developed synthetic polymers, including polyesters and polyamides, have gained widespread attention as suture materials in the medical field. These materials have been widely used in medical applications because of their physical and chemical properties, which are based on monomer units, polymerization reactions, and the formation of copolymers with various components at adjustable concentrations. Furthermore, well-established technologies are available for the synthesis and fabrication of complex devices. These polymers integrate oriented fibers of the same material into the matrix to demonstrate mechanical and structural features, including mechanical self-reinforcement. Additionally, they have highly sophisticated mechanical qualities, such as shape-memory polymers, which can easily deform and revert to their original form in response to stimuli such as pH and temperature. Medical fields in which these materials have been used include vascular stents, medication delivery devices, clot removal devices, sutures, aneurysm or ductus arteriosus occlusion, and orthodontic therapy [13–16].

15.2 FUTURE POTENTIAL OF SYNTHETIC POLYMER IN THE MANAGEMENT OF INFLAMMATORY DISORDER

Synthetic polymers are instrumental in shaping the landscape of drug delivery, resulting in significant breakthroughs in this field. Synthetic polymers have been extensively employed in the development of drug delivery platforms owing to their biocompatibility, capacity for controlled/sustained drug release, and adaptability. The current use of synthetic polymers in drug delivery underscores the

DOI: 10.1201/9781003477389-15

dynamic nature of this field, which includes biodegradable materials that enhance biocompatibility and smart polymers that enable precise control, contributing to the progress of drug delivery systems [12–18]. The use of synthetic polymers is poised to transform the administration of medicine and provide novel avenues for improved patient outcomes as researchers continue to explore new possibilities. Figure 15.1 shows the future of synthetic polymers in various fields.

The future of synthetic polymers is a topic of significant promise and potential as researchers and innovators continue to push the boundaries of materials science. Cutting-edge technologies, such as controlled polymerization methods and computational design, have produced synthetic polymers with precise and customizable properties, opening up unprecedented applications across various industries [14, 17, 19]. The integration of AI and materials science has accelerated the computational design of polymers, leading to the rapid discovery of novel materials with optimized performances. The potential of synthetic polymers in drug delivery systems and tissue engineering is particularly noteworthy, with the potential to revolutionize the treatment of various disorders. Furthermore, the next wave of inventive and sustainable technologies is anticipated to be significantly influenced by the incorporation of synthetic polymers into flexible electronics and energy storage technologies [3, 5, 15, 17]. The possibilities of using bioactive polymers in drug delivery are significant, with the potential to address many drug-related issues and unlock new opportunities for the advancement of health care. Table 15.1 shows the various studies conducted with novel synthetic polymers to combat inflammatory disorders.

15.2.1 Synthetic Polymers in Anti-Cancer Therapy

Cancer poses a serious threat to human health because of its high incidence and deadly nature. Many anti-cancer medications, including monoclonal antibodies, cytokines, and chemotherapeutic medicines, have been developed and have demonstrated high potency in cancer management. However, therapeutic outcomes are typically constrained by tumor heterogeneity and the emergence of medication resistance. Scientists are constantly striving to develop new anti-cancer therapies. Recently, anti-cancer approaches based on synthetic polymers have attracted great interest in pharmaceuticals [25–28].

Recently, a synthetic cationic lipo-peptide-mimicking lipo-polylysine (C_m-PLL_n) with structure-dependent anti-cancer activity was confirmed. Yi et al., in their work, rationally designed a series of C_m-PLL_n. Among them, C_{12}-PLL_{13} demonstrated optimal anti-cancer efficacy and selectivity, eliminating nine different tumor cell types, including those with drug-resistant and highly metastatic tumors, while having minimal negative effects on healthy human umbilical vein endothelial cells. Mechanistic investigations revealed that cationic C_{12}-PLL_{13} interacts with the negatively charged membranes of cancer cells to cause necrosis via membrane rupture. Finally, C_{12}-PLL_{13} successfully

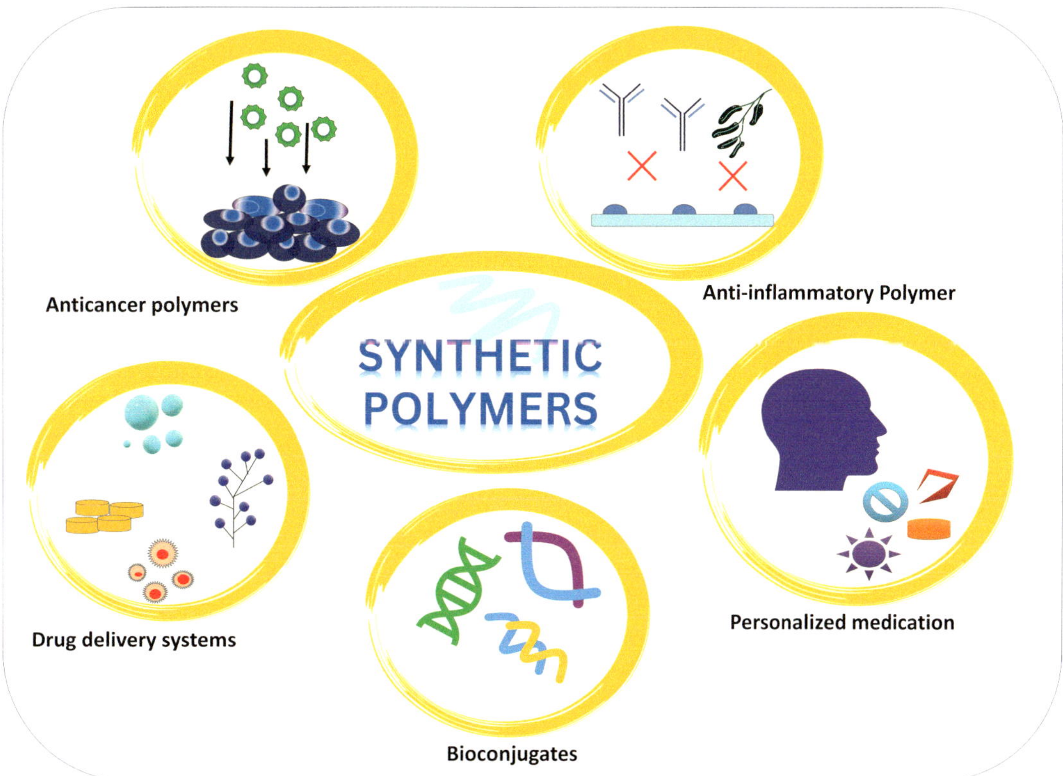

FIGURE 15.1 Future of synthetic polymers in various fields.

TABLE 15.1
Synthetic Polymers for Various Inflammatory Disease Applications

Polymer	Inflammatory Disease Condition	Outcome
PEG-PGC_{43}-$PLLA_{19}$ and PEG-PGC_{43}-$PDLA_{20}$	Cancer	The polymer proved highly effective against various cancer cell lines, and treatment with stereo complexes inhibited tumor growth and metastasis. Without causing resistance, these polymers show potential in cancer therapy [19, 20].
PCL-g-PAMAM cationic dendronized polymers	Rheumatoid arthritis	In rats with collagen-induced arthritis, the cationic dendronized polymers prevented bone deterioration, synovial hyperplasia, and joint edema [21].
Micronized porcine cartilage/poly (lactic-co-glycolic acid)	Tissue engineering, anti-inflammatory	*According to in vitro investigations, the amount of micronized porcine cartilage rose along with the rates of cell adhesion and proliferation* [22].
Polyaniline-Polypyrrolidone (PANi-PVP)	Tumor	In vivo, photothermal ablation of the tumor was accomplished with great therapeutic efficacy with the combination of well-dispersed PANi-PVP and near-infrared irradiation [22, 23].
Poly(ethylene glycol) - poly(propylene sulfide) copolymer	Atherosclerosis	The micelle quickly releases the drug, and the polymer consumes reactive oxygen species (ROS) at the site [24].
Polydopamine	Acute lung injury, acute inflammation-induced injury	Compared with healthy animals, polydopamine reduced the proinflammatory cytokinin in mice with acute lung injury [25].

Note: Polyethylene glycol: PEG, guanidinium-functionalized polycarbonate: PGC, poly(L-lactide): PLLA, poly(D-lactide): PDLA, polyamidoamine: PAMAM, polycaprolactone: PCL

inhibited tumor growth in a mouse C26 tumor model without negative side effects. Thus, lipopolylysine, which mimics lipopeptides, is a highly effective and biocompatible cancer treatment [29].

Takahashi et al. designed anti-cancer polymers to kill dormant prostate cancer cells. They demonstrated the molecular design of membrane-active co-polymers as synthetic anti-cancer peptide mimics. Prostate cancer cells are susceptible to the anti-cancer effects of these co-polymers. These co-polymers disrupt cell membranes to produce these anti-cancer effects. The anti-cancer activity and selectivity can be adjusted by modifying the cationic and hydrophobic monomer compositions of the co-polymers. Notably, the copolymer killed isolated dormant prostate cancer cells that were inherently docetaxel-resistant. The in vitro tumor spheroid models shrank or dispersed owing to the efficacy of the copolymer in eliminating dormant cells in the model [30, 31].

Shen et al. reported a host defense peptide (HDP)-mimicking synthetic polypeptide-based antineoplastic drug-free anti-cancer approach. This polypeptide demonstrated broad-spectrum anti-cancer action in drug-resistant and highly metastatic tumor cells. Detailed mechanistic studies have revealed that rapid necrosis of cancer cells within a short time can be achieved with the cationic anti-cancer polypeptide (ACPP) through a membrane-lytic mechanism. DA-ACPP, a pH-dependent zwitterionic derivative of ACPP, was synthesized for in vivo applications. When exposed to slightly acidic tumor microenvironments, DA-ACPP converts back to ACPP, which selectively kills cancer cells and exhibits negligible hemolysis under neutral physiological settings. These results demonstrate that artificial HDP-mimicking polypeptides are secure and efficient anti-tumor agents [31].

In addition to their involvement in a wide range of biological processes, saccharides also play a significant role in several diseases, including cancer [6]. Saccharides have garnered considerable interest in the biomaterial domain; however, their binding affinities to their corresponding protein receptors are frequently modest. When saccharide ligands are supplied as aggregates or clusters and lectins have numerous glycosylation binding sites, higher affinities can be attained. The "multivalent effect" or "cluster glycoside effect" causes multivalent carbohydrate-protein interactions, raising protein association constants and encouraging lectins to have a higher affinity for their multivalent glyco-ligands than the corresponding monovalent ligands. Owing to their availability of binding sites and potent glycan-based interactions, many research groups have explored polymers as ideal constructions that mimic natural saccharide ligands and drive higher activity. In addition, they also serve as delivery systems for nucleic acids [6].

While most research on glycopolymers focuses on their lectin interactions in cancer therapy, new findings have demonstrated that glycopolymers can also be employed in photodynamic and photothermal therapies and photoacoustic tumor imaging [32, 33]. Sun et al. developed brushes made of water-soluble glycopolymer for targeted photodynamic therapy. First, they created a poly(benzene-alt-fluorene) backbone by covalently attaching an Ir (III) complex photosensitizer. The second stage involved grafting a block copolymer based on polygalactose (PGal) and oligo (ethylene glycol) methacrylate onto the polymer backbone using atom transfer radical polymerization (ATRP). Under light irradiation, the glycopolymers effectively caused apoptosis of hepatocellular carcinoma cells and reduced tumor growth because PGal exhibited a specific identification and binding ability to these cells [34, 35].

Considering the available research data, synthetic polymers have great potential for anti-cancer therapy. Specific synthetic polymers can exhibit anti-cancer activity by suppressing tumor cells without drugs. Combining these polymers with anti-cancer drugs for drug delivery could enhance their anti-cancer effects. Concerns regarding drug resistance, targeting, and toxicity can be resolved through this novel approach. Although anti-cancer therapies based on synthetic polymers have gained great attention, only a limited number of studies have been performed on drug-free therapies. Several polymers with proven anti-cancer properties are yet to be explored in depth.

15.2.2 Synthetic Polymers as a Biomimetic Tool

Recently, biomimetic polymers have shown great promise as reliable building blocks for developing approaches to imitating/mimicking physiological molecules used by organisms to preserve homeostasis [36–42]. Biomimetic technologies maximize the therapeutic effect in regenerative medicine by exploiting the natural interactions between the implant and biological components of the host. Thus far, tissue engineering has produced encouraging results by developing delivery platforms that mimic the extracellular matrix to restore bone and cartilage [3]. In the early 1950s, Overberger et al.'s groundbreaking work opened new avenues for the study of bioactive polymers. Between 1960 and 1970, the period saw increased research intensity from researchers such as Donaruma, Samour, Vogl, Takemoto, Goodman, Ringsdorf, Levy, and others. By the 1970s, biomimetic polymers had been well established [42].

Selective extraction and purification of C-reactive protein (CRP) from complex bodily fluids are necessary to investigate the biological functions of CRP in various illnesses. [42]. However, clear nonspecific protein adsorption has been observed in the current affinity sorbents. As a result, additional purification processes are required to recover the target protein. Concerning the specific interplay between CRP and the cell outer membrane, a novel crosslinker, N, N′-methylenebisacrylamide, and a superior anti-biofouling agent, 2-(methacryloyloxy)ethyl phosphorylcholine, were combined to form a novel biomimetic polymer. After optimizing the washing and elution conditions methodically, an easy enrichment strategy was developed based on the polymer mentioned earlier. The new biomimetic polymer demonstrated favorable physical properties and inferior nonspecific protein binding for CRP compared to the marketed immobilized p-aminophenyl phosphoryl choline gel. Additionally, it has been effectively used to selectively enrich CRP in the serum of rats and humans. Effective CRP purification may be achieved with the help of a unique enrichment strategy based on an adaptable biomimetic polymer, which will advance the comprehensive investigation of CRP biological functions.

Poly(glycerol sebacate) (PGS) is a synthetic polymeric material with outstanding biocompatibility, high plasticity, and controlled degradation [43, 44]. Nevertheless, the use of PGSs in cartilage tissue engineering is restricted because their breakdown occurs faster than cartilage regeneration. Polycaprolactone (PCL) is a common synthetic polymer that is ideal for biodegradation and has high mechanical strength. Therefore, a scaffold made of a mixture of PCL, which degrades slowly, and PGS, which breaks down quickly, was developed, and its potential for use in artificial cartilage development. Scanning electron microscopy (SEM) revealed that the skeletons offered advantageous porosity microstructures. The PGS/PCL scaffolds exhibited superior mechanical strength and longer degradation period, as demonstrated by an in vitro degradation test. In PGS/PCL scaffolds, chondrocyte-like cells can be produced by bone marrow-derived mesenchymal stem cells (BMSCs), as demonstrated by the chondrogenic culture for cell scaffold construction. PGS/PCL scaffolds have favorable mechanical qualities, cytocompatibility, and adjustable biodegradation, making them potentially useful for cartilage tissue regeneration. This offers a fresh opportunity to address tissue engineering more effectively, and perhaps even treat other inflammatory diseases [45].

Heart failure (HF) is a pathological condition which affects 1%–2% of the global population. It is characterized by increased dysfunction and cardiomyocyte death. Although there is increased interest in cell-based therapeutics, the sustainable injection of cardiomyocytes remains challenging. An injectable biomimetic reverse thermal gel (RTG) specifically designed to facilitate long-term cardiomyocyte survival was developed. This RTG biopolymer was given in solution form to cardiomyocytes, which were quickly converted to a gel-based matrix at normal body temperature. The suitability of this biopolymer in maintaining cardiomyocyte viability was examined in this study. RTG was biofunctionalized using laminin or poly-l-lysine. Adult rat ventricular myocytes (ARVM) and neonatal rat ventricular myocytes (NRVM) were cultivated in 3D environments using plain-RTG and biomolecule-functionalized RTG. Dishes with conventional 2D biomolecule coatings served as the controls. It was discovered that RTG-lysine caused NRVM to proliferate and create functional syncytia that resembled hearts. After 21 days, beating cells were observed in both RTG and RTG-lysine groups in terms of cell contraction. Furthermore, after eight days, RTG-laminin contained more than 50% viable ARVMs identified by sarcomeric cross-striations, a well-defined cardiac phenotype. These results demonstrate the enormous potential of the developed RTG cell therapy platform for minimally invasive cardiomyocyte transplantation [46].

Biomaterials ideally mimic the mechanical, chemical, biological, and functional properties of living tissues or biomolecules. Phosphatidylcholine, a lipid found in mammalian cell membranes, is a model for the developing of

2-Methacryloyloxyethyl phosphorylcholine (MPC), which is one of the most effective biocompatible polymers. Medical devices use MPC polymers to develop artificial valves and joints [47].

Recently, apoptotic cell membrane-inspired polymers containing a phosphoryl serine (PS) group have been developed, as phosphatidyl-serine (PtdSer) mimetic polymers are particularly favorable materials against inflammation. Nevertheless, synthesizing PtdSer-mimetic polymers has proven difficult because of the presence of two reactive groups: the amino and carboxy groups in serine, which limit effective and selective modification approaches. Hence, the phosphoramidite technique is frequently used in solid-phase DNA synthesis. Using this reaction to produce a phosphate ester between two hydroxyl groups has the benefit of good selectivity under mild conditions [48]. A methacryloyloxyethyl phosphorylserine (MPS) monomer was synthesized, and poly (MPS) was obtained by a free radical polymerization reaction. The later deprotection and oxidation of PtdSer were also performed. The obtained poly(MPS) was subjected to lipopolysaccharide (LPS)-activated RAW 264.7 macrophages to investigate the immune-suppressive action. Under LPS-stimulated conditions, amoeboid macrophage cell extensions were found to be inhibited. In addition, effective NF- κB in macrophage suppression was also seen. Further investigations led to the finding that 50 pg or more of poly(MPS) within each macrophage is necessary to produce an inhibitory action. The time-dependent immunosuppression action of poly(MPS) study showed that after 3 hours of incubation, NF-κB expression was inhibited, indicating the requirement of a minimum time of three hours for immunosuppression [47].

In a previous study, a synthetic artificial β-cell (AβC) was created using multicompartmental vesicle-in-vesicle superstructures to modify insulin administration via a membrane fusion-mediated process [49]. The biochemical mechanisms of naturally occurring β-cells, which are well organized into compartments with well-defined boundaries, inspired this design. The large liposome vesicle on the outside functioned like a cytoplasmic membrane, but the insulin-loaded inner lipid nanovesicles resembled the granules found in normal β-cells. Additionally, this superstructure contains a system for glucose metabolism and a system for membrane fusion (pH-controlled reversible shielding polymers and complementary peptides that can form coiled coils). By varying the pH within the large vesicles to varying degrees, AβCs could differentiate between different glucose levels. This would lead to merging of the inner and outer large vesicles, which would cause insulin to be released under hyperglycemic conditions and retain basal low insulin release at normal blood glucose levels. All biochemical reactions occur inside large vesicles to prevent obstacles in the buffering environment. This made it possible to react quickly to dynamic variations in the blood sugar levels. With the help of the F127 thermogel, these bioinspired AβCs can be subcutaneously "transplanted," and unlike typical cell transplantation therapy, they can sustain normal blood glucose levels for several days without requiring immunosuppressive medications. Additionally, as these formulations and devices comprise sensible and labile chemicals such as ketals and aryl boronates, care should be taken to ensure their stability throughout the manufacturing and storage procedures.

15.2.3 Anti-inflammatory Potential of Synthetic Polymers

Infection or injury typically causes inflammation. It usually begins when immune cells are drawn to injured tissues by a pathogen or by damage-associated molecular patterns (PAMPs/DAMPs). While persistent irregular inflammation can be linked to negative health consequences in various circumstances, inflammation is frequently necessary for a balanced immune response. For example, autoimmune disorders, a subclass of autoinflammatory diseases defined by a detrimental immune response to self-antigens, have been identified in up to 4% of the global population. They often manifest in relapsing-remitting cycles, ranging in severity from mild to fatal, and produce crippling and persistent agony. The World Health Organization (WHO) generally believes that the most urgent threat to human health is chronic inflammatory diseases, which include autoimmune diseases that cause inflammation, heart conditions, cancer, and obesity. Synthetic polymers are promising for the improved management of these challenging conditions, where extensive research has been carried out for better therapeutic prospects in the future.

Poly(propylene sulfoxide) (PPSO) is a sulfoxide-based polymer that has recently attracted attention for its potential as a "stealth" macromolecule and its pharmacological activity as an antioxidant and anti-inflammatory [50]. The combination of these two ideas might initially seem strange because poly (ethylene glycol) (PEG) is practically biologically inert and is hence "stealth." Conversely, under homeostatic conditions, polysulfides may combine significant inertness toward biomolecules with the ability to scavenge ROS linked to inflammation. Thus, polysulfides are a particularly special kind of "active" or "stealth" polymer. Anionic ring-opening polymerization was used to produce poly (propylene sulfide) (PPS), which was subsequently used in the controlled oxidation of PPS to create PPSO. Low toxicity and somewhat reduced complement activation and macrophage absorption (in human plasma) were the in vitro characteristics of PPSO compared to PEG of the same size. Crucially, in contrast to PEG, PPSO demonstrated a potent and dose-dependent ROS (hydrogen peroxide and hypochlorite)scavenging action on LPS-activated macrophages, leading to a proportional decrease in cytokine output. Further thorough research on PPSO is needed to determine the mechanism of endocytosis, effectiveness of removing ROS, interactions with proteins, pharmacokinetics, and in vivo effects on inflammation models, among other things.

Synthetic polymer coatings are versatile and easily processed materials widely used in contemporary medical equipment and implants. These coatings can be shaped into a wide range of intricate patterns and forms and are specifically engineered for certain uses by manipulating their composition, physical characteristics, and chemical makeup. However, the integration and biological performance of these devices are severely limited when these materials are inserted into the body because they trigger strong inflammatory host reactions. The complicated process of biomaterial-mediated inflammation includes protein adsorption, leukocyte recruitment and activation, inflammatory mediator secretion, and fibrous encapsulation of implants. In-depth research is required to develop more biocompatible implants by altering the characteristics of the material using different anti-inflammatory polymeric surface coatings. These consist of passive and active techniques, such as anti-inflammatory drugs and nonfouling surface treatments [51]. New strategies are required to decrease the need for multiple implantation surgeries and to improve the in vivo longevity and performance of devices.

Although many patients with ulcerative colitis, an inflammatory colon illness that severely reduces the quality of life, respond to anti-inflammatory drugs for symptomatic relief, many of them require colectomy, a procedure in which the colon is removed entirely or in part. Several pharmaceutical drugs have been used to treat ulcerative colitis. However, their inefficiency is usually caused by several issues, including insufficient drug accumulation at the target site and deficiency in target-specific drug delivery [51]. Various adaptable polymeric biomaterials can be used to deliver medications in a target-specific manner through the oral route, improve drug concentration at the target site, and reduce chemotherapy-related side effects.

A pH-sensitive hydrogel was developed in a recent study using a heat-initiated free radical polymerization approach. The poly(methoxyl ethylene glycol) methyl ether methacrylateP(CE-MAA-MEG)-co-caprolactone-co-methacrylicacid-co-poly(ethyleneglycol)substantially alleviated inflammation that was detected in ulcerative colitis animal models, according to the results from the animal model [51]. Therefore, PEG-based hydrogels should be considered when designing drug delivery systems for ulcerative colitis in the future.

Numerous inflammatory illnesses are largely influenced by the pathophysiology of effector cells produced by inflammatory monocytes. However, no single therapeutic strategy can precisely alter these cells. Research has shown that inflammatory monocytes utilize the macrophage receptor with collagenous structure (MARCO) to absorb negatively charged immune-modifying microparticles (IMPs) infused with polystyrene, microdiamonds, or biodegradable poly(lactic-co-glycolic) acid in an opsonin-independent manner [52]. These monocytes then stopped migrating to inflammatory areas and were sequestered in the spleen by IMP infusion, followed by caspase-3-mediated apoptosis and apoptotic cell clearance processes. In murine models of lethal flavivirus encephalitis, experimental autoimmune encephalomyelitis, colitis induced by dextran sodium sulfate, peritonitis induced by thioglycolate, and myocardial infarction, administration of IMPs significantly decreased monocyte accumulation at inflammatory foci, decreased disease symptoms, and enhanced tissue repair. These results highlight the translation of IMPs for therapeutic application in illnesses aggravated by inflammatory monocytes.

Inhibiting tumor necrosis factor alpha (TNF-α) and interleukin-1 (IL-1) are examples of anti-cytokine employed in the current therapeutic management of rheumatoid arthritis, mainly targeting the inflammatory response. However, the response rate is still not equal. Lazar et al. developed a very interesting technique to fuse neutrophil membranes onto polymeric cores and inject these nanoparticles into the knee joint of a transgenic mouse collagen-induced arthritis. Neutrophils are important for reducing inflammation and for repairing tissue damage. Through the suppression of the overall arthritis severity and reduction of joint destruction, the results demonstrated strong therapeutic efficacy. Notably, the core of the nanoparticles was prepared using PLGA, an FDA-approved biodegradable polymer. In a different method, it was shown that polymeric nanoparticles coated with macrophage membranes could bind and neutralize endotoxins and reduce systemic inflammatory responses in vivo.[53, 54].

Given this range of applications, it is likely safe to say that synthetic polymers can potentially treat several inflammatory diseases. While certain polymers can produce anti-inflammatory effects on their own, others function as targeting agents for the delivery of drugs to particular sites. Further research on different polymers is required to improve their applications. Combinations of different synthetic polymers can occasionally produce miraculous effects, similar to those described in the preceding sections. Activity augmentation, targeting, and many other goals can be accomplished by the appropriate use of these polymers. The prospects of synthetic polymers hold great potential for treating various inflammatory disorders.

15.2.4 Theranostic Polymeric Nanoparticles

Rather than using a generalized technique, the theranostic approach is thought to be a new type of treatment that allows for individualized drug administration. This molecular-based therapeutic strategy efficiently reduces undesirable systemic side effects due to intersubject variability [55]. The theranostic approach combines diagnosis and treatment into one window [56, 57]. Theranostic polymeric nanoparticles (TNPs) are mainly used for molecular imaging, environmentally sensitive medication release, and active and passive targeting. [58].

Polymeric drug conjugates are new drug delivery systems with a size range of 1–10 nm. After several modifications, they improve biodistribution by avoiding renal and hepatic clearance processes. Conjugates with high molecular weights have high biodistribution and plasma half-life [59, 60]. The three major components of a polymer-drug conjugate are the therapeutic moiety, the polymer component, and the imaging or contrast agent. Occasionally, the drug may also function as a contrast agent. Therapeutic drugs such as bleomycin, ellipticine, paclitaxel, and doxorubicin possess fluorescence qualities [61–63].

Using stealth polymer nanoconjugates, drugs and contrast agents can be effectively delivered on the same platform. When internal or external stimuli are applied, a stimuli-responsive polymeric nanoparticle can efficiently release the medication [63]. Rheumatoid arthritis (RA) is an autoimmune disorder characterized by joint inflammation and muscle stiffness, and if untreated, it reaches the point where joint replacement is necessary. Methotrexate-based conventional therapy would result in significant systemic toxicities; therefore, to reduce their systemic toxicities and enhance their therapeutic index, it is imperative to administer the drug precisely and effectively to targeted inflammatory joints [64, 65]. Chitosan theranostic polymeric nanoparticles (CHNPs) loaded with dexamethasone and methotrexate have stronger anti-arthritic properties than the drug alone. Following the IV injection, more nanoparticles were found at the inflammatory site. Technetium-99m (99mTc) was conjugated to the nanoparticles for tracking for biodistribution of nanoparticles. The results revealed that the nanoparticles were distributed in the body after intraperitoneal injection [66].

Similarly, hyperbranched semiconducting polymer nanoparticles (MTX-HSP-PEG-NPs) were developed to deliver methotrexate. The nanoparticles were coupled with Technetium-99m (^{99m}Tc), which can be viewed by SPECT imaging. In vivo evaluation further revealed that compared to free methotrexate at the same dose, MTX-HSP-PEG-NPs had low systemic toxicity and a high therapeutic index for the considerable relief of RA. A biodistribution study showed that the EPR effect mediates nanoparticle uptake [66, 67].

15.2.5 Intrinsic Disordered Synthetic Polymers Against Inflammation

Polymers can significantly impact immune compatibility and are crucial in biomedical devices and drug delivery systems because immunological reactions may result in tissue damage, inflammation, and treatment failure [68]. Various techniques can be used to develop immunologically compatible polymers, such as surface modification with hydrophilic polymers, or can be designed to improve host immunity. Biodegradable polymers have been identified as promising candidates for meeting new requirements because of their unique capacity to modify physicochemical characteristics. Regulating the type and proportion of the starting materials used in polymer synthesis will help to achieve this [69, 70].

Protein polymers with inherently disordered sections are produced from elastomeric proteins and are known as intrinsically disordered polymers, including resilins and elastins [71]. Because of their low sequence complexity and similar biophysical properties to intrinsically disordered protein polymers (IDPs), a family of thermoresponsive bioengineered proteins, known as elastin-like polypeptides (ELPs), has become a fascinating model for IDP. ELPs comprise repetitive pentapeptide sequence units that go through Val-Pro-Gly-X-Gly, where X is the guest residue. In addition to proline, any residue of an amino acid can be it [72]. Many hydrophobic drugs, whose solubility restricts their therapeutic efficacy, are thought to be best suited for ELP nanoparticle delivery [73]. Consider the problems with doxorubicin (Dox), which has low solubility and serious toxic side effects, including myelosuppression, mucositis, and heart toxicity. Dox-ELPs nanoparticles can generate enhanced cytotoxicity during hyperthermia while retaining low cytotoxicity at 37°C by conjugating Dox to ELPs [74].

In addition, the reversible phase change of ELPs at a lower critical solution temperature (LCST), which is one of their main characteristics, occurs at certain solution temperatures. As ordered water molecules hydrate the hydrophobic side chains of ELPs, causing them to become soluble in aqueous buffers, ELPs display prolonged random coils below the LCST. An increase in the solution temperature above the LCST causes the ELPs to change from random coils to β-spiral structures. Hydrophobic side chains were revealed because ELPs release bound water molecules during this transition [75].

More and more polymers are now able to self-assemble in response to extrinsic stimuli, such as the high temperature used to treat solid tumors, or clinically significant intrinsic stimuli, such as the pH difference (pH6.5–6.8) between the tumor microenvironment and normal tissues, or the differences in calcium concentration between the extracellular and intracellular environments [75]. Dreher et al. developed thermally responsive ELPs using an N-terminal peptide ligand in a linear AB di-block architecture that self-assembled into spherical micelles when heated beyond the body temperature. They conjugated a peptide that specifically targeted the angiogenic tumor vasculature. The results showed that this can be used as a novel drug delivery tool for cancer management [76].

Single-chain polymeric nanoparticles (SCPNs) are IDPs that can change their conformation depending on the external environment. The morphologies exhibited by enzymes and IDPs have inspired considerable development of SCNPs. They can fold into a nanoparticle of between 2-20nm in size. They are synthesized via covalent or noncovalent interactions between individual polymers [77, 78]. Sathyan et al. developed SCPNs catalytic to deliver anti-cancer prodrugs. Pd II is a transition metal that is an excellent catalyst for activating prodrugs. Triphenylphosphine and bipyridine, Pd (II)-binding ligands, were used to functionalize the amphiphilic polymers. The resultant polymers disintegrate into 4–6 nm-sized nanoparticles with the Pd (II) catalyst in an interior hydrophobic domain. They concluded that this nanoparticle catalytic system could deliver more anti-cancer prodrugs in the future [79].

15.2.6 Personalized Medications and Synthetic Polymers in Inflammatory Diseases

Personalized medication is much more interconnected with circadian rhythm. Intersubject variability in this rhythm also affects treatment outcomes [80]. The patterns in the intensity of the symptoms demonstrate the need for precise therapies in which the precise dosage of medicines is released into the body at predetermined periods during the day. For example, RA will be active in the early morning, so the patient requires a delayed-release system to correct this [81]. Personalized medicine has become a popular strategy in drug delivery systems to improve treatment outcomes by developing individualized therapies based on each patient's needs [82]. More precisely, a release profile approximating the patient's circadian cycle is thought to be beneficial for novel approaches to cancer treatment [83]. The use of polymers is not restricted to nanoparticle-based drug delivery systems. It can also be used to develop an electronic device that can release drugs in a controlled manner.

For personalized medication, microelectronics and polymers have been integrated to control drug release based on individual variations. Microfluidic devices are widely employed as high-throughput platforms for drug screening and tissue-engineering applications [84]. Polymeric microcontainers and controlled-release microparticles have also been produced using microfluidics as drug delivery platforms. Therapeutic drugs can be delivered using polymeric programmed systems that can operate remotely or activate the entire polymeric system [85]. A pH-responsive and uPAR-targeted gemcitabine (Gem) drug delivery system made of polymeric micelles (Gem@TpHResMic) was developed using a microfluidic approach. These nanoparticles have a limited size distribution, strong colloidal stability, and high drug encapsulation efficiency (EE%). In an acidic environment, Gem@TpHResMic preferentially targets tumor cells that express uPA and releases gemcitabine in a controlled manner [86]. Hydrogels are polymeric materials that are extensively used in biomedical applications owing to their biocompatibility and adjustable mechanical and chemical properties [86]. Hydrogels can hold large amounts of water in their polymeric matrices. Darmawan et al. used doxorubicin and poly (ethylene glycol) diacrylate (PEGDA) hydrogels to develop a drug delivery system, and UV light patterning was used to alter the final geometry of the system. Following the in vitro cell tests, the synthesized system was characterized, and the drug release profile was assessed. Therefore, PEGDA is a promising option for the co-administration of drugs in personalized medicine because of its high biocompatibility and ease of customization [87, 88].

15.2.7 Artificial Intelligence and Machine Learning in Polymer-Based Drug Delivery

The techniques and effectiveness of research are changing dramatically in the fields of nanotechnology and personalized medicine. Many researchers have already used automated methods, but there is much potential for translation in the synthesis of nanoparticles. When combined with automation and artificial intelligence (AI), targeted therapeutic nanoparticles can be optimized for specific cell types and patients. AI is involved in different sectors such as personalized medication, controlled drug delivery, prediction of the stability of nanoparticles, and determination of synergistic drug combinations.

AI, which involves computer learning and eventually performing tasks, is called machine learning (ML). Artificial neural networks (ANNs) are ML techniques that are inspired by biological neurons. ANNs are effective ML instruments for modeling nonlinear interactions, which are commonly observed in industrial and various research contexts [89, 90]. AI can enhance synthetic models to produce biosynthetic models that resemble the natural membranes. Biomimicry helps develop potential drug delivery platforms and accelerate the systematic study of the complexities and divergences of biological membranes [91]. Drug efficacy for specific patients was predicted, and treatment regimens were optimized using computational models and AI algorithms. Majumder et al. reported that single-cell profile analysis using computational models can quickly identify drugs or pharmacological combinations for cancer treatment. With a sensitivity of up to 100%, the ML algorithm was utilized to predict the clinical response of new patients based on clinical responses to anti-cancer medications and their matching clinical outcomes [91].

On the other hand, AI is used to tune the size and shape of polymeric nanoparticles. It is difficult to generate homogenous polymeric micro-/nanoparticles with controlled characteristics. Microfluidic devices can produce microparticles with better monodispersity, precisely adjustable shapes, and excellent encapsulation efficiencies [92, 93, 94]. ML models have been developed and reported in several studies to forecast droplet microfluidic parameters, such as drug particle nanoprecipitation, droplet stability, and fluid and flow parameters [95, 96]. The size of PLGA droplets and the particles that form after drying are crucial factors in the microfluidic production of polymeric microparticles. Damiati et al. used ANNs to construct an in silico model for predicting these parameters. In microfluidics, PLGA concentrations and flow rates of the aqueous and dispersion phases are the most efficient experimental parameters for modifying droplet and particle sizes. These characteristics were considered when creating five distinct ANN models with simple structures. These models can forecast the sizes of PLGA particles produced by various microfluidic devices either separately or in combination [96, 97]. Haider et al. used an ANN technique to develop methotrexate phosphatidylserine-targeted nanoparticles to effectively manage cancer. The response surface approach was used to build and optimize methotrexate-loaded formulations, and an artificial neural network based on genetic algorithms (ANN-GA) was used to evaluate them. The prepared formulas were validated using ANN-GA modeling and optimized using

BBD for the response surface approach. The results were aligned with the RSM model, and the ANN-GA forecast of the optimized formulation fitted the experimental values well. They concluded that the Annexin V-conjugated nanoparticles showed more anti-cancer activity than the non-conjugated nanoparticles in MDA-MB 231 cells [98].

The next generation of polymeric nanocarriers must be intelligent, able to enter the bloodstream, identify cancer independently, and release the right amount of treatment while providing the greatest possible side effects. These artificially intelligent polymeric nanocontainers are administered orally, intravenously, or via inhalation into the lungs. Kordas et al. developed quadrupole-targeted polymeric containers that can recognize cancer microenvironments and release the drug doxorubicin with AI. These systems can release drugs in response to stimuli. The surfaces of the nanoparticles were decorated with folic acid for enhanced targeting. This new AI-assisted drug delivery system (Nano4Dox) was compared with Doxil in vivo and proved that the survival rate of mice was ten times higher than that of mice treated with Doxil alone [99].

15.2.8 Nucleic Acid–Binding Polymers

Mounting evidence suggests that immune cells are activated by cell-free nucleic acids (cfNAs) released by cells while dying through toll-like receptors (TLRs). TLR overactivation is linked to autoimmune and inflammatory diseases. Treatments for inflammation include the use of nucleic acid–binding polymers as cfNA scavengers. Lee et al. showed that these polymers can prevent all nucleic acid-sensing TLRs from activating, irrespective of whether they are designed to detect hypomethylated DNA. In an acute toxic shock mouse model, systemic treatment with these polymers could also avoid serious hepatic injury caused by proinflammatory nucleic acids. Consequently, by acting as molecular scavengers, these polymers offer an innovative class of anti-inflammatory agents to counteract the proinflammatory effects of different nucleic acids [100].

Recently, it has been demonstrated that natural polyphosphates, such as extracellular nucleic acids (RNA and DNA) and inorganic polyphosphates, stimulate blood coagulation. Developing safe and effective antithrombotic agents is challenging because of the increased risk of patient bleeding. The report analyzes the in vitro and in vivo anticoagulant and antithrombotic properties of NABPs. These polymers exhibit a significant affinity for binding to DNA, RNA, and inorganic polyphosphate molecules. They can also block clotting produced by RNA and polyphosphate, and activate the intrinsic route of coagulation in vitro [101].

It has been reported that the anti-inflammatory activity of cfNA-binding proteins is limited owing to their rapid elimination and poor biodistribution, whereas cfNA nanoparticles exhibit excellent biodistribution and renal clearance [101, 102]. Severe sepsis is a prevalent, costly, and fatal medical condition with few effective treatment alternatives. Developing innovative therapeutic approaches to combat severe sepsis would be aided by learning more about inflammatory dysregulation that results in sepsis. Dawulieti et al. determined the critical function of cell-free DNA (cfDNA) in controlling the proinflammatory pathway mediated by toll-like receptor 9 in inducing severe sepsis. The results showed that, compared to their nucleic acid–binding polymers, these nucleic acid–binding nanoparticles (NABNs) performed better in inhibiting cfDNA-induced inflammation and reducing organ damage caused by severe sepsis [103, 104]. Lohmann and coworkers designed a synthetic hydrogel for wound dressing to scavenge chemokines. Compared to the FDA-approved hydrogel Promogran, the developed hydrogel applied to skin wounds in diabetic mice enhanced wound healing and vascularization, and reduced inflammation. In chronic wounds, chemokine capture may be useful for promoting tissue regeneration.

15.3 CONCLUSION

Patients may respond differently to treatment depending on the intricacy, unpredictability, and heterogeneity of the inflammatory milieu and associated immune response. As inflammation is a very dynamic process, it would be ideal to have an anti-inflammatory therapy that can change the course of inflammation. The role of synthetic polymers in inflammatory disease conditions can be observed in different fields, such as drug delivery, implants, artificial membranes, and controlled release of circadian rhythms. In this context, several new inventions have been developed for synthetic polymers. Researchers have developed anti-cancer, anti-inflammatory, and inflammatory mediator scavenging polymers to improve patient compliance. The clinical success and translation of these novel synthetic polymers are major challenges. Extending preclinical achievements into clinical translation will lead to these new types of synthetic polymer-based anti-inflammatory treatments for various illnesses.

REFERENCES

1. Sung YK, Kim SW. Recent advances in polymeric drug delivery systems. *Biomater. Res.* 2020;24:12.
2. Kaur R, Kaur MS. Role of polymers in drug delivery. *J. Drug Deliv.* 2014; 4(3), 32–36.
3. Tong X, Pan W, Su T, Zhang M, Dong W, Qi X. Recent advances in natural polymer-based drug delivery systems. *React. Funct. Polym.* 2020;148:104501.
4. Bhatt P, Trehan S, Inamdar N, Mourya VK, Misra A. Polymers in Drug Delivery: An Update [Internet]. In: *Applications of Polymers in Drug Delivery*. Elsevier; 2021 [cited 2024 Jan 12]. page 1–42. Available from: https://linkinghub.elsevier.com/retrieve/pii/B978012819659500001X
5. Tavares MR, Pechar M, Chytil P, Etrych T. Polymer-Based Drug-Free Therapeutics for Anticancer, Anti-Inflammatory, and Antibacterial Treatment. *Macromol. Biosci.* 2021;21:2100135.
6. Espinosa-Cano E, Aguilar MR, Vázquez B, San Román J. Inflammation-Responsive Polymers [Internet]. In: *Smart Polymers and their Applications*. Elsevier; 2019 [cited 2024

Jan 12]. page 219–54. Available from: https://linkinghub.elsevier.com/retrieve/pii/B9780081024164000077
7. Langer R. Polymer-controlled drug delivery systems. *Acc. Chem. Res*. 1993;26:537–42.
8. Vilar G, Tulla-Puche J, Albericio F. Polymers and Drug Delivery Systems. *Curr. Drug Deliv*. 2012;9:367–94.
9. Liechty WB, Kryscio DR, Slaughter BV, Peppas NA. Polymers for Drug Delivery Systems. *Annu. Rev. Chem. Biomol. Eng*. 2010;1:149–73.
10. Wang H, Zhu H, Fu W, Zhang Y, Xu B, Gao F, et al. A high strength self-healable antibacterial and anti-inflammatory supramolecular polymer hydrogel. *Macromol. Rapid Commun*. 2017;38:1600695.
11. Schmaljohann D. Thermo- and pH-responsive polymers in drug delivery☆. *Adv. Drug Deliv. Rev*. 2006;58:1655–70.
12. Neuse EW. Synthetic Polymers as Drug-Delivery Vehicles in Medicine. *Met.-Based Drugs* 2008;2008:1–19.
13. Jung K, Corrigan N, Wong EHH, Boyer C. Bioactive Synthetic Polymers. *Adv. Mater*. 2022;34:2105063.
14. Bharadwaz A, Jayasuriya AC. Recent trends in the application of widely used natural and synthetic polymer nanocomposites in bone tissue regeneration. *Mater. Sci. Eng. C* 2020;110:110698.
15. Hacker MC, Krieghoff J, Mikos AG. Synthetic Polymers [Internet]. In: *Principles of Regenerative Medicine*. Elsevier; 2019 [cited 2024 Jan 12]. page 559–90. Available from: https://linkinghub.elsevier.com/retrieve/pii/B9780128098806000333
16. Maitz MF. Applications of synthetic polymers in clinical medicine. *Biosurface Biotribology* 2015;1:161–76.
17. Englert C, Brendel JC, Majdanski TC, Yildirim T, Schubert S, Gottschaldt M, et al. Pharmapolymers in the 21st century: Synthetic polymers in drug delivery applications. *Prog. Polym. Sci*. 2018;87:107–64.
18. Bracaglia LG, Winston S, Powell DA, Fisher JP. Synthetic polymer coatings diminish chronic inflammation risk in large ECM-based materials. *J. Biomed. Mater. Res. A* 2019;107:494–504.
19. Zhong G, Yang C, Liu S, Zheng Y, Lou W, Teo JY, et al. Polymers with distinctive anticancer mechanism that kills MDR cancer cells and inhibits tumor metastasis. *Biomaterials* 2019;199:76–87.
20. Peng B, Liang H, Li Y, Dong C, Shen J, Mao H, et al. Tuned Cationic Dendronized Polymer: Molecular Scavenger for Rheumatoid Arthritis Treatment. *Angew. Chem*. 2019;131: 4298–302.
21. Kim S, Jang JE, Lee JH, Khang G. Composite scaffold of micronized porcine cartilage/poly(lactic-co-glycolic acid) enhances anti-inflammatory effect. *Mater. Sci. Eng. C* 2018;88:46–52.
22. Wang L, Vivek R, Wu W, Wang G, Wang JY. Fabrication of Stable and Well-Dispersed Polyaniline–Polypyrrolidone Nanocomposite for Effective Photothermal Therapy. *ACS Biomater. Sci. Eng*. 2018;4(5), 1880–1890.
23. Wu T, Chen X, Wang Y, Xiao H, Peng Y, Lin L, et al. Aortic plaque-targeted andrographolide delivery with oxidation-sensitive micelle effectively treats atherosclerosis via simultaneous ROS capture and anti-inflammation. *Nanomedicine Nanotechnol. Biol. Med*. 2018;14:2215–26.
24. Zhao T, Zhang S, Guo Y, Wang Q. TiC_2: a new two-dimensional sheet beyond MXenes. *Nanoscale* 2016;8:233–42.
25. Maman S, Witz IP. A history of exploring cancer in context. *Nat. Rev. Cancer* 2018;18:359–76.
26. Balon K, Sheriff A, Jacków J, Łaczmański Ł. Targeting Cancer with CRISPR/Cas9-Based Therapy. *Int. J. Mol. Sci*. 2022;23:573.
27. Katzung BG, editor. *Basic & clinical pharmacology*. 14th edition. New York Chicago San Francisco Athens London Madrid Mexico City Milan New Delhi Singapore Sydney Toronto: McGraw-Hill Education; 2018.
28. Shen W, Zhang Y, Wan P, An L, Zhang P, Xiao C, et al. Antineoplastic Drug-Free Anticancer Strategy Enabled by Host-Defense-Peptides-Mimicking Synthetic Polypeptides. *Adv. Mater*. 2020;32:2001108.
29. Yi X, Wan P, Shen W, Zhang X, Zhang P, Xiao C. Synthetic lipo-polylysine with anti-cancer activity. *Biomater. Sci*. 2023;11:6611–8.
30. Takahashi H, Yumoto K, Yasuhara K, Nadres ET, Kikuchi Y, Buttitta L, et al. Anticancer polymers designed for killing dormant prostate cancer cells. *Sci. Rep*. 2019;9:1096.
31. Shen W, Zhang Y, Wan P, An L, Zhang P, Xiao C, et al. Antineoplastic Drug-Free Anticancer Strategy Enabled by Host-Defense-Peptides-Mimicking Synthetic Polypeptides. *Adv. Mater*. 2020;32:2001108.
32. Lee JM, Park DY, Yang L, Kim EJ, Ahrberg CD, Lee KB, et al. Generation of uniform-sized multicellular tumor spheroids using hydrogel microwells for advanced drug screening. *Sci. Rep*. 2018;8:17145.
33. Li X, Chen G. Glycopolymer-based nanoparticles: synthesis and application. *Polym. Chem*. 2015;6:1417–30.
34. Sun P, Wang G, Hou H, Yuan P, Deng W, Wang C, et al. A water-soluble phosphorescent conjugated polymer brush for tumor-targeted photodynamic therapy. *Polym. Chem*. 2017;8:5836–44.
35. Sharma AK, Arya A, Sahoo PK, Majumdar DK. Overview of biopolymers as carriers of antiphlogistic agents for treatment of diverse ocular inflammations. *Mater. Sci. Eng. C* 2016;67:779–91.
36. Drotleff S, Lungwitz U, Breunig M, Dennis A, Blunk T, Tessmar J, et al. Biomimetic polymers in pharmaceutical and biomedical sciences. *Eur. J. Pharm. Biopharm*. 2004;58:385–407.
37. Wu J, Xie L, Lin WZY, Chen Q. Biomimetic nanofibrous scaffolds for neural tissue engineering and drug development. *Drug Discov. Today* 2017;22:1375–84.
38. Brusini R, Varna M, Couvreur P. Advanced nanomedicines for the treatment of inflammatory diseases. *Adv. Drug Deliv. Rev*. 2020;157:161–78.
39. Bridges AW, García AJ. Anti-Inflammatory Polymeric Coatings for Implantable Biomaterials and Devices. *J. Diabetes Sci. Technol*. 2008;2:984–94.
40. Ain Q, Zeeshan M, Khan S, Ali H. Biomimetic hydroxyapatite as potential polymeric nanocarrier for the treatment of rheumatoid arthritis. *J. Biomed. Mater. Res. A* 2019;107:2595–600.
41. Peña B, Martinelli V, Jeong M, Bosi S, Lapasin R, Taylor MRG, et al. Biomimetic Polymers for Cardiac Tissue Engineering. *Biomacromolecules*. 2016 May 9;17(5):1593–601.
42. Wang Q, Jin H, Xia D, Shao H, Peng K, Liu X, et al. Biomimetic Polymer-Based Method for Selective Capture of C-Reactive Protein in Biological Fluids. *ACS Appl. Mater. Interfaces* 2018;10:41999–2008.
43. Liu Y, Tian K, Hao J, Yang T, Geng X, Zhang W. Biomimetic poly(glycerol sebacate)/polycaprolactone blend scaffolds for cartilage tissue engineering. *J. Mater. Sci. Mater. Med*. 2019;30:53.
44. Talouki PY, Tackallou SH, Shojaei S, Benisi SZ, Goodarzi V. The role of three-dimensional scaffolds based on polyglycerol sebacate/ polycaprolactone/ gelatin in the presence of Nanohydroxyapatite in promoting chondrogenic

differentiation of human adipose-derived mesenchymal stem cells. *Biol. Proced. Online* 2023;25:9.
45. Peña B, Martinelli V, Jeong M, Bosi S, Lapasin R, Taylor MRG, et al. Biomimetic Polymers for Cardiac Tissue Engineering. *Biomacromolecules* 2016;17:1593–601.
46. Ebara M. Apoptotic Cell-Mimetic Polymers for Anti-Inflammatory Therapy. *Chonnam Med. J.* 2019;55:1.
47. Nakagawa Y, Lee J, Liu Y, Abbasi S, Hong T, Cabral H, et al. Microglial Immunoregulation by Apoptotic Cellular Membrane Mimetic Polymeric Particles. *ACS Macro Lett.* 2022;11:270–5.
48. Chen Z, Wang Z, Gu Z. Bioinspired and Biomimetic Nanomedicines. *Acc. Chem. Res.* 2019;52:1255–64.
49. El Mohtadi F, d'Arcy R, Yang X, Turhan ZY, Alshamsan A, Tirelli N. Main Chain Polysulfoxides as Active 'Stealth' Polymers with Additional Antioxidant and Anti-Inflammatory Behaviour. *Int. J. Mol. Sci.* 2019;20:4583.
50. Li B, Xie J, Yuan Z, Jain P, Lin X, Wu K, et al. Mitigation of Inflammatory Immune Responses with Hydrophilic Nanoparticles. *Angew. Chem. Int. Ed.* 2018;57:4527–31.
51. Sohail M, Mudassir, Minhas MU, Khan S, Hussain Z, De Matas M, et al. Natural and synthetic polymer-based smart biomaterials for management of ulcerative colitis: a review of recent developments and future prospects. Drug Deliv. *Transl. Res.* 2019;9:595–614.
52. Getts DR, Terry RL, Getts MT, Deffrasnes C, Müller M, Van Vreden C, et al. Therapeutic Inflammatory Monocyte Modulation Using Immune-Modifying Microparticles. *Sci. Transl. Med. [Internet]* 2014 [cited 2024 Jan 12];6. Available from: https://doi.org/10.1126/scitranslmed.3007563
53. Lazar KM, Shetty S, Chilkoti A, Collier JH. Immune-active polymeric materials for the treatment of inflammatory diseases. *Curr. Opin. Colloid Interface Sci.* 2023;67:101726.
54. Pene F, Courtine E, Cariou A, Mira JP. Toward theragnostics. *Crit. Care Med.* 2009;37:S50–58.
55. Rai P, Mallidi S, Zheng X, Rahmanzadeh R, Mir Y, Elrington S, et al. Development and applications of photo-triggered theranostic agents. *Adv. Drug Deliv. Rev.* 2010;62:1094–124.
56. Mura S, Couvreur P. Nanotheranostics for personalized medicine. *Adv. Drug Deliv. Rev.* 2012;64:1394–416.
57. Janib SM. Imaging and drug delivery using theranostic nanoparticles. *Adv. Drug Deliv. Rev.* 2010;62:1052–3.
58. Seymour LW, Ferry DR, Kerr DJ, Rea D, Whitlock M, Poyner R, et al. Phase II studies of polymer-doxorubicin (PK1, FCE28068) in the treatment of breast, lung and colorectal cancer. *Int. J. Oncol.* 2009;34:1629–36.
59. Sadekar S, Ray A, Janàt-Amsbury M, Peterson CM, Ghandehari H. Comparative biodistribution of PAMAM dendrimers and HPMA copolymers in ovarian-tumor-bearing mice. *Biomacromolecules* 2011;12:88–96.
60. Mohan P, Rapoport N. Doxorubicin as a Molecular Nanotheranostic Agent: Effect of Doxorubicin Encapsulation in Micelles or Nanoemulsions on the Ultrasound-Mediated Intracellular Delivery and Nuclear Trafficking. *Mol. Pharm.* 2010;7:1959–73.
61. Heinrich AK, Lucas H, Schindler L, Chytil P, Etrych T, Mäder K, et al. Improved Tumor-Specific Drug Accumulation by Polymer Therapeutics with pH-Sensitive Drug Release Overcomes Chemotherapy Resistance. *Mol. Cancer Ther.* 2016;15:998–1007.
62. Motlagh NSH, Parvin P, Ghasemi F, Atyabi F. Fluorescence properties of several chemotherapy drugs: doxorubicin, paclitaxel and bleomycin. *Biomed. Opt. Express* 2016;7:2400–6.
63. Park H, Yang J, Seo S, Kim K, Suh J, Kim D, et al. Multifunctional nanoparticles for photothermally controlled drug delivery and magnetic resonance imaging enhancement. *Small Weinh. Bergstr. Ger.* 2008;4:192–6.
64. Syed A, Devi VK. Potential of targeted drug delivery systems in treatment of rheumatoid arthritis. *J. Drug Deliv. Sci. Technol.* 2019;53:101217.
65. Kumar V, Leekha A, Tyagi A, Kaul A, Mishra AK, Verma AK. Preparation and evaluation of biopolymeric nanoparticles as drug delivery system in effective treatment of rheumatoid arthritis. *Pharm. Res.* 2017;34:654–67.
66. Zhang Q, Li D, Zhong J, Wu Y, Shi Y, Yang H, et al. SPECT imaging and highly efficient therapy of rheumatoid arthritis based on hyperbranched semiconducting polymer nanoparticles. *Biomater. Sci.* 2021;9:1845–54.
67. DeFife KM, Grako K, Cruz-Aranda G, Price S, Chantung R, Macpherson K, et al. Poly(ester amide) co-polymers promote blood and tissue compatibility. *J. Biomater. Sci. Polym. Ed.* 2009;20:1495–511.
68. Jurak M, Wiącek AE, Ładniak A, Przykaza K, Szafran K. What affects the biocompatibility of polymers? *Adv. Colloid Interface Sci.* 2021;294:102451.
69. Huyer LD, Mandla S, Wang Y, Campbell S, Yee B, Euler C, et al. Macrophage immunomodulation through new polymers that recapitulate functional effects of itaconate as a power house of innate immunity. *Adv. Funct. Mater.* 2021;31:2003341.
70. Fuxreiter M. Fuzziness: linking regulation to protein dynamics. *Mol. Biosyst.* 2012;8:168–77.
71. Roberts S, Dzuricky M, Chilkoti A. Elastin-like Polypeptides as Models of Intrinsically Disordered Proteins. *FEBS Lett.* 2015;589:2477–86.
72. Clements AEB, Murphy WL. Injectable biomaterials for delivery of interleukin-1 receptor antagonist: Toward improving its therapeutic effect. *Acta Biomater.* 2019;93:123–34.
73. Bidwell GL, Fokt I, Priebe W, Raucher D. Development of elastin-like polypeptide for thermally targeted delivery of doxorubicin. *Biochem. Pharmacol.* 2007;73:620–31.
74. Doole FT, Camp CP, Kim M. Tailoring the formation and stability of self-assembled structures from precisely engineered intrinsically disordered protein polymers: A comprehensive review. *Giant* 2023;14:100158.
75. Dreher MR, Simnick AJ, Fischer K, Smith RJ, Patel A, Schmidt M, et al. Temperature Triggered Self-Assembly of Polypeptides into Multivalent Spherical Micelles. *J. Am. Chem. Soc.* 2008;130:687–94.
76. Latorre-Sánchez A, Pomposo JA. Recent bioinspired applications of single-chain nanoparticles. *Polym. Int.* 2016;65:855–60.
77. Arbe A, Pomposo JA, Moreno AJ, LoVerso F, González-Burgos M, Asenjo-Sanz I, et al. Structure and dynamics of single-chain nano-particles in solution. *Polymer* 2016;105:532–44.
78. Sathyan A, Croke S M, Pérez-López A, De Waal BFM, Unciti-Broceta A A, Palmans AR. Developing Pd(ii) based amphiphilic polymeric nanoparticles for pro-drug activation in complex media. *Mol. Syst. Des. Eng.* 2022;7:1736–48.
79. Greco CM, Sassone-Corsi P. Personalized medicine and circadian rhythms: Opportunities for modern society. *J. Exp. Med.* 2020;217:e20200702.
80. Sanka K, Pragada RR, Veerareddy PR. A pH-triggered delayed-release chronotherapeutic drug delivery system of aceclofenac for effective management of early morning symptoms of rheumatoid arthritis. *J. Microencapsul.* 2015;32:794–803.

81. Radhakrishnan A, Kuppusamy G, Ponnusankar S, Shanmukhan NK. Pharmacogenomic phase transition from personalized medicine to patient-centric customized delivery. *Pharmacogenomics J*. 2020;20:1–18.
82. Sulli G, Lam MTY, Panda S. Interplay between circadian clock and cancer: New frontiers for cancer treatment. *Trends Cancer* 2019;5:475–94.
83. Rezaei Kolahchi A, Khadem Mohtaram N, Pezeshgi Modarres H, Mohammadi MH, Geraili A, Jafari P, et al. Microfluidic-based multi-organ platforms for drug discovery. *Micromachines* 2016;7:162.
84. Geraili A, Xing M, Mequanint K. Design and fabrication of drug-delivery systems toward adjustable release profiles for personalized treatment. *VIEW* 2021;2:20200126.
85. Iacobazzi RM, Arduino I, Di Fonte R, Lopedota AA, Serratì S, Racaniello G, et al. Microfluidic-assisted preparation of targeted ph-responsive polymeric micelles improves gemcitabine effectiveness in PDAC: In vitro insights. *Cancers* 2022;14:5.
86. Li Y, Huang G, Zhang X, Li B, Chen Y, Lu T, et al. Magnetic Hydrogels and Their Potential Biomedical Applications. *Adv. Funct. Mater*. 2013;23:660–72.
87. Darmawan BA, Lee SB, Nan M, Nguyen VD, Park JO, Choi E. Shape-Tunable UV-Printed Solid Drugs for Personalized Medicine. *Polymers* 2022;14:2714.
88. Sutariya V. Artificial Neural Network in Drug Delivery and Pharmaceutical Research. *Open Bioinforma. J*. 2014;7:49–62.
89. Damiati SA, Alaofi AL, Dhar P, Alhakamy NA. Novel machine learning application for prediction of membrane insertion potential of cell-penetrating peptides. *Int. J. Pharm*. 2019;567:118453.
90. Parodi A, Molinaro R, Sushnitha M, Evangelopoulos M, Martinez JO, Arrighetti N, et al. Bio-inspired engineering of cell- and virus-like nanoparticles for drug delivery. *Biomaterials* 2017;147:155–68.
91. Majumder B, Baraneedharan U, Thiyagarajan S, Radhakrishnan P, Narasimhan H, Dhandapani M, et al. Predicting clinical response to anticancer drugs using an ex vivo platform that captures tumour heterogeneity. *Nat. Commun*. 2015;6:6169.
92. Damiati S, Kompella UB, Damiati SA, Kodzius R. Microfluidic Devices for Drug Delivery Systems and Drug Screening. *Genes* 2018;9:103.
93. Mahdi Y, Daoud K. Microdroplet size prediction in microfluidic systems via artificial neural network modeling for water-in-oil emulsion formulation. *J. Dispers. Sci. Technol*. 2017;38:1501–8.
94. Ali HSM, Blagden N, York P, Amani A, Brook T. Artificial neural networks modelling the prednisolone nanoprecipitation in microfluidic reactors. *Eur. J. Pharm. Sci*. 2009;37:514–22.
95. Khor JW, Jean N, Luxenberg ES, Ermon S, Tang SKY. Using machine learning to discover shape descriptors for predicting emulsion stability in a microfluidic channel. *Soft Matter* 2019;15:1361–72.
96. Damiati SA, Rossi D, Joensson HN, Damiati S. Artificial intelligence application for rapid fabrication of size-tunable PLGA microparticles in microfluidics. *Sci. Rep*. 2020;10:19517.
97. Haider T, Soni V. "Response surface methodology and artificial neural network-based modeling and optimization of phosphatidylserine targeted nanocarriers for effective treatment of cancer: In vitro and in silico studies." *J. Drug Deliv. Sci. Technol*. 2022;75:103663.
98. Kordas G. Quadrupole Stimuli-Responsive Targeted Polymeric Nanocontainers for Cancer Therapy: Artificial Intelligence in Drug Delivery Systems [Internet]. In: *Nanoengineering of Biomaterials*. John Wiley & Sons, Ltd; 2022 [cited 2024 Jan 3]. page 505–22. Available from: https://doi.org/10.1002/9783527832095.ch16
99. Lee J, Sohn JW, Zhang Y, Leong KW, Pisetsky D, Sullenger BA. Nucleic acid-binding polymers as anti-inflammatory agents. *Proc. Natl. Acad. Sci*. 2011;108:14055–60.
100. Jain S, Pitoc GA, Holl EK, Zhang Y, Borst L, Leong KW, et al. Nucleic acid scavengers inhibit thrombosis without increasing bleeding. *Proc. Natl. Acad. Sci*. 2012;109:12938–43.
101. Liang H, Peng B, Dong C, Liu L, Mao J, Wei S, et al. Cationic nanoparticle as an inhibitor of cell-free DNA-induced inflammation. *Nat. Commun*. 2018;9:4291.
102. Wu J, Liang H, Li Y, Shi Y, Bottini M, Chen Y, et al. Cationic Block Copolymer Nanoparticles with Tunable DNA Affinity for Treating Rheumatoid Arthritis. *Adv. Funct. Mater*. 2020;30:2000391.
103. Dawulieti J, Sun M, Zhao Y, Shao D, Yan H, Lao YH, et al. Treatment of severe sepsis with nanoparticulate cell-free DNA scavengers. *Sci. Adv*. 2020;6:eaay7148.
104. Lohmann N, Schirmer L, Atallah P, Wandel E, Ferrer RA, Werner C, et al. Glycosaminoglycan-based hydrogels capture inflammatory chemokines and rescue defective wound healing in mice. *Sci. Transl. Med*. 2017;9:eaai9044.

Index